Stochastic Differential Equations With Markovian Switching

Stochastic Differential Equations with Markovian Switching

Xuerong Mao
University of Strathclyde, UK

Chenggui Yuan
University of Wales Swansea, UK

Imperial College Press

Published by

Imperial College Press
57 Shelton Street
Covent Garden
London WC2H 9HE

Distributed by

World Scientific Publishing Co. Pte. Ltd.
5 Toh Tuck Link, Singapore 596224
USA office: 27 Warren Street, Suite 401-402, Hackensack, NJ 07601
UK office: 57 Shelton Street, Covent Garden, London WC2H 9HE

British Library Cataloguing-in-Publication Data
A catalogue record for this book is available from the British Library.

STOCHASTIC DIFFERENTIAL EQUATIONS WITH MARKOVIAN SWITCHING

Copyright © 2006 by Imperial College Press

All rights reserved. This book, or parts thereof, may not be reproduced in any form or by any means, electronic or mechanical, including photocopying, recording or any information storage and retrieval system now known or to be invented, without written permission from the Publisher.

For photocopying of material in this volume, please pay a copying fee through the Copyright Clearance Center, Inc., 222 Rosewood Drive, Danvers, MA 01923, USA. In this case permission to photocopy is not required from the publisher.

ISBN-13 978-1-86094-701-8
ISBN-10 1-86094-701-8

Printed in Singapore

Xuerong Mao would like to dedicate this book to his parents,
Mr Yiming Mao and Mrs Dezhen Chen

Chenggui Yuan would like to dedicate this book to his wife,
Mrs Wangdong Yang

Preface

Stochastic modelling has come to play an important role in many branches of science and industry. An area of particular interest has been the automatic control of stochastic systems, with consequent emphasis being placed on the analysis of stability in stochastic models.

The hybrid systems driven by continuous-time Markov chains have been used to model many practical systems where they may experience abrupt changes in their structure and parameters caused by phenomena such as component failures or repairs, changing subsystem interconnections, and abrupt environmental disturbances. The hybrid systems combine a part of the state that takes values continuously and another part of the state that takes discrete values. In 1971, [Kazangey and Sworder (1971)] presented a jump system, where a macroeconomic model of the national economy was used to study the effect of federal housing removal policies on the stabilization of the housing sector. The term describing the influence of interest rates was modelled by a finite-state Markov chain to provide a quantitative measure of the effect of interest rate uncertainty on optimal policy. [Athans (1987)] suggested that the hybrid systems would become a basic framework in posing and solving control-related issues in Battle Management Command, Control and Communications (BM/C^3) systems. The hybrid systems were also considered for the modelling of electric power systems by [Willsky and Rogers (1979)] as well as for the control of a solar thermal central receiver by [Sworder and Robinson (1973)]. In his book, [Mariton (1990)] explained that the hybrid systems had been emerging as a convenient mathematical framework for the formulation of various design problems in different fields such as target tracking (evasive target tracking problem), fault tolerant control and manufacturing processes.

One of the important classes of the hybrid systems is the linear jump systems

$$\dot{x}(t) = A(r(t))x(t)$$

which have been investigated for more than 30 years. The generalisation of the linear jump systems is the stochastic differential equations (SDEs) with Markovian switching

$$dx(t) = f(x(t), r(t))dt + g(x(t), r(t))dB(t),$$

which have recently received a great deal of attention. Here the state vector has two components $x(t)$ and $r(t)$: the first one is in general referred to as the state while the second one is regarded as the mode. In its operation, the system will switch from one mode to another in a random way, and the switching between the modes is governed by a Markov chain. The study of hybrid systems has included the optimal regulator, controllability, observability, stability and stabilization *etc*. For more information on the hybrid systems the reader is referred to [Basak *et al.* (1996); Bouks (1993); Costa and Boukas (1998); Dragan and Morozan (2002); Feng *et al.* (1992); Ji and Chizeck (1990); Lewin (1986); Mao *et al.* (2000); Morozan (1998); Pan and Bar-Shalom (1996); Souza and Fragoso (1993); Shaikhet (1996); Skorohod (1989); Sworder and Robinson (1973); Wonham (1970)]. In particular, the authors have made significant contributions to this area (see e.g. [Mao (1999a); Mao (1999b); Mao (2000b); Mao (2002b); Mao *et al.* (2000)] and [Yuan and Mao (2003a)]–[Yuan *et al.* (2003)]).

Although there is a book [Mariton (1990)] on the linear jump systems, there is so far no book on SDEs with Markovian switching and the papers on them were published in different journals in a form which is not convenient for the reader to understand the theory systematically. This book was therefore written. Some important features of this text are as follows:

- The text will be the first systematic presentation of the theory of SDEs with Markovian switching. It will present the basic principles of SDEs with Markovian switching at an introductory level but will emphasise the current research trends in the field of SDEs with Markovian switching at an advanced level.
- The text will cover various types of equations with Markovian switching from stochastic differential equations through interval systems to stochastic functional differential equations, all with Markovian switching.

- This text will discuss a number of approximation schemes including the Euler–Marayama and Carathedory under both global and local Lipschitz condition. Especially the numerical methods for SDEs under local Lipschitz condition are currently a very hot topic.
- This text will demonstrate the manifestations of the general Lyapunov method by showing how this effective technique can be adopted to study entirely different qualitative and quantitative properties of stochastic systems, e.g. asymptotic bounds, exponential stability and invariant measures.
- This text will emphasise the analysis of stability which is vital in the automatic control of stochastic systems.
- This text will be mainly based on the authors' recent research papers, for example, [Mao (1999a); Mao (1999b); Mao (2000b); Mao (2002b); Mao et al. (2000)] and [Yuan and Mao (2003a)]–[Yuan et al. (2003)]. It will hence discuss many hot topics including various applications to finance, population dynamics and control.

In other words, the text will take all the features of Itô equations, Markovian switching, interval systems as well as time-lag into account. The theory developed will be applicable in different and complicated situations in many branches of science and industry. In particular, we will discuss a number of important applications including population dynamics, financial modelling, stochastic stabilization and stochastic neural networks. All of them are currently hot topics in research.

We would like to thank the EPSRC/BBSRC, the Royal Society, the London Mathematics Society as well as the Edinburgh Mathematical Society for their financial support. We also wish to thank I.M. Davies, K.D. Elworthy, G. Gettinby, W.S.C. Gurney, D.J. Higham, Z. Hou, N. Jacob, X. Liao, E. Renshaw, A.M. Stuart, A. Truman, G.G. Yin, J. Zou, for their constant support, encouragement, assistance and collaboration. Thanks also to Lenore Betts at ICP who played a vital role in shaping the projet. Moreover, we should thank our families, in particular, Weihong and Wangdong, for their constant support.

Xuerong Mao and Chenggui Yuan

Contents

Preface vii

Notation xv

1. Brownian Motions and Stochastic Integrals 1
    1.1 Introduction 1
    1.2 Probability Theory 4
    1.3 Stochastic Processes 12
    1.4 Brownian Motions 18
    1.5 Stochastic Integrals 23
    1.6 Itô's Formula 38
    1.7 Markov Processes 43
    1.8 Generalised Itô's Formula 47
    1.9 Exercises 49

2. Inequalities 51
    2.1 Introduction 51
    2.2 Frequently Used Inequalities 51
    2.3 Gronwall-Type Inequalities 54
    2.4 Matrices and Inequalities 58
    2.5 Linear Matrix Inequalities 62
    2.6 M-Matrix Inequalities 67
    2.7 Stochastic Inequalities 69
    2.8 Exercises 75

3. Stochastic Differential Equations with Markovian Switching 77

3.1	Introduction	77
3.2	Stochastic Differential Equations	77
3.3	Existence and Uniqueness of Solutions	81
3.4	SDEs with Markovian Switching	88
3.5	L^p-Estimates	96
3.6	Almost Surely Asymptotic Estimates	101
3.7	Solutions as Markov Processes	104
3.8	Exercises	110

4. Approximate Solutions — 111

4.1	Introduction	111
4.2	Euler–Maruyama's Approximations	111
	4.2.1 Global Lipschitz Case	114
	4.2.2 Local Lipschitz Case	118
	4.2.3 More on Local Lipschitz Case	121
4.3	Caratheodory's Approximations	126
4.4	Split-Step Backward Euler Scheme	134
4.5	Backward Euler Scheme	146
4.6	Stochastic Theta Method	149
4.7	Exercises	154

5. Boundedness and Stability — 155

5.1	Introduction	155
5.2	Asymptotic Boundedness	157
5.3	Exponential Stability	164
	5.3.1 Nonlinear Jump Systems	178
	5.3.2 Multi-Dimensional Linear Equations	180
	5.3.3 Scalar Linear Equations	182
	5.3.4 Examples	187
5.4	Moment and Almost Sure Asymptotic Stability	191
5.5	Stability in Probability	204
5.6	Asymptotic Stability in Distribution	210
5.7	Exercises	226

6. Numerical Methods for Asymptotic Properties — 229

6.1	Introduction	229
6.2	Euler–Maruyama's Method and Exponential Stability	230
6.3	Euler–Maruyama's Method and Lyapunov Exponents	239

		6.4	Generalised Results and Stochastic Theta Method	241

    6.4 Generalised Results and Stochastic Theta Method 241
    6.5 Asymptotic Stability in Distribution of the EM Method:
         Constant Step Size 249
         6.5.1 Stability in Distribution of the EM Method 249
         6.5.2 Sufficient Criteria for Assumptions 6.16–6.18 256
         6.5.3 Convergence of Stationary Distributions 265
    6.6 Asymptotic Stability in Distribution of the EM Method:
         Variable Step Sizes 267
    6.7 Exercises 270

7. Stochastic Differential Delay Equations with Markovian Switching 271

    7.1 Introduction 271
    7.2 Stochastic Differential Delay Equations 273
    7.3 SDDEs with Markovian Switching 277
    7.4 Moment Properties 282
    7.5 Asymptotic Boundedness 285
    7.6 Exponential Stability 289
    7.7 Approximate Solutions 294
    7.8 Exercises 300

8. Stochastic Functional Differential Equations with Markovian Switching 301

    8.1 Introduction 301
    8.2 Stochastic Functional Differential Equations 301
    8.3 SFDEs with Markovian Switching 303
    8.4 Boundedness 305
    8.5 Asymptotic Stability 308
    8.6 Razumikhin-Type Theorems on Stability 311
    8.7 Examples 314
    8.8 Exercises 317

9. Stochastic Interval Systems with Markovian Switching 319

    9.1 Introduction 319
    9.2 Interval Matrices 320
    9.3 SDISs with Markovian Switching 322
    9.4 Razumikhin Technology on SDISs 328
         9.4.1 Delay Independent Criteria 328
         9.4.2 Delay Dependent Criteria 334

          9.4.3 Examples 341
    9.5 SISs with Markovian Switching 346
    9.6 Exercises 349

10. Applications   351

    10.1 Introduction 351
    10.2 Stochastic Population Dynamics 351
         10.2.1 Global Positive Solutions 353
         10.2.2 Ultimate Boundedness 356
         10.2.3 Moment Average in Time 358
    10.3 Stochastic Financial Modelling 360
         10.3.1 Non-Negative Solutions 361
         10.3.2 The EM Approximations 363
         10.3.3 Stochastic Volatility Model 370
         10.3.4 Options Under Stochastic Volatility 375
    10.4 Stochastic Stabilisation and Destabilisation 379
    10.5 Stochastic Neural Networks 387
    10.6 Exercises 394

Bibliographical Notes   395

Bibliography   397

Index   407

Notation

$$\begin{aligned}
\text{positive} :\ &> 0. \\
\text{nonpositive} :\ &\leq 0. \\
\text{negative} :\ &< 0. \\
\text{nonnegative} :\ &\geq 0.
\end{aligned}$$

a.s. : almost surely, or $\mathbb{P}$-almost surely, or with probability 1.

$A := B$: A is defined by B or A is denoted by B.

$A(x) \equiv B(x)$: $A(x)$ and $B(x)$ are identically equal, i.e. $A(x) = B(x)$ for all x.

$\emptyset$: the empty set.

I_A : the indicator function of a set A, i.e. $I_A(x) = 1$ if $x \in A$ or otherwise 0.

A^c : the complement of A in Ω, i.e. $A^c = \Omega - A$.

$A \subset B$: A is a subset of B, i.e. $A \cap B^c = \emptyset$.

$A \subset B$ a.s. : $\mathbb{P}(A \cap B^c) = 0$.

$\sigma(\mathcal{C})$: the σ-algebra generated by $\mathcal{C}$.

$a \vee b$: the maximum of a and b.

$a \wedge b$: the minimum of a and b.

$f : A \to B$: the mapping f from A to B.

$\mathbb{S}$: $= \{1, 2, \cdots, N\}$, the finite state space of a Markov chain.

$\mathbb{R} = \mathbb{R}^1$: the real line.

$\mathbb{R}_+$: $[0, \infty)$, the set of all nonnegative real numbers.

$\mathbb{R}^n$: the n-dimensional Euclidean space.

$\mathbb{R}^n_+$: $\{(x_1, \cdots, x_n) \in \mathbb{R}^n : x_i > 0,\ 1 \leq i \leq n\}$.

$\bar{\mathbb{R}}^n_+$: $\{(x_1, \cdots, x_n) \in \mathbb{R}^n : x_i \geq 0,\ 1 \leq i \leq n\}$.

$\mathcal{B}^n$: the Borel-σ-algebra on $\mathbb{R}^n$.

$\mathcal{B}$: $= \mathcal{B}^1$.

$\mathbb{R}^{n\times m}$:	the space of real $n \times m$-matrices.
$\mathcal{B}^{n\times m}$:	the Borel-σ-algebra on $\mathbb{R}^{n\times m}$.
$\mathbb{C}$:	the complex space.
$\mathbb{C}^n$:	the n-dimensional complex space.
$\mathbb{C}^{n\times m}$:	the space of complex $n \times m$-matrices.
$Z^{n\times n}$:	$= \{(a_{ij})_{n\times n} \in \mathbb{R}^{n\times n} : a_{ij} < 0,\ i \neq j\}$.
$\|x\|$:	the Euclidean norm of a vector x.
$\mathbb{S}_h$:	$= \{x \in \mathbb{R}^n : \|x\| < h\}$.
$\bar{\mathbb{S}}_h$:	$= \{x \in \mathbb{R}^n : \|x\| \leq h\}$.
A^T :	the transpose of a vector or matrix A.
$A = A^T$:	A is a symmetric matrix.
$A = A^T > 0$:	A is a symmetric positive-definite matrix.
$A = A^T \geq 0$:	A is a symmetric nonnegative-definite matrix.
$A \geq 0$:	each element of A is nonnegative.
$A > 0$:	$A \geq 0$ and at least one element of A is positive.
$A \gg 0$:	all element of A are positive.
$A_1 \geq A_2$:	if and only if $A_1 - A_2 \geq 0$.
$A_1 > A_2$:	if and only if $A_1 - A_2 > 0$.
$A_1 \gg A_2$:	if and only if $A_1 - A_2 \gg 0$.
$Z^{n\times n}$:	the space of matrices $A = (a_{ij})_{n\times n} \in \mathbb{R}^{n\times n}$ such that $a_{ij} \leq 0$ if $i \neq j$.
$\langle x, y \rangle$:	the inner product of vectors x and y, namely $\langle x, y \rangle = x^T y$.
$\mathrm{trace}(A)$:	the trace of a square matrix $A = (a_{ij})_{n\times n}$, i.e. $\mathrm{trace}(A) = \sum_{1\leq i \leq n} a_{ii}$.
$\lambda_{\min}(A)$:	the smallest eigenvalue of a symmetric matrix A.
$\lambda_{\max}(A)$:	the largest eigenvalue of a symmetric matrix A.
$\lambda_{\max}^+(A)$:	$= \sup_{x\in\mathbb{R}_+^n, |x|=1} x^T A x$ for a symmetric matrix A.
$\lambda(A)$:	the spectrum of A.
$\rho(A)$:	the spectral radius of A.
$\|A\|$:	$\|A\| = \sqrt{\mathrm{trace}(A^T A)}$, i.e. the trace norm of A.
$\|A\|$:	$\|A\| := \sup\{|Ax| : |x| = 1\} = \sqrt{\lambda_{\max}(A^T A)}$, i.e. the operator norm of A.
δ_{ij} :	Dirac's delta function, that is $\delta_{ij} = 1$ if $i = j$ or otherwise 0.
$C(D; \mathbb{R}^n)$:	the family of continuous $\mathbb{R}^n$-valued functions defined on D.
$C^m(D; \mathbb{R}^n)$:	the family of continuously m-times differentiable

Notation

	$\mathbb{R}^n$-valued functions defined on D.
$C_0^m(D;\mathbb{R}^n)$:	the family of functions in $C^m(D;\mathbb{R}^n)$ with compact support.
$C^{2,1}(D\times\mathbb{R}_+;R)$:	the family of all real-valued functions $V(x,t)$ defined on $D\times\mathbb{R}_+$ which are continuously twice differentiable in $x\in D$ and once differentiable in $t\in\mathbb{R}_+$.
∇ :	$\nabla = (\frac{\partial}{\partial x_1},\cdots,\frac{\partial}{\partial x_n})$.
Δ :	the Laplace operator, i.e. $\Delta = \sum_{i=1}^n \frac{\partial^2}{\partial x_i^2}$,
V_x :	$V_x = \nabla V = (V_{x_1},\cdots,V_{x_n}) = (\frac{\partial V}{\partial x_1},\cdots,\frac{\partial V}{\partial x_n})$.
V_{xx} :	$V_{xx} = (V_{x_ix_j})_{d\times d} = (\frac{\partial^2 V}{\partial x_i \partial x_j})_{n\times n}$.
$\|\xi\|_{L^p}$:	$= (\mathbb{E}|\xi|^p)^{1/p}$.
$L^p(\Omega;\mathbb{R}^n)$:	the family of $\mathbb{R}^n$-valued random variables ξ with $\mathbb{E}|\xi|^p < \infty$.
$L^p_{\mathcal{F}_t}(\Omega;\mathbb{R}^n)$:	the family of $\mathbb{R}^n$-valued $\mathcal{F}_t$-measurable random variables ξ with $\mathbb{E}|\xi|^p < \infty$.
$L_{\mathcal{F}_t}(\Omega;\mathbb{S})$:	the family of $\mathbb{S}$-valued $\mathcal{F}_t$-measurable random variables.
$C([-\tau,0];\mathbb{R}^n)$:	the space of continuous $\mathbb{R}^n$-valued functions φ defined on $[-\tau,0]$ with a norm $\|\varphi\| = \sup_{-\tau\le\theta\le 0}|\varphi(\theta)|$.
$L^p_{\mathcal{F}}([-\tau,0];\mathbb{R}^n)$:	the family of $C([-\tau,0];\mathbb{R}^n)$-valued random variables ϕ such that $\mathbb{E}\|\phi\|^p < \infty$.
$L^p_{\mathcal{F}_t}([-\tau,0];\mathbb{R}^n)$:	the family of $\mathcal{F}_t$-measurable $C([-\tau,0];\mathbb{R}^n)$-valued random variables ϕ such that $\mathbb{E}\|\phi\|^p < \infty$.
$C^b_{\mathcal{F}_t}([-\tau,0];\mathbb{R}^n)$:	the family of $\mathcal{F}_t$-measurable bounded $C([-\tau,0];\mathbb{R}^n)$-valued random variables.
$L^p([a,b];\mathbb{R}^n)$:	the family of Borel measurable functions $h:[a,b]\to\mathbb{R}^n$ such that $\int_a^b |h(t)|^p dt < \infty$.
$\mathcal{L}^p([a,b];\mathbb{R}^n)$:	the family of $\mathbb{R}^n$-valued $\mathcal{F}_t$-adapted processes $\{f(t)\}_{a\le t\le b}$ such that $\int_a^b |f(t)|^p dt < \infty$ a.s.
$\mathcal{M}^p([a,b];\mathbb{R}^n)$:	the family of processes $\{f(t)\}_{a\le t\le b}$ in $\mathcal{L}^p([a,b];\mathbb{R}^n)$ such that $\mathbb{E}\int_a^b |f(t)|^p dt < \infty$.
$\mathcal{L}^p(\mathbb{R}_+;\mathbb{R}^n)$:	the family of processes $\{f(t)\}_{t\ge 0}$ such that for every $T>0$, $\{f(t)\}_{0\le t\le T}\in \mathcal{L}^p([0,T];\mathbb{R}^n)$.
$\mathcal{M}^p(\mathbb{R}_+;\mathbb{R}^n)$:	the family of processes $\{f(t)\}_{t\ge 0}$ such that for every $T>0$, $\{f(t)\}_{0\le t\le T}\in \mathcal{M}^p([0,T];\mathbb{R}^n)$.
$\mathcal{K}$:	the family of all continuous increasing functions $\kappa:\mathbb{R}_+\to\mathbb{R}_+$ such that $\kappa(0)=0$ while $\kappa(u)>0$ for $u>0$.

κ^{-1} : the inverse function of $\kappa \in \mathcal{K}$.

$\mathcal{K}_\infty$: the family of all functions $\kappa \in \mathcal{K}$ with property that $\kappa(\infty) = \infty$.

$\mathcal{K}_\vee$: the family of all convex functions $\kappa \in \mathcal{K}$.

$\mathcal{K}_\wedge$: the family of all concave functions $\kappa \in \mathcal{K}$.

$L^1(\mathbb{R}_+; \mathbb{R}_+)$: the family of functions $\gamma : \mathbb{R}_+ \to \mathbb{R}_+$ such that $\int_0^\infty \gamma(t)dt < \infty$.

$D(\mathbb{R}_+; \mathbb{R}_+)$: the family of functions $\eta : \mathbb{R}_+ \to \mathbb{R}_+$ such that $\int_0^\infty \eta(t)dt = \infty$.

$\Psi(\mathbb{R}_+; \mathbb{R}_+)$: the family of continuous functions $\psi : \mathbb{R}_+ \to \mathbb{R}_+$ such that for any $\delta > 0$ and any increasing sequence $\{t_k\}_{k \geq 1}$, $\sum_{k=1}^\infty \int_{t_k}^{t_k+\delta} \psi(t)dt = \infty$.

$\mathcal{W}([-\tau, 0]; \mathbb{R}_+)$: the family of Borel-measurable functions w from $[-\tau, 0]$ to $\mathbb{R}_+$ such that $\int_{-\tau}^0 w(u)du = 1$.

$\mathcal{P}(\mathbb{R}^n \times \mathbb{S})$: the space of probability measures on $\mathbb{R}^n \times \mathbb{S}$.

$\mathbb{L}$: the family of mappings $f : \mathbb{R}^n \times \mathbb{S} \to \mathbb{R}$ satisfying $|f(x,i) - f(y,j)| \leq |x-y| + |i-j|$ and $|f(\cdot,\cdot)| \leq 1$.

Erf($\cdot$) : the error function given by Erf$(z) = (2\pi)^{-1/2} \int_0^z e^{-u^2/2}du$.

sign(x) : the sign function, namely sign$(x) = +1$ if $x \geq 0$ or otherwise -1.

Other notations will be explained where they first appear.

Chapter 1

Brownian Motions and Stochastic Integrals

1.1 Introduction

Systems in many branches of science and industry are often subject to various types of noise and uncertainty. For example, let us consider a simple model of an asset price. Suppose that at time t the asset price is $x(t)$. Consider a small subsequent time interval dt, during which $x(t)$ changes to $x(t) + dx(t)$. (We use the notation $d\cdot$ for the small change in any quantity over this time interval when we intend to consider it as an infinitesimal change.) By definition, the return of per unit of the asset price at time t is $dx(t)/x(t)$. How might we model this return?

To understand the modelling more easily, suppose that the asset is a bank deposit while the bank deposit interest rate is r. So $x(t)$ is the balance of the saving account at time t. Thus the return $dx(t)/x(t)$ of the saving at time t is rdt, that is

$$\frac{dx(t)}{x(t)} = rdt$$

or

$$\frac{dx(t)}{dt} = rx(t).$$

This ordinary differential equation can be solved exactly to give exponential growth in the value of the saving, i.e.

$$x(t) = x_0 e^{r(t-t_0)},$$

where x_0 is the initial deposit of the saving account at time t_0.

However asset prices do not move as money invested in a risk-free bank. It is often stated that asset prices must move randomly because of the *efficient market hypothesis* (see Chapter 10 for more details). The most common model decomposes the return $dx(t)/x(t)$ of the asset price into two parts. One is a predictable, deterministic and anticipated return akin to the return on money invested in a risk-free bank. It gives a contribution

$$\mu dt$$

to the return $dx(t)/x(t)$, where μ is a measure of the average rate of growth of the asset price, also known as the drift. The second contribution to $dx(t)/x(t)$ models the random change in the asset price in response to external effects, such as unexpected news. There are many external effects so by the well-known central limit theorem this second contribution can be represented by a random sample drawn from a normal distribution with mean zero and adds a term

$$\sigma dB(t)$$

to $dx(t)/x(t)$. Here σ is a number called the *volatility*, which measures the standard deviation of the returns. The quantity $dB(t)$ is the sample from a normal distribution with mean zero and variance dt. In other words, $dB(t)$ is the increment of a Brownian motion $B(t)$. Putting these contributions together, we obtain

$$\frac{dx(t)}{x(t)} = \mu dt + \sigma dB(t)$$

or

$$dx(t) = \mu x(t) dt + \sigma x(t) dB(t). \tag{1.1}$$

That is, in form of integration,

$$x(t) = x_0 + \int_{t_0}^{t} \mu x(u) du + \int_{t_0}^{t} \sigma x(u) dB(u). \tag{1.2}$$

The question is: what is the integration $\int_{t_0}^{t} \sigma x(u) dB(u)$? If the Brownian motion $B(t)$ were differentiable with its derivative $\dot{B}(t) = dB(t)/dt$, then the integral would have no problem at all as it could be done as the classical Lebesgue integral $\int_{t_0}^{t} \sigma x(u) \dot{B}(u) du$. Unfortunately, we shall see that the Brownian motion $B(t)$ is nowhere differentiable. Moreover, if the Brownian motion $B(t)$ were a process of finite variation, the integral $\int_{t_0}^{t} \sigma x(u) dB(u)$ could be regarded as the Lebesgue–Stieltjes one. However, we shall see that almost every sample path of the Brownian motion has infinite variation in any finite time interval. Hence the integral can not be defined in the ordinary way. It turns out that we need to make use of the stochastic nature of the Brownian motion in order to define the integral. This integral was first defined by one of the greatest Japanese mathematicians, K. Itô in 1949 and is now known as the *Itô stochastic integral*. Equation (1.1) is a linear stochastic differential equation (SDE) and is also known as the geometric Brownian motion which is a Nobel prize winning model in economics, namely the Black–Scholes model.

Let us now take one more step to see other random fluctuation. In their model (1.1), Black and Scholes assumed that the average rate of return μ and the volatility σ are constants. However, it has been proved by many authors that both of them, especially the volatility, are random processes in many situations. There is a strong evidence to indicate that the rate μ is a Markov jump process which can be modelled by a Markov chain. Of course, when the rate jumps, the volatility will jump accordingly. Taking these jumps into account, the classical model (1.1) has recently be generalised to form a new financial model

$$dx(t) = \mu(r(t))x(t)dt + \sigma(r(t))x(t)dB(t). \qquad (1.3)$$

Here $r(t)$ is a Markov chain with a finite state space $\mathbb{S} = \{1, 2, \ldots, N\}$ and μ, σ are now mappings from $\mathbb{S}$ to $[0, \infty)$. So, if the Markov chain is initially in state $r(0) = r_0 \in \mathbb{S}$, then before its first jump from r_0 to $r_1 \in \mathbb{S}$ at its first (random) jump time τ_1, the underlying asset price obeys the following geometric Brownian motion

$$dx(t) = \mu(r_0)x(t)dt + \sigma(r_0)x(t)dB(t)$$

with initial value $x(t_0) = x_0$. During this period from t_0 to τ_1, the rate and volatility are $\mu(r_0)$ and $\sigma(r_0)$, respectively. At time τ_1, the Markov chain jumps to r_1 where it will stay till the next jump at time τ_2. During the period from τ_1 to τ_2, the underlying asset price obeys another geometric Brownian motion

$$dx(t) = \mu(r_1)x(t)dt + \sigma(r_1)x(t)dB(t)$$

with initial value $x(\tau_1)$ at time τ_1, and the rate and volatility have been switched to $\mu(r_1)$ and $\sigma(r_1)$ from $\mu(r_0)$ and $\sigma(r_0)$, respectively. The underlying asset price will continue to switch from one geometric Brownian motion to other according to the Markovian switching. This type of random fluctuation, namely the Markovian switching, is one of the key features we are going to address in this book. Equation (1.3) is known as the geometric Brownian motion with Markovian switching or the hybrid geometric Brownian motion.

The main aims of this chapter are to introduce the stochastic nature of Brownian motion and to define the stochastic integral with respect to Brownian motion. To make this book self-contained, we shall briefly review the basic notations of probability theory and stochastic processes. We then give the mathematical definition of Brownian motions and introduce their important properties. Making use of these properties, we proceed to define the stochastic integral with respect to Brownian motion and establish the well-known Itô formula. To cope with the Markovian switching, which is the key feature of this book, we shall also review the essential notations and properties of Markov chains and establish the generalised Itô formula under Markovian switching.

1.2 Basic Notations of Probability Theory

Probability theory is concerned with the mathematical analysis of the intuitive notion of "chance" or "randomness," which, like all notions, is born of experience. The quantitative idea of randomness first took form at the gaming tables, and probability theory began, Pascal and Fermat (1654), as a theory of games of chance. Since then, the notion of chance has found its way into almost all branches of knowledge.

A theory becomes mathematical when it sets up a mathematical models of the phenomena with which it is concerned, that is, when, to describe the phenomena, it uses a collection of well-defined symbols and operations on the symbols. Probability theory deals with mathematical models of trials whose outcomes depend on chance. All the possible outcomes—the elementary events—are grouped together to form a set Ω with typical element $\omega \in \Omega$. Not every subset of Ω is in general an observable or interesting event. So we only group these observable or interesting events together as a family $\mathcal{F}$ of subsets of Ω. For the purpose of probability theory, such a family $\mathcal{F}$ should have the following properties:

- $\emptyset \in \mathcal{F}$, where $\emptyset$ denotes the empty set.
- If $A \in \mathcal{F}$, then its complement $A^C = \Omega - A \in \mathcal{F}$.
- If $\{A_i\}_{1 \leq i < \infty} \subset \mathcal{F}$, then $\bigcup_{i=1}^{\infty} A_i \in \mathcal{F}$.

A family $\mathcal{F}$ with these three properties is called a σ-*algebra*. The pair

$(\Omega, \mathcal{F})$ is called a *measurable space*, and the elements of $\mathcal{F}$ is henceforth called $\mathcal{F}$-*measurable sets* instead of events. If $\mathcal{C}$ is a family of subsets of Ω, then there exists a smallest σ-algebra $\sigma(\mathcal{C})$ on Ω which contains $\mathcal{C}$. This $\sigma(\mathcal{C})$ is called the σ-*algebra generated by* $\mathcal{C}$. If $\Omega = \mathbb{R}^n$ and $\mathcal{C}$ is the family of all open sets in $\mathbb{R}^n$, then $\mathcal{B}^n = \sigma(\mathcal{C})$ is called the *Borel* σ-*algebra* and the elements of $\mathcal{B}^n$ are called the *Borel sets*.

A real-valued function $X: \Omega \to \mathbb{R}$ is said to be $\mathcal{F}$-*measurable* if

$$\{\omega : X(\omega) \le a\} \in \mathcal{F} \quad \text{for all } a \in \mathbb{R}.$$

The function X is also called a real-valued ($\mathcal{F}$-measurable) *random variable*. An $\mathbb{R}^n$-valued function $X(\omega) = (X_1(\omega), \cdots, X_n(\omega))^T$ is said to be $\mathcal{F}$-*measurable* if all the elements X_i are $\mathcal{F}$-measurable. Similarly, an $n \times m$-matrix-valued function $X(\omega) = (X_{ij}(\omega))_{n \times m}$ is said to be $\mathcal{F}$-*measurable* if all the elements X_{ij} are $\mathcal{F}$-measurable. The *indicator function* I_A of a set $A \subset \Omega$ is defined by

$$I_A(\omega) = \begin{cases} 1 & \text{if } \omega \in A, \\ 0 & \text{if } \omega \notin A. \end{cases}$$

The indicator function I_A is $\mathcal{F}$-measurable if and only if A is an $\mathcal{F}$-measurable set, i.e. $A \in \mathcal{F}$. If the measurable space is $(\mathbb{R}^n, \mathcal{B}^n)$, a $\mathcal{B}^n$-measurable function is then called a *Borel measurable function*. More generally, let $(\Omega', \mathcal{F}')$ be another measurable space. A mapping $X: \Omega \to \Omega'$ is said to be $(\mathcal{F}, \mathcal{F}')$-*measurable* if

$$\{\omega : X(\omega) \in A'\} \in \mathcal{F} \quad \text{for all } A' \in \mathcal{F}'.$$

The mapping X is then called an Ω'-valued $(\mathcal{F}, \mathcal{F}')$-measurable (or simply, $\mathcal{F}$-measurable) random variable.

Let $X : \Omega \to \mathbb{R}^n$ be any function. The σ-*algebra* $\sigma(X)$ *generated by* X is the smallest σ-algebra on Ω containing all the sets $\{\omega : X(\omega) \in U\}$, $U \subset \mathbb{R}^n$ open. That is

$$\sigma(X) = \sigma(\{\omega : X(\omega) \in U\} : U \subset \mathbb{R}^n \text{ open}).$$

Clearly, X will then be $\sigma(X)$-measurable and $\sigma(X)$ is the smallest σ-algebra with this property. If X is $\mathcal{F}$-measurable, then $\sigma(X) \subset \mathcal{F}$, i.e. X generates a sub-σ-algebra of $\mathcal{F}$. If $\{X_i : i \in I\}$ is a collection of $\mathbb{R}^n$-valued functions, define

$$\sigma(X_i : i \in I) = \sigma\left(\bigcup_{i \in I} \sigma(X_i)\right)$$

which is called the *σ-algebra generated by* $\{X_i : i \in I\}$. It is the smallest σ-algebra with respect to which every X_i is measurable. The following result is useful. It is a special case of a result sometimes called the Doob–Dynkin lemma.

Lemma 1.1 *If $X, Y : \Omega \to \mathbb{R}^n$ are two given functions, then Y is $\sigma(X)$-measurable if and only if there exists a Borel measurable function $g : \mathbb{R}^n \to \mathbb{R}^n$ such that $Y = g(X)$.*

A *probability measure* $\mathbb{P}$ on a measurable space $(\Omega, \mathcal{F})$ is a function $\mathbb{P} : \mathcal{F} \to [0, 1]$ such that

(i) $\mathbb{P}(\Omega) = 1$;
(ii) for any disjoint sequence $\{A_i\}_{i \geq 1} \subset \mathcal{F}$ (i.e. $A_i \cap A_j = \emptyset$ if $i \neq j$)

$$\mathbb{P}\left(\bigcup_{i=1}^\infty A_i\right) = \sum_{i=1}^\infty \mathbb{P}(A_i).$$

The triple $(\Omega, \mathcal{F}, \mathbb{P})$ is called a *probability space*. If $(\Omega, \mathcal{F}, \mathbb{P})$ is a probability space, we set

$$\bar{\mathcal{F}} = \{A \subset \Omega : \exists\, B, C \in \mathcal{F} \text{ such that } B \subset A \subset C,\ \mathbb{P}(B) = \mathbb{P}(C)\}.$$

Then $\bar{\mathcal{F}}$ is a σ-algebra and is called the *completion* of $\mathcal{F}$. If $\mathcal{F} = \bar{\mathcal{F}}$, the probability space $(\Omega, \mathcal{F}, \mathbb{P})$ is said to be *complete*. If not, one can easily extend $\mathbb{P}$ to $\bar{\mathcal{F}}$ by defining $\mathbb{P}(A) = \mathbb{P}(B) = \mathbb{P}(C)$ for $A \in \bar{\mathcal{F}}$, where $B, C \in \mathcal{F}$ with the properties that $B \subset A \subset C$ and $\mathbb{P}(B) = \mathbb{P}(C)$. Now $(\Omega, \bar{\mathcal{F}}, \mathbb{P})$ is a complete probability space, called the *completion* of $(\Omega, \mathcal{F}, \mathbb{P})$.

In the remaining of this section, we let $(\Omega, \mathcal{F}, \mathbb{P})$ be a probability space. If X is a real-valued random variable and is *integrable* with respect to the probability measure $\mathbb{P}$, then the number

$$\mathbb{E}X = \int_\Omega X(\omega) d\mathbb{P}(\omega)$$

is called the *expectation* of X (with respect to $\mathbb{P}$). The number

$$Var(X) = \mathbb{E}(X - \mathbb{E}X)^2$$

is called the *variance* of X (here and in the sequel of this section we assume that all integrals concerned exist). The number $\mathbb{E}|X|^p$ ($p > 0$) is called the pth moment of X. If Y is another real-valued random variable,

$$Cov(X, Y) = \mathbb{E}[(X - \mathbb{E}X)(Y - \mathbb{E}Y)]$$

is called the *covariance* of X and Y. If $Cov(X,Y) = 0$, X and Y are said to be *uncorrelated*. For an $\mathbb{R}^n$-valued random variable $X = (X_1, \cdots, X_n)^T$, define $\mathbb{E}X = (\mathbb{E}X_1, \cdots, \mathbb{E}X_n)^T$. For an $n \times m$-matrix-valued random variable $X = (X_{ij})_{n \times m}$, define $\mathbb{E}X = (\mathbb{E}X_{ij})_{n \times m}$. If X and Y are both $\mathbb{R}^n$-valued random variables, the symmetric non-negative definite $n \times n$ matrix

$$Cov(X,Y) = \mathbb{E}[(X - \mathbb{E}X)(Y - \mathbb{E}Y)^T]$$

is called their *covariance matrix*.

Let X be an $\mathbb{R}^n$-valued random variable. Then X induces a probability measure μ_X on the Borel measurable space $(\mathbb{R}^n, \mathcal{B}^n)$, defined by

$$\mu_X(B) = \mathbb{P}\{\omega : X(\omega) \in B\} \quad \text{for } B \in \mathcal{B}^n,$$

and μ_X is called the *distribution* of X. The expectation of X can now be expressed as

$$\mathbb{E}X = \int_{\mathbb{R}^n} x d\mu_X(x).$$

More generally, if $g : \mathbb{R}^n \to \mathbb{R}^m$ is Borel measurable, we then have the following *transformation formula*

$$\mathbb{E}g(X) = \int_{\mathbb{R}^n} g(x) d\mu_X(x).$$

For $p \in (0, \infty)$, let $L^p = L^p(\Omega; \mathbb{R}^n)$ be the family of $\mathbb{R}^n$-valued random variables X with $\mathbb{E}|X|^p < \infty$. In L^1, we have $|\mathbb{E}X| \leq \mathbb{E}|X|$. Moreover, the following three inequalities are very useful:

- **Hölder's inequality**

$$|\mathbb{E}(X^T Y)| \leq (\mathbb{E}|X|^p)^{1/p} (\mathbb{E}|Y|^q)^{1/q}$$

if $p > 1$, $1/p + 1/q = 1$, $X \in L^p$, $Y \in L^q$;
- **Minkovski's inequality**

$$(\mathbb{E}|X + Y|^p)^{1/p} \leq (\mathbb{E}|X|^p)^{1/p} + (\mathbb{E}|Y|^p)^{1/p}$$

if $p > 1, X, Y \in L^p$;
- **Chebyshev's inequality**

$$\mathbb{P}\{\omega : |X(\omega)| \geq c\} \leq c^{-p}\mathbb{E}|X|^p$$

if $c > 0$, $p > 0$, $X \in L^p$.

A simple application of Hölder's inequality implies

$$(\mathbb{E}|X|^r)^{1/r} \leq (\mathbb{E}|X|^p)^{1/p}$$

if $0 < r < p < \infty$, $X \in L^p$.

Let X and X_k, $k \geq 1$, be $\mathbb{R}^n$-valued random variables. The following four convergence concepts are very important:

(a) If there exists a $\mathbb{P}$-null set $\Omega_0 \in \mathcal{F}$ such that for every $\omega \notin \Omega_0$, the sequence $\{X_k(\omega)\}$ converges to $X(\omega)$ in the usual sense in $\mathbb{R}^n$, then $\{X_k\}$ is said to converge to X *almost surely* or *with probability 1*, and we write $\lim_{k\to\infty} X_k = X$ a.s.

(b) If for every $\varepsilon > 0$, $\mathbb{P}\{\omega : |X_k(\omega) - X(\omega)| > \varepsilon\} \to 0$ as $k \to \infty$, then $\{X_k\}$ is said to converge to X *stochastically* or *in probability*.

(c) If X_k and X belong to L^p and $\mathbb{E}|X_k - X|^p \to 0$, then $\{X_k\}$ is said to converge to X *in pth moment* or *in L^p*.

(d) If for every real-valued continuous bounded function g defined on $\mathbb{R}^n$, $\lim_{k\to\infty} \mathbb{E}g(X_k) = \mathbb{E}g(X)$, then $\{X_k\}$ is said to converge to X *in distribution*.

These convergence concepts have the following relationship:

$$\text{convergence in } L^p$$
$$\Downarrow$$
$$\text{a.s. convergence} \Rightarrow \text{convergence in probability}$$
$$\Downarrow$$
$$\text{convergence in distribution}$$

Furthermore, a sequence converges in probability if and only if every subsequence of it contains an almost surely convergent subsequence. A sufficient condition for $\lim_{k\to\infty} X_k = X$ a.s. is the condition

$$\sum_{k=1}^{\infty} \mathbb{E}|X_k - X|^p < \infty \quad \text{for some } p > 0.$$

Let us now state two very important integration convergence theorems.

Theorem 1.1 *(Monotonic convergence theorem) If $\{X_k\}$ is an increasing sequence of non-negative random variables, then*

$$\lim_{k\to\infty} \mathbb{E}X_k = \mathbb{E}\left(\lim_{k\to\infty} X_k\right).$$

Theorem 1.2 *(Dominated convergence theorem) Let $p \geq 1$, $\{X_k\} \subset L^p(\Omega; \mathbb{R}^n)$ and $Y \in L^p(\Omega; \mathbb{R})$. Assume that $|X_k| \leq Y$ a.s. and $\{X_k\}$ converges to X in probability. Then $X \in L^p(\Omega; \mathbb{R}^n)$, $\{X_k\}$ converges to X in L^p, and*

$$\lim_{k \to \infty} \mathbb{E} X_k = \mathbb{E} X.$$

*When Y is bounded, this theorem is also referred to as the **bounded convergence theorem**.*

Two sets $A, B \in \mathcal{F}$ are said to be *independent* if $\mathbb{P}(A \cap B) = \mathbb{P}(A)\mathbb{P}(B)$. Three sets $A, B, C \in \mathcal{F}$ are said to be *independent* if

$$\mathbb{P}(A \cap B) = \mathbb{P}(A)\mathbb{P}(B), \quad \mathbb{P}(A \cap C) = \mathbb{P}(A)\mathbb{P}(C),$$

$$\mathbb{P}(B \cap C) = \mathbb{P}(B)\mathbb{P}(C) \quad \text{and} \quad \mathbb{P}(A \cap B \cap C) = \mathbb{P}(A)\mathbb{P}(B)\mathbb{P}(C).$$

Let I be an index set. A collection of sets $\{A_i : i \in I\} \subset \mathcal{F}$ is said to be *independent* if

$$\mathbb{P}(A_{i_1} \cap \cdots \cap A_{i_k}) = \mathbb{P}(A_{i_1}) \cdots \mathbb{P}(A_{i_k})$$

holds for arbitrary distinct finite indices $i_1, \cdots, i_k \in I$. Two sub-σ-algebras $\mathcal{F}_1$ and $\mathcal{F}_2$ of $\mathcal{F}$ are said to be *independent* if

$$\mathbb{P}(A_1 \cap A_2) = \mathbb{P}(A_1)\mathbb{P}(A_2) \quad \text{for all } A_1 \in \mathcal{F}_1,\ A_2 \in \mathcal{F}_2.$$

A collection of sub-σ-algebras $\{\mathcal{F}_i : i \in I\}$ is said to be *independent* if for arbitrary distinct finite indices $i_1, \cdots, i_k \in I$,

$$\mathbb{P}(A_{i_1} \cap \cdots \cap A_{i_k}) = \mathbb{P}(A_{i_1}) \cdots \mathbb{P}(A_{i_k})$$

holds for all $A_{i_1} \in \mathcal{F}_{i_1}, \cdots, A_{i_k} \in \mathcal{F}_{i_k}$. A family of random variables $\{X_i : i \in I\}$ (whose ranges may differ for different values of the index) is said to be *independent* if the σ-algebras $\sigma(X_i)$, $i \in I$ generated by them are independent. For example, two random variables $X : \Omega \to \mathbb{R}^n$ and $Y : \Omega \to \mathbb{R}^m$ are independent if and only if

$$\mathbb{P}\{\omega : X(\omega) \in A,\ Y(\omega) \in B\} = \mathbb{P}\{\omega : X(\omega) \in A\}\ \mathbb{P}\{\omega : Y(\omega) \in B\}$$

holds for all $A \in \mathcal{B}^n$, $B \in \mathcal{B}^m$. If X and Y are two independent real-valued integrable random variables, then XY is also integrable and

$$\mathbb{E}(XY) = \mathbb{E} X\ \mathbb{E} Y.$$

If $X, Y \in L^2(\Omega; \mathbb{R})$ are uncorrelated, then
$$Var(X+Y) = Var(X) + Var(Y).$$

If the X and Y are independent, they are uncorrelated; but the converse is not true. However, if (X, Y) has a joint normal distribution, then X and Y are independent if and only if they are uncorrelated.

Let $\{A_k\}$ be a sequence of sets in $\mathcal{F}$. The set of all those points which belong to almost all A_k (all but any finite number) is called the *inferior limit* of A_k, and is denoted by $\liminf_{k\to\infty} A_k$. Clearly,
$$\liminf_{k\to\infty} A_k = \bigcup_{i=1}^{\infty} \bigcap_{k=i}^{\infty} A_k.$$

The set of all those points which belong to infinitely many A_k is called the *superior limit* of A_k and is denoted by $\limsup_{k\to\infty} A_k$. It is easy to see
$$\limsup_{k\to\infty} A_k = \bigcap_{i=1}^{\infty} \bigcup_{k=i}^{\infty} A_k.$$

Moreover,
$$\liminf_{k\to\infty} A_k \subset \limsup_{k\to\infty} A_k.$$

With regard to their probabilities, we have the following well-known *Borel–Cantelli lemma*.

Lemma 1.2 *(Borel–Cantelli's lemma)*
(1) If $\{A_k\} \subset \mathcal{F}$ and $\sum_{k=1}^{\infty} \mathbb{P}(A_k) < \infty$, then
$$\mathbb{P}\left(\limsup_{k\to\infty} A_k\right) = 0.$$

That is, there exists a set $\Omega_1 \in \mathcal{F}$ with $\mathbb{P}(\Omega_1) = 1$ and an integer-valued random variable k_1 such that for every $\omega \in \Omega_1$ we have $\omega \notin A_k$ whenever $k \geq k_1(\omega)$.

(2) If the sequence $\{A_k\} \subset \mathcal{F}$ is independent and $\sum_{k=1}^{\infty} \mathbb{P}(A_k) = \infty$, then
$$\mathbb{P}\left(\limsup_{k\to\infty} A_k\right) = 1.$$

That is, there exists a set $\Omega_2 \in \mathcal{F}$ with $\mathbb{P}(\Omega_2) = 1$ such that for every $\omega \in \Omega_2$, there exists a sub-sequence $\{A_{k_i}\}$ such that the ω belongs to every A_{k_i}.

Let $A, B \in \mathcal{F}$ with $\mathbb{P}(B) > 0$. The *conditional probability of A under condition B* is

$$\mathbb{P}(A|B) = \frac{\mathbb{P}(A \cap B)}{\mathbb{P}(B)}.$$

However, we frequently encounter a family of conditions so we need the more general concept of *conditional expectation*. Let $X \in L^1(\Omega; \mathbb{R})$. Let $\mathcal{G} \subset \mathcal{F}$ be a sub-σ-algebra of $\mathcal{F}$ so $(\Omega, \mathcal{G})$ is a measurable space. In general, X is not $\mathcal{G}$-measurable. We now seek an integrable $\mathcal{G}$-measurable random variable Y such that it has the same values as X on the average in the sense that

$$\mathbb{E}(I_G Y) = \mathbb{E}(I_G X) \quad \text{i.e.} \quad \int_G Y(\omega) d\mathbb{P}(\omega) = \int_G X(\omega) d\mathbb{P}(\omega) \quad \forall G \in \mathcal{G}.$$

By the Radon–Nikodym theorem, there exists one such Y, almost surely unique. It is called the *conditional expectation of X under the condition $\mathcal{G}$*, and we write

$$Y = \mathbb{E}(X|\mathcal{G}).$$

If $\mathcal{G}$ is the σ-algebra generated by a random variable Y, we write

$$\mathbb{E}(X|\mathcal{G}) = \mathbb{E}(X|Y).$$

As an example, consider a collection of sets $\{A_k\} \subset \mathcal{F}$ with

$$\bigcup_k A_k = \Omega, \quad \mathbb{P}(A_k) > 0, \quad A_k \cap A_j = \emptyset \text{ if } k \neq j.$$

Let $\mathcal{G} = \sigma(\{A_k\})$, i.e. $\mathcal{G}$ is generated by $\{A_k\}$. Then $\mathbb{E}(X|\mathcal{G})$ is a *step function* on Ω given by

$$\mathbb{E}(X|\mathcal{G}) = \sum_k \frac{I_{A_k} \mathbb{E}(I_{A_k} X)}{\mathbb{P}(A_k)}.$$

In other words, if $\omega \in A_k$,

$$\mathbb{E}(X|\mathcal{G})(\omega) = \frac{\mathbb{E}(I_{A_k} X)}{\mathbb{P}(A_k)}.$$

It follows from the definition that

$$\mathbb{E}(\mathbb{E}(X|\mathcal{G})) = \mathbb{E}(X)$$

and

$$|\mathbb{E}(X|\mathcal{G})| \leq \mathbb{E}(|X| \,|\mathcal{G}) \quad a.s.$$

Other important properties of the conditional expectation are as follows (all the equalities and inequalities shown hold almost surely):

- $\mathbb{E}(X|\mathcal{G}) = \mathbb{E}X$ if $\mathcal{G} = \{\emptyset, \Omega\}$;
- $\mathbb{E}(X|\mathcal{G}) \geq 0$ if $X \geq 0$;
- $\mathbb{E}(X|\mathcal{G}) = X$ if X is $\mathcal{G}$-measurable;
- $\mathbb{E}(X|\mathcal{G}) = c$ if $X = c = const.$;
- $\mathbb{E}(aX + bY|\mathcal{G}) = a\mathbb{E}(X|\mathcal{G}) + b\mathbb{E}(Y|\mathcal{G})$ if $a, b \in \mathbb{R}$;
- $\mathbb{E}(X|\mathcal{G}) \leq \mathbb{E}(Y|\mathcal{G})$ if $X \leq Y$;
- $\mathbb{E}(XY|\mathcal{G}) = X\mathbb{E}(Y|\mathcal{G})$ if X is $\mathcal{G}$-measurable, in particular, $\mathbb{E}(\mathbb{E}(X|\mathcal{G})\,Y|\mathcal{G}) = \mathbb{E}(X|\mathcal{G})\,\mathbb{E}(Y|\mathcal{G})$;
- $\mathbb{E}(X|\mathcal{G}) = \mathbb{E}X$ if $\sigma(X)$ and $\mathcal{G}$ are independent, in particular, $\mathbb{E}(X|Y) = \mathbb{E}X$ if X, Y are independent;
- $\mathbb{E}(\mathbb{E}(X|\mathcal{G}_2)|\mathcal{G}_1) = \mathbb{E}(X|\mathcal{G}_1)$ if $\mathcal{G}_1 \subset \mathcal{G}_2 \subset \mathcal{F}$.

Finally, if $X = (X_1, \cdots, X_n)^T \in L^1(\Omega; \mathbb{R}^n)$, its *conditional expectation under* $\mathcal{G}$ is defined as

$$\mathbb{E}(X|\mathcal{G}) = (\mathbb{E}(X_1|\mathcal{G}), \cdots, \mathbb{E}(X_n|\mathcal{G}))^T.$$

1.3 Stochastic Processes

Let $(\Omega, \mathcal{F}, \mathbb{P})$ be a probability space. A *filtration* is a family $\{\mathcal{F}_t\}_{t \geq 0}$ of increasing sub-σ-algebras of $\mathcal{F}$ (i.e. $\mathcal{F}_t \subset \mathcal{F}_s \subset \mathcal{F}$ for all $0 \leq t < s < \infty$). The filtration is said to be *right continuous* if $\mathcal{F}_t = \bigcap_{s > t} \mathcal{F}_s$ for all $t \geq 0$. When the probability space is complete, the filtration is said to satisfy the *usual conditions* if it is right continuous and $\mathcal{F}_0$ contains all $\mathbb{P}$-null sets.

From now on, unless otherwise specified, we shall always work on a given complete probability space $(\Omega, \mathcal{F}, \mathbb{P})$ with a filtration $\{\mathcal{F}_t\}_{t \geq 0}$ satisfying the usual conditions. We also define $\mathcal{F}_\infty = \sigma(\bigcup_{t \geq 0} \mathcal{F}_t)$, i.e. the σ-algebra generated by $\bigcup_{t \geq 0} \mathcal{F}_t$.

A family $\{X_t\}_{t \in I}$ of $\mathbb{R}^n$-valued random variables is called a *stochastic process* with *parameter set* (or *index set*) I and *state space* $\mathbb{R}^n$. The parameter set I is usually (as in this book) the half line $\mathbb{R}_+ = [0, \infty)$, but it may also be an interval $[a, b]$, the non-negative integers or even subsets of

$\mathbb{R}^n$. Note that for each fixed $t \in I$ we have a random variable

$$\Omega \ni \omega \to X_t(\omega) \in \mathbb{R}^n.$$

On the other hand, for each fixed $\omega \in \Omega$ we have a function

$$I \ni t \to X_t(\omega) \in \mathbb{R}^n$$

which is called a *sample path* of the process, and we shall write $X_\cdot(\omega)$ for the path. Sometimes it is convenient to write $X(t,\omega)$ instead of $X_t(\omega)$, and the stochastic process may be regarded as a function of two variables (t,ω) from $I \times \Omega$ to $\mathbb{R}^n$. Similarly, one can define matrix-valued stochastic processes *etc.* We often write a stochastic process $\{X_t\}_{t\geq 0}$ as $\{X_t\}$, X_t or $X(t)$.

Let $\{X_t\}_{t\geq 0}$ be an $\mathbb{R}^n$-valued stochastic process. It is said to be *continuous* (resp. *right continuous*, *left continuous*) if for almost all $\omega \in \Omega$ function $X_t(\omega)$ is continuous (resp. right continuous, left continuous) on $t \geq 0$. It is said to be *cadlag* (*right continuous and left limit*) if it is right continuous and for almost all $\omega \in \Omega$ the left limit $\lim_{s \uparrow t} X_s(\omega)$ exists and is finite for all $t > 0$. It is said to be *integrable* if for every $t \geq 0$, X_t is an integrable random variable. It is said to be $\{\mathcal{F}_t\}$-*adapted* (or simply, *adapted*) if for every t, X_t is $\mathcal{F}_t$-measurable. It is said to be *measurable* if the stochastic process regarded as a function of two variables (t,ω) from $\mathbb{R}_+ \times \Omega$ to $\mathbb{R}^n$ is $\mathcal{B}(\mathbb{R}_+) \times \mathcal{F}$-measurable, where $\mathcal{B}(\mathbb{R}_+)$ is the family of all Borel sub-sets of $\mathbb{R}_+$. The stochastic process is said to be *progressively measurable* or *progressive* if for every $T \geq 0$, $\{X_t\}_{0 \leq t \leq T}$ regarded as a function of (t,ω) from $[0, T] \times \Omega$ to $\mathbb{R}^n$ is $\mathcal{B}([0, T]) \times \mathcal{F}_T$-measurable, where $\mathcal{B}([0, T])$ is the family of all Borel sub-sets of $[0, T]$. Let $\mathcal{O}$ (resp. $\mathcal{P}$) denote the smallest σ-algebra on $\mathbb{R}_+ \times \Omega$ with respect to which every cadlag adapted process (resp. left continuous process) is a measurable function of (t,ω). A stochastic process is said to be *optional* (resp. *predictable*) if the process regarded as a function of (t,ω) is $\mathcal{O}$-measurable (resp. $\mathcal{P}$-measurable). A real-valued stochastic process $\{A_t\}_{t\geq 0}$ is called an *increasing process* if for almost all $\omega \in \Omega$, $A_t(\omega)$ is non-negative nondecreasing right continuous on $t \geq 0$. It is called a *process of finite variation* if $A_t = \bar{A}_t - \hat{A}_t$ with $\{\bar{A}_t\}$ and $\{\hat{A}_t\}$ both increasing processes. It is obvious that the processes of finite variation are cadlag. Hence the adapted processes of finite variation are optional.

The relations among the various stochastic processes are summarised as follows:

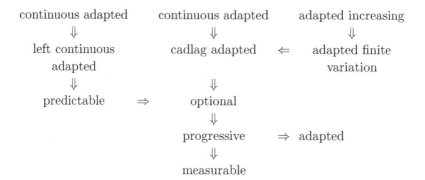

Let $\{X_t\}_{t\geq 0}$ be a stochastic process. Another stochastic process $\{Y_t\}_{t\geq 0}$ is called a *version* or *modification* of $\{X_t\}$ if for all $t \geq 0$, $X_t = Y_t$ a.s. (i.e. $\mathbb{P}\{\omega : X_t(\omega) = Y_t(\omega)\} = 1$). Two stochastic processes $\{X_t\}_{t\geq 0}$ and $\{Y_t\}_{t\geq 0}$ are said to be *indistinguishable* if for almost all $\omega \in \Omega$, $X_t(\omega) = Y_t(\omega)$ for all $t \geq 0$ (i.e. $\mathbb{P}\{\omega : X_t(\omega) = Y_t(\omega) \text{ for all } t \geq 0\} = 1$).

A random variable $\tau : \Omega \to [0, \infty]$ (it may take the value ∞) is called an $\{\mathcal{F}_t\}$-*stopping time* (or simply, *stopping time*) if $\{\omega : \tau(\omega) \leq t\} \in \mathcal{F}_t$ for any $t \geq 0$. Let τ and ρ be two stopping times with $\tau \leq \rho$ a.s. We define

$$[[\tau, \rho[[= \{(t, \omega) \in \mathbb{R}_+ \times \Omega : \tau(\omega) \leq t < \rho(\omega)\}$$

and call it a *stochastic interval*. Similarly, we can define stochastic intervals $[[\tau, \rho]]$, $]]\tau, \rho]]$ and $]]\tau, \rho[[$. If τ is a stopping time, define

$$\mathcal{F}_\tau = \{A \in \mathcal{F} : A \cap \{\omega : \tau(\omega) \leq t\} \in \mathcal{F}_t \text{ for all } t \geq 0\}$$

which is a sub-σ-algebra of $\mathcal{F}$. If τ and ρ are two stopping times with $\tau \leq \rho$ a.s., then $\mathcal{F}_\tau \subset \mathcal{F}_\rho$. The following two theorems are useful.

Theorem 1.3 *If $\{X_t\}_{t\geq 0}$ is a progressively measurable process and τ is a stopping time, then $X_\tau I_{\{\tau < \infty\}}$ is $\mathcal{F}_\tau$-measurable. In particular, if τ is finite, then X_τ is $\mathcal{F}_\tau$-measurable.*

Theorem 1.4 *Let $\{X_t\}_{t\geq 0}$ be an $\mathbb{R}^n$-valued cadlag $\{\mathcal{F}_t\}$-adapted process, and D an open subset of $\mathbb{R}^n$. Define*

$$\tau = \inf\{t \geq 0 : X_t \notin D\},$$

where we use the convention $\inf \emptyset = \infty$. Then τ is an $\{\mathcal{F}_t\}$-stopping time, and is called the first exit time from D. Moreover, if ρ is a stopping time, then

$$\theta = \inf\{t \geq \rho : X_t \notin D\}$$

is also an $\{\mathcal{F}_t\}$-stopping time, and is called the first exit time from D after ρ.

An $\mathbb{R}^n$-valued $\{\mathcal{F}_t\}$-adapted integrable process $\{M_t\}_{t\geq 0}$ is called a *martingale with respect to* $\{\mathcal{F}_t\}$ (or simply, *martingale*) if
$$\mathbb{E}(M_t|\mathcal{F}_s) = M_s \quad \text{a.s. for all } 0 \leq s < t < \infty.$$
It should be pointed out that every martingale has a cadlag modification since we always assume that the filtration $\{\mathcal{F}_t\}$ is right continuous. Therefore we can always assume that any martingale is cadlag in the sequel. If $X = \{X_t\}_{t\geq 0}$ is a progressively measurable process and τ is a stopping time, then $X^\tau = \{X_{\tau \wedge t}\}_{t\geq 0}$ is called a *stopped process* of X. The following is the well-known Doob martingale stopping theorem.

Theorem 1.5 *Let $\{M_t\}_{t\geq 0}$ be an $\mathbb{R}^n$-valued martingale with respect to $\{\mathcal{F}_t\}$, and let θ, ρ be two finite stopping times. Then*
$$\mathbb{E}(M_\theta|\mathcal{F}_\rho) = M_{\theta \wedge \rho} \quad a.s.$$
In particular, if τ is a stopping time, then
$$\mathbb{E}(M_{\tau \wedge t}|\mathcal{F}_s) = M_{\tau \wedge s} \quad a.s.$$
holds for all $0 \leq s < t < \infty$. That is, the stopped process $M^\tau = \{M_{\tau \wedge t}\}$ is still a martingale with respect to the same filtration $\{\mathcal{F}_t\}$.

A stochastic process $X = \{X_t\}_{t\geq 0}$ is called *square-integrable* if $\mathbb{E}|X_t|^2 < \infty$ for every $t \geq 0$. If $M = \{M_t\}_{t\geq 0}$ is a real-valued square-integrable continuous martingale, then there exists a unique continuous integrable adapted increasing process denoted by $\{\langle M, M \rangle_t\}$ such that $\{M_t^2 - \langle M, M \rangle_t\}$ is a continuous martingale vanishing at $t = 0$. The process $\{\langle M, M \rangle_t\}$ is called the *quadratic variation* of M. In particular, for any finite stopping time τ,
$$\mathbb{E}M_\tau^2 = \mathbb{E}\langle M, M \rangle_\tau.$$
If $N = \{N_t\}_{t\geq 0}$ is another real-valued square-integrable continuous martingale, we define
$$\langle M, N \rangle_t = \frac{1}{2}\Big(\langle M+N, M+N \rangle_t - \langle M, M \rangle_t - \langle N, N \rangle_t\Big),$$
and call $\{\langle M, N \rangle_t\}$ the *joint quadratic variation* of M and N. It is useful to know that $\{\langle M, N \rangle_t\}$ is the unique continuous integrable adapted process

of finite variation such that $\{M_t N_t - \langle M, N \rangle_t\}$ is a continuous martingale vanishing at $t = 0$. In particular, for any finite stopping time τ,

$$\mathbb{E} M_\tau N_\tau = \mathbb{E} \langle M, N \rangle_\tau.$$

A right continuous adapted process $M = \{M_t\}_{t \geq 0}$ is called a *local martingale* if there exists a nondecreasing sequence $\{\tau_k\}_{k \geq 1}$ of stopping times with $\tau_k \uparrow \infty$ a.s. such that every $\{M_{\tau_k \wedge t} - M_0\}_{t \geq 0}$ is a martingale. Every martingale is a local martingale (by Theorem 1.5), but the converse is not true. If $M = \{M_t\}_{t \geq 0}$ and $N = \{N_t\}_{t \geq 0}$ are two real-valued continuous local martingales, their *joint quadratic variation* $\{\langle M, N \rangle\}_{t \geq 0}$ is the unique continuous adapted process of finite variation such that $\{M_t N_t - \langle M, N \rangle_t\}_{t \geq 0}$ is a continuous local martingale vanishing at $t = 0$. When $M = N$, $\{\langle M, M \rangle\}_{t \geq 0}$ is called the *quadratic variation* of M. The following result is the useful strong law of large numbers.

Theorem 1.6 *(Strong law of large numbers)* Let $M = \{M_t\}_{t \geq 0}$ be a real-valued continuous local martingale vanishing at $t = 0$. Then

$$\lim_{t \to \infty} \langle M, M \rangle_t = \infty \quad \text{a.s.} \quad \Rightarrow \quad \lim_{t \to \infty} \frac{M_t}{\langle M, M \rangle_t} = 0 \quad \text{a.s.}$$

and also

$$\limsup_{t \to \infty} \frac{\langle M, M \rangle_t}{t} < \infty \quad \text{a.s.} \quad \Rightarrow \quad \lim_{t \to \infty} \frac{M_t}{t} = 0 \quad \text{a.s.}$$

More generally, if $A = \{A_t\}_{t \geq 0}$ is a continuous adapted increasing process such that

$$\lim_{t \to \infty} A_t = \infty \quad \text{and} \quad \int_0^\infty d\langle M, M \rangle_t (1 + A_t)^2 < \infty \quad \text{a.s.}$$

then

$$\lim_{t \to \infty} \frac{M_t}{A_t} = 0 \quad \text{a.s.}$$

A real-valued $\{\mathcal{F}_t\}$-adapted integrable process $\{M_t\}_{t \geq 0}$ is called a *supermartingale (with respect to $\{\mathcal{F}_t\}$)* if

$$\mathbb{E}(M_t | \mathcal{F}_s) \leq M_s \quad \text{a.s. for all } 0 \leq s < t < \infty.$$

It is called a *submartingale (with respect to $\{\mathcal{F}_t\}$)* if we replace the sign $\leq$ in the last formula with $\geq$. Clearly, $\{M_t\}$ is submartingale if and only if $\{-M_t\}$ is supermartingale. For a real-valued martingale $\{M_t\}$, $\{M_t^+ := \max(M_t, 0)\}$ and $\{M_t^- := \max(0, -M_t)\}$ are submartingales. For a supermartingale (resp. submartingale), $\mathbb{E}M_t$ is monotonically decreasing (resp. increasing). Moreover, if $p \geq 1$ and $\{M_t\}$ is an $\mathbb{R}^n$-valued martingale such that $M_t \in L^p(\Omega; \mathbb{R}^n)$, then $\{|M_t|^p\}$ is a non-negative submartingale. Moreover, Doob's stopping Theorem 1.5 holds for supermartingales and submartingales as well.

Theorem 1.7 *(Doob's martingale convergence theorem)*
(i) Let $\{M_t\}_{t\geq 0}$ be a real-valued right-continuous supermartingale. If
$$\sup_{0\leq t<\infty} \mathbb{E}M_t^- < \infty,$$
then M_t converges almost surely to a random variable $M_\infty \in L^1(\Omega; \mathbb{R})$. In particular, this holds if M_t is non-negative.

(ii) Let $\{M_t\}_{t\geq 0}$ be a real-valued right-continuous supermartingale. Then $\{M_t\}_{t\geq 0}$ is uniformly integrable, i.e.
$$\lim_{c\to\infty}\left[\sup_{t\geq 0}\mathbb{E}\Big(I_{\{|M_t|\geq c\}}|M_t|\Big)\right] = 0$$
if and only if there exists a random variable $M_\infty \in L^1(\Omega; \mathbb{R})$ such that $M_t \to M_\infty$ a.s. and in L^1 as well.

(iii) Let $X \in L^1(\Omega; \mathbb{R})$. Then
$$\mathbb{E}(X|\mathcal{F}_t) \to \mathbb{E}(X|\mathcal{F}_\infty) \quad \text{as } t\to\infty$$
a.s. and in L^1 as well.

Theorem 1.8 *(Supermartingale inequalities)* Let $\{M_t\}_{t\geq 0}$ be a real-valued supermartingale. Let $[a,b]$ be a bounded interval in $\mathbb{R}_+$. Then
$$c\, P\Big\{\omega : \sup_{a\leq t\leq b} M_t(\omega) \geq c\Big\} \leq \mathbb{E}M_a + \mathbb{E}M_b^-$$
and
$$c\, P\Big\{\omega : \inf_{a\leq t\leq b} M_t(\omega) \leq -c\Big\} \leq \mathbb{E}M_b^-$$
hold for all $c > 0$.

For submartingales we have the following well-known Doob inequality.

Theorem 1.9 *(Doob's submartingale inequalities)* Let $p > 1$. Let $\{M_t\}_{t\geq 0}$ be a real-valued non-negative submartingale such that $M_t \in$

$L^p(\Omega; \mathbb{R})$. Let $[a,b]$ be a bounded interval in $\mathbb{R}_+$. Then

$$\mathbb{E}\left(\sup_{a \le t \le b} M_t^p\right) \le \left(\frac{p}{p-1}\right)^p \mathbb{E} M_b^p.$$

Theorem 1.10 *(Non-negative semimartingale convergence theorem) Let $\{A_t\}_{t \ge 0}$ and $\{U_t\}_{t \ge 0}$ be two continuous adapted increasing processes with $A_0 = U_0 = 0$ a.s. Let $\{M_t\}_{t \ge 0}$ be a real-valued continuous local martingale with $M_0 = 0$ a.s. Let ξ be a non-negative $\mathcal{F}_0$-measurable random variable such that $\mathbb{E}\xi < \infty$. Define*

$$X_t = \xi + A_t - U_t + M_t \quad \text{for } t \ge 0.$$

If X_t is non-negative, then

$$\left\{\lim_{t \to \infty} A_t < \infty\right\} \subset \left\{\lim_{t \to \infty} X_t \text{ exists and is finite}\right\} \cap \left\{\lim_{t \to \infty} U_t < \infty\right\} \quad \text{a.s.}$$

where $B \subset D$ a.s. means $\mathbb{P}(B \cap D^c) = 0$. In particular, if $\lim_{t \to \infty} A_t < \infty$ a.s., then for almost all $\omega \in \Omega$

$$\lim_{t \to \infty} X_t(\omega) < \infty, \quad \lim_{t \to \infty} U_t(\omega) < \infty \quad \text{and} \quad -\infty < \lim_{t \to \infty} M_t(\omega) < \infty.$$

If we apply these results to an $\mathbb{R}^n$-valued martingale, we obtain the following Doob martingale inequalities.

Theorem 1.11 *(Doob's martingale inequalities) Let $\{M_t\}_{t \ge 0}$ be an $\mathbb{R}^n$-valued martingale. Let $[a,b]$ be a bounded interval in $\mathbb{R}_+$.*
(i) If $p \ge 1$, $c > 0$ and $M_t \in L^p(\Omega; \mathbb{R}^n)$, then

$$\mathbb{P}\left\{\omega : \sup_{a \le t \le b} |M_t(\omega)| \ge c\right\} \le \frac{\mathbb{E}|M_b|^p}{c^p}.$$

(ii) If $p > 1$ and $M_t \in L^p(\Omega; \mathbb{R}^n)$, then

$$\mathbb{E}\left(\sup_{a \le t \le b} |M_t|^p\right) \le \left(\frac{p}{p-1}\right)^p \mathbb{E}|M_b|^p.$$

1.4 Brownian Motions

Brownian motion is at the heart of most models in practice. Its name comes from the Scottish botanist Robert Brown who, in around 1827, reported experimental observations involving the erratic behaviour of a pollen grain when bombarded by (relatively small and effectively invisible) water molecules. A mathematical theory for Brownian motion has since been developed, with famous names such as Albert Einstein and Norbert Weiner making significant contributions. To describe the motion mathematically it is natural to use the concept of a stochastic process $B_t(\omega)$, interpreted as the position of the pollen grain ω at time t. Let us now give the mathematical definition of Brownian motion.

Definition 1.12 Let $(\Omega, \mathcal{F}, \mathbb{P})$ be a probability space with a filtration $\{\mathcal{F}_t\}_{t\geq 0}$. A (standard) one-dimensional Brownian motion is a real-valued continuous $\{\mathcal{F}_t\}$-adapted process $\{B_t\}_{t\geq 0}$ with the following properties:

(i) $B_0 = 0$ a.s.;
(ii) for $0 \leq s < t < \infty$, the increment $B_t - B_s$ is normally distributed with mean zero and variance $t - s$;
(iii) for $0 \leq s < t < \infty$, the increment $B_t - B_s$ is independent of $\mathcal{F}_s$.

We shall sometimes speak of a Brownian motion $\{B_t\}_{0\leq t\leq T}$ on $[0,T]$, for some $T > 0$, and the meaning of this terminology is apparent.

If $\{B_t\}_{t\geq 0}$ is a Brownian motion and $0 \leq t_0 < t_1 < \cdots < t_k < \infty$, then the increments $B_{t_i} - B_{t_{i-1}}$, $1 \leq i \leq k$ are independent, and we say that Brownian motion has *independent increments*. Moreover, the distribution of $B_{t_i} - B_{t_{i-1}}$ depends only on the difference $t_i - t_{i-1}$, and we say that Brownian motion has *stationary increments*.

The filtration $\{\mathcal{F}_t\}$ is a part of the definition of Brownian motion. However, we sometimes speak of a Brownian motion on a probability space $(\Omega, \mathcal{F}, \mathbb{P})$ without filtration. That is, $\{B_t\}_{t\geq 0}$ is a real-valued continuous process with properties (i) and (ii) but property (iii) is replaced by that it has independent increments. In this case, define $\mathcal{F}_t^B = \sigma(B_s : 0 \leq s \leq t)$ for $t \geq 0$, i.e. $\mathcal{F}_t^B$ is the σ-algebra generated by $\{B_s : 0 \leq s \leq t\}$. We call $\{\mathcal{F}_t^B\}_{t\geq 0}$ the *natural filtration* generated by $\{B_t\}$. Clearly, $\{B_t\}$ is a Brownian motion with respect to the natural filtration $\{\mathcal{F}_t^B\}$. Moreover, if $\{\mathcal{F}_t\}$ is a "larger" filtration in the sense that $\mathcal{F}_t^B \subset \mathcal{F}_t$ for $t \geq 0$, and $B_t - B_s$ is independent of $\mathcal{F}_s$ whenever $0 \leq s < t < \infty$, then $\{B_t\}$ is a Brownian motion with respect to the filtration $\{\mathcal{F}_t\}$.

In the definition we do not require the probability space $(\Omega, \mathcal{F}, \mathbb{P})$ be complete and the filtration $\{\mathcal{F}_t\}$ satisfy the usual conditions. However, it is often necessary to work on a complete probability space with a filtration satisfying the usual conditions. Let $\{B_t\}_{t\geq 0}$ be a Brownian motion defined on a probability space $(\Omega, \mathcal{F}, \mathbb{P})$. Let $(\Omega, \bar{\mathcal{F}}, \mathbb{P})$ be the completion of $(\Omega, \mathcal{F}, \mathbb{P})$. Clearly, $\{B_t\}$ is a Brownian motion on the complete

probability space $(\Omega, \bar{\mathcal{F}}, \mathbb{P})$. Let $\mathcal{N}$ be the collection of $\mathbb{P}$-null sets, i.e. $\mathcal{N} = \{A \in \bar{\mathcal{F}} : \mathbb{P}(A) = 0\}$. For $t \geq 0$, define
$$\bar{\mathcal{F}}_t = \sigma(\mathcal{F}_t^B \cup \mathcal{N}).$$
We called $\{\bar{\mathcal{F}}_t\}$ the *augmentation under $\mathbb{P}$ of the natural filtration* $\{\mathcal{F}_t^B\}$ *generated by* $\{B_t\}$. It is known that the augmentation $\{\bar{\mathcal{F}}_t\}$ is a filtration on $(\Omega, \bar{\mathcal{F}}, \mathbb{P})$ satisfying the usual condition. Moreover, $\{B_t\}$ is a Brownian motion on $(\Omega, \bar{\mathcal{F}}, \mathbb{P})$ with respect to $\{\bar{\mathcal{F}}_t\}$. This shows that given a Brownian motion $\{B_t\}_{t\geq 0}$ on a probability space $(\Omega, \mathcal{F}, \mathbb{P})$, one can construct a complete probability space with a filtration satisfying the usual conditions to work on.

However, throughout this book, unless otherwise specified, we would rather assume that $(\Omega, \mathcal{F}, \mathbb{P})$ is a complete probability space with a filtration $\{\mathcal{F}_t\}$ satisfying the usual conditions, and the one-dimensional Brownian motion $\{B_t\}$ is defined on it.

In Section 1.1 we mentioned that the integral $\int_{t_0}^{t} \sigma x(u) dB(u)$ cannot be defined as the classical Lebesgue integral since for almost every $\omega \in \Omega$, the Brownian sample path $B_{\cdot}(\omega)$ is nowhere differentiable. To begin, we note that Brownian motion has a remarkable scaling property: for any fixed $c \neq 0$,
$$X_t := \frac{B_{c^2 t}}{c}, \quad t \geq 0, \tag{1.4}$$
is a Brownian motion with respect to the filtration $\{\mathcal{F}_{c^2 t}\}$. Consider the quantity $|B_t|/t$. Since $B_0 = 0$, if B_t were differentiable then $|B_t|/t$ would converge to $|B_0'|$ as $t \to 0$. Let $t = 1/k^2$, where k is large, and set $c = k^2$ in (1.4). Since B_t and X_t have the same distributions we have
$$\mathbb{P}\left\{\frac{|B_{1/k^4}|}{1/k^4} > k\right\} = \mathbb{P}\left\{\frac{|X_{1/k^4}|}{1/k^4} > k\right\} = \mathbb{P}\left\{\frac{|B_1|}{1/k^2} > k\right\} = \mathbb{P}\left\{|B_1| > \frac{1}{k}\right\}.$$
Since B_1 is $N(0,1)$, we have
$$\lim_{k \to \infty} \mathbb{P}\left\{\frac{|B_{1/k^4}|}{1/k^4} > k\right\} = 1.$$
This shows that, with probability 1, B_t is not differentiable at $t = 0$. A similar argument can be used to show that B_t is nowhere differentiable, with probability 1.

Another way of examining the roughness of Brownian motion is to consider its variation. Recall that a continuously differentiable function,

$f \in C^1([0,T]; \mathbb{R})$, has finite variation. In fact, let k be any large integer and set $\Delta = T/k$ and $t_j = j\Delta$ for $0 \leq j \leq k$. The mean value theorem says that

$$f(t_j) - f(t_{j-1}) = \Delta f'(\theta_j), \quad \text{for some } \theta_j \in (t_{j-1}, t_j).$$

Thus

$$\sum_{j=1}^{k} |f(t_j) - f(t_{j-1})| = \Delta \sum_{j=1}^{k} |f'(\theta_j)| \leq T \max_{t \in [0,T]} |f'(t)|.$$

It follows that the variation of f obeys

$$\limsup_{k \to \infty} \sum_{j=1}^{k} |f(t_j) - f(t_{j-1})| \leq T \max_{t \in [0,T]} |f'(t)| < \infty.$$

To see whether Brownian motion has a similar property we use the inequality

$$\sum_{j=1}^{k} (B_{t_j} - B_{t_{j-1}})^2 \leq \left(\max_{1 \leq j \leq k} |B_{t_j} - B_{t_{j-1}}| \right) \sum_{j=1}^{k} |B_{t_j} - B_{t_{j-1}}|. \quad (1.5)$$

Note that the random variable $\sum_{j=1}^{k} (B_{t_j} - B_{t_{j-1}})^2$ has mean T and variance of $O(\Delta)$ (see Exercise 1.1). This implies

$$\lim_{k \to \infty} \sum_{j=1}^{k} (B_{t_j} - B_{t_{j-1}})^2 = T \quad a.s.$$

On the other hand, each $B_{t_j} - B_{t_{j-1}}$ has mean zero and variance Δ. It can then be shown that

$$\lim_{k \to \infty} \left(\max_{1 \leq j \leq k} |B_{t_j} - B_{t_{j-1}}| \right) = 0 \quad a.s.$$

In order for inequality (1.5) to hold it must therefore be true that, with probability 1, $\sum_{j=1}^{k} |B_{t_j} - B_{t_{j-1}}|$ is unbounded as $k \to \infty$. We thus say that Brownian motion has infinite variation in any finite time interval.

Although Brownian motion is rough, it has many important properties, and some of them are summarised below:

(a) $\{B_t\}$ is a continuous square-integrable martingale and its quadratic variation $\langle B, B \rangle_t = t$ for all $t \geq 0$.

(b) The strong law of large numbers states that
$$\lim_{t\to\infty} \frac{B_t}{t} = 0 \quad a.s.$$

(c) For almost every $\omega \in \Omega$, the Brownian sample path $B_\cdot(\omega)$ is locally Hölder continuous with exponent δ if $\delta \in (0, \frac{1}{2})$. However, for almost every $\omega \in \Omega$, the Brownian sample path $B_\cdot(\omega)$ is nowhere Hölder continuous with exponent $\delta > \frac{1}{2}$.

Besides, we have the following well-known law of the iterated logarithm.

Theorem 1.13 *(Law of the Iterated Logarithm) For almost every $\omega \in \Omega$, we have*

(i) $\limsup_{t\downarrow 0} \dfrac{B_t(\omega)}{\sqrt{2t \log \log(1/t)}} = 1$, (ii) $\liminf_{t\downarrow 0} \dfrac{B_t(\omega)}{\sqrt{2t \log \log(1/t)}} = -1$,

(iii) $\limsup_{t\to\infty} \dfrac{B_t(\omega)}{\sqrt{2t \log \log t}} = 1$, (iv) $\liminf_{t\to\infty} \dfrac{B_t(\omega)}{\sqrt{2t \log \log t}} = -1$.

This theorem shows that for any $\varepsilon > 0$ there exists a positive random variable ρ_ε such that for almost every $\omega \in \Omega$, the Brownian sample path $B_\cdot(\omega)$ is within the interval $\pm(1 + \varepsilon)\sqrt{2t \log \log t}$ whenever $t \geq \rho_\varepsilon(\omega)$, that is

$$-(1+\varepsilon)\sqrt{2t \log \log t} \leq B_t(\omega) \leq (1+\varepsilon)\sqrt{2t \log \log t} \quad \text{for all } t \geq \rho_\varepsilon(\omega).$$

On the other hand, the bounds $-(1-\varepsilon)\sqrt{2t \log \log t}$ and $(1-\varepsilon)\sqrt{2t \log \log t}$ (for $0 < \varepsilon < 1$) are exceeded in every t-neighbourhood of ∞ for every sample path.

Let us now define an n-dimensional Brownian motion.

Definition 1.14 An n-dimensional process $\{B_t = (B_t^1, \cdots, B_t^n)\}_{t\geq 0}$ is called an n-dimensional Brownian motion if every $\{B_t^i\}$ is a one-dimensional Brownian motion, and $\{B_t^1\}, \cdots, \{B_t^n\}$ are independent.

For an n-dimensional Brownian motion, we still have, for example,

$$\limsup_{t\to\infty} \frac{|B_t|}{\sqrt{2t \log \log t}} = 1 \quad a.s.$$

This is somewhat surprising because it means that the independent individual components of B_t are not simultaneously of the order $\sqrt{2t \log \log t}$, otherwise $\sqrt{n}$ instead of 1 would have appeared in the right-hand side of the above equality.

It is easy to see that an n-dimensional Brownian motion is an n-dimensional continuous martingale with the joint quadratic variations

$$\langle B^i, B^j \rangle_t = \delta_{ij} t \quad \text{for } 1 \leq i, j \leq n,$$

where δ_{ij} is the Dirac delta function, i.e.

$$\delta_{ij} = \begin{cases} 1 & \text{for } i = j, \\ 0 & \text{for } i \neq j. \end{cases}$$

It turns out that this property characterises Brownian motion among continuous local martingales. This is described by the following well-known Lévy theorem.

Theorem 1.15 *Let $\{M_t = (M_t^1, \cdots, M_t^n)\}_{t \geq 0}$ be an n-dimensional continuous local martingale with respect to the filtration $\{\mathcal{F}_t\}$ and $M_0 = 0$ a.s. If*

$$\langle M^i, M^j \rangle_t = \delta_{ij} t \quad \text{for } 1 \leq i, j \leq n,$$

then $\{M_t = (M_t^1, \cdots, M_t^n)\}_{t \geq 0}$ is an n-dimensional Brownian motion with respect to $\{\mathcal{F}_t\}$.

As an application of the Lévy theorem, one can show the following useful result.

Theorem 1.16 *Let $M = \{M_t\}_{t \geq 0}$ be a real-valued continuous local martingale such that $M_0 = 0$ and $\lim_{t \to \infty} \langle M, M \rangle_t = \infty$ a.s. For each $t \geq 0$, define the stopping time*

$$\tau_t = \inf\{s : \langle M, M \rangle_s > t\}.$$

Then $\{M_{\tau_t}\}_{t \geq 0}$ is a Brownian motion with respect to the filtration $\{\mathcal{F}_{\tau_t}\}_{t \geq 0}$.

1.5 Stochastic Integrals

In this section we shall define the stochastic integral

$$\int_0^t f(s) dB_s$$

with respect to an m-dimensional Brownian motion $\{B_t\}$ for a class of $n\times m$-matrix-valued stochastic processes $\{f(t)\}$. Since for almost all $\omega \in \Omega$, the Brownian sample path $B_\cdot(\omega)$ is not only nowhere differentiable but also has infinite variation in any finite time interval, the integral can not be defined in the ordinary way. However, we can define the integral for a large class of stochastic processes by making use of the stochastic nature of Brownian motion. This integral was first defined by K. Itô in 1949 and is now known as *Itô stochastic integral*. We shall now start to define the stochastic integral step by step.

Let $(\Omega, \mathcal{F}, \mathbb{P})$ be a complete probability space with a filtration $\{\mathcal{F}_t\}_{t\geq 0}$ satisfying the usual conditions. Let $B = \{B_t\}_{t\geq 0}$ be a one-dimensional Brownian motion defined on the probability space adapted to the filtration.

Definition 1.17 Let $0 \leq a < b < \infty$. Denote by $\mathcal{M}^2([a,b];\mathbb{R})$ the space of all real-valued measurable $\{\mathcal{F}_t\}$-adapted processes $f = \{f(t)\}_{a\leq t\leq b}$ such that

$$\|f\|_{a,b}^2 = \mathbb{E}\int_a^b |f(t)|^2 dt < \infty. \tag{1.6}$$

We identify f and $\bar{f}$ in $\mathcal{M}^2([a,b];\mathbb{R})$ if $\|f - \bar{f}\|_{a,b}^2 = 0$. In this case we say that f and $\bar{f}$ are equivalent and write $f = \bar{f}$.

Clearly, $\|\cdot\|_{a,b}$ defines a metric on $\mathcal{M}^2([a,b];\mathbb{R})$ and the space is complete under this metric. Let us point out that for every $f \in \mathcal{M}^2([a,b];\mathbb{R})$, there is a predictable $\bar{f} \in \mathcal{M}^2([a,b];\mathbb{R})$ such that $f = \bar{f}$. In fact, f has a progressively measurable modification $\hat{f}$ in $\mathcal{M}^2([a,b];\mathbb{R})$ and then we may take

$$\bar{f}(t) = \limsup_{h\downarrow 0} \frac{1}{h}\int_{t-h}^t \hat{f}(s)ds.$$

Thus, if necessary, we may assume that $f \in \mathcal{M}^2([a,b];\mathbb{R})$ is predictable without loss of generality. However, in this book we would rather follow the usual custom of not being very careful about the distinction between the equivalence processes.

For stochastic processes $f \in \mathcal{M}^2([a,b];\mathbb{R})$ we shall show how to define the Itô integral $\int_a^b f(t)dB_t$. The idea is natural: first define the integral $\int_a^b g(t)dB_t$ for a class of simple processes g. Then we show that each $f \in \mathcal{M}^2([a,b];\mathbb{R})$ can be approximated by such simple processes g's and we define the limit of $\int_a^b g(t)dB_t$ as the integral $\int_a^b f(t)dB_t$. Let us first introduce the concept of simple processes.

Definition 1.18 A real-valued stochastic process $g = \{g(t)\}_{a \le t \le b}$ is called a simple (or step) process if there exists a partition $a = t_0 < t_1 < \cdots < t_k = b$ of $[a, b]$, and bounded random variables ξ_i, $0 \le i \le k-1$ such that ξ_i is $\mathcal{F}_{t_i}$-measurable and

$$g(t) = \xi_0 I_{[t_0,\ t_1]}(t) + \sum_{i=1}^{k-1} \xi_i I_{(t_i,\ t_{i+1}]}(t). \tag{1.7}$$

Denote by $\mathcal{M}_0([a,b];\mathbb{R})$ the family of all such processes.

It is obvious that $\mathcal{M}_0([a,b];\mathbb{R}) \subset \mathcal{M}^2([a,b];\mathbb{R})$. We now give the definition of the Itô integral for such simple processes.

Definition 1.19 (Part 1 of the definition of Itô's integral) For a simple process g with the form of (1.7) in $\mathcal{M}_0([a,b];\mathbb{R})$, define

$$\int_a^b g(t) dB_t = \sum_{i=0}^{k-1} \xi_i (B_{t_{i+1}} - B_{t_i}) \tag{1.8}$$

and call it the stochastic integral of g with respect to the Brownian motion $\{B_t\}$ or the Itô integral.

Clearly, the stochastic integral $\int_a^b g(t) dB_t$ is $\mathcal{F}_b$-measurable. We shall now show that it belongs to $L^2(\Omega; \mathbb{R})$.

Lemma 1.3 *If $g \in \mathcal{M}_0([a,b];\mathbb{R})$, then*

$$\mathbb{E} \int_a^b g(t) dB_t = 0, \tag{1.9}$$

$$\mathbb{E} \left| \int_a^b g(t) dB_t \right|^2 = \mathbb{E} \int_a^b |g(t)|^2 dt. \tag{1.10}$$

Proof. Since ξ_i is $\mathcal{F}_{t_i}$-measurable whereas $B_{t_{i+1}} - B_{t_i}$ is independent of $\mathcal{F}_{t_i}$,

$$\mathbb{E} \int_a^b g(t) dB_t = \sum_{i=0}^{k-1} \mathbb{E}\big[\xi_i (B_{t_{i+1}} - B_{t_i})\big] = \sum_{i=0}^{k-1} \mathbb{E}\xi_i\, \mathbb{E}(B_{t_{i+1}} - B_{t_i}) = 0.$$

Moreover, note that $B_{t_{j+1}} - B_{t_j}$ is independent of $\xi_i \xi_j (B_{t_{i+1}} - B_{t_i})$ if $i < j$.

Thus

$$\mathbb{E}\left|\int_a^b g(t)dB_t\right|^2 = \sum_{0\leq i,j\leq k-1} \mathbb{E}\big[\xi_i\xi_j(B_{t_{i+1}} - B_{t_i})(B_{t_{j+1}} - B_{t_j})\big]$$

$$= \sum_{i=0}^{k-1} \mathbb{E}\big[\xi_i^2(B_{t_{i+1}} - B_{t_i})^2\big]$$

$$= \sum_{i=0}^{k-1} \mathbb{E}\xi_i^2 \mathbb{E}(B_{t_{i+1}} - B_{t_i})^2$$

$$= \sum_{i=0}^{k-1} \mathbb{E}\xi_i^2 (t_{i+1} - t_i) = \mathbb{E}\int_a^b |g(t)|^2 dt$$

as required. □

Lemma 1.4 *Let $g_1, g_2 \in \mathcal{M}_0([a,b]; \mathbb{R})$ and let c_1, c_2 be two real numbers. Then $c_1 g_1 + c_2 g_2 \in \mathcal{M}_0([a,b]; \mathbb{R})$ and*

$$\int_a^b [c_1 g_1(t) + c_2 g_2(t)]dB_t = c_1 \int_a^b g_1(t)dB_t + c_2 \int_a^b g_2(t)dB_t.$$

The proof is left to the reader as an exercise. We shall now use the properties shown in Lemmas 5.4 and 5.5 to extend the integral definition from simple processes to processes in $\mathcal{M}^2([a,b]; \mathbb{R})$. This is based on the following approximation result.

Lemma 1.5 *For any $f \in \mathcal{M}^2([a,b]; \mathbb{R})$, there exists a sequence $\{g_n\}$ of simple processes such that*

$$\lim_{k\to\infty} \mathbb{E}\int_a^b |f(t) - g_n(t)|^2 dt = 0. \tag{1.11}$$

Proof. We divide the whole proof into three steps.

Step 1. We first claim that for any $f \in \mathcal{M}^2([a,b]; \mathbb{R})$, there exists a sequence $\{\varphi_k\}_{k\geq 1}$ of bounded processes in $\mathcal{M}^2([a,b]; \mathbb{R})$ such that

$$\lim_{k\to\infty} \mathbb{E}\int_a^b |f(t) - \varphi_k(t)|^2 dt = 0. \tag{1.12}$$

In fact, for each k, put

$$\varphi_k(t) = [-k \vee f(t)] \wedge k.$$

Then (1.12) follows by the dominated convergence theorem (i.e. Theorem 1.2).

Step 2. We next claim that if $\varphi \in \mathcal{M}^2([a,b]; \mathbb{R})$ is bounded, say $|\varphi| \leq C = const.$, then there exists a sequence $\{\phi_k\}_{k \geq 1}$ of bounded continuous processes in $\mathcal{M}^2([a,b]; \mathbb{R})$ such that

$$\lim_{k \to \infty} \mathbb{E} \int_a^b |\varphi(t) - \phi_k(t)|^2 dt = 0. \tag{1.13}$$

In fact, for each k, let $\rho_k : \mathbb{R} \to \mathbb{R}_+$ be a continuous function such that $\rho_k(s) = 0$ for $s \leq -\frac{1}{k}$ and $s \geq 0$ and

$$\int_{-\infty}^{\infty} \rho_k(s) ds = 1.$$

Define

$$\phi_k(t) = \phi_k(t, \omega) = \int_a^b \rho_k(s - t) \varphi(s, \omega) ds.$$

Then for every ω, $\phi_k(\cdot, \omega)$ is continuous and $|\phi_k(t, \omega)| \leq C$. Also ϕ_k is a measurable $\{\mathcal{F}_t\}$-adapted process. Moreover, for all $\omega \in \Omega$,

$$\lim_{k \to \infty} \int_a^b |\varphi(t, \omega) - \phi_k(t, \omega)|^2 dt = 0.$$

So (1.13) follows by the bounded convergence theorem.

Step 3. We now claim that if $\phi \in \mathcal{M}^2([a,b]; \mathbb{R})$ is bounded and continuous, then there exists a sequence $\{g_k\}$ of simple processes such that

$$\lim_{k \to \infty} \mathbb{E} \int_a^b |\phi(t) - g_k(t)|^2 dt = 0. \tag{1.14}$$

In fact, for each k, let

$$g_k(t) = \phi(a)\, I_{[a,\, a+(b-a)/k]}(t)$$
$$+ \sum_{i=1}^{k-1} \phi(a + i(b-a)/k)\, I_{(a+i(b-a)/k,\, a+(i+1)(b-a)/k]}(t).$$

Then $g_k \in \mathcal{M}_0([a,b]; \mathbb{R})$, and for every ω,

$$\lim_{k \to \infty} \int_a^b |\phi(t, \omega) - g_k(t, \omega)|^2 dt = 0.$$

So (1.14) follows by the bounded convergence theorem once again. Finally, the conclusion of the lemma follows clearly from steps 1–3 and the proof is now complete. □

We can now explain how to define the Itô integral for a process $f \in \mathcal{M}^2([a,b]; \mathbb{R})$. By Lemma 5.6, there is a sequence $\{g_k\}_{k \geq 1}$ of simple processes such that
$$\lim_{k \to \infty} \mathbb{E} \int_a^b |f(t) - g_k(t)|^2 dt = 0.$$
Thus, by Lemmas 1.3 and 1.4,
$$\mathbb{E}\left|\int_a^b g_k(t) dB_t - \int_a^b g_j(t) dB_t\right|^2 = \mathbb{E}\left|\int_a^b [g_k(t) - g_k(t)] dB_t\right|^2$$
$$= \mathbb{E} \int_a^b |g_k(t) - g_k(t)|^2 dt \to 0 \quad \text{as } k, j \to \infty.$$
In other words, $\{\int_a^b g_k(t) dB_t\}$ is a Cauchy sequence in $L^2(\Omega; \mathbb{R})$. So the limit exists and we define the limit as the stochastic integral. This leads to the following definition.

Definition 1.20 (**Part 2 of the definition of Itô's integral**) Let $f \in \mathcal{M}^2([a,b]; \mathbb{R})$. The Itô integral of f with respect to $\{B_t\}$ is defined by
$$\int_a^b f(t) dB_t = \lim_{k \to \infty} \int_a^b g_k(t) dB_t \quad \text{in } L^2(\Omega; \mathbb{R}), \tag{1.15}$$
where $\{g_k\}$ is a sequence of simple processes such that
$$\lim_{k \to \infty} \mathbb{E} \int_a^b |f(t) - g_k(t)|^2 dt = 0. \tag{1.16}$$

The above definition is independent of the particular sequence $\{g_k\}$. Indeed, if $\{h_k\}$ is another sequence of simple processes converging to f in the sense that
$$\lim_{k \to \infty} \mathbb{E} \int_a^b |f(t) - h_k(t)|^2 dt = 0,$$
then the sequence $\{\varphi_k\}$, where $\varphi_{2k-1} = g_k$ and $\varphi_{2k} = h_k$, is also convergent to f in the same sense. Hence, by what we have proved, the sequence $\{\int_a^b \varphi_k(t) dB_t\}$ is convergent in $L^2(\Omega; \mathbb{R})$. It follows that the limits (in L^2) of $\int_a^b g_k(t) dB_t$ and of $\int_a^b h_k(t) dB_t$ are equal almost surely.

The stochastic integral has many nice properties. We first observe the following:

Theorem 1.21 *Let $f, g \in \mathcal{M}^2([a,b]; \mathbb{R})$, and let α, β be two real numbers. Then*

(i) $\int_a^b f(t)dB_t$ is $\mathcal{F}_b$-measurable;
(ii) $\mathbb{E}\int_a^b f(t)dB_t = 0$;
(iii) $\mathbb{E}|\int_a^b f(t)dB_t|^2 = \mathbb{E}\int_a^b |f(t)|^2 dt$;
(vi) $\int_a^b [\alpha f(t) + \beta g(t)]dB_t = \alpha \int_a^b f(t)dB_t + \beta \int_a^b g(t)dB_t$.

The proof is left to the reader as an exercise. The next theorem improves the results (ii) and (iii) of Theorem 1.21.

Theorem 1.22 Let $f \in \mathcal{M}^2([a,b]; \mathbb{R})$. Then

$$\mathbb{E}\left(\int_a^b f(t)dB(t)\Big|\mathcal{F}_a\right) = 0, \tag{1.17}$$

$$\mathbb{E}\left(\left|\int_a^b f(t)dB(t)\right|^2 \Big|\mathcal{F}_a\right) = \mathbb{E}\left(\int_a^b |f(t)|^2 dt\Big|\mathcal{F}_a\right)$$

$$= \int_a^b \mathbb{E}(|f(t)|^2|\mathcal{F}_a)dt. \tag{1.18}$$

We need a simple lemma.

Lemma 1.6 If $f \in \mathcal{M}^2([a,b]; \mathbb{R})$ and ξ is a real-valued bounded $\mathcal{F}_a$-measurable random variable, then $\xi f \in \mathcal{M}^2([a,b]; \mathbb{R})$ and

$$\int_a^b \xi f(t)dB_t = \xi \int_a^b f(t)dB_t. \tag{1.19}$$

Proof. It is clear that $\xi f \in \mathcal{M}^2([a,b]; \mathbb{R})$. If f is a simple processes, then (1.19) follows from the definition of the stochastic integral. For general $f \in \mathcal{M}^2([a,b]; \mathbb{R})$, let $\{g_k\}$ be a sequence of simple processes satisfying (1.16). Applying (1.19) to each g_k and taking $k \to \infty$, the required assertion (1.19) follows. □

Proof of Theorem 1.22. By the definition of conditional expectation, (1.17) holds if and only if

$$\mathbb{E}\left(I_A \int_a^b f(t)dB(t)\right) = 0$$

for all sets $A \in \mathcal{F}_a$. But by Lemma 1.6 and Theorem 1.21,

$$\mathbb{E}\left(I_A \int_a^b f(t)dB(t)\right) = \mathbb{E}\int_a^b I_A f(t)dB(t) = 0$$

as required. The proof of (1.18) is similar. □

Let $T > 0$ and $f \in \mathcal{M}^2([0,T];\mathbb{R})$. Clearly, for any $0 \le a < b \le T$, $\{f(t)\}_{a \le t \le b} \in \mathcal{M}^2([a,b];\mathbb{R})$ so $\int_a^b f(t)dB_t$ is well defined. It is easy to show that

$$\int_a^b f(t)dB_t + \int_b^c f(t)dB_t = \int_a^c f(t)dB_t \tag{1.20}$$

if $0 \le a < b < c \le T$.

Definition 1.23 Let $f \in \mathcal{M}^2([0,T];\mathbb{R})$. Define

$$I(t) = \int_0^t f(s)dB_s \quad \text{for } 0 \le t \le T,$$

where, by definition, $I(0) = \int_0^0 f(s)dB_s = 0$. We call $I(t)$ the indefinite Itô integral of f.

Clearly, $\{I(t)\}$ is $\{\mathcal{F}_t\}$-adapted. We now show the very important martingale property of the indefinite Itô integral.

Theorem 1.24 *If $f \in \mathcal{M}^2([0,T];\mathbb{R})$, then the indefinite integral $\{I(t)\}_{0 \le t \le T}$ is a square-integrable martingale with respect to the filtration $\{\mathcal{F}_t\}$. In particular,*

$$\mathbb{E}\left[\sup_{0 \le t \le T}\left|\int_0^t f(s)dB_s\right|^2\right] \le 4\mathbb{E}\int_0^T |f(s)|^2 ds. \tag{1.21}$$

Proof. Clearly, $\{I(t)\}_{0 \le t \le T}$ is square-integrable. To show the martingale property, let $0 \le s < t \le T$. By (1.20) and Theorem 1.22

$$\mathbb{E}(I(t)|\mathcal{F}_s) = \mathbb{E}(I(s)|\mathcal{F}_s) + \mathbb{E}\left(\int_s^t f(r)dB_r \Big| \mathcal{F}_s\right) = I(s)$$

as desired. The inequality (1.21) now follows from Doob's martingale inequality (i.e. Theorem 1.11). □

Theorem 1.25 *If $f \in \mathcal{M}^2([0,T];\mathbb{R})$, then the indefinite integral $\{I(t)\}_{0 \le t \le T}$ has a continuous version.*

Proof. Let $\{g_k\}$ be a sequence of simple processes such that

$$\lim_{k \to \infty} \mathbb{E}\int_0^T |f(s) - g_k(s)|^2 ds = 0. \tag{1.22}$$

Note from the definition of the stochastic integral and the continuity of the Brownian motion that the indefinite integrals
$$I_k(t) = \int_0^t g_k(s)dB_s, \quad 0 \le t \le T$$
are continuous. By Theorem 1.24, $\{I_k(t) - I_j(t)\}$ is a martingale, for each pair of integers k, j. Hence, by Doob's martingale inequality (namely, Theorem 1.11), for any $\varepsilon > 0$
$$\mathbb{P}\left\{\sup_{0 \le t \le T} |I_k(t) - I_j(t)| \ge \varepsilon\right\} \le \frac{1}{\varepsilon^2}\mathbb{E}|I_k(T) - I_j(T)|^2$$
$$= \frac{1}{\varepsilon^2}\mathbb{E}\int_0^T |g_k(s) - g_j(s)|^2 ds \to 0 \quad \text{as } k, j \to \infty.$$
For each $i = 1, 2, \cdots$, taking $\varepsilon = i^{-2}$, it follows that for some k_i sufficiently large,
$$\mathbb{P}\left\{\sup_{0 \le t \le T} |I_{k_i}(t) - I_j(t)| \ge \frac{1}{i^2}\right\} \le \frac{1}{i^2} \quad \text{if } j \ge k_i.$$
One can then choose the k_i in such a way that $k_i \uparrow \infty$ as $i \to \infty$ and
$$\mathbb{P}\left\{\sup_{0 \le t \le T} |I_{k_i}(t) - I_{k_{i+1}}(t)| \ge \frac{1}{i^2}\right\} \le \frac{1}{i^2}, \quad i \ge 1.$$
Since $\sum i^{-2} < \infty$, the Borel–Cantelli lemma (i.e. Lemma 1.2) implies that there exists a set $\Omega_0 \in \mathcal{F}$ with $\mathbb{P}(\Omega_0) = 0$ and an integer-valued random variable i_0 such that for every $\omega \in \Omega_0$,
$$\sup_{0 \le t \le T} |I_{k_i}(t, \omega) - I_{k_{i+1}}(t, \omega)| < \frac{1}{i^2} \quad \text{if } i \ge i_0(\omega).$$
In other words, with probability 1, $\{I_{k_i}(t)\}_{i \ge 1}$ is uniformly convergent in $t \in [0, T]$, and therefore the limit, denoted by $J(t)$, is continuous in $t \in [0, T]$ for almost all $\omega \in \Omega$. Since (1.22) implies
$$\lim_{i \to \infty} I_{k_i}(t) = \int_0^t f(s)dB_s \quad \text{in } L^2(\Omega, \mathbb{R}),$$
it follows that
$$J(t) = \int_0^t f(s)dB_s \quad \text{a.s.}$$
That is, the indefinite integral has a continuous version.

From now on, when we speak of the indefinite integral we always mean a continuous version of it.

Theorem 1.26 Let $f \in \mathcal{M}^2([0,T];\mathbb{R})$. Then the indefinite integral $I = \{I(t)\}_{0 \le t \le T}$ is a square-integrable continuous martingale and its quadratic variation is given by

$$\langle I, I \rangle_t = \int_0^t |f(s)|^2 ds, \quad 0 \le t \le T. \tag{1.23}$$

Proof. Obviously we need only to show (1.23). By the definition of the quadratic variation we need to show that $\{I^2(t) - \langle I, I \rangle_t\}$ is a continuous martingale vanishing at $t = 0$. But, obviously $I^2(0) - \langle I, I \rangle_0 = 0$. Moreover, if $0 \le r < t \le T$, by Theorem 1.22,

$$\mathbb{E}(I^2(t) - \langle I, I \rangle_t | \mathcal{F}_r)$$
$$= I^2(r) - \langle I, I \rangle_r + 2I(r)\mathbb{E}\left(\int_r^t f(s)dB_s \Big| \mathcal{F}_r\right)$$
$$+ \mathbb{E}\left(\left|\int_r^t f(s)dB_s\right|^2 \Big| \mathcal{F}_r\right) - \mathbb{E}\left(\int_r^t |f(s)|^2 ds \Big| \mathcal{F}_r\right)$$
$$= I^2(r) - \langle I, I \rangle_r$$

as desired. □

Let us now proceed to define the stochastic integrals with stopping time. We observe that if τ is an $\{\mathcal{F}_t\}$-stopping time, then $\{I_{[\![0,\,\tau]\!]}(t)\}_{t \ge 0}$ is a bounded right continuous $\{\mathcal{F}_t\}$-adapted process. In fact, the boundedness and right continuity are obvious. Moreover, for each $t \ge 0$,

$$\{\omega : I_{[\![0,\,\tau]\!]}(t,\omega) \le r\} = \begin{cases} \emptyset \in \mathcal{F}_t & \text{if } r < 0, \\ \{\omega : \tau(\omega) < t\} \in \mathcal{F}_t & \text{if } 0 \le r < 1, \\ \Omega \in \mathcal{F}_t & \text{if } r \ge 1, \end{cases}$$

that is, $I_{[\![0,\,\tau]\!]}(t)$ is $\mathcal{F}_t$-measurable. Therefore, $\{I_{[\![0,\,\tau]\!]}(t)\}_{t \ge 0}$ is also predictable.

Definition 1.27 Let $f \in \mathcal{M}^2([0,T];\mathbb{R})$, and let τ be an $\{\mathcal{F}_t\}$-stopping time such that $0 \le \tau \le T$. Then, $\{I_{[\![0,\,\tau]\!]}(t)f(t)\}_{0 \le t \le T} \in \mathcal{M}^2([0,T];\mathbb{R})$ clearly, and we define

$$\int_0^\tau f(s)dB_s = \int_0^T I_{[\![0,\,\tau]\!]}(s)f(s)dB_s.$$

Furthermore, if ρ is another stopping time with $0 \le \rho \le \tau$, we define
$$\int_\rho^\tau f(s)dB_s = \int_0^\tau f(s)dB_s - \int_0^\rho f(s)dB_s.$$
It is easy to see that
$$\int_\rho^\tau f(s)dB_s = \int_0^T I_{]\!]\rho,\,\tau]\!]}(s)f(s)dB_s. \tag{1.24}$$
If applying Theorem 1.21 to this we immediately obtain:

Theorem 1.28 *Let $f \in \mathcal{M}^2([0,T];\mathbb{R})$, and let ρ, τ be two stopping times such that $0 \le \rho \le \tau \le T$. Then*
$$\mathbb{E}\int_\rho^\tau f(s)dB_s = 0,$$
$$\mathbb{E}\left|\int_\rho^\tau f(s)dB_s\right|^2 = \mathbb{E}\int_\rho^\tau |f(s)|^2 ds.$$

However, the next theorem improves these results and is also a generalisation of Theorem 1.22.

Theorem 1.29 *Let $f \in \mathcal{M}^2([0,T];\mathbb{R})$, and let ρ, τ be two stopping times such that $0 \le \rho \le \tau \le T$. Then*
$$\mathbb{E}\left(\int_\rho^\tau f(s)dB_s\Big|\mathcal{F}_\rho\right) = 0, \tag{1.25}$$
$$\mathbb{E}\left(\left|\int_\rho^\tau f(s)dB_s\right|^2\Big|\mathcal{F}_\rho\right) = \mathbb{E}\left(\int_\rho^\tau |f(s)|^2 ds\Big|\mathcal{F}_\rho\right). \tag{1.26}$$

We need a useful lemma.

Lemma 1.7 *Let $f \in \mathcal{M}^2([0,T];\mathbb{R})$, and let τ be a stopping time such that $0 \le \tau \le T$. Then*
$$\int_0^\tau f(s)dB_s = I(\tau),$$
where $\{I(t)\}_{0 \le t \le T}$ is the indefinite integral of f given by Definition 1.23.

We leave the proof of this lemma to the reader, but prove Theorem 1.29.

Proof of Theorem 1.29. By Theorem 1.26 and the Doob martingale stopping theorem (i.e. Theorem 1.5),
$$\mathbb{E}(I(\tau)|\mathcal{F}_\rho) = I(\rho) \tag{1.27}$$

and
$$\mathbb{E}(I^2(\tau) - \langle I, I\rangle_\tau | \mathcal{F}_r) = I^2(\rho) - \langle I, I\rangle_\rho, \tag{1.28}$$

where $\{\langle I, I\rangle_t\}$ is defined by (1.23). Applying Lemma 1.7 one then sees from (1.27) that
$$\mathbb{E}\Big(\int_\rho^\tau f(s)dB_s\Big|\mathcal{F}_\rho\Big) = \mathbb{E}(I(\tau) - I(\rho)|\mathcal{F}_\rho) = 0$$

which is (1.25). Moreover, by (1.27) and (1.28),
$$\mathbb{E}(|I(\tau) - I(\rho)|^2|\mathcal{F}_\rho) = \mathbb{E}(I^2(\tau)|\mathcal{F}_\rho) - 2I(\rho)\mathbb{E}(I(\tau)|\mathcal{F}_\rho) + I^2(\rho)$$
$$= \mathbb{E}(I^2(\tau)|\mathcal{F}_\rho) - I^2(\rho) = \mathbb{E}(\langle I,I\rangle_\tau - \langle I,I\rangle_\rho|\mathcal{F}_\rho) = \mathbb{E}\Big(\int_\rho^\tau |f(s)|^2 ds\Big|\mathcal{F}_\rho\Big)$$

which, by Lemma 1.7, is the required (5.21). □

Corollary 1.30 *Let $f, g \in \mathcal{M}^2([0,T]; \mathbb{R})$, and let ρ, τ be two stopping times such that $0 \le \rho \le \tau \le T$. Then*
$$\mathbb{E}\Big(\int_\rho^\tau f(s)dB_s \int_\rho^\tau g(s)dB_s\Big|\mathcal{F}_\rho\Big) = \mathbb{E}\Big(\int_\rho^\tau f(s)g(s)ds\Big|\mathcal{F}_\rho\Big).$$

Proof. By Theorem 1.29,
$$4\mathbb{E}\Big(\int_\rho^\tau f(s)dB_s \int_\rho^\tau g(s)dB_s\Big|\mathcal{F}_\rho\Big)$$
$$= \mathbb{E}\Big(\Big|\int_\rho^\tau (f(s) + g(s))dB_s\Big|^2\Big|\mathcal{F}_\rho\Big) - \mathbb{E}\Big(\Big|\int_\rho^\tau (f(s) - g(s))dB_s\Big|^2\Big|\mathcal{F}_\rho\Big)$$
$$= \mathbb{E}\Big(\int_\rho^\tau (f(s) + g(s))^2 ds\Big|\mathcal{F}_\rho\Big) - \mathbb{E}\Big(\int_\rho^\tau (f(s) - g(s))^2 ds\Big|\mathcal{F}_\rho\Big)$$
$$= 4\mathbb{E}\Big(\int_\rho^\tau f(s)g(s)ds\Big|\mathcal{F}_\rho\Big)$$

as desired. □

Let us now begin to extend the Itô stochastic integral to the multidimensional case. Let $\{B_t = (B_t^1, \cdots, B_t^m)^T\}_{t\ge 0}$ be an m-dimensional Brownian motion defined on the complete probability space $(\Omega, \mathcal{F}, \mathbb{P})$ adapted to the filtration $\{\mathcal{F}_t\}$. Let $\mathcal{M}^2([0,T]; \mathbb{R}^{n\times m})$ denote the family of all $n \times m$-matrix-valued measurable $\{\mathcal{F}_t\}$-adapted processes $f =$

$\{(f_{ij}(t))_{n\times m}\}_{0\le t\le T}$ such that

$$\mathbb{E}\int_0^T |f(s)|^2 dt < \infty.$$

Here, and throughout this book, $|A|$ will denote the trace norm for matrix A, i.e. $|A| = \sqrt{\text{trace}(A^T A)}$.

Definition 1.31 Let $f \in \mathcal{M}^2([0,T];\mathbb{R}^{n\times m})$. Using matrix notation, we define the multi-dimensional indefinite Itô integral

$$\int_0^t f(s)dB_s = \int_0^t \begin{pmatrix} f_{11}(s) & \cdots & f_{1m}(s) \\ \vdots & & \vdots \\ f_{d1}(s) & \cdots & f_{dm}(s) \end{pmatrix} \begin{pmatrix} dB_s^1 \\ \vdots \\ dB_s^m \end{pmatrix}$$

to be the n-column-vector-valued process whose i'th component is the following sum of one-dimensional Itô integrals

$$\sum_{j=1}^m \int_0^t f_{ij}(s)dB_s^j.$$

Clearly, the Itô integral is an $\mathbb{R}^n$-valued continuous martingale with respect to $\{\mathcal{F}_t\}$. Besides, it has the following important properties.

Theorem 1.32 Let $f \in \mathcal{M}^2([0,T];\mathbb{R}^{n\times m})$, and let ρ, τ be two stopping times such that $0 \le \rho \le \tau \le T$. Then

$$\mathbb{E}\Big(\int_\rho^\tau f(s)dB_s \Big| \mathcal{F}_\rho\Big) = 0, \qquad (1.29)$$

$$\mathbb{E}\Big(\Big|\int_\rho^\tau f(s)dB_s\Big|^2 \Big| \mathcal{F}_\rho\Big) = \mathbb{E}\Big(\int_\rho^\tau |f(s)|^2 ds \Big| \mathcal{F}_\rho\Big). \qquad (1.30)$$

Assertion (1.29) follows from the definition of multi-dimensional Itô integral and Theorem 1.29, while (1.30) follows from Theorem 1.29 and the following lemma.

Lemma 1.8 Let $\{B_t^1\}_{t\ge 0}$ and $\{B_t^2\}_{t\ge 0}$ be two independent one-dimensional Brownian motions. Let $f, g \in \mathcal{M}^2([0,T];\mathbb{R})$, and let ρ, τ be two stopping times such that $0 \le \rho \le \tau \le T$. Then

$$\mathbb{E}\Big(\int_\rho^\tau f(s)dB_s^1 \int_\rho^\tau g(s)dB_s^2 \Big| \mathcal{F}_\rho\Big) = 0. \qquad (1.31)$$

Proof. We first claim that if $\varphi, \phi \in \mathcal{M}^2([a,b]; \mathbb{R})$. Then

$$\mathbb{E}\left(\int_a^b \varphi(s)dB_s^1 \int_a^b \phi(s)dB_s^2\right) = 0. \tag{1.32}$$

In fact, let φ, ϕ be simple processes with the forms

$$\varphi(t) = \xi_0 I_{[t_0,\, t_1]}(t) + \sum_{i=1}^{k-1} \xi_i I_{(t_i,\, t_{i+1}]}(t)$$

and

$$\phi(t) = \zeta_0 I_{[\bar{t}_0,\, \bar{t}_1]}(t) + \sum_{j=1}^{m-1} \zeta_i I_{(\bar{t}_j,\, \bar{t}_{j+1}]}(t).$$

Then

$$\mathbb{E}\left(\int_a^b \varphi(s)dB_s^1 \int_a^b \phi(s)dB_s^2\right) = \sum_{i=0}^{k-1}\sum_{j=0}^{m-1} \mathbb{E}\left[\xi_i\zeta_j(B_{t_{i+1}}^1 - B_{t_i}^1)(B_{\bar{t}_{j+1}}^2 - B_{\bar{t}_j}^2)\right].$$

But for every pair of i, j, if $t_i \leq \bar{t}_j$, then $B_{\bar{t}_{j+1}}^2 - B_{\bar{t}_j}^2$ is independent of $\xi_i\zeta_j(B_{t_{i+1}}^1 - B_{t_i}^1)$ and hence

$$\mathbb{E}\left[\xi_i\zeta_j(B_{t_{i+1}}^1 - B_{t_i}^1)(B_{\bar{t}_{j+1}}^2 - B_{\bar{t}_j}^2)\right] = 0.$$

Similarly, it still holds if $t_i > \bar{t}_j$. In other words, we have shown that (1.32) holds for simple processes φ, ϕ, but the general case follows by the approximation procedure.

We next observe that for any $0 \leq r < t \leq T$

$$\mathbb{E}\left(\int_r^t f(s)dB_s^1 \int_r^t g(s)dB_s^2 \Big| \mathcal{F}_r\right) = 0, \tag{1.33}$$

since, by (1.32) and Lemma 1.6, for any $A \in \mathcal{F}_r$

$$\mathbb{E}\left(I_A \int_r^t f(s)dB_s^1 \int_r^t g(s)dB_s^2\right) = \mathbb{E}\left(\int_r^t I_A f(s)dB_s^1 \int_r^t g(s)dB_s^2\right) = 0.$$

Therefore

$$\mathbb{E}\Big(\int_0^t f(s)dB_s^1 \int_0^t g(s)dB_s^2 \Big| \mathcal{F}_r\Big)$$
$$= \int_0^r f(s)dB_s^1 \int_0^r g(s)dB_s^2 + \int_0^r f(s)dB_s^1 \mathbb{E}\Big(\int_r^t g(s)dB_s^2 \Big| \mathcal{F}_r\Big)$$
$$+ \int_0^r g(s)dB_s^2 \mathbb{E}\Big(\int_r^t f(s)dB_s^1 \Big| \mathcal{F}_r\Big) + \mathbb{E}\Big(\int_r^t f(s)dB_s^1 \int_r^t g(s)dB_s^2 \Big| \mathcal{F}_r\Big)$$
$$= \int_0^r f(s)dB_s^1 \int_0^r g(s)dB_s^2.$$

That is, $\{\int_0^t f(s)dB_s^1 \int_0^t g(s)dB_s^2\}_{0 \le t \le T}$ is a martingale with respect to $\{\mathcal{F}_t\}$. Hence, by the Doob martingale stopping theorem,

$$\mathbb{E}\Big(\int_0^\tau f(s)dB_s^1 \int_0^\tau g(s)dB_s^2 \Big| \mathcal{F}_\rho\Big) = \int_0^\rho f(s)dB_s^1 \int_0^\rho g(s)dB_s^2. \quad (1.34)$$

Now the required assertion (1.31) follows from (1.34) easily. The proof of the lemma, hence of Theorem 1.32 is now complete. $\square$

We shall finally extend the stochastic integral to a larger class of stochastic processes. Let $\mathcal{L}^2(\mathbb{R}_+; \mathbb{R}^{n \times m})$ denote the family of all $n \times m$-matrix-valued measurable $\{\mathcal{F}_t\}$-adapted processes $f = \{f(t)\}_{t \ge 0}$ such that

$$\int_0^T |f(t)|^2 dt < \infty \quad \text{a.s. for every } T > 0.$$

Let $\mathcal{M}^2(\mathbb{R}_+; \mathbb{R}^{n \times m})$ denote the family of all processes $f \in \mathcal{L}^2(\mathbb{R}_+; \mathbb{R}^{n \times m})$ such that

$$\mathbb{E}\int_0^T |f(t)|^2 dt < \infty \quad \text{for every } T > 0.$$

Clearly, if $f \in \mathcal{M}^2(\mathbb{R}_+; \mathbb{R}^{n \times m})$, then $\{f(t)\}_{0 \le t \le T} \in \mathcal{M}^2([0,T]; \mathbb{R}^{n \times m})$ for every $T > 0$. Hence, the indefinite integral $\int_0^t f(s)dB_s$, $t \ge 0$ is well defined, and it is an $\mathbb{R}^n$-valued continuous square-integrable martingale. However, we aim to define the integral for all processes in $\mathcal{L}^2(\mathbb{R}_+; \mathbb{R}^{n \times m})$. Let $f \in \mathcal{L}^2(\mathbb{R}_+; \mathbb{R}^{n \times m})$. For each integer $k \ge 1$, define the stopping time

$$\tau_k = k \wedge \inf\{t \ge 0 : \int_0^t |f(s)|^2 ds \ge k\}.$$

Clearly, $\tau_k \uparrow \infty$ a.s. Moreover, $\{f(t)I_{[[0,\tau_k]]}(t)\}_{t\geq 0} \in \mathcal{M}^2(\mathbb{R}_+;\mathbb{R}^{n\times m})$ so the integral

$$I_k(t) = \int_0^t f(s)I_{[[0,\tau_k]]}(s)dB_s, \quad t \geq 0$$

is well defined. Note that for $1 \leq k \leq j$ and $t \geq 0$,

$$I_j(t \wedge \tau_k) = \int_0^{t\wedge\tau_k} f(s)I_{[[0,\tau_j]]}(s)dB_s = \int_0^t f(s)I_{[[0,\tau_j]]}(s)I_{[[0,\tau_k]]}(s)dB_s$$
$$= \int_0^t f(s)I_{[[0,\tau_k]]}(s)dB_s = I_k(t),$$

which implies

$$I_j(t) = I_k(t), \quad 0 \leq t \leq \tau_k.$$

So we may define the indefinite stochastic integral $\{I(t)\}_{t\geq 0}$ as

$$I(t) = I_k(t) \quad \text{on } 0 \leq t \leq \tau_k. \tag{1.35}$$

Definition 1.33 Let $f = \{f(t)\}_{t\geq 0} \in \mathcal{L}^2(\mathbb{R}_+;\mathbb{R}^{n\times m})$. The indefinite Itô integral of f with respect to $\{B_t\}$ is the $\mathbb{R}^n$-valued process $\{I(t)\}_{t\geq 0}$ defined by (1.35). As before, we usually write $\int_0^t f(s)dB_s$ instead of $I(t)$.

It is clear that the Itô integral $\int_0^t f(s)dB_s$, $t \geq 0$ is an $\mathbb{R}^n$-valued continuous local martingale.

1.6 Itô's Formula

In the previous section we defined the Itô stochastic integrals. However the basic definition of the integrals is not very convenient in evaluating a given integral. This is similar to the situation for classical Lebesgue integrals, where we do not use the basic definition but rather the fundamental theorem of calculus plus the chain rule in the explicit calculations. For example, it is very easy to use the chain rule to calculate $\int_0^t \cos(s)ds = \sin(t)$ but not so if you use the basic definition. In this section we shall establish the stochastic version of the chain rule for the Itô integrals, which is known as Itô's formula. We shall see in this book that Itô's formula is not only useful in evaluating the Itô integrals but, more importantly, it plays a key role in stochastic analysis.

Let $B(t) = (B_1(t), \cdots, B_m(t))^T$, $t \geq 0$ be an m-dimensional Brownian motion defined on the complete probability space $(\Omega, \mathcal{F}, \mathbb{P})$ adapted to the filtration $\{\mathcal{F}_t\}_{t \geq 0}$.

Definition 1.34 An n-dimensional Itô process is an $\mathbb{R}^n$-valued continuous adapted process $x(t) = (x_1(t), \cdots, x_n(t))^T$ on $t \geq 0$ of the form

$$x(t) = x(0) + \int_0^t f(s)ds + \int_0^t g(s)dB(s),$$

where $f = (f_1, \cdots, f_n)^T \in \mathcal{L}^1(\mathbb{R}_+; \mathbb{R}^n)$ and $g = (g_{ij})_{n \times m} \in \mathcal{L}^2(\mathbb{R}_+; \mathbb{R}^{n \times m})$. We shall say that $x(t)$ has a stochastic differential $dx(t)$ on $t \geq 0$ given by

$$dx(t) = f(t)dt + g(t)dB(t). \tag{1.36}$$

We shall sometimes speak of Itô process $x(t)$ and its stochastic differential $dx(t)$ on $t \in [a, b]$, and the meaning is apparent.

Let $C^{2,1}(\mathbb{R}^n \times \mathbb{R}_+; \mathbb{R})$ denote the family of all real-valued functions $V(x, t)$ defined on $\mathbb{R}^n \times \mathbb{R}_+$ such that they are continuously twice differentiable in x and once in t. If $V \in C^{2,1}(\mathbb{R}^n \times \mathbb{R}_+; \mathbb{R})$, we set

$$V_t = \frac{\partial V}{\partial t}, \quad V_x = \left(\frac{\partial V}{\partial x_1}, \cdots, \frac{\partial V}{\partial x_n}\right),$$

$$V_{xx} = \left(\frac{\partial^2 V}{\partial x_i \partial x_j}\right)_{n \times n} = \begin{pmatrix} \frac{\partial^2 V}{\partial x_1 \partial x_1} & \cdots & \frac{\partial^2 V}{\partial x_1 \partial x_n} \\ \vdots & & \vdots \\ \frac{\partial^2 V}{\partial x_n \partial x_1} & \cdots & \frac{\partial^2 V}{\partial x_n \partial x_n} \end{pmatrix}.$$

Clearly, when $V \in C^{2,1}(\mathbb{R} \times \mathbb{R}_+; \mathbb{R})$, we have $V_x = \frac{\partial V}{\partial x}$ and $V_{xx} = \frac{\partial^2 V}{\partial x^2}$.

We are now ready to state the well-known Itô formula.

Theorem 1.35 *(Itô's formula)* Let $x(t)$ be an n-dimensional Itô process on $t \geq 0$ with the stochastic differential

$$dx(t) = f(t)dt + g(t)dB(t),$$

where $f \in \mathcal{L}^1(\mathbb{R}_+; \mathbb{R}^n)$ and $g \in \mathcal{L}^2(\mathbb{R}_+; \mathbb{R}^{n \times m})$. Let $V \in C^{2,1}(\mathbb{R}^n \times \mathbb{R}_+; \mathbb{R})$. Then $V(x(t), t)$ is a real-valued Itô process with its stochastic differential given by

$$dV(x(t), t) = \Big[V_t(x(t), t) + V_x(x(t), t)f(t)$$
$$+ \frac{1}{2}\text{trace}\big(g^T(t)V_{xx}(x(t), t)g(t)\big)\Big]dt + V_x(x(t), t)g(t)dB(t) \quad a.s. \tag{1.37}$$

The proof can be found in many books e.g. [Mao (1997)] so is left to the reader as an exercise. Let us now introduce formally a multiplication table:

$$dtdt = 0, \quad dB_i dt = 0,$$
$$dB_i dB_i = dt, \quad dB_i dB_j = 0 \quad \text{if } i \neq j.$$

Then, for example,

$$dx_i(t)dx_j(t) = \sum_{k=1}^{m} g_{ik}(t)g_{jk}(t)dt. \tag{1.38}$$

Moreover, the Itô formula can be written as

$$dV(x(t),t) = V_t(x(t),t)dt + V_x(x(t),t)dx(t)$$
$$+ \frac{1}{2}dx^T(t)V_{xx}(x(t),t)dx(t). \tag{1.39}$$

Note that if $x(t)$ were continuously differentiable in t, then (by the classical calculus formula for total derivatives) the term $\frac{1}{2}dx^T(t)V_{xx}(x(t),t)dx(t)$ would not appear. For example, let $V(x,t) = x_1 x_2$, then (1.38) and (1.39) yield

$$d[x_1(t)x_2(t)] = x_1(t)dx_2(t) + x_2(t)dx_1(t) + dx_1 dx_2 \tag{1.40}$$
$$= x_1(t)dx_2(t) + x_2(t)dx_1(t) + \sum_{k=1}^{m} g_{1k}(t)g_{2k}(t)dt,$$

which is different from the classical formula of integration by parts $d(uv) = vdu + udv$ if both u, v are differentiable. More clearly, we have

$$d[\sin^2(t)] = 2\sin(t)d\sin(t)$$

but we don't have, if $B(t)$ is a scalar Brownian motion,

$$d[B^2(t)] = 2B(t)dB(t),$$

instead, we have

$$d[B^2(t)] = 2B(t)dB(t) + dt.$$

However we do have the stochastic version of integration by parts formula which is similar to the classical one.

Theorem 1.36 *(**Integration by parts formula**) Let $x(t)$, $t \geq 0$ be a one-dimensional Itô process with the stochastic differential*

$$dx(t) = f(t)dt + g(t)dB(t),$$

where $f \in \mathcal{L}^1(\mathbb{R}_+; \mathbb{R})$ and $g \in \mathcal{L}^2(\mathbb{R}_+; R^{1\times m})$. Let $y(t)$, $t \geq 0$ be a real-valued continuous adapted process of finite variation. Then

$$d[x(t)y(t)] = y(t)dx(t) + x(t)dy(t), \quad (1.41)$$

that is

$$x(t)y(t) - x(0)y(0) = \int_0^t y(s)[f(s)ds + g(s)dB(s)] + \int_0^t x(s)dy(s), \quad (1.42)$$

where the last integral is the Lebesgue–Stieltjes integral.

Given $V \in C^{2,1}(\mathbb{R}^n \times \mathbb{R}_+; \mathbb{R})$, define an operator $LV : \mathbb{R}^n \times \mathbb{R}_+ \to \mathbb{R}$ by

$$LV(x,t) = V_t(x,t) + V_x(x,t)f(t) + \frac{1}{2}\text{trace}\big(g^T(t)V_{xx}(x,t)g(t)\big), \quad (1.43)$$

which is called the *diffusion operator* of the Itô process (1.36) associated with the $C^{2,1}$-function V. With this diffusion operator, the Itô formula (1.37) can be written as

$$dV(x(t),t) = LV(x(t),t)dt + V_x(x(t),t)g(t)dB(t) \quad \text{a.s.} \quad (1.44)$$

If $0 \leq \tau \leq \rho < \infty$ are two stopping times such that

$$\mathbb{E}\int_\tau^\rho |V_x(x(t),t)g(t)|^2 dt < \infty$$

and the integrations involved below are all finite, then

$$\mathbb{E}V(x(\rho),\rho) - \mathbb{E}V(x(\tau),\tau) = \mathbb{E}\int_\tau^\rho LV(x(t),t)dt. \quad (1.45)$$

Let us now give a number of examples to illustrate the use of Itô's formula in evaluating the stochastic integrals.

Example 1.37 Let $B(t)$ be a one-dimensional Brownian motion. To compute the stochastic integral

$$\int_0^t e^{-s/2 + B(s)} dB(s),$$

we let $V(x,t) = e^{-t/2+x}$ and $x(t) = B(t)$, and then, by the Itô formula, we compute

$$d\left[e^{-t/2+B(t)}\right] = -\frac{1}{2}e^{-t/2+B(t)}dt + e^{-t/2+B(t)}dB(t) + \frac{1}{2}e^{-t/2+B(t)}dt$$
$$= e^{-t/2+B(t)}dB(t).$$

That yields

$$\int_0^t e^{-s/2+B(s)}dB(s) = e^{-t/2+B(t)} - 1.$$

Rewrite this as

$$e^{-\frac{1}{2}p^2 t + pB(t)} = 1 + \int_0^t e^{-\frac{1}{2}p^2 s + pB(s)}dB(s).$$

It can be shown (see Exercise 1.10) that

$$\mathbb{E}\int_0^t |e^{-\frac{1}{2}p^2 s + pB(s)}|^2 ds = \frac{1}{p^2}[e^{p^2 t} - 1], \quad \forall t \geq 0.$$

So $\int_0^t e^{-\frac{1}{2}p^2 s + pB(s)}dB(s)$ is a martingale on $t \geq 0$ vanishing at $t = 0$. We hence obtain the following important result.

Theorem 1.38 *(The exponential martingale formula)* Let $B(t)$ be a one-dimensional Brownian motion and $p > 0$. Then $e^{-\frac{1}{2}p^2 t + pB(t)}$ is a martingale on $t \geq 0$ with initial value 1 and hence for any bounded stopping time τ,

$$\mathbb{E}\left[e^{-\frac{1}{2}p^2 \tau + pB(\tau)}\right] = 1.$$

Example 1.39 Let $B(t)$ be a one-dimensional Brownian motion. What is the integration of the Brownian sample path over the time interval $[0, t]$, i.e. $\int_0^t B(s)ds$? The integration by parts formula yields

$$d[tB(t)] = B(t)dt + t\,dB(t).$$

Therefore

$$\int_0^t B(s)ds = tB(t) - \int_0^t s\,dB(s).$$

On the other hand, we may apply Itô's formula to $B^3(t)$ to obtain

$$dB^3(t) = 3B^2(t)dB(t) + 3B(t)dt,$$

which gives the alternative

$$\int_0^t B(s)ds = \frac{1}{3}B^3(t) - \int_0^t B^2(s)dB(s).$$

Example 1.40 Let $B(t)$ be an m-dimensional Brownian motion. Let $V : \mathbb{R}^m \to \mathbb{R}$ be C^2. Then Itô's formula implies

$$V(B(t)) = V(0) + \frac{1}{2}\int_0^t \Delta V(B(s))ds + \int_0^t V_x(B(s))dB(s),$$

where $\Delta = \sum_{i=1}^m \frac{\partial^2}{\partial x_i^2}$ is the Laplace operator. In particular, let V be a quadratic function, namely $V(x) = x^T Q x$, where Q is an $m \times m$ matrix. Then

$$B^T(t)QB(t) = \text{trace}(Q)t + \int_0^t B^T(s)(Q + Q^T)dB(s).$$

Example 1.41 Let $x(t)$ be an n-dimensional Itô process as given by Definition 1.34. Let Q be an $n \times n$ matrix. Then

$$x^T(t)Qx(t) - x^T(0)Qx(0)$$
$$= \int_0^t \left(x^T(s)(Q + Q^T)f(s) + \frac{1}{2}\text{trace}[g^T(s)(Q + Q^T)g(s)]\right)ds$$
$$+ \int_0^t x^T(s)(Q + Q^T)g(s)dB(s).$$

1.7 Markov Processes

In this section we will recall some basic facts about a Markov process. An n-dimensional $\mathcal{F}_t$-adapted process $X = \{X_t\}_{t\geq 0}$ is called a *Markov process* if the following *Markov property* is satisfied: for all $0 \leq s \leq t < \infty$ and $A \in \mathcal{B}(\mathbb{R}^n)$,

$$\mathbb{P}(X(t) \in A|\mathcal{F}_s) = \mathbb{P}(X(t) \in A|X(s)).$$

This is equivalent to the following one: for any bounded Borel measurable function $\varphi : \mathbb{R}^n \to \mathbb{R}$ and $0 \leq s \leq t < \infty$,

$$\mathbb{E}(\varphi(X(t))|\mathcal{F}_s) = \mathbb{E}(\varphi(X(t))|X(s)).$$

The *transition probability or function* of the Markov process is a function $P(s, x; t, A)$, defined on $0 \leq s \leq t < \infty$, $x \in \mathbb{R}^n$ and $A \in \mathcal{B}(\mathbb{R}^n)$, with the following properties:

(1) For every $0 \leq s \leq t < \infty$ and $A \in \mathcal{B}(\mathbb{R}^n)$,
$$P(s, X(s); t, A) = \mathbb{P}(X(t) \in A | X(s)).$$

(2) $P(s, x; t, \cdot)$ is a probability measure on $\mathcal{B}(\mathbb{R}^n)$ for every $0 \leq s \leq t < \infty$ and $x \in \mathbb{R}^n$.

(3) $P(s, \cdot; t, A)$ is Borel measurable for every $0 \leq s \leq t < \infty$ and $A \in \mathcal{B}(\mathbb{R}^n)$.

(4) The Kolmogorov–Chapman equation
$$P(s, x; t, A) = \int_{\mathbb{R}^n} P(u, y; t, A) P(s, x; u, dy)$$
holds for any $0 \leq s \leq u \leq t < \infty$, $x \in \mathbb{R}^n$ and $A \in \mathcal{B}(\mathbb{R}^n)$.

Clearly, in terms of transition probability, the Markov property becomes
$$\mathbb{P}(X(t) \in A | \mathcal{F}_s) = P(s, X(s); t, A).$$

We shall use the notion
$$\mathbb{P}(X(t) \in A | X(s) = x) = P(s, x; t, A)$$
and
$$\mathbb{E}_{s,x} \varphi(X(t)) = \int_{\mathbb{R}^n} \varphi(y) P(s, x; t, dy).$$

A Markov process $X = \{X(t)\}_{t \geq 0}$ is said to be *homogeneous* if its transition probability $P(s, x; t, A)$ is stationary, namely
$$P(s + u, x; t + u, A) = P(s, x; t, A)$$
for all $0 \leq s \leq t < \infty$, $x \in \mathbb{R}^n$, $u \geq 0$ and $A \in \mathcal{B}(\mathbb{R}^n)$. In this case, the transition probability $P(s, x; t, A)$ depends only on $t - s$ and it can be simply written as $P(0, x; t, A) = P(t, x, A)$. Moreover, the Kolmogorov–Chapman equation becomes
$$P(t + s, x, A) = \int_{\mathbb{R}^n} P(s, y, A) P(t, x, dy).$$

Furthermore, with the notation

$$\mathbb{E}_x(\varphi(X(t))) = \int_{\mathbb{R}^n} \varphi(y) P(t, x, dy),$$

the Markov property becomes

$$\mathbb{E}(\varphi(X(t))|\mathcal{F}_s) = \mathbb{E}_{X(s)} \varphi(X(t-s)).$$

An n-dimensional process $\{X_t\}_{t \geq 0}$ is called a *strong Markov process* if the following *strong Markov property* is satisfied: for any bounded Borel measurable function $\varphi : \mathbb{R}^n \to \mathbb{R}$, any finite $\{\mathcal{F}_t\}$-stopping time τ and $t \geq 0$,

$$\mathbb{E}(\varphi(X(t+\tau))|\mathcal{F}_\tau) = \mathbb{E}(\varphi(X(t+\tau))|X(\tau)).$$

Especially, in the homogeneous case, this becomes

$$\mathbb{E}(\varphi(X(t+\tau))|\mathcal{F}_\tau) = \mathbb{E}_{X(\tau)} \varphi(X(t)).$$

A stochastic process $X = \{X(t)\}_{t \geq 0}$, defined on a probability space $(\Omega, \mathcal{F}, \mathbb{P})$, with values in a countable set Ξ (to be called the *state space* of the process), is called a *continuous-time Markov chain* if for any finite set $0 \leq t_1 < t_2 < \cdots < t_n < t_{n+1}$ of "times", and corresponding set $i_1, i_2, \ldots, i_{n-1}, i, j$ of states in Ξ such that $\mathbb{P}\{X(t_n) = i, X(t_{n-1}) = i_{n-1}, \ldots, X(t_1) = i_1\} > 0$, we have

$$\mathbb{P}\{X(t_{n+1}) = j|X(t_n) = i, X(t_{n-1}) = i_{n-1}, \ldots, X(t_1) = i_1\}$$
$$= \mathbb{P}\{X(t_{n+1}) = j|X(t_n) = i\}.$$

If for all s, t such that $0 \leq s \leq t < \infty$ and all $i, j \in \Xi$ the conditional probability $\mathbb{P}\{X(t) = j|X(s) = i\}$ depends only on $t - s$, we say that the process $X = \{X(t)\}_{t \geq 0}$ is *homogeneous*. In this case, then, $\mathbb{P}\{X(t) = j|X(s) = i\} = \mathbb{P}\{X(t-s) = j|X(0) = i\}$, and the function

$$P_{ij}(t) =: \mathbb{P}\{X(t) = j|X(s) = i\}, \quad i, j \in \Xi, t \geq 0,$$

is called the *transition function* or *transition probability* of the process. The function $P_{ij}(t)$ is called *standard* if $\lim_{t \to 0} P_{ii}(t) = 1$ for all $i \in \Xi$.

Theorem 1.42 [Anderson (1991)] *Let $P_{ij}(t)$ be a standard transition function, then $\gamma_i := \lim_{t \to 0}[1 - P_{ii}(t)]/t$ exists (but may be ∞) for all $i \in \Xi$.*

A state $i \in \Xi$ is said to be *stable* if $\gamma_i < \infty$.

Theorem 1.43 [Anderson (1991)] *Let $P_{ij}(t)$ be a standard transition function, and let j be a stable state. Then $\gamma_{ij} = P'_{ij}(0)$ exists and is finite for all $i \in \Xi$.*

Let $\gamma_{ii} = -\gamma_i$ and $\Gamma = (\gamma_{ij})_{i,j \in \Xi}$. Γ is called the *generator* of the Markov chain. If the state space is *finite* which we can take to be $\mathbb{S} = \{1, 2, \ldots, N\}$, then the process is called a continuous-time *finite* Markov chain. Throughout this book, we assume that all Markov chains are finite and all states are stable. For such a Markov chain, almost every sample path is a right continuous step function.

Theorem 1.44 [Anderson (1991)] *Let $P(t) = (P_{ij}(t))_{N \times N}$ be the transition probability matrix and $\Gamma = (\gamma_{ij})_{N \times N}$ be the generator of a finite Markov chain. Then*

$$P(t) = e^{t\Gamma}.$$

It is useful to emphasise that a continuous-time Markov chain $X(t)$ with generator $\Gamma = \{\gamma_{ij}\}_{N \times N}$ can be represented as a stochastic integral with respect to a Poisson random measure (see [Skorohod (1989)] and [Ghosh et al. (1997)]). Indeed, let Δ_{ij} be consecutive, left closed, right open intervals of the real line each having length γ_{ij} such that

$$\Delta_{12} = [0, \gamma_{12}),$$
$$\Delta_{13} = [\gamma_{12}, \gamma_{12} + \gamma_{13}),$$
$$\vdots$$
$$\Delta_{1N} = \left[\sum_{j=2}^{N-1} \gamma_{1j}, \sum_{j=2}^{N} \gamma_{1j} \right),$$
$$\Delta_{21} = \left[\sum_{j=2}^{N} \gamma_{1j}, \sum_{j=2}^{N} \gamma_{1j} + \gamma_{21} \right),$$
$$\Delta_{23} = \left[\sum_{j=2}^{N} \gamma_{1j} + \gamma_{21}, \sum_{j=2}^{N} \gamma_{1j} + \gamma_{21} + \gamma_{23} \right),$$
$$\vdots$$
$$\Delta_{2N} = \left[\sum_{j=2}^{N} \gamma_{1j} + \sum_{j=1, j \neq 2}^{N-1} \gamma_{2j}, \sum_{j=2}^{N} \gamma_{1j} + \sum_{j=1, j \neq 2}^{N} \gamma_{2j} \right)$$

and so on. Define a function
$$h : \mathbb{S} \times \mathbb{R} \to \mathbb{R}$$
by
$$h(i, y) = \begin{cases} j - i & \text{if } y \in \Delta_{ij}, \\ 0 & \text{otherwise.} \end{cases} \tag{1.46}$$

Then
$$dX(t) = \int_{\mathbb{R}} h(X(t-), y) \nu(dt, dy),$$

with initial condition $X(0) = i_0$, where $\nu(dt, dy)$ is a Poisson random measure with intensity $dt \times \mu(dy)$, in which μ is the Lebesgue measure on $\mathbb{R}$.

1.8 Generalised Itô's Formula

Let $(\Omega, \mathcal{F}, \{\mathcal{F}_t\}_{t \geq 0}, \mathbb{P})$ be a complete probability space with a filtration $\{\mathcal{F}_t\}_{t \geq 0}$ satisfying the usual conditions (i.e. it is increasing and right continuous while $\mathcal{F}_0$ contains all $\mathbb{P}$-null sets). Let $B(t) = (B_t^1, \ldots, B_t^m)^T$ be an m-dimensional Brownian motion defined on the probability space. Let $r(t), t \geq 0$, be a right-continuous Markov chain on the probability space taking values in a finite state space $\mathbb{S} = \{1, 2, \ldots, N\}$ with generator $\Gamma = (\gamma_{ij})_{N \times N}$ given by

$$\mathbb{P}\{r(t + \delta) = j | r(t) = i\} = \begin{cases} \gamma_{ij} \delta + o(\delta) & \text{if } i \neq j, \\ 1 + \gamma_{ii} \delta + o(\delta) & \text{if } i = j, \end{cases}$$

where $\delta > 0$. Here $\gamma_{ij} \geq 0$ is transition rate from i to j if $i \neq j$ while

$$\gamma_{ii} = -\sum_{j \neq i} \gamma_{ij}.$$

We assume that the Markov chain $r(\cdot)$ is independent of the Brownian motion $B(\cdot)$.

Let $x(t)$ be an n-dimensional Itô process on $t \geq 0$ with the stochastic differential

$$dx(t) = f(t)dt + g(t)dB(t),$$

where $f \in \mathcal{L}^1(\mathbb{R}_+; \mathbb{R}^n)$ and $g \in \mathcal{L}^2(\mathbb{R}_+; \mathbb{R}^{n \times m})$. The Itô formula established in Section 1.6 shows that a $C^{2,1}(\mathbb{R}^n \times \mathbb{R}_+; \mathbb{R}_+)$-function V maps the Itô process $x(t)$ into another Itô process $V(x(t), t)$. On the other hand, we will consider the paired process $(x(t), r(t))$ in this book and we need to know how a function $V : \mathbb{R}^n \times \mathbb{R}_+ \times \mathbb{S} \to \mathbb{R}$ will map $(x(t), r(t))$ into another process $V(x(t), t, r(t))$. For this purpose, let $C^{2,1}(\mathbb{R}^n \times \mathbb{R}_+ \times \mathbb{S}; \mathbb{R})$ denote the family of all real-valued functions $V(x, t, i)$ on $\mathbb{R}^n \times \mathbb{R}_+ \times \mathbb{S}$ which are continuously twice differentiable in x and once in t. If $V \in C^{2,1}(\mathbb{R}^n \times \mathbb{R}_+ \times \mathbb{S}; \mathbb{R})$, define an operator LV from $\mathbb{R}^n \times \mathbb{R}_+ \times \mathbb{S}$ to $\mathbb{R}$ by

$$LV(x,t,i) = V_t(x,t,i) + V_x(x,t,i)f(t)$$
$$+ \frac{1}{2}\text{trace}[g^T(t)V_{xx}(x,t,i)g(t)], + \sum_{j=1}^{N} \gamma_{ij} V(x,t,j), \quad (1.47)$$

where

$$V_t(x,t,i) = \frac{\partial V(x,t,i)}{\partial t}, \quad V_x(x,t,i) = \left(\frac{\partial V(x,t,i)}{\partial x_1}, \ldots, \frac{\partial V(x,t,i)}{\partial x_n} \right)$$

and

$$V_{xx}(x,t,i) = \left(\frac{\partial^2 V(x,t,i)}{\partial x_i \partial x_j} \right)_{n \times n}.$$

The following formula, known as the generalised Itô formula, reveals how V maps the paired process $(x(t), r(t))$ into a new process $V(x(t), t, r(t))$.

Theorem 1.45 *If $V \in C^{2,1}(\mathbb{R}^n \times \mathbb{R}_+ \times \mathbb{S}; \mathbb{R})$, then for any $t \geq 0$*

$$V(x(t), t, r(t))$$
$$= V(x(0), 0, r(0)) + \int_0^t LV(x(s), s, r(s))ds$$
$$+ \int_0^t V_x(x(s), s, r(s))g(x(s), s, r(s))dB(s)$$
$$+ \int_0^t \int_\mathbb{R} (V(x(s), s, i_0 + h(r(s), l)) - V(x(s), s, r(s)))\mu(ds, dl), \quad (1.48)$$

where the function h is defined by (1.46) and $\mu(ds, dl) = \nu(ds, dl) - \mu(dl)ds$ is a martingale measure while ν and μ have been defined in the end of Section 1.7.

The proof can be found in [Skorohod (1989)] on page 104 (namely, the proof of Lemma 3 there). In particular, taking the expectation on both sides of (1.48), we get the following useful lemma.

Lemma 1.9 Let $V \in C^{2,1}(\mathbb{R}^n \times \mathbb{R}_+ \times \mathbb{S}; \mathbb{R}_+)$ and τ_1, τ_2 be bounded stopping times such that $0 \leq \tau_1 \leq \tau_2$ a.s. If $V(x(t), t, r(t))$ and $LV(x(t), t, r(t))$ etc, are bounded on $t \in [\tau_1, \tau_2]$ with probability 1, then

$$\mathbb{E}V(x(\tau_2), \tau_2, r(\tau_2)) = \mathbb{E}V(x(\tau_1), \tau_1, r(\tau_1))$$
$$+ \mathbb{E}\int_{\tau_1}^{\tau_2} LV(x(s), s, r(s))ds. \quad (1.49)$$

1.9 Exercises

1.1 Let B_t, $0 \leq t \leq T$ be a scalar Brownian motion and let k be a positive integer. Set $\Delta = T/k$ and $t_j = j\Delta$ for $0 \leq j \leq k$. Show

$$\mathbb{E}(B_{t_j} - B_{t_{j-1}})^2 = \Delta \quad \text{and} \quad \mathbb{E}(B_{t_j} - B_{t_{j-1}})^4 = 3\Delta^2.$$

Deduce that

$$\mathbb{E}[(B_{t_j} - B_{t_{j-1}})(B_{t_i} - B_{t_{j-i}})] = 0 \quad \text{for } i \neq j.$$

Hence show that $\sum_{j=1}^{k}(B_{t_j} - B_{t_{j-1}})^2$ has mean T. Next, show that

$$\mathbb{E}\left(\sum_{j=1}^{k}(B_{t_j} - B_{t_{j-1}})^2\right)^2 = T^2 + 2T\Delta$$

and hence deduce that $\sum_{j=1}^{k}(B_{t_j} - B_{t_{j-1}})^2$ has variance of $O(\Delta)$.

1.2 Show that a scalar Brownian motion B_t has a scaling property: for any fixed $c \neq 0$,

$$X_t := \frac{B_{c^2 t}}{c}, \quad t \geq 0,$$

is a Brownian motion with respect to the filtration $\{\mathcal{F}_{c^2 t}\}$.

1.3 Prove properties (a)–(c) listed on page 21 for a Brownian motion.

1.4 Prove Lemma 1.4.

1.5 In Step 2 of the proof of Lemma 1.5, $\phi_k(t)$ is defined for the given process $\varphi(t)$. Show that for every ω, $\phi_k(\cdot, \omega)$ is continuous and and

$$\lim_{k \to \infty} \int_a^b |\varphi(t, \omega) - \phi_k(t, \omega)|^2 dt = 0.$$

1.6 Prove Theorem 1.21.

1.7 Prove Lemma 1.7.

1.8 Prove the Itô formula. (You may refer to [Mao (1997)].)

1.9 Let Q be an $m \times m$ matrix and define $V(x) = x^T Q x$ for $x \in \mathbb{R}^m$. Show that $V_x(x) = x^T(Q + Q^T)$ and $V_{xx} = Q + Q^T$ and hence show by the Itô formula that

$$B^T(t)QB(t) = \text{trace}(Q)t + \int_0^t B^T(s)(Q + Q^T)dB(s),$$

where $B(t)$ is an m-dimensional Brownian motion.

1.10 Let $\xi \sim N(0, \sigma^2)$ (i.e. a normal distribution with mean 0 and variance σ^2). Show

$$\mathbb{E}(e^\xi) = e^{\frac{1}{2}\sigma^2}.$$

Hence show that if $B(t)$ is a scalar Brownian motion and $p > 0$, then

$$\int_0^t \mathbb{E}|e^{-\frac{1}{2}p^2 s + pB(s)}|^2 ds = \frac{1}{p^2}[e^{p^2 t} - 1], \quad \forall t \geq 0.$$

Chapter 2

Inequalities

2.1 Introduction

Inequalities appear much more frequently than equalities in mathematics. For example, inequalities appear on almost every page from this page on in this book. This is because mathematics has become more and more complicated and equalities can often not be driven so we are forced to estimate lots of quantities leading to many inequalities. In this chapter we will present some important inequalities which will be used throughout this book. We will begin with some elementary but frequently-used inequalities and then establish several Gronwall-type integral inequalities. We will also review several useful matrix inequalities including linear matrix inequalities and M-matrix inequalities. We will finally establish a number of the most important inequalities in stochastic analysis including the well-known Burkholder–Davis–Gundy inequality and the exponential martingale inequality.

2.2 Frequently Used Inequalities

There are several inequalities which are used frequently in this book. Although they are elementary, they may be used in different ways to create various useful inequalities.

Let us begin with the simplest inequality

$$2ab \leq a^2 + b^2, \quad \forall a, b \in \mathbb{R}. \tag{2.1}$$

From this follows

$$2ab \leq \varepsilon a^2 + \frac{1}{\varepsilon} b^2, \quad \forall a, b \in \mathbb{R} \text{ and } \forall \varepsilon > 0. \tag{2.2}$$

This is because

$$2ab = 2(\sqrt{\varepsilon}\, a)\left(\frac{b}{\sqrt{\varepsilon}}\right) \le \varepsilon a^2 + \frac{1}{\varepsilon} b^2$$

as required. The reader will see that this slightly modified inequality (2.2) turns out to be used much more frequently in this book than its original inequality (2.1).

Let us now proceed to the Yong inequality

$$|a|^\beta |b|^{(1-\beta)} \le \beta |a| + (1-\beta)|b|, \quad \forall a, b \in \mathbb{R} \text{ and } \forall \beta \in [0,1]. \tag{2.3}$$

Using the same trick as (2.2) was obtained, we derive that for any $\varepsilon > 0$

$$|a|^\beta |b|^{(1-\beta)} = (\varepsilon |a|)^\beta \left(\frac{|b|}{\varepsilon^{\beta/(1-\beta)}}\right)^{(1-\beta)} \le \varepsilon \beta |a| + \frac{1-\beta}{\varepsilon^{\beta/(1-\beta)}} |b|, \tag{2.4}$$

which is an alternative bound for $|a|^\beta |b|^{(1-\beta)}$ and is more flexible for the free parameter ε may be chosen to suit various situations. Another variation of the Yong inequality is the following:

$$|a|^p |b|^q \le |a|^{(p+q)} + \frac{q}{p+q} \left[\frac{p}{\varepsilon(p+q)}\right]^{p/q} |b|^{(p+q)} \tag{2.5}$$

for $\forall a, b \in \mathbb{R}$ and $\forall p, q, \varepsilon > 0$. In fact,

$$|a|^p |b|^q = \left(\frac{\varepsilon(p+q)}{p}|a|^{(p+q)}\right)^{p/(p+q)} \left(\left[\frac{p}{\varepsilon(p+q)}\right]^{p/q} |b|^{(p+q)}\right)^{q/(p+q)}$$

$$\le \varepsilon |a|^{(p+q)} + \frac{q}{p+q}\left[\frac{p}{\varepsilon(p+q)}\right]^{p/q} |b|^{(p+q)}$$

as stated. A more general Yong inequality is

$$\prod_{i=1}^k |a_i|^{\beta_i} \le \sum_{i=1}^k \beta_i |a_i|, \tag{2.6}$$

where $a_i \in \mathbb{R}$ and $\beta_i \ge 0$ obeying $\sum_{i=1}^k \beta_i = 1$.

Another inequality frequently-used in this book is the discrete Hölder inequality

$$\left|\sum_{i=1}^k a_i b_i\right| \le \left(\sum_{i=1}^k |a_i|^p\right)^{1/p} \left(\sum_{i=1}^k |b_i|^q\right)^{1/q} \tag{2.7}$$

if $p, q > 1$, $1/p + 1/q = 1$, $a_i, b_i \in \mathbb{R}$ and $k \geq 2$. In particular, if $b_i = 1$ for all $1 \leq i \leq k$, we have

$$\left|\sum_{i=1}^{k} a_i\right| \leq k^{(p-1)/p} \left(\sum_{i=1}^{k} |a_i|^p\right)^{1/p}.$$

Of course, this holds for $p = 1$ obviously. So we always have

$$\left|\sum_{i=1}^{k} a_i\right|^p \leq k^{(p-1)} \sum_{i=1}^{k} |a_i|^p \qquad (2.8)$$

if $p \geq 1$. On the other hand, when $p \in (0, 1)$, we have

$$\left|\sum_{i=1}^{k} a_i\right|^p \leq \left(k \max_{1 \leq i \leq k} |a_i|\right)^p = k^p \max_{1 \leq i \leq k} |a_i|^p \leq k^p \sum_{i=1}^{k} |a_i|^p. \qquad (2.9)$$

One more useful inequality is

$$|a^p - b^p| \leq p|a - b|(a^{p-1} + b^{p-1}) \quad \forall a, b \geq 0 \text{ and } p \geq 1. \qquad (2.10)$$

In fact, in the case $a \geq b \geq 0$, the mean value theorem gives

$$a^p - b^p = p\xi^{p-1}(a - b)$$

for some $\xi \in [b, a]$. Thus

$$a^p - b^p \leq p a^{p-1}(a - b) \leq p(a - b)(a^{p-1} + b^{p-1}).$$

The case $b > a$ follows by symmetry.

In this book we will also often use convex or concave functions and their associated inequalities. A function $\varphi : \mathbb{R} \to \mathbb{R}$ is said to be *convex* if

$$\varphi(\tfrac{1}{2}(x+y)) \leq \tfrac{1}{2}(\varphi(x) + \varphi(y)) \quad \forall x, y \in \mathbb{R}.$$

It said to be *concave* if

$$\varphi(\tfrac{1}{2}(x+y)) \geq \tfrac{1}{2}(\varphi(x) + \varphi(y)) \quad \forall x, y \in \mathbb{R}.$$

Clearly, φ is convex if and only if $-\varphi$ is concave so we shall only state a couple of useful inequalities for a convex function φ. First of all, we have

$$\varphi(\alpha x + (1-\alpha)y) \leq \alpha \varphi(x) + (1-\alpha)\varphi(y) \quad \forall x, y \in \mathbb{R}$$

if $\alpha \in (0, 1)$. More generally, if $\alpha_1, \cdots, \alpha_k$ are positive constants such that $\sum_{i=1}^{k} \alpha_i = 1$, then

$$\varphi\Big(\sum_{i=1}^{k}\alpha_i x_i\Big) \le \sum_{i=1}^{k}\alpha_i\varphi(x_i) \quad \forall x_1,\cdots,x_k \in \mathbb{R}.$$

Even more generally, we have the following well-known Jensen inequality.

Theorem 2.1 *(Jensen's inequality) If $\varphi:\mathbb{R}\to\mathbb{R}$ is a convex function while $\xi:\Omega\to\mathbb{R}$ is a random variable on a probability space $(\Omega,\mathcal{F},\mathbb{P})$ such that $\mathbb{E}|\xi|<\infty$, then*

$$\varphi(\mathbb{E}\xi) \le \mathbb{E}\varphi(\xi).$$

2.3 Gronwall-Type Inequalities

The integral inequalities of Gronwall-type have been widely applied in the theory of ordinary differential equations and stochastic differential equations to prove the results on existence, uniqueness, boundedness, comparison, continuous dependence, perturbation and stability *etc*. Naturally, Gronwall-type inequalities will play an important role in this book. For the convenience of the reader, we establish a number of well-known inequalities of this type in this section.

Theorem 2.2 *(Gronwall's inequality) Let $T>0$ and $c\ge 0$. Let $u(\cdot)$ be a Borel measurable bounded non-negative function on $[0,T]$, and let $v(\cdot)$ be a non-negative integrable function on $[0,T]$. If*

$$u(t) \le c + \int_0^t v(s)u(s)ds \quad \text{for all } 0\le t\le T, \tag{2.11}$$

then

$$u(t) \le c\exp\Big(\int_0^t v(s)ds\Big) \quad \text{for all } 0\le t\le T. \tag{2.12}$$

Proof. Without loss of generality we may assume that $c>0$. Set

$$z(t) = c + \int_0^t v(s)u(s)ds \quad \text{for } 0\le t\le T.$$

Then $u(t) \le z(t)$. Moreover, by the chain rule of classical calculus, we have

$$\log(z(t)) = \log(c) + \int_0^t \frac{v(s)u(s)}{z(s)}ds \le \log(c) + \int_0^t v(s)ds.$$

This implies
$$z(t) \le c\exp\left(\int_0^t v(s)ds\right) \quad \text{for } 0 \le t \le T,$$
and the required inequality (2.12) follows since $u(t) \le z(t)$. □

Theorem 2.3 *(Bihari's inequality) Let $T > 0$ and $c > 0$. Let $K : \mathbb{R}_+ \to \mathbb{R}_+$ be a continuous nondecreasing function such that $K(t) > 0$ for all $t > 0$. Let $u(\cdot)$ be a Borel measurable bounded non-negative function on $[0,T]$, and let $v(\cdot)$ be a non-negative integrable function on $[0,T]$. If*

$$u(t) \le c + \int_0^t v(s)K(u(s))ds \quad \text{for all } 0 \le t \le T, \tag{2.13}$$

then

$$u(t) \le G^{-1}\left(G(c) + \int_0^t v(s)ds\right) \tag{2.14}$$

holds for all such $t \in [0,T]$ that

$$G(c) + \int_0^t v(s)ds \in Dom(G^{-1}), \tag{2.15}$$

where

$$G(r) = \int_1^r \frac{ds}{K(s)} \quad \text{on } r > 0,$$

and G^{-1} is the inverse function of G.

Proof. Set
$$z(t) = c + \int_0^t v(s)K(u(s))ds \quad \text{for } 0 \le t \le T.$$

Then $u(t) \le z(t)$. By the chain rule of classical calculus, one can derive that

$$G(z(t)) = G(c) + \int_0^t \frac{v(s)K(u(s))}{K(z(s))}ds \le G(c) + \int_0^t v(s)ds \tag{2.16}$$

for all $t \in [0,T]$. Hence, for $t \in [0,T]$ satisfying (2.15) one sees from (2.16) that

$$z(t) \le G^{-1}\left(G(c) + \int_0^t v(s)ds\right),$$

and the desired inequality (2.14) follows since $u(t) \leq z(t)$. □

Theorem 2.4 Let $T > 0$, $\alpha \in [0,1)$ and $c > 0$. Let $u(\cdot)$ be a Borel measurable bounded non-negative function on $[0,T]$, and let $v(\cdot)$ be a non-negative integrable function on $[0,T]$. If

$$u(t) \leq c + \int_0^t v(s)[u(s)]^\alpha ds \quad \text{for all } 0 \leq t \leq T, \tag{2.17}$$

then

$$u(t) \leq \left(c^{1-\alpha} + (1-\alpha)\int_0^t v(s)ds\right)^{\frac{1}{1-\alpha}} \tag{2.18}$$

holds for all $t \in [0,T]$.

Proof. Without loss of generality, we may assume $c > 0$. Set

$$z(t) = c + \int_0^t v(s)[u(s)]^\alpha ds \quad \text{for } 0 \leq t \leq T.$$

Then $u(t) \leq z(t)$ and $z(t) > 0$. By the fundamental differential formula, one can show that

$$[z(t)]^{1-\alpha} = c^{1-\alpha} + (1-\alpha)\int_0^t \frac{v(s)[u(s)]^\alpha}{[z(s)]^\alpha}ds$$

$$\leq c^{1-\alpha} + (1-\alpha)\int_0^t v(s)ds$$

for all $t \in [0,T]$, and the required inequality (2.18) follows. □

The discrete inequalities of Gronwall-type have also been widely applied in the theory of difference equations, SDEs driven by discontinous semi-martingales, numerical analysis *etc.* We will only establish a discrete Gronwall inequality below but more discrete inequalities of Gronwall-type can be found in [Mao (1991)].

Theorem 2.5 *(Discrete Gronwall's inequality)* Let M be a positive integer. Let u_k and v_k be non-negative numbers for $k = 0, 1, \cdots, M$. If

$$u_k \leq u_0 + \sum_{j=0}^{k-1} v_j u_j, \quad \forall k = 1, 2, \cdots, M, \tag{2.19}$$

then

$$u_k \leq u_0 \exp\left(\sum_{j=0}^{k-1} v_j\right), \quad \forall k = 1, 2, \cdots, M, \quad (2.20)$$

Proof. Define, for $t \in [0, M+1)$,

$$u(t) = \sum_{j=0}^{M} u_j I_{[j,j+1)}(t), \quad v(t) = \sum_{j=0}^{M} v_j I_{(j,j+1]}(t),$$

and

$$A(t) = \sum_{j=1}^{M} I_{[j,M+1]}(t).$$

It then follows from (2.19) that

$$u(t) \leq u(0) + \int_0^t v(s)u(s-)dA(s), \quad \forall t \in [0, T],$$

where the integral is of the Lebesgue–Stieltjes type and $u(s-) = \lim_{t \uparrow s} u(t)$ (i.e. the left limit). Without loss of generality we may assume $u(0) = u_0 > 0$; otherwise replace u_0 by $\varepsilon > 0$ and then let $\varepsilon \to 0$. Set

$$w(t) = u(0) + \int_0^t v(s)u(s-)dA(s), \quad t \in [0, M+1).$$

Then $w(t)$ is positive and non-decreasing and $w(t) \geq u(t)$ whence $w(t-) \geq u(t-)$ as well. By the fundamental differential formula we deduce that

$$\log(w(t)) = \log(u(0)) + \int_0^t \frac{v(s)u(s-)}{w(s-)} dA(s)$$
$$+ \sum_{0 < s \leq t} \left[\log(w(s)) - \log(w(s-)) - \frac{w(s) - w(s-)}{w(s-)}\right]. \quad (2.21)$$

But

$$\log(w(s)) - \log(w(s-)) = \int_{w(s-)}^{w(s)} \frac{dz}{z}$$

$$\leq \int_{w(s-)}^{w(s)} \frac{dz}{w(s-)} \leq \frac{w(s) - w(s-)}{w(s-)}.$$

Substituting this into (2.21) and making use of the fact that $w(s-) \geq u(s-)$ we obtain

$$\log(w(t)) = \log(u(0)) + \int_0^t v(s) dA(s).$$

In particular, for any $k = 1, 2, \cdots, M$, we have

$$\log(w(k)) \leq \log(u_0) + \int_0^k v(s) dA(s) = \log(u_0) + \sum_{j=0}^{k-1} v_j.$$

This implies

$$u_k \leq w(k) \leq u_0 \exp\left(\sum_{j=0}^{k-1} v_j\right)$$

as required. $\square$

2.4 Matrices and Inequalities

Let us first present some well-known results from matrix calculus which are widely used in this book. Let A, C, D, F be matrices of appropriate dimensions and let x, y be column vectors of appropriate dimensions.

Denote by det.(A) the determinant of a square matrix A. If det.$(A) \neq 0$, then A is said to be nonsingular. In this case A is invertible and its inverse is denoted by A^{-1}. To compute the inverse of a sum of matrices, the following Sherman–Morrison–Woodbury formula can be applied:

$$(A + CDF)^{-1} = A^{-1} - A^{-1}C(D^{-1} + FA^{-1}C)^{-1}FA^{-1}. \quad (2.22)$$

The following rule inverts a nonsingular matrix in block form:

$$\begin{bmatrix} A & C \\ F & D \end{bmatrix}^{-1} = \begin{bmatrix} (A - CD^{-1}F)^{-1} & -A^{-1}C(D - FA^{-1}C)^{-1} \\ -D^{-1}F(A - CD^{-1}F)^{-1} & (D - FA^{-1}C)^{-1} \end{bmatrix}. \quad (2.23)$$

For a square matrix A, its trace denoted by trace(A) is the sum of its diagonal elements. The following are three useful properties:

$$\text{trace}(A + C) = \text{trace}(A) + \text{trace}(C), \quad (2.24)$$

$$\text{trace}(AC) = \text{trace}(CA), \quad (2.25)$$

$$x^T A y = \text{trace}(A y x^T). \tag{2.26}$$

For a matrix $A \in \mathbb{R}^{n \times m}$, its induced operator norm is defined by

$$\|A\| = \sup_{x \in \mathbb{R}^m, |x|=1} |Ax| \tag{2.27}$$

while its trace norm is given by

$$|A| = \sqrt{\text{trace}(A^T A)}. \tag{2.28}$$

We always have

$$|Ax| \leq \|A\| |x| \quad \text{and} \quad |Ax| \leq |A| |x|, \quad \forall x \in \mathbb{R}^m.$$

For a square $n \times n$ matrix A, its eigenvalue is a root λ to the eigen equation $\det.(\lambda I - A) = 0$, where I is the identity matrix. Clearly, an eigenvalue may be a complex number and A has n eigenvalues in the complex space $\mathbb{C}$. Denote by $\lambda(A)$ the set of all eigenvalues of A which is called the spectrum of A. Define $\rho(A) = \max\{|\lambda| : \lambda \in \lambda(A)\}$ which is called the spectral radius of A.

A square matrix A is said to be symmetric if $A = A^T$. For a symmetric matrix $A \in \mathbb{R}^{n \times n}$, its eigenvalues are all real numbers, namely $\lambda(A) \subset \mathbb{R}$. Denoted its largest and smallest eigenvalue by $\lambda_{\max}(A)$ and $\lambda_{\min}(A)$, respectively. In this symmetric case, the spectral radius of A becomes $\rho(A) = |\lambda_{\max}(A)| \vee |\lambda_{\min}(A)|$. Moreover,

$$\lambda_{\max}(A) = \sup_{x \in \mathbb{R}^n, |x|=1} x^T A x \quad \text{and} \quad \lambda_{\min}(A) = \inf_{x \in \mathbb{R}^n, |x|=1} x^T A x. \tag{2.29}$$

The following inequality is also useful

$$\lambda_{\min}(A) |x|^2 \leq x^T A x \leq \lambda_{\max}(A) |x|^2, \quad \forall x \in \mathbb{R}^n. \tag{2.30}$$

If $\lambda_{\min}(A) > 0$, A is said to be positive-definite and we sometimes write $A > 0$. In this case, we always have

$$x^T A x > 0, \quad \forall x \in \mathbb{R}^n - \{0\}.$$

If $\lambda_{\min}(A) \geq 0$, A is said to be non-negative-definite and we sometimes write $A \geq 0$. In this case, we always have

$$x^T A x \geq 0, \quad \forall x \in \mathbb{R}^n.$$

In terms of the eigenvalue, the operator norm of a matrix $A \in \mathbb{R}^{n \times m}$ can be written as

$$\|A\| = \sqrt{\lambda_{\max}(A^T A)}. \tag{2.31}$$

In the remaining of this section, we let $A = (a_{ij})_{n \times n} \in \mathbb{R}^{n \times n}$ be symmetric. Define $\mathbb{R}_+^n = \{(x_1, \cdots, x_n)^T \in \mathbb{R}^n : x_i > 0,\ 1 \le i \le n\}$, namely the positive cone of $\mathbb{R}^n$, and $\bar{\mathbb{R}}_+^n = \{(x_1, \cdots, x_n)^T \in \mathbb{R}^n : x_i \ge 0,\ 1 \le i \le n\}$, the non-negative cone of $\mathbb{R}^n$. Define

$$\lambda_{\max}^+(A) = \sup_{x \in \bar{\mathbb{R}}_+^n,\, |x|=1} x^T A x.$$

Let us emphasise that this is different from the largest eigenvalue $\lambda_{\max}(A)$ of the matrix A. Indeed, recalling (2.29) we observe that

$$\lambda_{\max}^+(A) \le \lambda_{\max}(A).$$

In many situations we even have $\lambda_{\max}^+(A) < \lambda_{\max}(A)$. For example, for

$$A = \begin{bmatrix} -1 & -1 \\ -1 & -1 \end{bmatrix},$$

we have $\lambda_{\max}^+(A) = -1 < \lambda_{\max}(A) = 0$. On the other hand, $\lambda_{\max}^+(A)$ does have many similar properties as $\lambda_{\max}(A)$ has. For example, it follows straightforward from the definition that

$$x^T A x \le \lambda_{\max}^+(A)|x|^2 \quad \forall x \in \bar{\mathbb{R}}_+^n$$

and

$$\lambda_{\max}^+(A) \le \|A\|.$$

Moreover

$$\lambda_{\max}^+(A + C) \le \lambda_{\max}^+(A) + \lambda_{\max}^+(C)$$

if C is another symmetric $n \times n$ matrix.

The results established in this book require to verify $\lambda_{\max}^+(A) \le 0$ or $\lambda_{\max}^+(A) < 0$. It is easy to see from the definition that $\lambda_{\max}^+(A) \le 0$ if

$$a_{ij} \le 0 \quad \text{for all } 1 \le i, j \le n.$$

But the following two theorems give better results.

Theorem 2.6 *We always have*

$$\lambda_{\max}^+(A) \le \max_{1\le i\le n}\left(a_{ii} + \sum_{j\ne i}(0\vee a_{ij})\right). \tag{2.32}$$

Consequently, $\lambda_{\max}^+(A) \le 0$ *if*

$$a_{ii} \le -\sum_{j\ne i}(0\vee a_{ij}), \quad 1\le i\le n;$$

while $\lambda_{\max}^+(A) < 0$ *if*

$$a_{ii} < -\sum_{j\ne i}(0\vee a_{ij}), \quad 1\le i\le n.$$

Proof. For any $x\in\mathbb{R}_+^n$ with $|x|=1$, compute

$$x^T A x = \sum_{i=1}^n \sum_{j=1}^n a_{ij} x_i x_j$$
$$\le \sum_{i=1}^n a_{ii} x_i^2 + \sum_{i=1}^n \sum_{j\ne i}(0\vee a_{ij}) x_i x_j$$
$$\le \sum_{i=1}^n a_{ii} x_i^2 + \frac{1}{2}\sum_{i=1}^n \sum_{j\ne i}(0\vee a_{ij})(x_i^2 + x_j^2)$$
$$= \sum_{i=1}^n a_{ii} x_i^2 + \sum_{i=1}^n \left(\sum_{j\ne i}(0\vee a_{ij})\right)x_i^2$$
$$= \sum_{i=1}^n \left(a_{ii} + \sum_{j\ne i}(0\vee a_{ij})\right)x_i^2$$
$$\le \max_{1\le i\le n}\left(a_{ii} + \sum_{j\ne i}(0\vee a_{ij})\right).$$

Thus the required assertion (2.32) follows. □

Theorem 2.7 *Given a symmetric matrix* $A = (a_{ij})_{n\times n}$, *we define its associated matrix* $\tilde{A} = (\tilde{a}_{ij})_{n\times n}$ *by*

$$\tilde{a}_{ii} = a_{ii}, \quad 1\le i\le n$$

while

$$\tilde{a}_{ij} = 0\vee a_{ij}, \quad 1\le i,j\le n,\ i\ne j.$$

Then $\lambda^+_{\max}(A) \le \lambda_{\max}(\tilde{A})$.

Proof. Clearly,
$$a_{ij} \le \tilde{a}_{ij}, \quad 1 \le i, j \le n.$$
By definition, it is easy to see that
$$\lambda^+_{\max}(A) \le \lambda^+_{\max}(\tilde{A}).$$
But we always have
$$\lambda^+_{\max}(\tilde{A}) \le \lambda_{\max}(\tilde{A}).$$
The required assertion hence follows. □

2.5 Linear Matrix Inequalities

The history of linear matrix inequalities (LMIs) in the analysis of dynamical systems goes back more than 100 years. The story began in about 1890, when Lyapunov published his seminal work introducing what we now call Lyapunov theory. He showed that the linear differential equation

$$\frac{dx(t)}{dt} = Ax(t) \tag{2.33}$$

is asymptotically stable (i.e. all trajectories converge to zero) if and only if there exists a positive-definite symmetric matrix P such that

$$A^T P + PA < 0. \tag{2.34}$$

The requirement that $A^T P + PA < 0$ subject to $P = P^T > 0$ is what we now call a Laypunov inequality on P, which is a special form of an LMI. Lyapunov also showed that this first LMI could be solved explicitly. In fact, we can pick up any $Q = Q^T > 0$ and then solve the linear equation $A^T P + PA = -Q$ for the matrix P, which is guaranteed to be positive-definite if system (2.33) is asymptotically stable. In summary, the first LMI used to analyse stability of dynamical system was the Lyapunov inequality (2.34) which can be solved analytically (by solving a set of linear equations).

The next major milestone occurred in the 1940s. Lur'e, Postnikov, and others applied Lyapunov's methods to some specific practical problems in control engineering, especially, the problem of stability of a control system with a nonlinearity in the actuator. Although they did not explicitly form matrix inequalities, their stability criteria have the form of LMIs. These inequalities were reduced to polynomial inequalities which were then checked "by hand" (for, needless to say, small systems).

The third major breakthrough came in the early 1960s, when Yakubovich, Popov, Kalman, and other researchers succeeded in reducing the solution of the LMIs that arose in the problem of Lur'e to simple graphical criteria, using what we now call the positive-real (PR) lemma. By 1970, it was known that the LMI appearing in the PR lemma could be solved not only by graphical means, but also by solving a certain algebraic Riccati equation (ARE). In 1971, when he studied the problem of quadratic optimal control, J.C. Willems was led to the LMI

$$\begin{bmatrix} A^T P + PA + Q & PB + C^T \\ B^T P + C & R \end{bmatrix} < 0, \qquad (2.35)$$

and pointed out that this could be solved by studying the symmetric solutions of the ARE

$$A^T P + PA - (PB + C^T) R^{-1} (B^T P + C) + Q = 0. \qquad (2.36)$$

which in turn could be found by an eigen-decomposition of a related Hamiltonian matrix.

The next major advance was the simple observation that the LMIs which arise in system and control theory can be formulated as convex optimisation problems that are amenable to computer solution. Although this is a simple observation, it has some important consequences, the most important of which is that we can reliably solve many LMIs for which no "analytic solution" has been found (or is likely to be found). This observation was made explicitly by several researchers. Pyatnitskii and Skorodinskii were perhaps the first researchers to make this point, clearly and completely, in 1982. In the late 1980s, the powerful and efficient interior-point methods were developed to solve the LMIs that arise in system and control theory by several researchers including Karmarkar, Nesterov, Nemirovskii.

The theory of LMIs has become very rich but we shall only cite a few LMIs here for the use of this book. Many more LMIs can be found, for example, in [Boyd et al. (1994)]. The key idea here is to convert nonlinear (convex) matrix inequalities to LMIs which can then be solved efficiently by numerical methods e.g. the interior-point algorithms. (We will not discuss the algorithms in this book due to the page limit.) The Schur complements play a key role in this conversion and we state them as a theorem.

Theorem 2.8 *(The Schur complements)* The LMI

$$\begin{bmatrix} Q & S \\ S^T & R \end{bmatrix} > 0, \tag{2.37}$$

where $Q = Q^T$ and $R = R^T$, is equivalent to

$$R > 0, \quad Q - SR^{-1}S^T > 0. \tag{2.38}$$

This theorem shows that the set of nonlinear inequalities (2.38) can be represented as the LMI (2.37). Indeed, the Schur complements enable us to represent many convex constraints and quadratic matrix inequalities as LMIs.

As an example, the (maximum singular value) matrix norm constraint $\|F\| < 1$, where $F \in \mathbb{R}^{n \times m}$, can be represented as the LMI

$$\begin{bmatrix} I_n & F \\ F & I_m \end{bmatrix} > 0,$$

where I_n and I_m are $n \times n$ and $m \times m$ identity matrix, respectively.

The constraint $x^T P^{-1} x < 1$, $P > 0$, where $x \in \mathbb{R}^n$ and $P = P^T \in \mathbb{R}^{n \times n}$, is expressed as the LMI

$$\begin{bmatrix} P & x \\ x^T & 1 \end{bmatrix} > 0.$$

More generally, the constraint

$$\text{trace}(S^T P^{-1} S) < 1, \quad P > 0,$$

where $P = P^T \in \mathbb{R}^{n \times n}$, $S \in \mathbb{R}^{n \times m}$, can be represented, by introducing a new slack matrix variable $X = X^T \in \mathbb{R}^{m \times m}$, as the LMI

$$\text{trace}(X) < 1, \quad \begin{bmatrix} X & S \\ S^T & P \end{bmatrix} > 0.$$

As another related example, consider the quadratic matrix inequality in P (by "in P" we mean P is a matrix variable):

$$A^T P + PA + PDR^{-1}D^T P + Q < 0,$$

where A, D, $Q = Q^T$, $R = R^T > 0$ are given matrices of appropriate sizes, and $P = P^T$ is the variable. Note that this is a *quadratic* matrix inequality in the variable P. However, the Schur complements allow us to express it as the LMI

$$\begin{bmatrix} -A^T P - PA - Q & PD \\ D^T P & R \end{bmatrix} > 0.$$

Although many LMIs have to be solved numerically by computer, there are several LMI problems with analytic solutions. Let us begin with the simplest LMI arising in control theory, the Lyapunov inequality:

$$A^T P + PA < 0, \quad P = P^T > 0, \qquad (2.39)$$

where $A \in \mathbb{R}^{n \times n}$ is given and P is the variable. Lyapunov showed that this LMI is feasible if and only if the matrix A is stable, i.e. all trajectories of $\dot{x} = Ax$ converge to zero, or equivalently, all eigenvalues of A have negative real part. To solve this LMI, we simply pick up any $Q = Q^T > 0$ and solve the Lyapunov equation $A^T P + PA = -Q$, which is nothing but a set of $\frac{1}{2}n(n+1)$ linear equations for the $\frac{1}{2}n(n+1)$ scalar variables in P. This set of linear equations will be solvable and result in $P = P^T > 0$ if and only if the LMI is feasible.

One more important example is given by the *positive-real* (PR) lemma, which yields a "frequence-domain" interpretation for a certain LMI problem, and under some additional assumptions, a numerical solution procedure via Riccati equations as well. The LMI considered is

$$\begin{bmatrix} A^T P + PA & PF - C^T \\ F^T P - C & -D^T - D \end{bmatrix} \leq 0, \quad P = P^T > 0, \qquad (2.40)$$

where $A \in \mathbb{R}^{n \times n}$, $F \in \mathbb{R}^{n \times m}$, $C \in \mathbb{R}^{m \times n}$ and $D \in \mathbb{R}^{m \times m}$ are given, and the matrix $P = P^T \in \mathbb{R}^{n \times n}$ is the variable. We assume that A is stable and $D + D^T > 0$. Note that the LMI (2.40) is equivalent to the quadratic matrix inequality

$$A^T P + PA + (PF - C^T)(D + D^T)^{-1}(PF - C^T)^T \leq 0. \qquad (2.41)$$

The link with system and control theory is given by the following result. The LMI (2.40) is feasible if and only if the linear system

$$\dot{x}(t) = Ax(t) + Fu(t), \quad y(t) = Cx(t) + Du(t), \quad x(0) = 0 \qquad (2.42)$$

is *passive*, i.e., satisfies

$$\int_0^T u(t)^T y(t) dt \geq 0$$

for all u and $T \geq 0$. Passivity can also be expressed in terms of the *transfer matrix* in $\mathbb{C}^{m \times m}$ of the linear system (2.42), defined as

$$H(s) := C(sI - A)^{-1}F + D$$

for $s \in \mathbb{C}$. Passivity is equivalent to the transfer matrix H being *positive-real*, which means that

$$H(s) + H(s)^* \geq 0 \quad \forall s \in \mathbb{C} \text{ with } \text{Re}(s) \geq 0,$$

where $H(s)^*$ is the conjugate transpose of $H(s)$. With a few further technical assumptions, including $D + D^T > 0$, the LMI (2.40) can be solved by a method based on Riccati equations and Hamiltonian matrices. With these assumptions the LMI (2.40) is feasible if and only if there exists a real matrix $P = P^T > 0$ satisfying the ARE

$$A^T P + PA + (PF - C^T)(D + D^T)^{-1}(PF - C^T)^T = 0, \quad (2.43)$$

which is just the quadratic matrix inequality (2.41) with equality substituted for inequality. To this ARE, we first form the associated Hamiltonian matrix

$$M = \begin{bmatrix} A - F(D + D^T)^{-1}C & F(D + D^T)^{-1}F^T \\ -C^T(D + D^T)^{-1}C & -A^T + C^T(D + D^T)^{-1}F^T \end{bmatrix}.$$

Then the LMI (2.40) is feasible if and only if M has no pure imaginary eigenvalues.

The same results appear in another important form, the *bounded-real* (BR) lemma. The LMI considered is

$$\begin{bmatrix} A^T P + PA + C^T C & PF + C^T D \\ F^T P + D^T C & D^T D - I \end{bmatrix} \leq 0, \quad P = P^T > 0, \quad (2.44)$$

where $A \in \mathbb{R}^{n \times n}$, $F \in \mathbb{R}^{n \times m}$, $C \in \mathbb{R}^{m \times n}$ and $D \in \mathbb{R}^{m \times m}$ are given, and the matrix $P = P^T \in \mathbb{R}^{n \times n}$ is the variable. We assume that A is stable. This LMI is feasible if and only if the linear system (2.42) is *nonexpansive*, i.e.,

$$\int_0^T y(t)^T y(t) dt \leq \int_0^T u(t)^T u(t) dt$$

for all solution of (2.42) with $x(0) = 0$. This nonexpansivity is equivalent to the transfer matrix H defined as above obeying the BR condition

$$H(s)H(s)^* \leq I \quad \forall s \in \mathbb{C} \text{ with } \text{Re}(s) > 0.$$

This is sometimes expressed as $\|H\|_\infty \leq 1$, where

$$\|H\|_\infty := \sup\{\|H(s)\| : s \in \mathbb{C} \text{ with } \mathrm{Re}(s) > 0\}$$

is called the $\mathbb{H}_\infty$ norm of the transfer matrix H. Once again we can relate the LMI (2.44) to an ARE. With some appropriate conditions including $D^T D < I$, the LMI (2.44) is feasible if and only if the ARE

$$A^T P + PA + C^T C + (PF + C^T D)(I - D^T D)^{-1}(PF + C^T D)^T = 0 \quad (2.45)$$

has a real solution $P = P^T$. We can solve this ARE by forming the associated Hamiltonian matrix

$$M = \begin{bmatrix} A + F(I - D^T D)^{-1} D^T C & F(I - D^T D)^{-1} B^T \\ -C^T (I - DD^T)^{-1} C & -A^T - C^T D(I - D^T D)^{-1} F^T \end{bmatrix}.$$

Then the LMI (2.44) is feasible if and only if M has no pure imaginary eigenvalues.

2.6 M-Matrix Inequalities

The theory of M-matrices has played an important role in the study of stability, stabilisation, control *etc.* We shall cite some useful results on M-matrices for the use of this book but more details can be found in [Berman and Plemmons (1994)].

Let us introduce a few more new notations. Let G be a vector or matrix. By $G \geq 0$ we mean each element of G is non-negative. By $G > 0$ we mean $G \geq 0$ and at least one element of G is positive. By $G \gg 0$ we mean all element of G are positive. Let G_1 and G_2 be two vectors or matrices with same dimensions. We write $G_1 \geq G_2$, $G_1 > G_2$ and $G_1 \gg G_2$ if and only if $G_1 - G_2 \geq 0$, $G_1 - G_2 > 0$ and $G_1 - G_2 \gg 0$, respectively. Moreover, we also adopt here the traditional notation by letting

$$Z^{n \times n} = \{A = (a_{ij})_{n \times n} : a_{ij} \leq 0,\ i \neq j\}.$$

Definition 2.9 A square matrix $A = (a_{ij})_{n \times n}$ is called a nonsingular M-matrix if A can be expressed in the form $A = sI - G$ with some $G \geq 0$ and $s > \rho(G)$, where I is the identity $n \times n$ matrix and $\rho(G)$ the spectral radius of G.

It is easy to see that a nonsingular M-matrix A has nonpositive off-diagonal and positive diagonal entries, that is
$$a_{ii} > 0 \text{ while } a_{ij} \leq 0, \ i \neq j.$$
In particular, $A \in Z^{n \times n}$. There are many conditions which are equivalent to the statement that A is a nonsingular M-matrix and we now cite some of them for the use of this book

Theorem 2.10 *If $A \in Z^{n \times n}$, then the following statements are equivalent:*

(1) *A is a nonsingular M-matrix.*
(2) *All of the principal minors of A are positive; that is*
$$\begin{vmatrix} a_{11} & \cdots & a_{1k} \\ \vdots & & \vdots \\ a_{k1} & \cdots & a_{kk} \end{vmatrix} > 0 \quad \text{for every } k = 1, 2, \cdots, n.$$
(3) *Every real eigenvalue of each principal submatrix of A is positive.*
(4) *Every real eigenvalue of A is positive.*
(5) *For each $x \neq 0$ there exists a positive diagonal matrix D such that $x^T A D x > 0$.*
(6) *$A + D$ is nonsingular for each non-negative diagonal matrix D.*
(7) *A is positive stable; that is, the real part of each eigenvalue of A is positive.*
(8) *There exists a symmetric positive-definite matrix P such that $AP + PA^T$ is positive-definite.*
(9) *A is semipositive; that is, there exists $x \gg 0$ in $\mathbb{R}^n$ such that $Ax \gg 0$.*
(10) *A is inverse-positive; that is, A^{-1} exists and $A^{-1} \geq 0$.*
(11) *A is monotone; that is, for $x \in \mathbb{R}^{n \times n}$,*
$$Ax \geq 0 \text{ implies } x \geq 0.$$
(12) *The new matrix by switching the ith row of A with its jth row and then the ith column with the jth column is a nonsingular M-matrix.*
(13) *For any $y \gg 0$ in $\mathbb{R}^n$, the linear equation $Ax = y$ has a unique solution $x \gg 0$.*
(14) *For each $y \geq 0$ the set*
$$S_y := \{x \geq 0 : A^T x \leq y\}$$
is bounded, and A is nonsingular.

(15) $S_0 = \{0\}$; that is, the inequalities $A^T x \leq 0$ and $x \geq 0$ have only the trivial solution $x = 0$, and A is nonsingular.

2.7 Stochastic Inequalities

The examples in Section 1.6 have already shown how useful the Itô formula is in computing stochastic integration. In this section will further demonstrate the powerfulness of the Itô formula by applying it to establish several very important moment inequalities for stochastic integrals.

Throughout this section, we let $B(t) = (B_1(t), \cdots, B_m(t))^T$, $t \geq 0$ be an m-dimensional Brownian motion defined on the complete probability space $(\Omega, \mathcal{F}, \mathbb{P})$ adapted to the filtration $\{\mathcal{F}_t\}_{t \geq 0}$.

Theorem 2.11 Let $p \geq 2$. Let $g \in \mathcal{M}^2([0,T]; \mathbb{R}^{n \times m})$ such that

$$\mathbb{E} \int_0^T |g(s)|^p ds < \infty.$$

Then

$$\mathbb{E} \left| \int_0^T g(s) dB(s) \right|^p \leq \left(\frac{p(p-1)}{2} \right)^{\frac{p}{2}} T^{\frac{p-2}{2}} \mathbb{E} \int_0^T |g(s)|^p ds. \qquad (2.46)$$

In particular, for $p = 2$, there is equality.

Proof. For $p = 2$ the required result follows from Theorem 1.32 so we only need to show the theorem for the case of $p > 2$. For $0 \leq t \leq T$, set

$$x(t) = \int_0^t g(s) dB(s).$$

By Itô's formula and Theorem 1.32,

$$\mathbb{E}|x(t)|^p$$
$$= \frac{p}{2} \mathbb{E} \int_0^t \left(|x(s)|^{p-2} |g(s)|^2 + (p-2)|x(s)|^{p-4} |x^T(s) g(s)|^2 \right) ds \qquad (2.47)$$
$$\leq \frac{p(p-1)}{2} \mathbb{E} \int_0^t |x(s)|^{p-2} |g(s)|^2 ds. \qquad (2.48)$$

By Hölder's inequality, we compute

$$\mathbb{E}|x(t)|^p \le \frac{p(p-1)}{2}\left(\mathbb{E}\int_0^t |x(s)|^p ds\right)^{\frac{p-2}{p}} \left(\mathbb{E}\int_0^t |g(s)|^p ds\right)^{\frac{2}{p}}$$

$$= \frac{p(p-1)}{2}\left(\int_0^t \mathbb{E}|x(s)|^p ds\right)^{\frac{p-2}{p}} \left(\mathbb{E}\int_0^t |g(s)|^p ds\right)^{\frac{2}{p}}.$$

Note from (2.47) that $\mathbb{E}|x(t)|^p$ is nondecreasing in t. It then follows

$$\mathbb{E}|x(t)|^p \le \frac{p(p-1)}{2}\left[t\mathbb{E}|x(t)|^p\right]^{\frac{p-2}{p}} \left(\mathbb{E}\int_0^t |g(s)|^p ds\right)^{\frac{2}{p}}.$$

This yields

$$\mathbb{E}|x(t)|^p \le \left(\frac{p(p-1)}{2}\right)^p 2t^{\frac{p-2}{2}} \mathbb{E}\int_0^t |g(s)|^p ds,$$

and the required assertion (2.46) follows by replacing t with T. □

Theorem 2.12 *Under the same assumptions as Theorem 2.11,*

$$\mathbb{E}\left(\sup_{0\le t\le T}\left|\int_0^t g(s)dB(s)\right|^p\right) \le \left(\frac{p^3}{2(p-1)}\right)^{p/2} T^{\frac{p-2}{2}} \mathbb{E}\int_0^T |g(s)|^p ds. \quad (2.49)$$

Proof. Recall that the stochastic integral $\int_0^t g(s)dB(s)$ is an $\mathbb{R}^n$-valued continuous martingale. Hence, by Theorem 1.10, we have

$$\mathbb{E}\left(\sup_{0\le t\le T}\left|\int_0^t g(s)dB(s)\right|^p\right) \le \left(\frac{p}{p-1}\right)^p \mathbb{E}\left|\int_0^T g(s)dB(s)\right|^p.$$

In view of Theorem 2.11, we then obtain assertion (2.49). □

The following theorem is known as the Burkholder–Davis–Gundy inequality.

Theorem 2.13 *Let $g \in \mathcal{L}^2(\mathbb{R}_+; \mathbb{R}^{n\times m})$. Define, for $t \ge 0$,*

$$x(t) = \int_0^t g(s)dB(s) \quad and \quad A(t) = \int_0^t |g(s)|^2 ds.$$

Then for every $p > 0$, there exist universal positive constants c_p, C_p, which are only dependent on p, such that

$$c_p \mathbb{E}|A(t)|^{\frac{p}{2}} \le \mathbb{E}\left(\sup_{0\le s\le t} |x(s)|^p\right) \le C_p \mathbb{E}|A(t)|^{\frac{p}{2}} \quad (2.50)$$

for all $t \ge 0$. In particular, one may take

$$c_p = (p/2)^p, \quad C_p = (32/p)^{p/2} \qquad \text{if } 0 < p < 2;$$
$$c_p = 1, \quad C_p = 4 \qquad \text{if } p = 2;$$
$$c_p = (2p)^{-p/2}, \quad C_p = \left[p^{p+1}/2(p-1)^{p-1}\right]^{p/2} \qquad \text{if } p > 2.$$

Proof. We may assume without loss of generality that both $x(t)$ and $A(t)$ are bounded. Otherwise, for each integer $k \geq 1$, define the stopping time

$$\tau_k = \inf\{t \geq 0 : |x(t)| \vee A(t) \geq k\}.$$

If we can show (2.50) for the stopped processes $x(t \wedge \tau_k)$ and $A(t \wedge \tau_k)$, then the general case follows upon letting $k \to \infty$. Besides, for convenience, we set $x^*(t) = \sup_{0 \leq s \leq t} |x(s)|$.

<u>Case 1</u> : $p = 2$. The required (2.50) follows from Theorem 1.32 and the Doob martingale inequality immediately.

<u>Case 2</u> : $p > 2$. It follows from (2.48) that

$$\mathbb{E}|x(t)|^p \leq \frac{p(p-1)}{2} \mathbb{E}\left[|x^*(t)|^{p-2} A(t)\right]$$
$$\leq \frac{p(p-1)}{2} \left[\mathbb{E}|x^*(t)|^p\right]^{\frac{p-2}{p}} \left[\mathbb{E}|A(t)|^{\frac{p}{2}}\right]^{\frac{2}{p}}, \qquad (2.51)$$

where the Hölder inequality has been used. But, by the Doob martingale inequality,

$$\mathbb{E}|x^*(t)|^p \leq \left(\frac{p}{p-1}\right)^p \mathbb{E}|x(t)|^p.$$

Substituting this into (2.51) yields

$$\mathbb{E}|x^*(t)|^p \leq \left(\frac{p^{(p+1)}}{2(p-1)^{p-1}}\right)^{\frac{p}{2}} \mathbb{E}|A(t)|^{\frac{p}{2}}$$

which is the right-hand-side inequality of (2.50). To show the left-hand-side one, we set

$$y(t) = \int_0^t |A(s)|^{\frac{p-2}{4}} g(s) dB(s).$$

Then

$$\mathbb{E}|y(t)|^2 = \mathbb{E}\int_0^t |A(s)|^{\frac{p-2}{2}} |g(s)|^2 ds$$
$$= \mathbb{E}\int_0^t |A(s)|^{\frac{p-2}{2}} dA(s) = \frac{2}{p}\mathbb{E}|A(t)|^{\frac{p}{2}}. \qquad (2.52)$$

On the other hand, the integration by parts formula yields

$$x(t)|A(t)|^{\frac{p-2}{4}} = \int_0^t |A(s)|^{\frac{p-2}{4}} dx(s) + \int_0^t x(s)d\Big(|A(s)|^{\frac{p-2}{4}}\Big)$$

$$= y(t) + \int_0^t x(s)d\Big(|A(s)|^{\frac{p-2}{4}}\Big).$$

Thus

$$|y(t)| \le |x(t)||A(t)|^{\frac{p-2}{4}} + \int_0^t |x(s)|d\Big(|A(s)|^{\frac{p-2}{4}}\Big) \le 2x^*(t)|A(t)|^{\frac{p-2}{4}}.$$

Substituting this into (2.52) one sees that

$$\frac{2}{p}\mathbb{E}|A(t)|^{\frac{p}{2}} \le 4\mathbb{E}\Big[|x^*(t)|^2|A(t)|^{\frac{p-2}{2}}\Big] \le 4\Big[\mathbb{E}|x^*(t)|^p\Big]^{\frac{2}{p}}\Big[\mathbb{E}|A(t)|^{\frac{p}{2}}\Big]^{\frac{p-2}{p}}.$$

This implies

$$\frac{1}{(2p)^{p/2}}\mathbb{E}|A(t)|^{\frac{p}{2}} \le \mathbb{E}|x^*(t)|^p$$

as desired.

<u>Case 3</u>: $0 < p < 2$. Fix $\varepsilon > 0$ arbitrarily and define

$$\eta(t) = \int_0^t [\varepsilon + A(s)]^{\frac{p-2}{4}} g(s)dB(s) \quad \text{and} \quad \eta^*(t) = \sup_{0 \le s \le t} |\eta(s)|.$$

Then

$$\mathbb{E}|\eta(t)|^2 = E\int_0^t [\varepsilon + A(s)]^{\frac{p-2}{2}} dA(s) \le 2pE[\varepsilon + A(t)]^{\frac{p}{2}}. \qquad (2.53)$$

On the other hand, the integration by parts formula gives

$$\eta(t)[\varepsilon + A(t)]^{\frac{2-p}{4}} = \int_0^t g(s)dB(s) + \int_0^t \eta(s)d\Big([\varepsilon + A(s)]^{\frac{2-p}{4}}\Big)$$

$$= x(t) + \int_0^t \eta(s)d\Big([\varepsilon + A(s)]^{\frac{2-p}{4}}\Big).$$

Thus

$$|x(t)| \le |\eta(t)|[\varepsilon + A(t)]^{\frac{2-p}{4}} + \int_0^t |\eta(s)|d\Big([\varepsilon + A(s)]^{\frac{2-p}{4}}\Big)$$

$$\le 2\eta^*(t)[\varepsilon + A(t)]^{\frac{2-p}{4}}.$$

Since this holds for all $t \geq 0$ and the right-hand side is nondecreasing, we must have

$$\mathbb{E}|x^*(t)|^p \leq 2^p \mathbb{E}\left[|\eta^*(t)|^p [\varepsilon + A(t)]^{\frac{p(2-p)}{4}}\right]$$
$$\leq 2^p \left[\mathbb{E}|\eta^*(t)|^2\right]^{\frac{p}{2}} \left[E[\varepsilon + A(t)]^{\frac{p}{2}}\right]^{\frac{2-p}{2}}. \quad (2.54)$$

But, by Doob's martingale inequality and (2.53),

$$\mathbb{E}|\eta^*(t)|^2 \leq 4\mathbb{E}|\eta(t)|^2 \leq \frac{8}{p}\mathbb{E}[\varepsilon + A(t)]^{\frac{p}{2}}.$$

Substituting this into (2.54) one sees that

$$\mathbb{E}|x^*(t)|^p \leq \left(\frac{32}{p}\right)^{\frac{p}{2}} \mathbb{E}[\varepsilon + A(t)]^{\frac{p}{2}}.$$

Letting $\varepsilon \to 0$ we obtain the right-hand-side inequality of (2.50). To show the left-hand-side one, we write, for any fixed $\varepsilon > 0$,

$$|A(t)|^{\frac{p}{2}} = \left(|A(t)|^{\frac{p}{2}}[\varepsilon + x^*(t)]^{\frac{-p(2-p)}{2}}\right)[\varepsilon + x^*(t)]^{\frac{p(2-p)}{2}}.$$

Then, applying Hölder's inequality, one sees that

$$\mathbb{E}|A(t)|^{\frac{p}{2}} \leq \left[\mathbb{E}\big(A(t)[\varepsilon + x^*(t)]^{p-2}\big)\right]^{\frac{p}{2}} \left(\mathbb{E}[\varepsilon + x^*(t)]^p\right)^{\frac{2-p}{2}}. \quad (2.55)$$

Define

$$\xi(t) = \int_0^t [\varepsilon + x^*(s)]^{\frac{p-2}{2}} g(s) dB(s).$$

Then

$$\mathbb{E}|\xi(t)|^2 = E\int_0^t [\varepsilon + x^*(s)]^{p-2} dA(s) \geq \mathbb{E}\big([\varepsilon + x^*(t)]^{p-2} A(t)\big). \quad (2.56)$$

On the other hand, the integration by parts formula gives

$$x(t)[\varepsilon + x^*(t)]^{\frac{p-2}{2}} = \xi(t) + \int_0^t x(s) d\Big([\varepsilon + x^*(s)]^{\frac{p-2}{2}}\Big)$$
$$= \xi(t) + \frac{p-2}{2}\int_0^t x(s)[\varepsilon + x^*(s)]^{\frac{p-4}{2}} d[\varepsilon + x^*(s)].$$

Thus

$$|\xi(t)| \leq x^*(t)[\varepsilon + x^*(t)]^{\frac{p-2}{2}} + \frac{2-p}{2}\int_0^t x^*(s)[\varepsilon + x^*(s)]^{\frac{p-4}{2}} d[\varepsilon + x^*(s)]$$

$$\leq [\varepsilon + x^*(t)]^{\frac{p}{2}} + \frac{2-p}{2}\int_0^t [\varepsilon + x^*(s)]^{\frac{p-2}{2}} d[\varepsilon + x^*(s)]$$

$$\leq \frac{2}{p}[\varepsilon + x^*(t)]^{\frac{p}{2}}.$$

This, together with (2.56), implies

$$\mathbb{E}\Big([\varepsilon + x^*(t)]^{p-2} A(t)\Big) \leq \left(\frac{2}{p}\right)^2 \mathbb{E}[\varepsilon + x^*(t)]^p.$$

Substituting this into (2.55) we get that

$$\mathbb{E}|A(t)|^{\frac{p}{2}} \leq \left(\frac{2}{p}\right)^p \mathbb{E}[\varepsilon + x^*(t)]^p.$$

Finally, letting $\varepsilon \to 0$ we have

$$\left(\frac{p}{2}\right)^p \mathbb{E}|A(t)|^{\frac{p}{2}} \leq \mathbb{E}|x^*(t)|^p$$

as required. The proof is now complete. □

The following theorem is known as the exponential martingale inequality which will play an important role in this book.

Theorem 2.14 Let $g = (g_1, \cdots, g_m) \in \mathcal{L}^2(\mathbb{R}_+; \mathbb{R}^{1 \times m})$, and let T, α, β be any positive numbers. Then

$$\mathbb{P}\left\{\sup_{0 \leq t \leq T}\left[\int_0^t g(s)dB(s) - \frac{\alpha}{2}\int_0^t |g(s)|^2 ds\right] > \beta\right\} \leq e^{-\alpha\beta}. \tag{2.57}$$

Proof. For every integer $k \geq 1$, define the stopping time

$$\tau_k = \inf\left\{t \geq 0 : \left|\int_0^t g(s)dB(s)\right| + \int_0^t |g(s)|^2 ds \geq k\right\},$$

and the Itô process

$$x_k(t) = \alpha \int_0^t g(s) I_{[[0,\tau_k]]}(s) dB(s) - \frac{\alpha^2}{2}\int_0^t |g(s)|^2 I_{[[0,\tau_k]]}(s) ds.$$

Clearly, $x_k(t)$ is bounded and $\tau_k \uparrow \infty$ a.s. Applying the Itô formula to $\exp[x_k(t)]$ we obtain that

$$\exp[x_k(t)] = 1 + \int_0^t \exp[x_k(s)]dx_k(s) + \frac{\alpha^2}{2}\int_0^t \exp[x_k(s)]|g(s)|^2 I_{[[0,\tau_k]]}(s)ds$$
$$= 1 + \alpha \int_0^t \exp[x_k(s)]g(s)I_{[[0,\tau_k]]}(s)dB(s).$$

In view of Theorem 1.32, one sees that $\exp[x_k(t)]$ is a non-negative martingale on $t \geq 0$ with $\mathbb{E}\big(\exp[x_k(t)]\big) = 1$. Hence, by Theorem 1.10, we get that

$$\mathbb{P}\left\{\sup_{0 \leq t \leq T} \exp[x_k(t)] \geq e^{\alpha\beta}\right\} \leq e^{-\alpha\beta}\mathbb{E}\big(\exp[x_k(T)]\big) = e^{-\alpha\beta}.$$

That is,

$$\mathbb{P}\left\{\sup_{0 \leq t \leq T}\left[\int_0^t g(s)I_{[[0,\tau_k]]}(s)dB(s) - \frac{\alpha}{2}\int_0^t |g(s)|^2 I_{[[0,\tau_k]]}(s)ds\right] > \beta\right\} \leq e^{-\alpha\beta}.$$

Now the required assertion (2.57) follows by letting $k \to \infty$. □

2.8 Exercises

2.1 Prove the Yong inequality (2.3) and hence by induction show its general form (2.6).

2.2 Show that a function $\varphi: \mathbb{R} \to \mathbb{R}$ is convex if and only if

$$\varphi(\alpha x + (1-\alpha)y) \leq \alpha\varphi(x) + (1-\alpha)\varphi(y) \quad \forall x, y \in \mathbb{R}$$

holds for any $\alpha \in (0, 1)$.

2.3 From the definition (2.27) of the matrix operator norm and the property (2.29) of the largest eigenvalue, show, for a matrix $A \in \mathbb{R}^{n \times m}$, that

$$\|A\| = \sqrt{\lambda_{\max}(A^T A)}.$$

And hence show $\|A\| \leq |A|$.

2.4 Let $F \in \mathbb{R}^{n \times m}$. Show that $\|F\| < 1$ is equivalent to $I_n - FF^T > 0$, where where I_n is the $n \times n$ identity matrix. Hence show that the matrix norm constraint $\|F\| < 1$ can be represented as the LMI

$$\begin{bmatrix} I_n & F \\ F & I_m \end{bmatrix} > 0.$$

2.5 The universal positive constants c_p and C_p stated in Theorem 2.13 are not optimal. For example, the formula there gives $C_1 = 4\sqrt{2}$ but in fact we can have $C_1 = 3$. That is, if $g \in \mathcal{L}^2(\mathbb{R}_+; \mathbb{R}^{n \times m})$ and define, for $t \geq 0$,

$$x(t) = \int_0^t g(s)dB(s) \quad \text{and} \quad A(t) = \int_0^t |g(s)|^2 ds,$$

then show

$$\mathbb{E}\left(\sup_{0 \leq s \leq t} |x(s)|\right) \leq 3\mathbb{E}(|A(t)|^{\frac{1}{2}}) \quad \forall t \geq 0.$$

(You may refer to [Da Prato and Zabczyk (1992)].)

Chapter 3

Stochastic Differential Equations with Markovian Switching

3.1 Introduction

In Section 1.1 we introduced, as an example, the geometric Brownian motion with Markovian switching or the hybrid geometric Brownian motion

$$dx(t) = \mu(r(t))x(t)dt + \sigma(r(t))x(t)dB(t).$$

In this chapter we shall turn to finding the solution to the equation. In general, we shall investigate the solution to a non-linear stochastic differential equation (SDE) with Markovian switching

$$dx(t) = f(x(t), t, r(t))dt + g(x(t), t, r(t))dB(t) \quad \text{on } t \in [t_0, T]. \quad (3.1)$$

The questions are:

- What is the solution?
- Are there existence-and-uniqueness theorems for such a solution?
- What are the properties of the solution?
- How can the solution be obtained in practice?

In this chapter and the next we shall answer these questions one by one.

3.2 Stochastic Differential Equations

To make the theory more understandable, let us first discuss stochastic differential equations (SDEs) without Markovian switching.

Let $(\Omega, \mathcal{F}, \mathbb{P})$ be a complete probability space with a filtration $\{\mathcal{F}_t\}_{t \geq 0}$ satisfying the usual conditions. Throughout this chapter, unless otherwise specified, we let $B(t) = (B_1(t), \cdots, B_m(t))^T$, $t \geq 0$ be an m-dimensional

Brownian motion defined on the space. Let $0 \leq t_0 < T < \infty$. Let $x_0 \in L^2_{\mathcal{F}_{t_0}}(\Omega; \mathbb{R}^n)$, i.e. an $\mathcal{F}_{t_0}$-measurable $\mathbb{R}^n$-valued random variable such that $\mathbb{E}|x_0|^2 < \infty$. Let $f : \mathbb{R}^n \times [t_0, T] \to \mathbb{R}^n$ and $g : \mathbb{R}^n \times [t_0, T] \to \mathbb{R}^{n \times m}$ be both Borel measurable. Consider the n-dimensional stochastic differential equation of Itô type

$$dx(t) = f(x(t), t)dt + g(x(t), t)dB(t), \quad t_0 \leq t \leq T \qquad (3.2)$$

with initial value $x(t_0) = x_0$. By the definition of stochastic differential, this equation is equivalent to the following stochastic integral equation

$$x(t) = x_0 + \int_{t_0}^{t} f(x(s), s)ds + \int_{t_0}^{t} g(x(s), s)dB(s) \quad \forall t \in [t_0, T]. \qquad (3.3)$$

Let us first give the definition of the solution.

Definition 3.1 An $\mathbb{R}^n$-valued stochastic process $\{x(t)\}_{t_0 \leq t \leq T}$ is called a *solution* of equation (3.2) if it has the following properties:

(1) $\{x(t)\}$ is continuous and $\mathcal{F}_t$-adapted;
(2) $\{f(x(t), t)\} \in \mathcal{L}^1([t_0, T]; \mathbb{R}^n)$ and $\{g(x(t), t)\} \in \mathcal{L}^2([t_0, T]; \mathbb{R}^{n \times m})$;
(3) equation (3.3) holds with probability 1.

A solution $\{x(t)\}$ is said to be *unique* if any other solution $\{\bar{x}(t)\}$ is indistinguishable from $\{x(t)\}$, that is

$$\mathbb{P}\{x(t) = \bar{x}(t) \text{ for all } t_0 \leq t \leq T\} = 1.$$

Remark 3.2 Denote the solution of equation (3.2) by $x(t; t_0, x_0)$. Note from equation (3.3) that for any $s \in [t_0, T]$,

$$x(t) = x(s) + \int_s^t f(x(u), u)du + \int_s^t g(x(u), u)dB(u) \quad \forall s \leq t \leq T. \qquad (3.4)$$

But, this is a stochastic differential equation on $[s, T]$ with initial value $x(s) = x(s; t_0, x_0)$, whose solution is $x(t; s, x(s; t_0, x_0))$. Therefore, we see that the solution of equation (3.2) satisfies the following *flow* or *semigroup property*

$$x(t; t_0, x_0) = x(t; s, x(s; t_0, x_0)), \quad t_0 \leq s \leq t \leq T.$$

We shall now give some examples of stochastic differential equations.

Example 3.3 Let $B(t)$, $t \geq 0$ be a one-dimensional Brownian motion. Define the two-dimensional stochastic process

$$x(t) = (x_1(t), x_2(t))^T = (\cos(B(t)), \sin(B(t)))^T \quad \forall t \geq 0. \tag{3.5}$$

The process $x(t)$ is called *Brownian motion on the unit circle*. We now show that $x(t)$ satisfies a linear stochastic differential equation. By Itô's formula,

$$dx_1(t) = -\sin(B(t))dB(t) - \frac{1}{2}\cos(B(t))dt = -\frac{1}{2}x_1(t)dt - x_2(t)dB(t),$$

$$dx_2(t) = \cos(B(t))dB(t) - \frac{1}{2}\sin(B(t))dt = -\frac{1}{2}x_2(t)dt + x_1(t)dB(t).$$

That is, in matrix form,

$$dx(t) = -\frac{1}{2}x(t)dt + Kx(t)dB(t), \tag{3.6}$$

where

$$K = \begin{bmatrix} 0 & -1 \\ 1 & 0 \end{bmatrix}.$$

Example 3.4 The charge $Q(t)$ at time t at a fixed point in an electrical circuit satisfies the second order differential equation

$$L\ddot{Q}(t) + R\dot{Q}(t) + \frac{1}{C}Q(t) = F(t), \tag{3.7}$$

where L is the inductance, R the resistance, C the capacitance and $F(t)$ the potential source. Suppose that the potential source is subject to the environmental noise and is described by $F(t) = G(t) + \alpha \dot{B}(t)$, where $\dot{B}(t)$ is a one-dimensional white noise (i.e. $B(t)$ is a Brownian motion) and α is the intensity of the noise. Then equation (3.7) becomes

$$L\ddot{Q}(t) + R\dot{Q}(t) + \frac{1}{C}Q(t) = G(t) + \alpha \dot{B}(t). \tag{3.8}$$

Introduce the two-dimensional process $x(t) = (x_1(t), x_2(t))^T = (Q(t), \dot{Q}(t))^T$. Then equation (3.8) can be expressed as an Itô equation

$$dx_1(t) = x_2(t)dt,$$

$$dx_2(t) = \frac{1}{L}\left(-\frac{1}{C}x_1(t) - Rx_2(t) + G(t)\right)dt + \frac{\alpha}{L}dB(t).$$

That is,
$$dx(t) = [Ax(t) + H(t)]dt + K dB(t), \qquad (3.9)$$
where
$$A = \begin{bmatrix} 0 & 1 \\ -1/CL & -R/L \end{bmatrix}, \quad H(t) = \begin{bmatrix} 0 \\ G(t)/L \end{bmatrix}, \quad K = \begin{bmatrix} 0 \\ \alpha/L \end{bmatrix}.$$

Example 3.5 More generally, consider the nth-order differential equation with white noise of the form
$$y^{(n)}(t) = F(y(t), \cdots, y^{(n-1)}(t), t) + G(y(t), \cdots, y^{(n-1)}(t), t)\dot{B}(t), \quad (3.10)$$
where $y^{(i)} = d^i y(t)/dt^i$, $F : \mathbb{R}^n \times \mathbb{R}_+ \to \mathbb{R}$, $G : \mathbb{R}^n \times \mathbb{R}_+ \to \mathbb{R}^{1 \times m}$, and $\dot{B}(t)$ is an m-dimensional white noise, i.e. $B(t)$ is an m-dimensional Brownian motion. Introducing the $\mathbb{R}^n$-valued stochastic process $x(t) = (x_1(t), \cdots, x_k(t))^T = (y(t), \cdots, y^{(n-1)}(t))^T$, we can then convert equation (3.10) into an n-dimensional Itô equation
$$dx(t) = \begin{bmatrix} x_2(t) \\ \vdots \\ x_k(t) \\ F(x(t), t) \end{bmatrix} dt + \begin{bmatrix} 0 \\ \vdots \\ 0 \\ G(x(t), t) \end{bmatrix} dB(t). \qquad (3.11)$$

Example 3.6 If $g(x, t) \equiv 0$, equation (3.2) becomes the ordinary differential equation
$$\dot{x}(t) = f(x(t), t), \quad t \in [t_0, T] \qquad (3.12)$$
with initial value $x(t_0) = x_0$. In this case, the random influence can only show up in the initial value x_0. As a special case, consider the one-dimensional equation
$$\dot{x}(t) = 3[x(t)]^{2/3}, \quad t \in [t_0, T] \qquad (3.13)$$
with initial value $x(t_0) = 1_A$, where $A \in \mathcal{F}_{t_0}$. It is easy to verify that for any $0 < \alpha < T - t_0$, the stochastic process
$$x(t) = x(t, \omega) = \begin{cases} (t - t_0 + 1)^3 & \text{for } t_0 \le t \le T, \ \omega \in A, \\ 0 & \text{for } t_0 \le t \le t_0 + \alpha, \ \omega \notin A, \\ (t - t_0 - \alpha)^3 & \text{for } t_0 + \alpha < t \le T, \ \omega \notin A \end{cases}$$

is a solution to equation (3.13). In other words, equation (3.13) has infinitely many solutions. As another special case, consider the one-dimensional equation

$$\dot{x}(t) = [x(t)]^2, \quad t \in [t_0, T] \tag{3.14}$$

with initial value $x(t_0) = x_0$, which is a random variable taking values larger than $1/[T - t_0]$. It is easy to verify that equation (3.14) has the (unique) solution

$$x(t) = \left[\frac{1}{x_0} - (t - t_0)\right]^{-1} \quad \text{only on } t_0 \leq t < t_0 + \frac{1}{x_0}(< T),$$

but there is no solution defined for all $t \in [t_0, T]$ in this case.

Example 3.7 Let $B(t)$ be a one-dimensional Brownian motion. [Girsanov (1962)] has shown that the one-dimensional Itô equation

$$x(t) = \int_{t_0}^{t} |x(s)|^\alpha dB(s)$$

has a unique solution when $\alpha \geq 1/2$, but it has infinitely many solutions when $0 < \alpha < 1/2$.

3.3 Existence and Uniqueness of Solutions

Examples 3.6 and 3.7 show that an Itô equation may not have a unique solution defined on the whole interval $[t_0, T]$. Let us now turn to finding the conditions that guarantee the existence and uniqueness of the solution to equation (3.2).

Theorem 3.8 *Assume that there exist two positive constants $\bar{K}$ and K such that*

(Lipschitz condition) *for all $x, y \in \mathbb{R}^n$ and $t \in [t_0, T]$*

$$|f(x,t) - f(y,t)|^2 \vee |g(x,t) - g(y,t)|^2 \leq \bar{K}|x-y|^2; \tag{3.15}$$

(Linear growth condition) *for all $(x,t) \in \mathbb{R}^n \times [t_0, T]$*

$$|f(x,t)|^2 \vee |g(x,t)|^2 \leq K(1 + |x|^2). \tag{3.16}$$

Then there exists a unique solution $x(t)$ to equation (3.2) and the solution belongs to $\mathcal{M}^2([t_0, T]; \mathbb{R}^n)$.

The whole proof is rather technical and long. To make it more understandable, we first prepare a lemma which gives a bound for the solution.

Lemma 3.1 *Assume that the linear growth condition (3.16) holds. If $x(t)$ is a solution of equation (3.2), then*

$$\mathbb{E}\left(\sup_{t_0 \leq t \leq T} |x(t)|^2\right) \leq (1 + 3\mathbb{E}|x_0|^2)e^{3K(T-t_0)(T-t_0+4)}. \tag{3.17}$$

In particular, $x(t)$ belongs to $\mathcal{M}^2([t_0, T]; \mathbb{R}^n)$.

Proof. For every integer $k \geq 1$, define the stopping time

$$\tau_k = T \wedge \inf\{t \in [t_0, T] : |x(t)| \geq k\}.$$

Clearly, $\tau_k \uparrow T$ a.s. Set $x_k(t) = x(t \wedge \tau_k)$ for $t \in [t_0, T]$. Then $x_k(t)$ satisfies the equation

$$x_k(t) = x_0 + \int_{t_0}^t f(x_k(s), s) I_{[[t_0, \tau_k]]}(s) ds + \int_{t_0}^t g(x_k(s), s) I_{[[t_0, \tau_k]]}(s) dB(s).$$

Using the elementary inequality $|a+b+c|^2 \leq 3(|a|^2 + |b|^2 + |c|^2)$, the Hölder inequality and condition (3.16), one can show that

$$|x_k(t)|^2 \leq 3|x_0|^2 + 3K(t - t_0) \int_{t_0}^t (1 + |x_k(s)|^2) ds$$

$$+ 3 \left| \int_{t_0}^t g(x_k(s), s) I_{[[t_0, \tau_k]]}(s) dB(s) \right|^2.$$

Hence, by the Doob martingale inequality and condition (3.16) one can further show that

$$\mathbb{E}\left(\sup_{t_0 \leq s \leq t} |x_k(s)|^2\right) \leq 3\mathbb{E}|x_0|^2 + 3K(T - t_0) \int_{t_0}^t (1 + \mathbb{E}|x_k(s)|^2) ds$$

$$+ 12\mathbb{E} \int_{t_0}^t |g(x_k(s), s)|^2 I_{[[t_0, \tau_k]]}(s) ds$$

$$\leq 3\mathbb{E}|x_0|^2 + 3K(T - t_0 + 4) \int_{t_0}^t (1 + \mathbb{E}|x_k(s)|^2) ds.$$

Consequently

$$1 + \mathbb{E}\left(\sup_{t_0 \leq s \leq t} |x_k(s)|^2\right)$$

$$\leq 1 + 3\mathbb{E}|x_0|^2 + 3K(T - t_0 + 4) \int_{t_0}^t \left[1 + \mathbb{E}\left(\sup_{t_0 \leq r \leq s} |x_k(r)|^2\right)\right] ds.$$

Now the Gronwall inequality yields that

$$1 + \mathbb{E}\Big(\sup_{t_0 \leq t \leq T} |x_k(t)|^2\Big) \leq (1 + 3\mathbb{E}|x_0|^2)e^{3K(T-t_0)(T-t_0+4)}.$$

Thus

$$\mathbb{E}\Big(\sup_{t_0 \leq t \leq \tau_k} |x(t)|^2\Big) \leq (1 + 3\mathbb{E}|x_0|^2)e^{3K(T-t_0)(T-t_0+4)}.$$

Finally the required inequality (3.17) follows by letting $k \to \infty$. □

Proof of Theorem 3.8. Uniqueness. Let $x(t)$ and $\bar{x}(t)$ be two solutions of equation (3.2). By Lemma 3.1, both of them belong to $\mathcal{M}^2([t_0,T];\mathbb{R}^n)$. Note that

$$x(t) - \bar{x}(t) = \int_{t_0}^t [f(x(s),s) - f(\bar{x}(s),s)]ds + \int_{t_0}^t [g(x(s),s) - g(\bar{x}(s),s)]dB(s).$$

Using the Hölder inequality, the Doob martingale inequality and the Lipschitz condition (3.15) one can show in the same way as in the proof of Lemma 3.1 that

$$\mathbb{E}\Big(\sup_{t_0 \leq s \leq t} |x(s) - \bar{x}(s)|^2\Big) \leq 2\bar{K}(T+4)\int_{t_0}^t \mathbb{E}\Big(\sup_{t_0 \leq r \leq s} |x(r) - \bar{x}(r)|^2\Big)ds.$$

The Gronwall inequality then yields that

$$\mathbb{E}\Big(\sup_{t_0 \leq t \leq T} |x(t) - \bar{x}(t)|^2\Big) = 0.$$

Hence, $x(t) = \bar{x}(t)$ for all $t_0 \leq t \leq T$ almost surely. The uniqueness has been proved.

Existence. Set $x_0(t) \equiv x_0$, and for $k = 1, 2, \cdots$, define the Picard iterations

$$x_k(t) = x_0 + \int_{t_0}^t f(x_{k-1}(s),s)ds + \int_{t_0}^t g(x_{k-1}(s),s)dB(s) \qquad (3.18)$$

for $t \in [t_0, T]$. Obviously, $x_0(\cdot) \in \mathcal{M}^2([t_0,T];\mathbb{R}^n)$. Moreover, it is easy to see by induction that $x_k(\cdot) \in \mathcal{M}^2([t_0,T];\mathbb{R}^n)$, because we have from (3.18) that

$$\mathbb{E}|x_k(t)|^2 \leq c_1 + 3K(T+1)\int_{t_0}^t \mathbb{E}|x_{k-1}(s)|^2 ds, \qquad (3.19)$$

where $c_1 = 3\mathbb{E}|x_0|^2 + 3KT(T+1)$. It also follows from (3.19) that for any $j \geq 1$,

$$\max_{1 \leq k \leq j} \mathbb{E}|x_k(t)|^2 \leq c_1 + 3K(T+1) \int_{t_0}^{t} \max_{1 \leq k \leq j} \mathbb{E}|x_{k-1}(s)|^2 ds$$

$$\leq c_1 + 3K(T+1) \int_{t_0}^{t} \left(\mathbb{E}|x_0|^2 + \max_{1 \leq k \leq j} \mathbb{E}|x_k(s)|^2 \right) ds$$

$$\leq c_2 + 3K(T+1) \int_{t_0}^{t} \max_{1 \leq k \leq j} \mathbb{E}|x_k(s)|^2 ds,$$

where $c_2 = c_1 + 3KT(T+1)\mathbb{E}|x_0|^2$. The Gronwall inequality implies

$$\max_{1 \leq k \leq j} \mathbb{E}|x_k(t)|^2 \leq c_2 e^{3KT(T+1)}.$$

Since j is arbitrary, we must have

$$\mathbb{E}|x_k(t)|^2 \leq c_2 e^{3KT(T+1)} \quad \forall t \in [t_0, T], \ k \geq 1. \tag{3.20}$$

Next, we note that

$$|x_1(t) - x_0(t)|^2 = |x_1(t) - x_0|^2$$

$$\leq 2 \left| \int_{t_0}^{t} f(x_0, s) ds \right|^2 + 2 \left| \int_{t_0}^{t} g(x_0, s) dB(s) \right|^2.$$

Taking the expectation and using (3.16), we get

$$\mathbb{E}|x_1(t) - x_0(t)|^2$$
$$\leq 2K(t - t_0)^2(1 + \mathbb{E}|x_0|^2) + 2K(t - t_0)(1 + \mathbb{E}|x_0|^2) \leq C, \tag{3.21}$$

where $C = 2K(T - t_0 + 1)(T - t_0)(1 + \mathbb{E}|x_0|^2)$. We now claim that for $k \geq 0$,

$$\mathbb{E}|x_{k+1}(t) - x_k(t)|^2 \leq \frac{C[M(t - t_0)]^k}{k!} \quad \forall t \in [t_0, T], \tag{3.22}$$

where $M = 2\bar{K}(T - t_0 + 1)$. We shall show this by induction. In view of (3.21) we see that (3.22) holds when $k = 0$. Under the inductive assumption that (3.22) holds for some $k \geq 0$, we shall show that (3.22) still holds for

$k+1$. Note that

$$|x_{k+2}(t) - x_{k+1}(t)|^2 \leq 2\left|\int_{t_0}^t [f(x_{k+1}(s), s) - f(x_k(s), s)]ds\right|^2$$
$$+ 2\left|\int_{t_0}^t [g(x_{k+1}(s), s) - g(x_k(s), s)]dB(s)\right|^2. \quad (3.23)$$

Taking the expectation and using (3.15) as well as the inductive assumption, we derive that

$$\mathbb{E}|x_{k+2}(t) - x_{k+1}(t)|^2 \leq 2\bar{K}(t - t_0 + 1)\mathbb{E}\int_{t_0}^t |x_{k+1}(s) - x_k(s)|^2 ds$$
$$\leq M \int_{t_0}^t \mathbb{E}|x_{k+1}(s) - x_k(s)|^2 ds$$
$$\leq M \int_{t_0}^t \frac{C[M(s-t_0)]^k}{k!} ds = \frac{C[M(t-t_0)]^{k+1}}{(k+1)!}.$$

That is, (3.22) holds for $k+1$. Hence, by induction, (3.22) holds for all $k \geq 0$. Furthermore, replacing k in (3.23) with $k-1$ we see that

$$\sup_{t_0 \leq t \leq T} |x_{k+1}(t) - x_k(t)|^2 \leq 2\bar{K}(T - t_0) \int_{t_0}^T |x_k(s) - x_{k-1}(s)|^2 ds$$

$$+ 2 \sup_{t_0 \leq t \leq T} \left|\int_{t_0}^T [g(x_k(s), s) - g(x_{k-1}(s), s)]dB(s)\right|^2.$$

Taking the expectation and using the Doob martingale inequality and (3.22), we find that

$$\mathbb{E}\left(\sup_{t_0 \leq t \leq T} |x_{k+1}(t) - x_k(t)|^2\right) \leq 2\bar{K}(T - t_0 + 4) \int_{t_0}^T \mathbb{E}|x_k(s) - x_{k-1}(s)|^2 ds$$
$$\leq 4M \int_{t_0}^T \frac{C[M(s-t_0)]^{k-1}}{(k-1)!} ds = \frac{4C[M(T-t_0)]^k}{k!}.$$

Hence

$$\mathbb{P}\left\{\sup_{t_0 \leq t \leq T} |x_{k+1}(t) - x_k(t)| > \frac{1}{2^k}\right\} \leq \frac{4C[4M(T-t_0)]^k}{k!}.$$

Since $\sum_{k=0}^{\infty} 4C[4M(T-t_0)]^k/k! < \infty$, the Borel–Cantelli lemma yields that for almost all $\omega \in \Omega$ there exists a positive integer $k_0 = k_0(\omega)$ such that

$$\sup_{t_0 \leq t \leq T} |x_{k+1}(t) - x_k(t)| \leq \frac{1}{2^n} \quad \text{whenever } k \geq k_0.$$

It follows that, with probability 1, the partial sums

$$x_0(t) + \sum_{i=0}^{k-1} [x_{i+1}(t) - x_i(t)] = x_k(t)$$

are convergent uniformly in $t \in [0, T]$. Denote the limit by $x(t)$. Clearly, $x(t)$ is continuous and $\mathcal{F}_t$-adapted. On the other hand, one sees from (3.22) that for every t, $\{x_k(t)\}_{k \geq 1}$ is a Cauchy sequence in L^2 as well. Hence we also have $x_k(t) \to x(t)$ in L^2. Letting $k \to \infty$ in (3.20) gives

$$\mathbb{E}|x(t)|^2 \leq c_2 e^{3KT(T+1)} \quad \forall t \in [t_0, T].$$

Therefore, $x(\cdot) \in \mathcal{M}^2([t_0, T]; \mathbb{R}^n)$. It remains to show that $x(t)$ satisfies equation (3.3). Note that

$$\mathbb{E}\left| \int_{t_0}^t f(x_k(s), s)ds - \int_{t_0}^t f(x(s), s)ds \right|^2$$
$$+ \mathbb{E}\left| \int_{t_0}^t g(x_k(s), s)dB(s) - \int_{t_0}^t g(x(s), s)dB(s) \right|^2$$
$$\leq \bar{K}(T - t_0 + 1) \int_{t_0}^T \mathbb{E}|x_k(s) - x(s)|^2 ds \to 0 \quad \text{as } n \to \infty.$$

Hence we can let $k \to \infty$ in (3.18) to obtain that

$$x(t) = x_0 + \int_{t_0}^t f(x(s), s)ds + \int_{t_0}^t g(x(s), s)dB(s) \quad \text{on } t_0 \leq t \leq T$$

as desired. □

In the proof above we have shown that the Picard iterations $x_k(t)$ converge to the unique solution $x(t)$ of equation (3.2). The following theorem gives an estimate on error of the convergence.

Theorem 3.9 *Let the assumptions of Theorem 3.8 hold. Let $x(t)$ be the unique solution of equation (3.2) and $x_k(t)$ be the Picard iterations defined* by (3.18). Then

$$\mathbb{E}\left(\sup_{t_0\leq t\leq T}|x_k(t)-x(t)|^2\right) \leq \frac{8C[M(T-t_0)]^k}{k!}e^{8M(T-t_0)} \quad (3.24)$$

for all $n \geq 1$, where C and M are the same as defined in the proof of Theorem 3.8, that is $C = 2K(T-t_0+1)(T-t_0)(1+\mathbb{E}|x_0|^2)$ and $M = 2\bar{K}(T-t_0+1)$.

Proof. From

$$x_k(t) - x(t) = \int_{t_0}^t [f(x_{k-1}(s),s) - f(x(s),s)]ds$$
$$+ \int_{t_0}^t [g(x_{k-1}(s),s) - g(x(s),s)]dB(s),$$

we can derive that

$$\mathbb{E}\left(\sup_{t_0\leq s\leq t}|x_k(s)-x(s)|^2\right) \leq 2\bar{K}(T-t_0+4)\int_{t_0}^t \mathbb{E}|x_{k-1}(s)-x(s)|^2 ds$$

$$\leq 8M\int_{t_0}^t \mathbb{E}|x_k(s)-x_{k-1}(s)|^2 ds + 8M\int_{t_0}^t \mathbb{E}|x_k(s)-x(s)|^2 ds.$$

Substituting (3.22) into this yields that

$$\mathbb{E}\left(\sup_{t_0\leq s\leq t}|x_k(s)-x(s)|^2\right)$$
$$\leq 8M\int_{t_0}^T \frac{C[M(s-t_0)]^{k-1}}{(k-1)!}ds + 8M\int_{t_0}^t \mathbb{E}|x_k(s)-x(s)|^2 ds$$
$$\leq \frac{8C[M(T-t_0)]^k}{k!} + 8M\int_{t_0}^t \mathbb{E}\left(\sup_{t_0\leq r\leq s}|x_k(r)-x(r)|^2\right)ds.$$

Consequently, the required inequality (3.24) follows by applying the Gronwall inequality. □

This theorem shows that one can use the Picard iteration procedure to obtain the approximate solutions of equation (3.2), and (3.24) gives the estimate for the error of the approximation. We shall discuss other approximation procedures later.

To close this section let us make a number of important remarks.

Remark 3.10 (a) The coefficients f and g can depend on ω in a general manner as long as they are adapted. It is easy to observe from the proofs above that in this case, the Lipschitz and the linear growth condition still guarantee the existence and uniqueness of the solution.

(b) Both initial time t_0 and final time T can be random variables provided they are stopping times.

(c) In the above we require the initial value x_0 be L^2, but in general it is enough for x_0 to be a random variable as long as it is $\mathcal{F}_{t_0}$-measurable. For further details, please see [Gihman and Skorohod (1972)].

3.4 SDEs with Markovian Switching

After establishing the fundamental existence-and-uniqueness theorem of solutions to SDEs we can now begin to discuss SDEs with Markovian switching.

Let $r(t)$, $t \geq t_0$, be a right-continuous Markov chain on the probability space taking values in a finite state space $\mathbb{S} = \{1, 2, \cdots, N\}$ with generator $\Gamma = (\gamma_{ij})_{N \times N}$ given by

$$\mathbb{P}\{r(t+\Delta) = j | r(t) = i\} = \begin{cases} \gamma_{ij}\Delta + o(\Delta) & \text{if } i \neq j, \\ 1 + \gamma_{ii}\Delta + o(\Delta) & \text{if } i = j, \end{cases}$$

where $\Delta > 0$. Here $\gamma_{ij} \geq 0$ is the transition rate from i to j if $i \neq j$ while

$$\gamma_{ii} = -\sum_{j \neq i} \gamma_{ij}.$$

We assume that the Markov chain $r(\cdot)$ is $\mathcal{F}_t$-adapted but independent of the Brownian motion $B(\cdot)$. Consider an SDE with Markovian switching of the form

$$dx(t) = f(x(t), t, r(t))dt + g(x(t), t, r(t))dB(t), \quad t_0 \leq t \leq T \quad (3.25)$$

with initial data $x(t_0) = x_0 \in L^2_{\mathcal{F}_{t_0}}(\Omega; \mathbb{R}^n)$ and $r(t_0) = r_0$, where r_0 is an $\mathbb{S}$-valued $\mathcal{F}_{t_0}$-measurable random variable and

$$f : \mathbb{R}^n \times \mathbb{R}_+ \times \mathbb{S} \to \mathbb{R}^n \quad \text{and} \quad g : \mathbb{R}^n \times \mathbb{R}_+ \times \mathbb{S} \to \mathbb{R}^{n \times m}.$$

Definition 3.11 An $\mathbb{R}^n$-valued stochastic process $\{x(t)\}_{t_0 \leq t \leq T}$ is called a *solution* of equation (3.25) if it has the following properties:

(1) $\{x(t)\}_{t_0 \leq t \leq T}$ is continuous and $\mathcal{F}_t$-adapted;
(2) $\{f(x(t), t, r(t))\}_{t_0 \leq t \leq T} \in \mathcal{L}^1([t_0, T]; \mathbb{R}^n)$ while $\{g(x(t), t, r(t))\}_{t_0 \leq t \leq T} \in \mathcal{L}^2([t_0, T]; \mathbb{R}^{n \times m})$;

(3) for any $t \in [t_0, T]$, equation

$$x(t) = x(t_0) + \int_{t_0}^t f(x(s), s, r(s))ds + \int_{t_0}^t g(x(s), s, r(s))dB(s)$$

holds with probability 1.

A solution $\{x(t)\}_{t_0 \leq t \leq T}$ is said to be *unique* if any other solution $\{\bar{x}(t)\}_{t_0 \leq t \leq T}$ is indistinguishable from $\{x(t)\}_{t_0 \leq t \leq T}$.

Remark 3.12 Denote the solution of equation (3.25) by $x(t; t_0, x_0, r_0)$ and the Markov chain by $r(t; t_0, r_0)$. Note

$$x(t) = x(s) + \int_s^t f(x(u), u, r(u))du + \int_s^t g(x(u), u, r(u))dB(u) \quad \forall s \leq t \leq T.$$

But, this is a stochastic differential equation on $[s, T]$ with initial data $x(s) = x(s; t_0, x_0, r_0)$ and $r(s) = r(s; t_0, r_0)$, whose solution is $x(t; s, x(s), r(s))$. Therefore, we see that the pair $(x(t; t_0, x_0, r_0), r(t; t_0, r_0))$ has the following *flow* or *semigroup property*

$$(x(t; t_0, x_0, r_0), r(t; t_0, r_0))$$
$$= (x(t; s, x(s; t_0, x_0), r(s; t_0, r_0)), r(t; s, r(s; t_0, r_0)))$$

for $t_0 \leq s \leq t \leq T$.

Theorem 3.13 *Assume that there exist two positive constants $\bar{K}$ and K such that*

(Lipschitz condition) *for all $x, y \in \mathbb{R}^n$, $t \in [t_0, T]$ and $i \in \mathbb{S}$*

$$|f(x, t, i) - f(y, t, i)|^2 \vee |g(x, t, i) - g(y, t, i)|^2 \leq \bar{K}|x - y|^2; \quad (3.26)$$

(Linear growth condition) *for all $(x, t, i) \in \mathbb{R}^n \times [t_0, T] \times \mathbb{S}$*

$$|f(x, t, i)|^2 \vee |g(x, t, i)|^2 \leq K(1 + |x|^2). \quad (3.27)$$

Then there exists a unique solution $x(t)$ to equation (3.25) and, moreover,

$$\mathbb{E}\left(\sup_{t_0 \leq t \leq T} |x(t)|^2\right) \leq (1 + 3\mathbb{E}|x_0|^2)e^{3K(T-t_0)(T-t_0+4)} \quad (3.28)$$

so the solution belongs to $\mathcal{M}^2([t_0, T]; \mathbb{R}^n)$.

Proof. Recall that almost every sample path of $r(\cdot)$ is a right-continuous step function with a finite number of simple jumps on $[t_0, T]$. So there is a sequence $\{\tau_k\}_{k \geq 0}$ of stopping times such that

- for almost every $\omega \in \Omega$ there is a finite $\bar{k} = \bar{k}(\omega)$ for $t_0 = \tau_0 < \tau_1 < \cdots < \tau_{\bar{k}} = T$ and $\tau_k = T$ if $k > \bar{k}$;
- $r(\cdot)$ is a random constant on every interval $[[\tau_k, \tau_{k+1}[[$, namely

$$r(t) = r(\tau_k) \text{ on } \tau_k \leq t < \tau_{k+1} \text{ for } \forall k \geq 0.$$

We first consider equation (3.25) on $t \in [[\tau_0, \tau_1]]$ which becomes

$$dx(t) = f(x(t), t, r_0)dt + g(x(t), t, r_0)dB(t) \qquad (3.29)$$

with initial data x_0 and r_0. By Theorem 3.8 and Remark 3.10, equation (3.29) has a unique solution which belongs to $\mathcal{M}^2([[\tau_0, \tau_1]]; \mathbb{R}^n)$. In particular, $x(\tau_1) \in L^2_{\mathcal{F}_{\tau_1}}(\Omega; \mathbb{R}^n)$. We next consider equation (3.25) on $t \in [[\tau_1, \tau_2]]$ which becomes

$$dx(t) = f(x(t), t, r(\tau_1))dt + g(x(t), t, r(\tau_1))dB(t) \qquad (3.30)$$

with initial data $x(\tau_1)$ and $r(\tau_1)$. Again by Theorem 3.8 and Remark 3.10, equation (3.30) has a unique solution which belongs to $\mathcal{M}^2([[\tau_1, \tau_2]]; \mathbb{R}^n)$. Repeating this procedure we see that equation (3.25) has a unique solution $x(t)$ on $[t_0, T]$. In the same way as Lemma 3.1 was proved we can show (3.28). $\square$

The Lipschitz condition (3.26) means that the coefficients $f(x, t, i)$ and $g(x, t, i)$ do not change faster than a linear function of x as change in x. This implies in particular the continuity of $f(x, t, i)$ and $g(x, t, i)$ in x for all $t \in [t_0, T]$. Thus, functions that are discontinuous with respect to x are excluded as the coefficients. Besides, functions like $\sin(x^2)$ do not satisfy the Lipschitz condition. These indicate that the Lipschitz condition is too restrictive. To remove this (uniform) Lipschitz condition let us introduce the concept of local solution.

Definition 3.14 Let σ_∞ be a stopping time such that $t_0 \leq \sigma_\infty \leq T$ a.s. An $\mathbb{R}^n$-valued $\mathcal{F}_t$-adapted continuous stochastic process $\{x(t) : t_0 \leq t < \sigma_\infty\}$ is called a *local solution* of equation (3.25) if $x(t_0) = x_0$ and, moreover, there is a nondecreasing sequence $\{\sigma_k\}_{k \geq 1}$ of stopping times such that $t_0 \leq \sigma_k \uparrow \sigma_\infty$ a.s. and

$$x(t) = x(t_0) + \int_{t_0}^{t \wedge \sigma_k} f(x(s), s, r(s))ds + \int_{t_0}^{t \wedge \sigma_k} g(x(s), s, r(s))dB(s)$$

holds for any $t \in [t_0, T)$ and $k \geq 1$ with probability 1. If, furthermore,

$$\limsup_{t \to \sigma_\infty} |x(t)| = \infty \text{ whenever } \sigma_\infty < T,$$

then it is called a *maximal local solution* and σ_∞ is called the *explosion time*. A maximal local solution $\{x(t) : t_0 \le t < \sigma_\infty\}$ is said to be *unique* if any other maximal solution $\{\bar{x}(t) : t_0 \le t < \bar{\sigma}_\infty\}$ is indistinguishable from it, namely $\sigma_\infty = \bar{\sigma}_\infty$ and $x(t) = \bar{x}(t)$ for $t_0 \le t < \sigma_\infty$ with probability 1.

The following theorem shows the existence of unique maximal local solution under the local Lipschitz condition without the linear growth condition.

Theorem 3.15 *Assume that* (local Lipschitz condition) *for every integer $k \ge 1$, there exists a positive constant h_k such that, for all $t \in [t_0, T]$, $i \in \mathbb{S}$ and those $x, y \in \mathbb{R}^n$ with $|x| \vee |y| \le k$,*

$$|f(x,t,i) - f(y,t,i)|^2 \vee |g(x,t,i) - g(y,t,i)|^2 \le \bar{h}_k |x-y|^2. \quad (3.31)$$

Then there exists a unique maximal local solution to equation (3.25).

Proof. This theorem is proved by a truncation procedure. We only outline the proof but leave the details to the reader. For each $k \ge 1$, define the truncation function

$$f_k(x,t,i) = \begin{cases} f(x,t,i) & \text{if } |x| \le k, \\ f(kx/|x|,t,i) & \text{if } |x| > k, \end{cases}$$

and $g_k(x,t,i)$ similarly. Then f_k and g_k satisfy the Lipschitz condition and the linear growth condition. Hence by Theorem 3.13, there is a unique solution $x_k(\cdot)$ in $\mathcal{M}^2([t_0, T]; \mathbb{R}^n)$ to the equation

$$dx_k(t) = f_k(x_k(t), t, r(t))dt + g_k(x_k(t), t, r(t))dB(t), \quad t \in [t_0, T] \quad (3.32)$$

with initial data $x_k(t_0) = x_0$ and $r(t_0) = r_0$. Define the stopping time

$$\sigma_k = T \wedge \inf\{t \in [t_0, T] : |x_k(t)| \ge k\}.$$

It is not difficult to show that

$$x_k(t) = x_{k+1}(t) \quad \text{if } t_0 \le t \le \sigma_k. \quad (3.33)$$

This implies that σ_k is increasing so has its limit $\sigma_\infty = \lim_{k\to\infty} \sigma_k$. Define $\{x(t) : t_0 \le t < \sigma_\infty\}$ by

$$x(t) = x_k(t), \quad t \in [[\sigma_{k-1}, \sigma_k[[, \quad k \ge 1,$$

where $\sigma_0 = t_0$. By (3.33), $x(l \wedge \sigma_k) = x_k(t \wedge \sigma_k)$. It therefore follows from (3.32) that

$$x(t \wedge \sigma_k) = x_0 + \int_{t_0}^{t \wedge \sigma_k} f_k(x(s), s, r(s))ds + \int_{t_0}^{t \wedge \sigma_k} g_k(x(s), s, r(s))dB(s)$$

$$= x_0 + \int_{t_0}^{t \wedge \sigma_k} f(x(s), s, r(s))ds + \int_{t_0}^{t \wedge \sigma_k} g(x(s), s, r(s))dB(s)$$

for any $t \in [t_0, T)$ and $k \geq 1$. It is also easy to see that if $\sigma_\infty < T$, then

$$\limsup_{t \to \sigma_\infty} |x(t)| \geq \limsup_{k \to \infty} |x(\sigma_k)| = \limsup_{k \to \infty} |x_k(\sigma_k)| = \infty.$$

Hence $\{x(t) : t_0 \leq t < \sigma_\infty\}$ is a maximal local solution. To show the uniqueness, let $\{\bar{x}(t) : t_0 \leq t < \bar{\sigma}_\infty\}$ be another maximal local solution. Define

$$\bar{\sigma}_k = \bar{\sigma}_\infty \wedge \inf\{t \in [[t_0, \bar{\sigma}_\infty[[: |\bar{x}(t)| \geq k\}.$$

It is easy to show that $\bar{\sigma}_k \to \bar{\sigma}_\infty$ a.s. and

$$\mathbb{P}\{x(t) = \bar{x}(t) \; \forall t \in [[t_0, \sigma_k \wedge \bar{\sigma}_k]]\} = 1 \quad \forall k \geq 1.$$

Letting $k \to \infty$ yields that

$$\mathbb{P}\{x(t) = \bar{x}(t) \; \forall t \in [[t_0, \sigma_\infty \wedge \bar{\sigma}_\infty[[\} = 1.$$

To complete the proof, we need to show that $\sigma_\infty = \bar{\sigma}_\infty$ a.s. In fact, for almost any $\omega \in \{\sigma_\infty < \bar{\sigma}_\infty\}$, we have

$$|\bar{x}(\sigma_\infty, \omega)| = \lim_{k \to \infty} |\bar{x}(\sigma_k, \omega)| = \lim_{k \to \infty} |x(\sigma_k, \omega)| = \infty$$

which contradicts the fact that $\bar{x}(t, \omega)$ is continuous on $t \in [t_0, \bar{\sigma}_\infty(\omega))$. We must hence have $\sigma_\infty \geq \bar{\sigma}_\infty$ a.s. Similarly, we can show $\sigma_\infty \leq \bar{\sigma}_\infty$ a.s. Therefore we must have $\sigma_\infty = \bar{\sigma}_\infty$ a.s. □

The following theorem is a generalisation of Theorem 3.13— the (uniform) Lipschitz condition is replaced by the local Lipschitz condition.

Theorem 3.16 *Assume that the local Lipschitz condition (3.31) and the linear growth condition (3.27) are satisfied. Then the conclusions of Theorem 3.13 still hold.*

Proof. We use the same notations as in the proof of Theorem 3.15. By (3.27) and the definition of f_k and g_k we have that

$$|f_k(x, t, i)|^2 \vee |g_k(x, t, i)|^2 \leq K(1 + |x|^2) \quad \forall (x, t, i) \in \mathbb{R}^n \times [t_0, T] \times \mathbb{S}.$$

So by Theorem 3.13,
$$\mathbb{E}\left(\sup_{t_0\leq t\leq T}|x_k(t)|^2\right) \leq (1+3\mathbb{E}|x_0|^2)e^{3K(T-t_0)(T-t_0+4)}.$$
That is,
$$\mathbb{E}\left(\sup_{t_0\leq t\leq \sigma_k}|x(t)|^2\right) \leq (1+3\mathbb{E}|x_0|^2)e^{3K(T-t_0)(T-t_0+4)}.$$
Letting $k \to \infty$ yields
$$\mathbb{E}\left(\sup_{t_0\leq t< \sigma_\infty}|x(t)|^2\right) \leq (1+3\mathbb{E}|x_0|^2)e^{3K(T-t_0)(T-t_0+4)}.$$
This implies easily that $\sigma_\infty = T$ a.s. and the assertions follow. □

The local Lipschitz condition allows us to include many functions as the coefficients $f(x,t,i)$ and $g(x,t,i)$ e.g. functions that have continuous partial derivatives of first order with respect to x on $\mathbb{R}^n \times [t_0,T] \times \mathbb{S}$. However, the linear growth condition still excludes some important functions like $-|x|^2 x$ as the coefficients. The following result improves the situation.

Theorem 3.17 *Assume that the local Lipschitz condition (3.31) holds, but the linear growth condition (3.27) is replaced with the following monotone condition: There exists a positive constant K such that for all $(x,t,i) \in \mathbb{R}^n \times [t_0,T] \times \mathbb{S}$*

$$x^T f(x,t,i) + \frac{1}{2}|g(x,t,i)|^2 \leq K(1+|x|^2). \quad (3.34)$$

Then there exists a unique solution $x(t)$ to equation (3.2) in $\mathcal{M}^2([t_0,T];\mathbb{R}^n)$.

This theorem can be proved in a similar way as Theorem 3.16 was proved— the local Lipschitz condition guarantees that the solution exists in $[t_0, \sigma_\infty)$, but the monotone condition instead of the linear growth condition guarantees that $\sigma_\infty = T$ a.s. so the solution exists on the whole interval $[t_0, T]$. We leave the details to the reader as an exercise. It should be stressed that if the linear growth condition (3.27) holds then the monotone condition (3.34) is satisfied, but not conversely. For example, consider the one-dimensional stochastic differential equation

$$dx(t) = a(r(t))[x(t) - x^3(t)]dt + b(r(t))x^2(t)dB(t), \quad t \in [t_0,T],$$

where $B(t)$ is a one-dimensional Brownian motion and $a,b: \mathbb{S} \to \mathbb{R}_+$ such that $b^2(i) \leq 2a(i)$ for $\forall i \in \mathbb{S}$. Clearly, the local Lipschitz condition is satisfied but the linear growth condition is not. On the other hand, note that

$$a(i)x[x-x^3] + \frac{1}{2}b^2(i)x^4 \leq a(i)x^2 \leq K(1+x^2)$$

with $K = \max_{i \in \mathbb{S}} a(i)$. That is, the monotone condition is fulfilled. Hence Theorem 3.17 guarantees that this equation has a unique solution on $[t_0, T]$.

In this book we often discuss a stochastic differential equation on $[t_0, \infty)$, that is

$$dx(t) = f(x(t), t, r(t))dt + g(x(t), t, r(t))dB(t), \quad t \in [t_0, \infty) \quad (3.35)$$

with initial data $x(t_0) = x_0$ and $r(t_0) = r_0$. If the assumptions of the existence-and-uniqueness theorem hold on every finite subinterval $[t_0, T]$ of $[t_0, \infty)$, then equation (3.35) has a unique solution $x(t)$ on the entire interval $[t_0, \infty)$. Such a solution is called a *global* solution. For example, the following result follows from Theorem 3.17.

Theorem 3.18 *Assume that for every real number $T > t_0$ and integer $k \geq 1$, there exists a positive constant $K_{T,k}$ such that for all $t \in [t_0, T]$, $i \in \mathbb{S}$ and all $x, y \in \mathbb{R}^n$ with $|x| \vee |y| \leq k$,*

$$|f(x,t,t) - f(y,t,i)|^2 \vee |g(x,t,i) - g(y,t,i)|^2 \leq K_{T,k}|x-y|^2. \quad (3.36)$$

Assume also that for every $T > t_0$, there exists a positive constant K_T such that for all $(x, t, i) \in \mathbb{R}^n \times [t_0, T] \times \mathbb{S}$,

$$x^T f(x,t,i) + \frac{1}{2}|g(x,t,i)|^2 \leq K_T(1+|x|^2). \quad (3.37)$$

Then there exists a unique global solution $x(t)$ to equation (3.35) and the solution belongs to $\mathcal{M}^2([t_0, \infty); \mathbb{R}^n)$.

To establish a more general result we need more notations. Let $C^{2,1}(\mathbb{R}^n \times \mathbb{R}_+ \times \mathbb{S}; \mathbb{R}_+)$ denote the family of all non-negative functions $V(x, t, i)$ on $\mathbb{R}^n \times \mathbb{R}_+ \times \mathbb{S}$ which are continuously twice differentiable in x and once in t. If $V \in C^{2,1}(\mathbb{R}^n \times \mathbb{R}_+ \times \mathbb{S}; \mathbb{R}_+)$, define an operator LV from $\mathbb{R}^n \times \mathbb{R}_+ \times \mathbb{S}$ to $\mathbb{R}$ by

$$LV(x,t,i) = V_t(x,t,i) + V_x(x,t,i)f(x,t,i)$$
$$+ \frac{1}{2}\text{trace}[g^T(x,t,i)V_{xx}(x,t,i)g(x,t,i)] + \sum_{j=1}^{N}\gamma_{ij}V(x,t,j),$$

where

$$V_t(x,t,i) = \frac{\partial V(x,t,i)}{\partial t}, \quad V_x(x,t,i) = \left(\frac{\partial V(x,t,i)}{\partial x_1}, \cdots, \frac{\partial V(x,t,i)}{\partial x_n}\right),$$

$$V_{xx}(x,i) = \left(\frac{\partial^2 V(x,i)}{\partial x_i \partial x_j}\right)_{n \times n}.$$

Theorem 3.19 *Assume that the local Lipschitz condition (3.36) holds. Assume also that there is a function $V \in C^{2,1}(\mathbb{R}^n \times \mathbb{R}_+ \times \mathbb{S}; \mathbb{R}_+)$ and a constant $\alpha > 0$ such that*

$$\lim_{|x| \to \infty} \left(\inf_{(t,i) \in \mathbb{R}_+ \times \mathbb{S}} V(x,t,i) \right) = \infty \tag{3.38}$$

and

$$LV(x,t,i) \leq \alpha(1 + V(x,t,i)), \quad \forall (x,t,i) \in \mathbb{R}^n \times \mathbb{R}_+ \times \mathbb{S}. \tag{3.39}$$

Then there exists a unique global solution $x(t)$ to equation (3.35).

Proof. It is sufficient to prove the theorem for any initial constant values $x_0 \in \mathbb{R}^n$ and $r_0 \in \mathbb{S}$. By Theorem 3.15, the local Lipschitz condition guarantees the existence of the unique maximal solution $x(t)$ on $[t_0, \sigma_\infty)$, where σ_∞ is the explosion time. We need to show $\sigma_\infty = \infty$ a.s. If this is not true, then we can find a pair of positive constants ε and T such that

$$\mathbb{P}\{\sigma_\infty \leq T\} > 2\varepsilon.$$

For each integer $k \geq 1$, define the stopping time

$$\sigma_k = \inf\{t \geq t_0 : |x(t)| \geq k\}.$$

Since $\sigma_k \to \sigma_\infty$ almost surely, we can find a sufficiently large integer k_0 for

$$\mathbb{P}\{\sigma_k \leq T\} > \varepsilon, \quad \forall k \geq k_0. \tag{3.40}$$

Fix any $k \geq k_0$. For any $t_0 \leq t \leq T$, by the generalised Itô formula,

$$\mathbb{E}V(x(t \wedge \sigma_k), t \wedge \sigma_k, r(t \wedge \sigma_k))$$
$$= V(x_0, t_0, r_0) + \mathbb{E}\int_{t_0}^{t \wedge \sigma_k} LV(x(s), s, r(s))ds$$
$$\leq V(x_0, t_0, r_0) + \alpha(T - t_0) + \alpha \int_{t_0}^{t} \mathbb{E}V(x(s \wedge \sigma_k), t \wedge \sigma_k, s(t \wedge \sigma_k))ds.$$

The Gronwall inequality implies

$$\mathbb{E}V(x(T \wedge \sigma_k), T \wedge \sigma_k, r(T \wedge \sigma_k)) \leq [V(x_0, t_0, r_0) + \alpha(T - t_0)]e^{\alpha(T-t_0)}.$$

So

$$\mathbb{E}\Big(I_{\{\sigma_k \leq T\}} V(x(\sigma_k), \sigma_k, r(\sigma_k))\Big) \leq [V(x_0, t_0, r_0) + \alpha(T - t_0)]e^{\alpha(T-t_0)}. \tag{3.41}$$

On the other hand, if we define

$$h_k = \inf\{V(x, t, i) : |x| \geq k, \ t \in [t_0, T], \ i \in \mathbb{S}\},$$

then $h_k \to \infty$ by (3.38). It now follows from (3.40) and (3.41) that

$$[V(x_0, t_0, r_0) + \alpha(T - t_0)]e^{\alpha(T-t_0)} \geq h_k \mathbb{P}\{\sigma_k \leq T\} \geq \varepsilon h_k.$$

Letting $k \to \infty$ yields a contradiction so we must have $\sigma_\infty = \infty$ a.s. $\square$

3.5 L^p-Estimates

In this section, we let $x(t)$, $t_0 \leq t \leq T$ be the unique solution of equation (3.25) with initial data $x(t_0) = x_0$ and $r(t_0) = r_0$, and we shall investigate the pth moment of the solution.

Theorem 3.20 *Assume that there is a function $V \in C^{2,1}(\mathbb{R}^n \times [t_0, T] \times \mathbb{S}; \mathbb{R}_+)$ and positive constants p, α, β such that for all $(x, t, i) \in \mathbb{R}^n \times [t_0, T] \times \mathbb{S}$,*

$$\beta |x|^p \leq V(x, t, i) \tag{3.42}$$

and

$$LV(x, t, i) \leq \alpha V(x, t, i). \tag{3.43}$$

If initial data $x(t_0) = x_0$ and $r(t_0) = r_0$ satisfy $\mathbb{E}V(x_0, t_0, r_0) < \infty$, then

$$\mathbb{E}|x(t)|^p \leq \frac{1}{\beta}\mathbb{E}V(x_0, t_0, r_0)e^{\alpha(t-t_0)}, \quad \forall t \in [t_0, T]. \tag{3.44}$$

Proof. For each integer $k \geq 1$, define the stopping time

$$\sigma_k = T \wedge \inf\{t \geq t_0 : |x(t)| \geq k\}.$$

Then $\sigma_k \to T$ almost surely. By the generalised Itô formula and condition (3.43) we can show that for $t \in [t_0, T]$,

$$\mathbb{E}V(x(t \wedge \sigma_k), t \wedge \sigma_k, r(t \wedge \sigma_k))$$
$$\leq \mathbb{E}V(x_0, t_0, r_0) + \alpha \int_{t_0}^{t} \mathbb{E}V(x(s \wedge \sigma_k), t \wedge \sigma_k, s(t \wedge \sigma_k))ds.$$

The Gronwall inequality implies

$$\mathbb{E}V(x(t \wedge \sigma_k), t \wedge \sigma_k, r(t \wedge \sigma_k)) \leq \mathbb{E}V(x_0, t_0, r_0)e^{\alpha(t-t_0)}$$

for all $t \in [t_0, T]$. Making use of condition (3.42) and then letting $k \to \infty$ yields the required assertion (3.44). □

Corollary 3.21 *Let $p \geq 2$ and $x_0 \in L^p_{\mathcal{F}_{t_0}}(\Omega; \mathbb{R}^n)$. Assume that there exists a constant $\alpha > 0$ such that for all $(x, t, i) \in \mathbb{R}^n \times [t_0, T] \times \mathbb{S}$,*

$$x^T f(x, t, i) + \frac{p-1}{2}|g(x, t, i)|^2 \leq \alpha(1 + |x|^2). \tag{3.45}$$

Then

$$\mathbb{E}|x(t)|^p \leq 2^{\frac{p-2}{2}}(1 + \mathbb{E}|x_0|^p)e^{p\alpha(t-t_0)} \quad \forall t \in [t_0, T]. \tag{3.46}$$

Proof. Let $V(x, t, i) = (1 + |x|^2)^{p/2}$. Compute the operator

$$LV(x, t, i) = p(1 + |x|^2)^{\frac{p-2}{2}} x^T f(x, t, i) + \frac{p}{2}(1 + |x|^2)^{\frac{p-2}{2}}|g(x, t, i)|^2$$
$$+ \frac{p(p-2)}{2}(1 + |x|^2)^{\frac{p-4}{2}}|x^T g(x, t, i)|^2$$
$$\leq p(1 + |x|^2)^{\frac{p-2}{2}}\left[x^T f(x, t, i) + \frac{p-1}{2}|g(x, t, i)|^2\right]$$
$$\leq p\alpha(1 + |x|^2)^{\frac{p}{2}}.$$

By Theorem 3.20,

$$\mathbb{E}(1 + |x(t)|^2)^{p/2} \leq \mathbb{E}(1 + |x_0|^2)^{p/2} e^{p\alpha(t-t_0)} \quad \forall t \in [t_0, T] \tag{3.47}$$

and the required assertion (3.46) follows. □

We now verify that if the linear growth condition (3.27) is fulfilled, then (3.45) is satisfied with $\alpha = \sqrt{K} + K(p-1)/2$. In fact, using (3.27) and the elementary inequality $2ab \leq a^2 + b^2$ we derive that for any $\varepsilon > 0$,

$$2x^T f(x, t, i) \leq 2|x||f(x, t, i)| = 2(\sqrt{\varepsilon}|x|)(|f(x, t, i)|/\sqrt{\varepsilon})$$

$$\le \varepsilon|x|^2 + \frac{1}{\varepsilon}|f(x,t,i)|^2 \le \varepsilon|x|^2 + \frac{K}{\varepsilon}(1+|x|^2).$$

Letting $\varepsilon = \sqrt{K}$ yields

$$x^T f(x,t,i) \le \sqrt{K}(1+|x|^2).$$

Consequently

$$x^T f(x,t,i) + \frac{p-1}{2}|g(x,t,i)|^2 \le \left[\sqrt{K} + \frac{K(p-1)}{2}\right](1+|x|^2).$$

We therefore obtain the following useful corollary.

Corollary 3.22 *Let $p \ge 2$ and $x_0 \in L^p_{\mathcal{F}_{t_0}}(\Omega;\mathbb{R}^n)$. Assume that the linear growth condition (3.27) holds. Then inequality (3.46) holds with $\alpha = \sqrt{K} + K(p-1)/2$.*

We now apply these results to show one of the important properties of the solution.

Theorem 3.23 *Let $p \ge 2$ and $x_0 \in L^p_{\mathcal{F}_{t_0}}(\Omega;\mathbb{R}^n)$. Assume that the linear growth condition (3.27) holds. Then*

$$\mathbb{E}|x(t) - x(s)|^p \le C(t-s)^{\frac{p}{2}}, \quad t_0 \le s < t \le T, \tag{3.48}$$

where

$$C = 2^{p-2}(1+\mathbb{E}|x_0|^p)e^{p\alpha(T-t_0)}\left([2(T-t_0)]^{\frac{p}{2}} + [p(p-1)]^{\frac{p}{2}}\right)$$

and $\alpha = \sqrt{K} + K(p-1)/2$. In particular, the pth moment of the solution is continuous on $[t_0, T]$.

Proof. By the elementary inequality $|a+b|^p \le 2^{p-1}(|a|^p + |b|^p)$, it is easy to see that

$$\mathbb{E}|x(t)-x(s)|^p \le 2^{p-1}\mathbb{E}\left|\int_s^t f(x(u),u,r(u))du\right|^p$$
$$+ 2^{p-1}\mathbb{E}\left|\int_s^t g(x(u),u,r(u))dB(u)\right|^p.$$

Using the Hölder inequality, Theorem 1.7.1 and the linear growth condition, one can then derive that

$$\mathbb{E}|x(t) - x(s)|^p \leq [2(t-s)]^{p-1}\mathbb{E}\int_s^t |f(x(u),u,r(u))|^p du$$

$$+ \frac{1}{2}[2p(p-1)]^{\frac{p}{2}}(t-s)^{\frac{p-2}{2}}\mathbb{E}\int_s^t |g(x(u),u,r(u))|^p du$$

$$\leq c_1(t-s)^{\frac{p-2}{2}}\int_s^t \mathbb{E}(1+|x(u)|^2)^{\frac{p}{2}} du,$$

where $c_1 = 2^{\frac{p-2}{2}}K^{\frac{p}{2}}\left([2(T-t_0)]^{\frac{p}{2}} + [p(p-1)]^{\frac{p}{2}}\right)$. Applying (3.47) one sees that

$$\mathbb{E}|x(t) - x(s)|^p \leq c_1(t-s)^{\frac{p-2}{2}}\int_s^t 2^{\frac{p-2}{2}}(1+\mathbb{E}|x_0|^p)e^{p\alpha(r-t_0)} du$$

$$\leq c_1 2^{\frac{p-2}{2}}(1+\mathbb{E}|x_0|^p)e^{p\alpha(T-t_0)}(t-s)^{\frac{p}{2}},$$

which is the required inequality (3.48). □

The following theorem will be useful in the next chapter when we discuss approximate solutions.

Theorem 3.24 *Let $p \geq 2$ and $x_0 \in L^p_{\mathcal{F}_{t_0}}(\Omega;\mathbb{R}^n)$. Assume that there is a $K > 0$ such that*

$$x^T f(x,t,i) \vee |g(x,t,i)|^2 \leq K(1+|x|^2), \quad \forall (x,t,i) \in \mathbb{R}^n \times [t_0,T] \times \mathbb{S}. \tag{3.49}$$

Then

$$\mathbb{E}\left(\sup_{t_0 \leq t \leq T} |x(t)|^p\right) \leq (1+2\mathbb{E}|x_0|^p)e^{2p(10p+1)K(T-t_0)}. \tag{3.50}$$

Proof. By the generalised Itô formula and condition (3.49),

$$|x(t)|^p \leq |x_0|^p + M(t)$$

$$+ \int_{t_0}^t p|x(s)|^{p-2}\left[x^T(s)f(x(s),s,r(s)) + \frac{p-1}{2}|g(x(s),s,r(s))|^2\right]ds$$

$$\leq |x_0|^p + M(t) + \frac{1}{2}p(p+1)K\int_{t_0}^t |x(s)|^{p-2}(1+|x(s)|^2)ds$$

$$\leq |x_0|^p + M(t) + p(p+1)K\int_{t_0}^t (1+|x(s)|^p)ds,$$

where
$$M(t) = \int_{t_0}^{t} p|x(s)|^{p-2}x^T(s)g(x(s),s,r(s))dB(s).$$

So, for any $t_1 \in [t_0, T]$,

$$\mathbb{E}\left(\sup_{t_0 \leq t \leq t_1} |x(t)|^p\right) \leq \mathbb{E}|x_0|^p + \mathbb{E}\left(\sup_{t_0 \leq t \leq t_1} |M(t)|\right)$$
$$+ p(p+1)K\mathbb{E}\int_{t_0}^{t_1}(1+|x(s)|^p)ds. \qquad (3.51)$$

On the other hand, by the Burkholder–Davis–Gundy inequality and Exercise 2.5,

$$\mathbb{E}\left(\sup_{t_0 \leq t \leq t_1}|M(t)|\right) \leq 3\mathbb{E}\left(\int_{t_0}^{t_1} p^2|x(s)|^{2p-2}|g(x(s),s,r(s))|^2 ds\right)^{\frac{1}{2}}$$
$$\leq 3\mathbb{E}\left(\sup_{t_0 \leq t \leq t_1}|x(t)|^p \int_{t_0}^{t_1} p^2|x(s)|^{p-2}|g(x(s),s,r(s))|^2 ds\right)^{\frac{1}{2}}$$
$$\leq 0.5\mathbb{E}\left(\sup_{t_0 \leq t \leq t_1}|x(t)|^p\right) + 4.5\int_{t_0}^{t_1} p^2 K|x(s)|^{p-2}(1+|x(s)|^2)ds$$
$$\leq 0.5\mathbb{E}\left(\sup_{t_0 \leq t \leq t_1}|x(t)|^p\right) + 9p^2 K\mathbb{E}\int_{t_0}^{t_1}(1+|x(s)|^p)ds.$$

Substituting this into (3.51) gives

$$\mathbb{E}\left(\sup_{t_0 \leq t \leq t_1}|x(t)|^p\right) \leq 2\mathbb{E}|x_0|^p + 2p(10p+1)K\int_{t_0}^{t_1}(1+|x(s)|^p)ds.$$

Consequently

$$1 + \mathbb{E}\left(\sup_{t_0 \leq t \leq t_1}|x(t)|^p\right)$$
$$\leq 1 + 2\mathbb{E}|x_0|^p + 2p(10p+1)K\int_{t_0}^{t_1}\left[1+\mathbb{E}\left(\sup_{t_0 \leq t \leq s}|x(t)|^p\right)\right]ds.$$

By the Gronwall inequality, the desired assertion (3.50) follows. □

In the above we have discussed the L^p-estimates for the solutions in the case when $p \geq 2$. Let us now consider the case of $0 < p < 2$. This is rather easy if we note that the Hölder inequality implies

$$\mathbb{E}|x(t)|^p \leq \left[\mathbb{E}|x(t)|^2\right]^{\frac{p}{2}}.$$

In other words, the estimate for $\mathbb{E}|x(t)|^p$ can be done via the estimate for the second moment. For example, the following result follows from Corollary 3.21.

Corollary 3.25 *Let $0 < p < 2$ and $x_0 \in L^2_{\mathcal{F}_{t_0}}(\Omega; \mathbb{R}^n)$. Assume that there exists a constant $\alpha > 0$ such that for all $(x, t, i) \in \mathbb{R}^n \times [t_0, T] \times \mathbb{S}$,*

$$x^T f(x,t,i) + \frac{1}{2}|g(x,t,i)|^2 \leq \alpha(1+|x|^2).$$

Then

$$\mathbb{E}|x(t)|^p \leq (1+\mathbb{E}|x_0|^2)^{\frac{p}{2}} e^{p\alpha(t-t_0)} \quad \forall t \in [t_0, T].$$

3.6 Almost Surely Asymptotic Estimates

Let us now consider the n-dimensional stochastic differential equation with Markovian switching

$$dx(t) = f(x(t), t, r(t))dt + g(x(t), t, r(t))dB(t) \quad \text{on } t \in [t_0, \infty) \quad (3.52)$$

with initial value $x(t_0) = x_0$. Assume that the equation has a unique global solution $x(t)$ on $[t_0, \infty)$. Besides, we shall impose the linear growth condition: There is a positive constant K such that,

$$|f(x,t,i)|^2 \vee |g(x,t)|^2 \leq K(1+|x|^2). \quad (3.53)$$

If $p \geq 2$ and $x_0 \in L^p_{\mathcal{F}_{t_0}}(\Omega; \mathbb{R}^n)$, Corollary 3.22 shows that

$$\mathbb{E}|x(t)|^p \leq 2^{\frac{p-2}{2}}(1+\mathbb{E}|x_0|^p)^{\frac{p}{2}} e^{p\alpha(t-t_0)} \quad \forall t \geq t_0,$$

where $\alpha = \sqrt{K} + K(p-1)/2$. This means that the pth moment will grow at most exponentially with exponent $p\alpha$. This can also be expressed as

$$\limsup_{t \to \infty} \frac{1}{t} \log(\mathbb{E}|x(t)|^p) \leq p\alpha. \quad (3.54)$$

The left-hand side of (3.54) is called the *pth moment Lyapunov exponent* (for $p \leq 2$ too), and (3.54) shows that the pth moment Lyapunov exponent should not be greater than $p\alpha$. In this section, we shall establish the almost surely asymptotic estimate for the solution. More precisely, we shall estimate

$$\limsup_{t \to \infty} \frac{1}{t} \log|x(t)| \quad (3.55)$$

almost surely, which is called the *sample Lyapunov exponent*, or simply *Lyapunov exponent*.

Theorem 3.26 Let (3.53) hold. If $x_0 \in L^2_{\mathcal{F}_{t_0}}(\Omega; \mathbb{R}^n)$, then

$$\limsup_{t \to \infty} \frac{1}{t} \log |x(t)| \leq \sqrt{K} + \frac{K}{2} \quad a.s. \tag{3.56}$$

Proof. Set $\alpha = \sqrt{K} + K/2$. By Corollary 3.22,

$$\mathbb{E}|x(t)|^2 \leq (1 + \mathbb{E}|x_0|^2)e^{2\alpha(t-t_0)} \quad \forall t \geq t_0. \tag{3.57}$$

Let $k = 1, 2, \cdots$. For $t_0 + k \leq t \leq t_0 + k + 1$,

$$x(t) = x(t_0 + k) + \int_{t_0+k}^{t} f(x(s), s, r(s))ds$$
$$+ \int_{t_0+k}^{t} g(x(s), s, r(s))dB(s).$$

Hence

$$\mathbb{E}\left(\sup_{t_0+k \leq t \leq t_0+k+1} |x(t)|^2\right)$$
$$\leq 3\mathbb{E}|x(t_0+k)|^2 + 3\mathbb{E}\left(\sup_{t_0+k \leq t \leq t_0+k+1} \left|\int_{t_0+k}^{t} f(x(s), s, r(s))ds\right|^2\right)$$
$$+ 3\mathbb{E}\left(\sup_{t_0+k \leq t \leq t_0+k+1} \left|\int_{t_0+k}^{t} g(x(s), s, r(s))dB(s)\right|^2\right). \tag{3.58}$$

Using the Hölder inequality and the linear growth condition we compute

$$\mathbb{E}\left(\sup_{t_0+k \leq t \leq t_0+k+1} \left|\int_{t_0+k}^{t} f(x(s), s, r(s))ds\right|^2\right)$$
$$\leq \mathbb{E}\left|\int_{t_0+k}^{t_0+k+1} f(x(s), s, r(s))ds\right|^2 \leq K \int_{t_0+k}^{t_0+k+1} (1 + \mathbb{E}|x(s)|^2)ds, \tag{3.59}$$

while using the Doob martingale inequality we have

$$\mathbb{E}\left(\sup_{t_0+k \leq t \leq t_0+k+1} \left|\int_{t_0+k}^{t} g(x(s), s, r(s))dB(s)\right|^2\right).$$
$$\leq 4\mathbb{E}\int_{t_0+k}^{t_0+k+1} |g(x(s), s, r(s))|^2 ds \leq 4K \int_{t_0+k}^{t_0+k+1} (1 + \mathbb{E}|x(s)|^2)ds. \tag{3.60}$$

Substituting (3.59) and (3.60) into (3.58) yields

$$\mathbb{E}\Big(\sup_{t_0+k\le t\le t_0+k+1} |x(t)|^2\Big)$$
$$\le 3\mathbb{E}|x(t_0+k)|^2 + 15K\int_{t_0+k}^{t_0+k+1}(1+\mathbb{E}|x(s)|^2)ds.$$

Making use of (3.57) we obtain

$$\mathbb{E}\Big(\sup_{t_0+k\le t\le t_0+k+1} |x(t)|^2\Big) \le Ce^{2\alpha k}, \tag{3.61}$$

where $C = (30K+3)(1+\mathbb{E}|x_0|^2)$. Now, let $\varepsilon > 0$ be arbitrary. By the well-known Chebyshev inequality, we have

$$\mathbb{P}\Big\{\sup_{t_0+k\le t\le t_0+k+1} |x(t)| > e^{(\alpha+\varepsilon)k}\Big\}$$
$$\le e^{-2(\alpha+\varepsilon)k}\mathbb{E}\Big(\sup_{t_0+k\le t\le t_0+k+1} |x(t)|^2\Big)$$
$$\le Ce^{2\varepsilon k} \quad \forall k \ge 1.$$

In view of the well-known Borel–Cantelli lemma, we sees that for almost all $\omega \in \Omega$,

$$\sup_{t_0+k\le t\le t_0+k+1} |x(t)| \le e^{(\alpha+\varepsilon)k} \tag{3.62}$$

holds for all but finitely many k. Hence, there exists a $k_0(\omega)$, for almost all $\omega \in \Omega$, for which (3.62) holds whenever $k \ge k_0$. Consequently, for almost all $\omega \in \Omega$,

$$\frac{1}{t}\log(|x(t)|) \le \frac{(\alpha+\varepsilon)k}{t_0+k}$$

if $t_0+k \le t \le t_0+k+1$ and $k \ge k_0$. Therefore

$$\limsup_{t\to\infty}\frac{1}{t}\log(|x(t)|) \le \alpha+\varepsilon \quad a.s.$$

The assertion follows since $\varepsilon > 0$ is arbitrary. $\square$

A slightly more general result than Theorem 3.26 is stated as Exercise 3.2 below.

3.7 Solutions as Markov Processes

In this section we shall discuss the Markov property of the solutions.

Theorem 3.27 *Let $\xi(t)$ be a solution of the equation*

$$d\xi(t) = f(\xi(t), t, r(t))dt + g(\xi(t), t, r(t))dB(t) \quad \text{on } t \geq 0, \tag{3.63}$$

whose coefficients satisfy the conditions of the existence-and-uniqueness theorem. Then $(\xi(t), r(t))$ is a Markov process whose transition probability is defined by

$$P(s, (x, i); t, A \times \{j\}) = \mathbb{P}\{\xi_s^{x,i}(t) \in A \times \{j\}\} \tag{3.64}$$

for $(x, i) \in \mathbb{R}^n \times \mathbb{S}$, $A \in \mathcal{B}(\mathbb{R}^n)$ and $j \in \mathbb{S}$, where $\xi_s^{x,i}(t)$ is the solution of equation (3.63) on $t \geq s$ with initial data $\xi(s) = x \in \mathbb{R}^n$ and $r(s) = i \in \mathbb{S}$. That is

$$\xi_s^{x,i}(t) = x + \int_s^t f(\xi_s^{x,i}(u), u, r_s^i(u))du + \int_s^t g(\xi_s^{x,i}(u), u, r_s^i(u))dB(u) \tag{3.65}$$

on $t \geq s$, where $r_s^i(t)$ stands for the Markov chain on $t \geq s$ starting from state i at time $t = s$.

To prove this theorem, let us cite a classical result (see e.g. [Mao (1997), Lemma 9.2 on p.87]).

Lemma 3.2 *Let $\bar{h}(x, \omega)$ be a scalar bounded measurable random function of x, independent of $\mathcal{F}_s$. Let ζ be an $\mathcal{F}_s$-measurable random variable. Then*

$$\mathbb{E}(\bar{h}(\zeta, \omega)|\mathcal{F}_s) = H(\zeta), \tag{3.66}$$

where $H(x) = \mathbb{E}\bar{h}(x, \omega)$.

Proof of Theorem 3.27. Recall (see Chapter 1)

$$dr(t) = \int_{\mathbb{R}} h(r(t-), y)\nu(dt, dy),$$

where $\nu(dt, dy)$ is a Poisson random measure and h is defined by (1.46). This yields

$$r_s^i(t) = i + \int_s^t \int_{\mathbb{R}} h(r_s^i(u-), y)\nu(du, dy)$$

for $0 \leq s < t < \infty$ and $i \in \mathbb{S}$.

Let $\mathcal{G}_s = \sigma\{B(u) - B(s), \nu(u,U) - \nu(s,U) : u \geq s, U \in \mathcal{B}(\mathbb{R})\}$. Clearly, $\mathcal{G}_s$ is independent of $\mathcal{F}_s$. Moreover, the value of $(\xi_s^{x,i}(t)), r_s^i(t))$ depends on the increments $B(u) - B(s), \nu(u, dy) - \nu(s, dy)$ for $u \geq s$ and so is $\mathcal{G}_s$-measurable. Hence $(\xi_s^{x,i}(t), r_s^i(t))$ is independent of $\mathcal{F}_s$. On the other hand, note that $(\xi(t), r(t)) = (\xi_s^{\xi(s), r(s)}(t), r_s^{r(s)}(t))$, since $(\xi(t), r(t))$ satisfy the equations

$$\xi(t) = \xi(s) + \int_s^t f(\xi(u), u, r(u))du + \int_s^t g(\xi(u), u, r(u))dB(u) \quad (3.67)$$

and

$$r(t) = r(s) + \int_s^t \int_{\mathbb{R}} h(r(u-), y)\nu(du, dy).$$

Now, for any $A \in \mathcal{B}(\mathbb{R}^n), j \in \mathbb{S}$, we can apply Lemma 3.2 with $\bar{h}((x,i), \omega) = I_{A \times \{j\}}(\xi_s^{x,i}(t), r_s^i(t))$ to compute

$$\mathbb{P}((\xi(t), r(t)) \in A \times \{j\} | \mathcal{F}_s) = \mathbb{E}(I_{A \times \{j\}}(\xi(t), r(t)) | \mathcal{F}_s)$$
$$= \mathbb{E}(I_{A \times \{j\}}(\xi_s^{\xi(s), r(s)}(t), r_s^{r(s)}(t)) | \mathcal{F}_s)$$
$$= \mathbb{E}[I_{A \times \{j\}}(\xi_s^{x,i}(t), r_s^i(t))]|_{x=\xi(s), i=r(s)}$$
$$= P(s, (\xi(s), r(s)); t, A \times \{j\}).$$

This proves the Markov property. □

Generally speaking, a Markov process is not a strong one. The conditions that guarantee a Markov process possesses the strong Markov property are the right continuity of the sample paths plus the so-called *Feller property*. Given that the 2nd component of the Markov process $(\xi(t), r(t))$ takes its value in the discrete space $\mathbb{S}$, the Feller property for this process can be stated as follows: For any $i \in \mathbb{S}, \lambda > 0$ and any bounded continuous function $\varphi : \mathbb{R}^n \to \mathbb{R}$, the mapping

$$(x, s) \to \sum_{j \in \mathbb{S}} \int_{\mathbb{R}^n} \varphi(y) P(s, (x, i); s + \lambda, dy \times \{j\})$$

is continuous. In this case, the transition function $P(s, (x, i); t, A \times \{j\})$, or the corresponding Markov process, is said to obey the Feller property.

For the strong property of the solution we need to strengthen the conditions slightly.

Theorem 3.28 *Let $(\xi(t), r(t))$ be the solution of the equations*

$$d\xi(t) = f(\xi(t), t, r(t))dt + g(\xi(t), t, r(t))dB(t)$$

and

$$dr(t) = \int_{\mathbb{R}} h(r(t-), y)\nu(dt, dy).$$

Assume the coefficients are uniformly Lipschitz continuous and satisfy the linear growth condition, that is there are two constants K and $\bar{K}$ such that

$$|f(x,t,i) - f(y,t,i)|^2 \vee |g(x,t,i) - g(y,t,i)|^2 \leq \bar{K}|x-y|^2 \tag{3.68}$$

and

$$|f(x,t,i)|^2 \vee |g(x,t,i)|^2 \leq K(1+|x|^2) \tag{3.69}$$

for all $x, y \in \mathbb{R}^n, i \in \mathbb{S}$ and $t \geq 0$. Then $(\xi(t), r(t))$ is a strong Markov process.

We again need to prepare a lemma in order to prove this theorem.

Lemma 3.3 *Let (3.68) and (3.69) hold. For $(x,i) \in \mathbb{R}^n \times \mathbb{S}$ and $0 \leq s \leq t < \infty$, let $\xi_s^{x,i}(t)$ and $r_s^i(t)$ be the same as defined in Theorem 3.27. Then for any $T > 0$ and $\delta > 0$,*

$$\mathbb{E}\left(\sup_{u \leq t \leq T} |\xi_s^{x,i}(t) - \xi_u^{y,i}(t)|^2\right) \leq C(|x-y|^2 + |u-s| + o(|u-s|)) \tag{3.70}$$

if $0 \leq s, u \leq T$, $|x| \vee |y| \leq \delta$ and $i \in \mathbb{S}$, where C is a positive constant dependent on T, δ, K and $\bar{K}$.

Proof. Without any loss of generality we may assume that $s \leq u$. Clearly, for $u \leq t \leq T$,

$$\xi_s^{x,i}(t) - \xi_u^{y,i}(t) = \xi_s^{x,i}(u) - y$$
$$+ \int_u^t [f(\xi_s^{x,i}(v), v, r_s^i(v)) - f(\xi_u^{y,i}(v), v, r_u^i(v))]dv$$
$$+ \int_u^t [g(\xi_s^{x,i}(v), v, r_s^i(v)) - g(\xi_u^{y,i}(v), v, r_u^i(v))]dB(v). \tag{3.71}$$

By (3.68), we compute

$$\mathbb{E}\int_u^t |f(\xi_s^{x,i}(v),v,r_s^i(v)) - f(\xi_u^{y,i}(v),v,r_u^i(v))|^2 dv$$

$$\le 2\mathbb{E}\int_u^t |f(\xi_s^{x,i}(v),v,r_s^i(v)) - f(\xi_u^{y,i}(v),v,r_s^i(v))|^2 dv$$

$$+ 2\mathbb{E}\int_u^t |f(\xi_u^{y,i}(v),v,r_s^i(v)) - f(\xi_u^{y,i}(v),v,r_u^i(v))|^2 dv$$

$$\le 2\bar{K}\int_u^t \mathbb{E}|\xi_s^{x,i}(v) - \xi_u^{y,i}(v)|^2 dv$$

$$+ 2\mathbb{E}\int_u^t |f(\xi_u^{y,i}(v),v,r_s^i(v)) - f(\xi_u^{y,i}(v),v,r_u^i(v))|^2 dv. \tag{3.72}$$

Noting that $\xi_u^{y,i}(v)$ and $I_{\{r_s^i(v)\ne r_u^i(v)\}}$ are conditionally independent with respect to the σ-algebra generated by $r_s^i(v)$, we compute, using (3.68), that

$$\mathbb{E}\int_u^t |f(\xi_u^{y,i}(v),v,r_s^i(v)) - f(\xi_u^{y,i}(v),v,r_u^i(v))|^2 dv$$

$$\le 2\mathbb{E}\int_u^t \left[|f(\xi_u^{y,i}(v),v,r_s^i(v))|^2 + |f(\xi_u^{y,i}(v),v,r_u^i(v))|^2\right] I_{\{r_s^i(v)\ne r_u^i(v)\}} dv$$

$$\le 4K\mathbb{E}\int_u^t [1 + |\xi_u^{y,i}(v)|^2] I_{\{r_s^i(v)\ne r_u^i(v)\}} dv$$

$$\le 4K\int_u^t \mathbb{E}\left[\mathbb{E}[(1+|\xi_u^{y,i}(v)|^2)I_{\{r_s^i(v)\ne r_u^i(v)\}}|r_s^i(v)]\right] dv$$

$$\le 4K\int_u^t \mathbb{E}\left[\mathbb{E}[1+|\xi_u^{y,i}(v)|^2|r_s^i(v)]\mathbb{E}[I_{\{r_s^i(v)\ne r_u^i(v)\}}|r_s^i(v)]\right] dv. \tag{3.73}$$

But, by the Markov property,

$$\mathbb{E}[I_{\{r_s^i(v)\ne r_u^i(v)\}}|r_s^i(v)] = \sum_{j\in\mathbb{S}} I_{\{r_s^i(v)=j\}}\mathbb{P}\{r_u^i(v)\ne j|r_s^i(v)=j\}$$

$$= \sum_{j\in\mathbb{S}} I_{\{r_s^i(v)=j\}}\sum_{k\ne j}(\gamma_{jk}(u-s) + o(u-s))$$

$$\le \max_{i\in\mathbb{S}}(-\gamma_{ii})(u-s) + o(u-s). \tag{3.74}$$

Therefore, applying Theorem 3.13,

$$\mathbb{E}\int_u^t |f(\xi_u^{y,i}(v),v,r_s^i(v)) - f(\xi_u^{y,i}(v),v,r_u^i(v))|^2 dv$$

$$\le C|u-s| + o(|u-s|), \tag{3.75}$$

where C is a positive constant dependent on K, T etc, but independent of u, s and, in particular, in the remainder of the proof it may change line by line. Substitute (3.75) to (3.72) yields

$$\mathbb{E} \int_u^t |f(\xi_u^{y,i}(v), v, r_s^i(v)) - f(\xi_u^{y,i}(v), v, r_u^i(v))|^2 dv$$
$$\leq 2\bar{K} \int_u^t \mathbb{E}|\xi_s^{x,i}(v) - \xi_u^{y,i}(v)|^2 dv + C|u-s| + o(|u-s|). \qquad (3.76)$$

Similarly, we can show that

$$\mathbb{E} \int_u^t |g(\xi_u^{y,i}(v), v, r_s^i(v)) - g(\xi_u^{y,i}(v), v, r_u^i(v))|^2 dv$$
$$\leq 2\bar{K} \int_u^t \mathbb{E}|\xi_s^{x,i}(v) - \xi_u^{y,i}(v)|^2 dv + C|u-s| + o(|u-s|). \qquad (3.77)$$

Note from Theorem 3.23 (condition (3.69) is used here) that

$$\mathbb{E}|\xi_s^{x,i}(u) - y|^2 \leq 2\mathbb{E}|\xi_s^{x,i}(u) - x|^2 + 2|x-y|^2$$
$$\leq C|u-s|^2 + 2|x-y|^2. \qquad (3.78)$$

Using (3.76)–(3.78), by the Hölder inequality and the Burkholder–Davis–Gundy inequality, we can show (the details are left to the reader as an exercise) from (3.71) that for any $\bar{u} \in [u, T]$,

$$\mathbb{E}\left(\sup_{u \leq t \leq \bar{u}} |\xi_s^{x,i}(t) - \xi_u^{y,i}(t)|^2\right)$$
$$\leq C|u-s|^2 + o(|u-s|) + 6|x-y|^2$$
$$+ 6\bar{K}(T+4) \int_u^{\bar{u}} \mathbb{E}\left(\sup_{u \leq t \leq v} |\xi_s^{x,i}(t) - \xi_u^{y,i}(t)|^2\right) dv. \qquad (3.79)$$

The required assertion finally follows from the Gronwall equality. $\square$

We can now show the strong Markov property of the solution.

Proof of Theorem 3.28. We need only to verify the Feller property, this is, to show the mapping

$$(x, s) \to \sum_{j=1}^N \int_{\mathbb{R}^n} \varphi(y) P(s, (x, i); s+\lambda, dy \times \{j\}) = \mathbb{E}\varphi(\xi_s^{x,i}(s+\lambda))$$

is continuous, for any bounded continuous function $\varphi : \mathbb{R}^n \to \mathbb{R}$ and any fixed $i \in \mathbb{S}$, $\lambda > 0$. Note that

$$\mathbb{E}\varphi(\xi_s^{x,i}(s+\lambda)) - \mathbb{E}\varphi(\xi_u^{y,i}(u+\lambda))$$
$$= \mathbb{E}\varphi(\xi_s^{x,i}(s+\lambda)) - \mathbb{E}\varphi(\xi_s^{x,i}(u+\lambda))$$
$$+ \mathbb{E}\varphi(\xi_s^{x,i}(u+\lambda)) - \mathbb{E}\varphi(\xi_u^{y,i}(u+\lambda)).$$

But, by Lemma 3.3 and the bounded convergence theorem,

$$\mathbb{E}\varphi(\xi_s^{x,i}(u+\lambda)) - \mathbb{E}\varphi(\xi_u^{y,i}(u+\lambda)) \to 0 \text{ as } (y,u) \to (x,s)$$

and, by Theorem 3.23,

$$\mathbb{E}\varphi(\xi_s^{x,i}(s+\lambda)) - \mathbb{E}\varphi(\xi_s^{x,i}(u+\lambda)) \to 0 \text{ as } u \to s.$$

In consequence,

$$\mathbb{E}\varphi(\xi_s^{x,i}(s+\lambda)) - \mathbb{E}\varphi(\xi_u^{y,i}(u+\lambda)) \to 0 \text{ as } (y,u) \to (x,s).$$

In other words, $\mathbb{E}\varphi(\xi_s^{x,i}(s+\lambda))$ as a function of (x,s) is continuous, and that is the Feller property. $\square$

The following theorem takes the global Lipschitz condition off.

Theorem 3.29 *The conclusion of Theorem 3.28 still holds if the global Lipschitz condition (3.68) is replaced by the local Lipschitz condition (3.31).*

Proof. We use the same notations as above. For each integer $k \geq 1$, define $\xi_{k,s}^{x,i}(t) = \xi_s^{x,i}(t \wedge \tau_k)$ for $t \geq s$, where $\tau_k = \inf\{t \geq s : |x_s^{x,i}(t)| \geq k\}$. Recalling the proof of Theorem 3.15, we observe from Theorem 3.28 that $(\xi_{k,s}^{x,i}(t), r_s^i(t))$ is a strong Markov process. Hence, fix any bounded continuous function $\varphi : \mathbb{R}^n \to \mathbb{R}$ and any $i \in \mathbb{S}$, $\lambda > 0$, the mapping

$$(x,s) \to \mathbb{E}\varphi(\xi_{k,s}^{x,i}(s+\lambda))$$

is continuous, namely

$$\lim_{(y,u)\to(x,s)} \mathbb{E}\varphi(\xi_{k,u}^{y,i}(s+\lambda)) = \mathbb{E}\varphi(\xi_{k,s}^{x,i}(s+\lambda)). \quad (3.80)$$

But, by the bounded convergence theorem, we have

$$\lim_{k\to\infty} \mathbb{E}\varphi(\xi_{k,s}^{x,i}(s+\lambda)) = \lim_{k\to\infty} \mathbb{E}\varphi(\xi_s^{x,i}((s+\lambda)\wedge\tau_k)) = \lim_{k\to\infty} \mathbb{E}\varphi(\xi_s^{x,i}((s+\lambda))).$$

We can hence take $k \to \infty$ in (3.80) to obtain

$$\lim_{(y,u)\to(x,s)} \mathbb{E}\varphi(\xi_u^{y,i}(s+\lambda)) = \mathbb{E}\varphi(\xi_s^{x,i}(s+\lambda)),$$

which is the Feller property of the process $(\xi(t), r(t))$ and hence its strong Markov property follows. □

3.8 Exercises

3.1 Prove Theorem 3.17 as hinted on page 93.

3.2 Consider equation (3.52) with initial condition $x(t_0) = x_0 \in L^2_{\mathcal{F}_{t_0}}(\Omega; \mathbb{R}^n)$. Assume that there exists a constant $\alpha > 0$ such that for all $(x, t, i) \in \mathbb{R}^n \times [t_0, T] \times \mathbb{S}$,

$$x^T f(x,t,i) + \frac{1}{2}|g(x,t,i)|^2 \le \alpha(1 + |x|^2).$$

Show that the solution of equation (3.52) obeys

$$\limsup_{t \to \infty} \frac{1}{t} \log |x(t)| \le \alpha \quad a.s.$$

3.3 Assume that the coefficients of equation (3.52) obey

$$|f(x,t,i)|^2 \vee |g(x,t,i)|^2 \le K(e^{\alpha t} + |x|^2) \quad \forall (x,t,i) \in \mathbb{R}^n \times \mathbb{R}_+ \times \mathbb{S}$$

for some $K, \alpha > 0$. Let $x(t)$ be a solution to equation (3.52) such that

$$\limsup_{t \to \infty} \frac{1}{t} \log(\mathbb{E}|x(t)|^2) \le \alpha.$$

Show that

$$\limsup_{t \to \infty} \frac{1}{t} \log |x(t)| \le \frac{\alpha}{2} \quad a.s.$$

3.4 Using (3.76)–(3.78), by the Hölder inequality and the Burkholder–Davis–Gundy inequality, show (3.79) from (3.71).

Chapter 4

Approximate Solutions

4.1 Introduction

In the previous chapter we established a number of criteria on the existence and uniqueness of the solution to an SDE with Markovian switching and discussed the properties of the solution. However, most SDEs with Markovian switching do not have explicit solutions. It is therefore important and necessary to know how to obtain the solutions in practice. By solutions in practice we mean the approximate solutions which can be computed numerically. In this chapter we shall discuss a number of methods to obtain the approximate solutions. We will emphasize the Euler–Maruyama method, one of the most powerful numerical schemes, but will discuss the other methods too.

It is useful to point out that this chapter presents several new techniques to obtain approximate solutions in the case when the coefficients of equations satisfy the local Lipschitz condition rather than the global Lipschitz condition.

4.2 Euler–Maruyama's Approximations

We will use the same notations as in the previous chapter. But to make the theory more understandable we will mainly discuss approximate solutions to autonomous SDEs with Markovian switching. Moreover, we will, without loss of any generality, set the initial time $= 0$ while letting the initial data x_0 and r_0 be non-random. That is, let us consider the n-dimensional SDE with Markovian switching

$$dx(t) = f(x(t), r(t))dt + g(x(t), r(t))dB(t), \quad 0 \le t \le T, \quad (4.1)$$

with initial data $x(0) = x_0 \in \mathbb{R}^n$ and $r(0) = r_0 \in \mathbb{S}$, where $f : \mathbb{R}^n \times \mathbb{S} \to \mathbb{R}^n$ and $g : \mathbb{R}^n \times \mathbb{S} \to \mathbb{R}^{n \times m}$. As a standing hypothesis we assume that both f and g are sufficiently smooth so that equation (4.1) has a unique solution (see the previous chapter).

The aim of this section is to develop the Euler–Maruyama method to obtain the approximate solutions to equation (4.1). To define the Euler–Maruyama approximate solution, let us recall the property of the embedded discrete-time Markov chain: Given a step size $\Delta > 0$, let $r_k^\Delta = r(k\Delta)$ for $\Delta > 0$ and $k \geq 0$. Then $\{r_k^\Delta,\ k = 0, 1, 2, \cdots\}$ is a discrete-time Markov chain with the one-step transition probability matrix

$$P(\Delta) = (P_{ij}(\Delta))_{N \times N} = e^{\Delta \Gamma}. \tag{4.2}$$

The discrete-time Markov chain $\{r_k^\Delta,\ k = 0, 1, 2, \cdots\}$ can be simulated as follows: compute the one-step transition probability matrix

$$P(\Delta) = (P_{ij}(\Delta))_{N \times N} = e^{\Delta \Gamma}.$$

Let $r_0^\Delta = r_0$ and generate a random number ξ_1 which is uniformly distributed in $[0, 1]$. If $\xi_1 = 1$ then let $r_1^\Delta = r_1 = N$ or otherwise find the unique integer $r_1 \in \mathbb{S}$ for

$$\sum_{j=1}^{r_1-1} P_{r_0,j}(\Delta) \leq \xi_1 < \sum_{j=1}^{r_1} P_{r_0,j}(\Delta)$$

and let $r_1^\Delta = r_1$, where we set $\sum_{i=1}^{0} P_{r_0,j}(\Delta) = 0$ as usual. Generate independently a new random number ξ_2 which is again uniformly distributed in $[0, 1]$. If $\xi_2 = 1$ then let $r_2^\Delta = r_2 = N$ or otherwise find the unique integer $r_2 \in \mathbb{S}$ for

$$\sum_{j=1}^{r_2-1} P_{r_1,j}(\Delta) \leq \xi_2 < \sum_{j=1}^{r_2} P_{r_1,j}(\Delta)$$

and let $r_2^\Delta = r_2$. Repeating this procedure a trajectory of $\{r_k^\Delta,\ k = 0, 1, 2, \cdots\}$ can be generated. This procedure can be carried out independently to obtain more trajectories.

After explaining how to simulate the discrete-time Markov chain $\{r_k^\Delta\}$, we can now define the Euler–Maruyama (EM) approximate solution to equation (4.1). Given a step size $\Delta > 0$, let $t_k = k\Delta$ for $k \geq 0$. Compute the discrete approximations $X_k \approx x(t_k)$ by setting $X_0 = x_0$, $r_0^\Delta = r_0$ and

forming
$$X_{k+1} = X_k + f(X_k, r_k^\Delta)\Delta + g(X_k, r_k^\Delta)\Delta B_k, \quad (4.3)$$

where $\Delta B_k = B(t_{k+1}) - B(t_k)$. Let
$$\bar{X}(t) = X_k, \quad \bar{r}(t) = r_k^\Delta \quad \text{for } t \in [t_k, t_{k+1}) \quad (4.4)$$

and define the continuous EM approximate solution
$$X(t) = X_0 + \int_0^t f(\bar{X}(s), \bar{r}(s))ds + \int_0^t g(\bar{X}(s), \bar{r}(s))dB(s). \quad (4.5)$$

Note that $X(t_k) = \bar{X}(t_k) = X_k$, that is $X(t)$ and $\bar{X}(t)$ coincide with the discrete solution at the gridpoints.

Let us now present a useful lemma.

Lemma 4.1 *Assume that f and g satisfy the linear growth condition: There is a constant $K > 0$ such that*
$$|f(x,i)| \vee |g(x,i)| \leq K(1+|x|) \quad \forall (x,i) \in \mathbb{R}^n \times \mathbb{S}. \quad (4.6)$$

Then for any $p \geq 2$ there is a constant H, which is dependent on only p, T, K, x_0 but independent of Δ, such that the exact solution and the EM approximate solution to equation (4.1) have the property that
$$\mathbb{E}\left[\sup_{0 \leq t \leq T} |x(t)|^p\right] \vee \mathbb{E}\left[\sup_{0 \leq t \leq T} |X(t)|^p\right] \leq H. \quad (4.7)$$

Proof. It follows from (4.5) that
$$|X(t)|^p$$
$$\leq 3^{p-1}\left[|x_0|^p + |\int_0^t f(\bar{X}(s), \bar{r}(s))ds|^p + |\int_0^t g(\bar{X}(s), \bar{r}(s))dB(s)|^p\right]$$
$$\leq 3^{p-1}\left[|x_0|^p + T^{p-1}\int_0^t |f(\bar{X}(s), \bar{r}(s))|^p ds + |\int_0^t g(\bar{X}(s), \bar{r}(s))dB(s)|^p\right].$$

This implies that for any $0 \leq t_1 \leq T$,
$$\mathbb{E}\left[\sup_{0 \leq t \leq t_1} |X(t)|^p\right] \leq 3^{p-1}\left[|x_0|^p + T^{p-1}\int_0^{t_1} \mathbb{E}|f(\bar{X}(s), \bar{r}(s))|^p ds\right.$$
$$\left. + \mathbb{E}\left[\sup_{0 \leq t \leq t_1} |\int_0^t g(\bar{X}(s), \bar{r}(s))dB(s)|^p\right]\right]. \quad (4.8)$$

By the Burkholder–Davis–Gundy inequality (i.e. Theorem 2.13) and the Hölder inequality we compute that

$$\mathbb{E}\Big[\sup_{0\le t\le t_1}|\int_0^t g(\bar X(s),\bar r(s))dB(s)|^p\Big] \le C_p \mathbb{E}\Big[\int_0^{t_1}|g(\bar X(s),\bar r(s))|^2 ds\Big]^{p/2}$$

$$\le C_p T^{p/2-1}\mathbb{E}\int_0^{t_1}|g(\bar X(s),\bar r(s))|^p ds,$$

where C_p is a constant. Substituting this into (4.6) and then using the linear growth condition (4.8) we obtain

$$\mathbb{E}\Big[\sup_{0\le t\le t_1}|X(t)|^p\Big]$$
$$\le 3^{p-1}\Big[|x_0|^p + 2^{p-1}h^p(T^{p-1}+T^{p/2-1})\mathbb{E}\int_0^{t_1}(1+|\bar X(s)|^p)ds$$
$$\le H_1 + H_1\int_0^{t_1}\mathbb{E}\Big[\sup_{0\le t\le s}|X(r)|^p\Big]ds, \qquad (4.9)$$

where $H_1 = H_1(p,T,h,x_0)$ is a constant independent of Δ. Applying the well-known Gronwall inequality to (4.9) yields

$$\mathbb{E}\Big[\sup_{0\le t\le T}|X(t)|^p\Big] \le H_1 e^{T H_1} := H.$$

Similarly, we can show that

$$\mathbb{E}\Big[\sup_{0\le t\le T}|x(t)|^p\Big] \le H.$$

(This follows from Theorem 3.24 too but the bound H may differ.) So the required assertion follows. □

4.2.1 Global Lipschitz Case

We can now begin to discuss the strong convergence of the EM approximate solution to the exact solution under the following global Lipschitz condition: There is a constant $\bar K > 0$ such that

$$|f(x,i)-f(y,i)| \vee |g(x,i)-g(y,i)| \le \bar K|x-y| \qquad (4.10)$$

for all $x,y \in \mathbb{R}^n$ and $i \in \mathbb{S}$.

Note from this global Lipschitz condition that

$$|f(x,i)| \le |f(x,i) - f(0,i)| + |f(0,i)| \le \bar{K}|x| + |f(0,i)|$$

and

$$|g(x,i)| \le |g(x,i) - g(0,i)| + |g(0,i)| \le \bar{K}|x| + |g(0,i)|.$$

Hence

$$|f(x,i)| \vee |g(x,i)| \le K(1+|x|) \qquad (4.11)$$

with $K = \bar{K} \vee \max\{|f(0,i)| \vee |g(0,i)| : i \in \mathbb{S}\}$. In other words, the global Lipschitz condition (4.10) implies the linear growth condition (4.6). Lemma 4.1 then shows that under condition (4.10) any pth moments, especially the 2nd moments, of the exact solution and the EM approximate solution to equation (4.1) are finite.

Theorem 4.1 *Under the global Lipschitz condition (4.10),*

$$\mathbb{E}\left[\sup_{0 \le t \le T} |X(t) - x(t)|^2\right] \le C\Delta + o(\Delta), \qquad (4.12)$$

where C is a positive constant independent of Δ.

Proof. By the Hölder inequality and the Doob martingale inequality, it is not difficult to show that for $0 \le t \le T$,

$$\mathbb{E}\left(\sup_{0 \le s \le t} |X(s) - x(s)|^2\right) \le 2T\mathbb{E}\int_0^t |f(\bar{X}(s), \bar{r}(s)) - f(x(s), r(s))|^2 ds$$

$$+ 8\mathbb{E}\int_0^t |g(\bar{X}(s), \bar{r}(s)) - g(x(s), r(s))|^2 ds. \qquad (4.13)$$

Note from (4.10) that

$$\mathbb{E}\int_0^t |f(\bar{X}(s), \bar{r}(s)) - f(x(s), r(s))|^2 ds$$

$$\le 2\mathbb{E}\int_0^t |f(\bar{X}(s), r(s)) - f(x(s), r(s))|^2 ds$$

$$+ 2\mathbb{E}\int_0^t |f(\bar{X}(s), \bar{r}(s)) - f(\bar{X}(s), r(s))|^2 ds$$

$$\leq 2\bar{K}^2 \int_0^t \mathbb{E}|\bar{X}(s) - x(s)|^2 ds$$
$$+ 2\mathbb{E}\int_0^T |f(\bar{X}(s), \bar{r}(s)) - f(\bar{X}(s), r(s))|^2 ds. \tag{4.14}$$

Let $j = [T/\Delta]$, the integer part of T/Δ. Then

$$\mathbb{E}\int_0^T |f(\bar{X}(s), \bar{r}(s)) - f(\bar{X}(s), r(s))|^2 ds$$
$$= \sum_{k=0}^{j} \mathbb{E}\int_{t_k}^{t_{k+1}} |f(\bar{X}(t_k), r(t_k)) - f(\bar{X}(t_k), r(s))|^2 ds \tag{4.15}$$

with t_{j+1} being now set to be T. Moreover, in the remainder of the proof C is a positive constant independent of Δ which may change line by line. With these notations we derive, using (4.10), that

$$\mathbb{E}\int_{t_k}^{t_{k+1}} |f(\bar{X}(t_k), r(t_k)) - f(\bar{X}(t_k), r(s))|^2 ds$$
$$\leq 2\mathbb{E}\int_{t_k}^{t_{k+1}} \left[|f(\bar{X}(t_k), r(t_k))|^2 + |f(\bar{X}(t_k), r(s))|^2\right] I_{\{r(s) \neq r(t_k)\}} ds$$
$$\leq C\mathbb{E}\int_{t_k}^{t_{k+1}} \left[1 + |\bar{X}(t_k)|^2\right] I_{\{r(s) \neq r(t_k)\}} ds$$
$$\leq C\int_{t_k}^{t_{k+1}} \mathbb{E}\Big[\mathbb{E}\big[(1 + |\bar{X}(t_k)|^2) I_{\{r(s) \neq r(t_k)\}} | r(t_k)\big]\Big] ds$$
$$= C\int_{t_k}^{t_{k+1}} \mathbb{E}\Big[\mathbb{E}\big[(1 + |\bar{X}(t_k)|^2) | r(t_k)\big]\mathbb{E}\big[I_{\{r(s) \neq r(t_k)\}} | r(t_k)\big]\Big] ds,$$

where in the last step we use the fact that $\bar{X}(t_k)$ and $I_{\{r(s) \neq r(t_k)\}}$ are conditionally independent with respect to the σ-algebra generated by $r(t_k)$. But, by the Markov property,

$$\mathbb{E}\big[I_{\{r(s) \neq r(t_k)\}} | r(t_k)\big] = \sum_{i \in \mathbb{S}} I_{\{r(t_k) = i\}} \mathbb{P}(r(s) \neq i | r(t_k) = i)$$
$$= \sum_{i \in \mathbb{S}} I_{\{r(t_k) = i\}} \sum_{j \neq i} (\gamma_{ij}(s - t_k) + o(s - t_k))$$
$$\leq \left(\max_{1 \leq i \leq N} (-\gamma_{ii})\Delta + o(\Delta)\right) \sum_{i \in \mathbb{S}} I_{\{r(t_k) = i\}}$$
$$\leq C\Delta + o(\Delta). \tag{4.16}$$

So, by Lemma 4.1,

$$\mathbb{E} \int_{t_k}^{t_{k+1}} |f(\bar{X}(t_k), r(t_k)) - f(\bar{X}(t_k), r(s))|^2 ds$$

$$\leq (C\Delta + o(\Delta)) \int_{t_k}^{t_{k+1}} [1 + \mathbb{E}|\bar{X}(t_k)|^2] ds \leq \Delta(C\Delta + o(\Delta)).$$

Substituting this into (4.15) gives

$$\mathbb{E} \int_0^T |f(\bar{X}(s), \bar{r}(s)) - f(\bar{X}(s), r(s))|^2 ds \leq C\Delta + o(\Delta). \quad (4.17)$$

Combining (4.14) and (4.17) we obtain that

$$\mathbb{E} \int_0^t |f(\bar{X}(s), \bar{r}(s)) - f(x(s), r(s))|^2 ds$$

$$\leq 2\bar{K}^2 \int_0^t \mathbb{E}|\bar{X}(s) - x(s)|^2 ds + C\Delta + o(\Delta). \quad (4.18)$$

Similarly, we can show that

$$\mathbb{E} \int_0^t |g(\bar{X}(s), \bar{r}(s)) - g(x(s), r(s))|^2 ds$$

$$\leq 2\bar{K}^2 \int_0^t \mathbb{E}|\bar{X}(s) - x(s)|^2 ds + C\Delta + o(\Delta). \quad (4.19)$$

Substituting (4.18) and (4.19) into (4.13) yields that

$$\mathbb{E}\left(\sup_{0 \leq s \leq t} |X(s) - x(s)|^2 \right) \leq C \int_0^t \mathbb{E}|\bar{X}(s) - x(s)|^2 ds + C\Delta + o(\Delta). \quad (4.20)$$

Note

$$\mathbb{E}|\bar{X}(s) - x(s)|^2 \leq 2\mathbb{E}|X(s) - x(s)|^2 + 2\mathbb{E}|X(s) - \bar{X}(s)|^2. \quad (4.21)$$

But, for $s \in [0, t]$, let $k_s = [s/\Delta]$, the integer part of s/Δ. It then follows from (4.5) and (4.11) as well as Lemma 4.1 that

$$\mathbb{E}|\bar{X}(s) - X(s)|^2 \leq C\mathbb{E}\Big[(1 + |X_{k_s}|^2)(\Delta^2 + |B(s) - B(t_{k_s})|^2)\Big]$$

$$\leq C\Delta. \quad (4.22)$$

Putting (4.21) and (4.22) into (4.20) we see that

$$\mathbb{E}\left(\sup_{0\le s\le t} |X(s) - x(s)|^2\right) \le C \int_0^t \mathbb{E}|X(s) - x(s)|^2 ds + C(\Delta + o(\Delta))$$

$$\le C \int_0^t \mathbb{E}\left(\sup_{0\le r\le s} |X(r) - x(r)|^2\right) ds + C\Delta + o(\Delta)$$

and the required result (4.12) follows from the Gronwall inequality. □

4.2.2 Local Lipschitz Case

We have just shown the strong convergence of the EM method on equation (4.1) under the global Lipschitz condition. But in many situations, the coefficients f and g are only locally Lipschitz continuous. It is therefore useful to establish the strong convergence of the EM method under the local Lipschitz condition. By the local Lipschitz condition we mean that for each $R = 1, 2, \cdots$, there is a constant $K_R > 0$ such that

$$|f(x,i) - f(y,i)| \vee |g(x,i) - g(y,i)| \le K_R |x - y| \qquad (4.23)$$

for all $i \in \mathbb{S}$ and those $x, y \in \mathbb{R}^n$ with $|x| \vee |y| \le R$.

Theorem 4.2 *Under the local Lipschitz condition (4.23) and the linear growth condition (4.6), the EM approximate solution converges to the exact solution of equation (4.1) in the sense that*

$$\lim_{\Delta \to 0} \mathbb{E}\left[\sup_{0\le t\le T} |X(t) - x(t)|^2\right] = 0. \qquad (4.24)$$

Proof. Fix a $p > 2$. By Lemma 4.1, there is a positive constant H independent of Δ such that

$$\mathbb{E}\left[\sup_{0\le t\le T} |X(t)|^p\right] \vee \mathbb{E}\left[\sup_{0\le t\le T} |x(t)|^p\right] \le H. \qquad (4.25)$$

For sufficiently large integer R, define the stopping times

$$\tau_R = T \wedge \inf\{t \in [0,T] : |X(t)| \ge R\},$$
$$\rho_R = T \wedge \inf\{t \in [0,T] : |x(t)| \ge R\},$$
$$\theta_R = \tau_R \wedge \rho_R.$$

Let

$$e(t) = X(t) - x(t).$$

Recall the Young inequality: for $r^{-1} + q^{-1} = 1$ and $\forall a, b, \delta$

$$ab \leq \frac{\delta}{r}a^r + \frac{1}{q\delta^{q/r}}b^q.$$

Thus, for any $\delta > 0$,

$$\mathbb{E}\left[\sup_{0\leq t\leq T} |e(t)|^2\right]$$
$$=\mathbb{E}\left[\sup_{0\leq t\leq T} |e(t)|^2 I_{\{\tau_R > T, \rho_R > T\}}\right] + \mathbb{E}\left[\sup_{0\leq t\leq T} |e(t)|^2 I_{\{\tau_R \leq T \text{ or } \rho_R \leq T\}}\right]$$
$$\leq \mathbb{E}\left[\sup_{0\leq t\leq T} |e(t \wedge \theta_R)|^2 I_{\{\theta_R > T\}}\right] + \frac{2\delta}{p}\mathbb{E}\left[\sup_{0\leq t\leq T} |e(t)|^p\right]$$
$$+ \frac{1-2/p}{\delta^{2/(p-2)}}\mathbb{P}(\tau_R \leq T \text{ or } \rho_R \leq T). \tag{4.26}$$

Now, by (4.25),

$$\mathbb{P}(\tau_R \leq T) = \mathbb{E}\left[I_{\{\tau_R \leq T\}} \frac{|X(\tau_R)|^p}{R^p}\right] \leq \frac{1}{R^p}\mathbb{E}\left[\sup_{0\leq t\leq T}|X(t)|^p\right] \leq \frac{H}{R^p}.$$

A similar result can be derived for ρ_R, so that

$$\mathbb{P}(\tau_R \leq T \text{ or } \rho_R \leq T) \leq \frac{2H}{R^p}.$$

Note also from (4.25) that

$$\mathbb{E}\left[\sup_{0\leq t\leq T}|e(t)|^p\right] \leq 2^{p-1}\left(\mathbb{E}\left[\sup_{0\leq t\leq T}|X(t)|^p\right] + \mathbb{E}\left[\sup_{0\leq t\leq T}|x(t)|^p\right]\right) \leq 2^p H.$$

Using these bounds gives

$$\mathbb{E}\left[\sup_{0\leq t\leq T}|e(t)|^2\right] \leq \mathbb{E}\left[\sup_{0\leq t\leq T}|X(t \wedge \theta_R) - x(t \wedge \theta_R)|^2\right]$$
$$+ \frac{2^{p+1}\delta H}{p} + \frac{2(p-2)H}{p\delta^{2/(p-2)}R^p}. \tag{4.27}$$

In a similar way as Theorem 4.1 was proved, we can show that

$$\mathbb{E}\left[\sup_{0\leq t\leq T}|X(t\wedge\theta_R) - x(t\wedge\theta_R)|^2\right] \leq C_R \Delta + o(\Delta), \tag{4.28}$$

where C_R is a constant independent of Δ. Substituting this into (4.27) gives

$$\mathbb{E}\left[\sup_{0\le t\le T}|e(t)|^2\right] \le C_R\Delta + o(\Delta) + \frac{2^{p+1}\delta H}{p} + \frac{2(p-2)H}{p\delta^{2/(p-2)}R^p}. \qquad (4.29)$$

Now, given any $\varepsilon > 0$, we can choose δ so that

$$\frac{2^{p+1}\delta H}{p} < \frac{\epsilon}{3},$$

then choose R sufficiently large for

$$\frac{2(p-2)H}{p\delta^{2/(p-2)}R^p} < \frac{\epsilon}{3},$$

and finally choose Δ sufficiently small for

$$C_R\Delta + o(\Delta) < \frac{\epsilon}{3},$$

so that, in (4.29),

$$\mathbb{E}\left[\sup_{0\le t\le T}|e(t)|^2\right] < \epsilon$$

as required. $\square$

We remark that although Theorem 4.2 does not reveal the order of convergence as Theorem 4.1 did, its proof is optimal in the sense that in the globally Lipschitz case ($L_R \le L$ for all R) we have $C_R = C$ in (4.28) and hence may take $\delta = \Delta$ and $R = \Delta^{-1/(p-2)}$ in (4.29) to recover the result

$$\mathbb{E}\left[\sup_{0\le t\le T}|e(t)|^2\right] \le C\Delta + o(\Delta)$$

obtained in Theorem 4.1.

We observe that the proof of Theorem 4.2 uses only the local Lipschitz condition (4.23) and the bounded pth moment property (4.25), namely

$$\mathbb{E}\left[\sup_{0\le t\le T}|X(t)|^p\right] \vee \mathbb{E}\left[\sup_{0\le t\le T}|x(t)|^p\right] \le H \qquad (4.30)$$

for some $p > 2$ and a positive constant H independent of Δ. So the following general statement holds.

Theorem 4.3 *Under the local Lipschitz condition (4.23) and the bounded pth moment condition (4.30) (p > 2), the EM approximate solution converges to the exact solution of equation (4.1) in the sense that*

$$\lim_{\Delta \to 0} \mathbb{E}\left[\sup_{0 \le t \le T} |X(t) - x(t)|^2\right] = 0.$$

Lemma 4.1 shows that the linear growth condition (4.6) implies the bounded pth moment property (4.30).

In Theorem 4.3 we require $p > 2$ in order to have (4.29) so that we can control the error

$$\mathbb{E}\left[\sup_{0 \le t \le T} |X(t) - x(t)|^2\right] < \varepsilon$$

for any prescribed ε by choosing the step size Δ sufficiently small as shown in the proof of Theorem 4.2. However, in the case when $p = 2$ we still have the following strong convergence theorem.

Theorem 4.4 *Under the local Lipschitz condition (4.23) and the bounded 2nd moment condition (4.30) (p = 2), the EM approximate solution converges to the exact solution of equation (4.1) in the sense that*

$$\lim_{\Delta \to 0} \mathbb{E}\left[\sup_{0 \le t \le T} |X(t) - x(t)|^2\right] = 0.$$

Of course we will not be able to control the error so precisely as in the case when $p > 2$. We leave the proof of Theorem 4.4 as an exercise.

4.2.3 *More on Local Lipschitz Case*

Let us now concentrate on the equation with only the local Lipschitz condition (4.23) but without the linear growth condition (4.6) nor the bounded pth moment property (4.30). The following theorem describes the convergence in probability, instead of L^2, of the EM solutions to the exact solution under some additional conditions in terms of Lyapunov-type functions.

Theorem 4.5 *Let the local Lipschitz condition (4.23) hold. Assume that there exists a C^2 function $V : \mathbb{R}^n \times \mathbb{S} \to \mathbb{R}_+$ satisfying the following three conditions:*

(i) $\lim_{|x| \to \infty} V(x, i) = \infty$ *for any* $i \in \mathbb{S}$;

(ii) *for some $h > 0$,*

$$LV(x,i) \leq h(1 + V(x,i)) \quad \forall (x,i) \in \mathbb{R}^n \times S,$$

where

$$LV(x,i) = V_x(x,i)f(x,i) + \frac{1}{2}\text{trace}\left[g^T(x,i)V_{xx}g(x,i)\right] + \sum_{j=1}^{N}\gamma_{ij}V(x,j);$$

(iii) *for each $R > 0$ there exists a positive constant K_R such that for all $i \in \mathbb{S}$ and those $x, y \in \mathbb{R}^n$ with $|x| \vee |y| \leq R$,*

$$|V(x,i) - V(y,i)| \vee |V_x(x,i) - V_x(y,i)| \vee |V_{xx}(x,i) - V_{xx}(y,i)| \leq K_R|x-y|.$$

Then

$$\lim_{\Delta \to 0}\left(\sup_{0 \leq t \leq T}|X(t) - x(t)|^2\right) = 0 \quad \text{in probability.} \tag{4.31}$$

Proof. We divide the whole proof into three steps.

Step 1. For sufficiently large R, define the stopping time

$$\theta = T \wedge \inf\{t \in [0,T]: |x(t)| \geq R\}.$$

Applying the generalized Itô formula to $V(x(t), r(t))$ yields

$$\mathbb{E}[V(x(t \wedge \theta), r(t \wedge \theta))] = V(x_0, r_0) + \mathbb{E}\int_0^{t \wedge \theta} LV(x(s), r(s))ds.$$

By condition (ii)

$$\mathbb{E}[V(x(t \wedge \theta), r(t \wedge \theta))] \leq V(x_0, r_0) + h\mathbb{E}\int_0^{t \wedge \theta}(1 + V(x(s), r(s)))ds$$

$$\leq V(x_0, r_0) + hT + h\int_0^t \mathbb{E}V(\bar{X}(s \wedge \theta), r(s \wedge \theta))ds.$$

Using the Gronwall inequality, we obtain

$$\mathbb{E}[V(x(T \wedge \theta), r(T \wedge \theta))] \leq (V(x_0, r_0) + hT)e^{hT}. \tag{4.32}$$

Let

$$v_R = \inf\{V(x,i): |x| \geq R, i \in \mathbb{S}\}.$$

By condition (i), $v_R \to \infty$ as $R \to \infty$. Noting that $|x(\theta)| = R$ whenever $\theta < T$, we derive from (4.32) that

$$(V(x_0, r_0) + hT)e^{hT} \geq \mathbb{E}[V(x(\theta), r(\theta))I_{\{\theta < T\}}]$$
$$\geq v_R \mathbb{P}(\theta < T).$$

That is

$$\mathbb{P}(\theta < T) \leq \frac{e^{hT}}{v_R}(V(x_0, r_0) + hT). \tag{4.33}$$

Step 2. For sufficiently large R define the stopping time

$$\rho = T \wedge \inf\{t \in [0, T] : |X(t)| \geq R\}.$$

Using (4.5) and applying the generalized Itô's formula to $V(X(t), r(t))$ yields

$$\mathbb{E}[V(X(\rho \wedge t), r(\rho \wedge t))] = V(x_0, r_0)$$
$$+ \mathbb{E} \int_0^{\rho \wedge t} \bigg[V_x(X(s), r(s)) f(\bar{X}(s), \bar{r}(s))$$
$$+ \frac{1}{2}\text{trace}[g^T(\bar{X}(s), \bar{r}(s)) V_{xx}(X(s), r(s)) g(\bar{X}(s), \bar{r}(s))]$$
$$+ \sum_{j=1}^{N} \gamma_{r(s)j} V(X(s), j) \bigg] ds.$$

Rearranging the terms on the right-hand side by plus-and-minus technique and using condition (ii) we obtain that

$$\mathbb{E}[V(X(\rho \wedge t), r(\rho \wedge t))]$$
$$\leq V(x_0, r_0) + h\mathbb{E} \int_0^{\rho \wedge t} (1 + V(X(s), r(s))) ds$$
$$+ h\mathbb{E} \int_0^{\rho \wedge t} [V(\bar{X}(s), \bar{r}(s)) - V(X(s), \bar{r}(s))] ds$$
$$+ h\mathbb{E} \int_0^{\rho \wedge t} [V(X(s), \bar{r}(s)) - V(X(s), r(s))] ds$$

$$+ \mathbb{E} \int_0^{\rho \wedge t} |V_x(X(s), r(s)) - V_x(\bar{X}(s), r(s))| |f(\bar{X}(s), \bar{r}(s))| ds$$

$$+ \mathbb{E} \int_0^{\rho \wedge t} |V_x(\bar{X}(s), r(s)) - V_x(\bar{X}(s), \bar{r}(s))| |f(\bar{X}(s), \bar{r}(s))| ds$$

$$+ \frac{1}{2} \mathbb{E} \int_0^{\rho \wedge t} |V_{xx}(X(s), r(s)) - V_{xx}(\bar{X}(s), r(s))| |g(\bar{X}(s), \bar{r}(s))|^2 ds$$

$$+ \frac{1}{2} \mathbb{E} \int_0^{\rho \wedge t} |V_{xx}(\bar{X}(s), r(s)) - V_{xx}(\bar{X}(s), \bar{r}(s))| |g(\bar{X}(s), \bar{r}(s))|^2 ds$$

$$+ \mathbb{E} \int_0^{\rho \wedge t} \sum_{j=1}^{N} |\gamma_{r(s)j}| |V(X(s), j) - V(\bar{X}(s), j)| ds$$

$$+ \mathbb{E} \int_0^{\rho \wedge t} \sum_{j=1}^{N} |\gamma_{r(s)j} - \sum_{j=1}^{N} \gamma_{\bar{r}(s)j}| |V(\bar{X}(s), j)| ds. \tag{4.34}$$

By condition (iii) we have

$$\mathbb{E} \int_0^{\rho \wedge t} [V(\bar{X}(s), \bar{r}(s)) - V(X(s), \bar{r}(s))] ds$$

$$\leq \mathbb{E} \int_0^{\rho \wedge t} K_R |\bar{X}(s) - X(s)| ds$$

$$\leq K_R \int_0^T \mathbb{E} |\bar{X}(\rho \wedge s) - X(\rho \wedge s)| ds$$

$$\leq K_R \int_0^T \left(\mathbb{E} |\bar{X}(\rho \wedge s) - X(\rho \wedge s)|^2 \right)^{\frac{1}{2}} ds.$$

Using (4.16) we can show, in the same way as (4.17) was proved, that

$$\mathbb{E} \int_0^{\rho \wedge t} [V(X(s), \bar{r}(s)) - V(X(s), r(s))] ds \leq 2 V_R T \beta (\Delta + o(\Delta)),$$

where $\beta = \max_{1 \leq i \leq N}(-\gamma_{ii})$, $V_R = \max\{V(x, i) : |x| \leq R, \ i \in \mathbb{S}\}$. We can similarly estimate the other terms on the right-hand side of (4.34) to get that

$$\mathbb{E}[V(X(\rho \wedge t), r(\rho \wedge t))] \leq V(x_0, r_0) + hT + h \mathbb{E} \int_0^{\rho \wedge t} V(X(s), r(s)) ds$$

$$+ C_1(R) \int_0^T \left(\mathbb{E} |\bar{X}(\rho \wedge s) - X(\rho \wedge s)|^2 \right)^{\frac{1}{2}} ds$$

$$+ C_1(R)(\Delta + o(\Delta)), \tag{4.35}$$

where $C_1(R)$ and the following $C_2(R)$, $C_3(R)$, $\cdots$ are all constants dependent of R but independent of Δ. But, in the same way as (4.22) was proved, we can show that

$$\mathbb{E}|\bar{X}(\rho \wedge s) - X(\rho \wedge s)|^2 \leq C_2(R)\Delta \quad \forall s \in [0, T].$$

Substituting this into (4.35) yields that

$$\mathbb{E}[V(X(\rho \wedge t), r(\rho \wedge t))] \leq V(x_0, r_0) + hT + C_3(R)(\Delta^{\frac{1}{2}} + o(\Delta))$$
$$+ h \int_0^t \mathbb{E}V(X(\rho \wedge s), r(\rho \wedge s))ds.$$

By the Gronwall inequality,

$$\mathbb{E}[V(X(\rho \wedge T), r(\rho \wedge T))]$$
$$\leq e^{hT}\left[V(x_0, r_0) + hT + C_3(R)(\Delta^{\frac{1}{2}} + o(\Delta))\right]. \quad (4.36)$$

In the same way as (4.33) was obtained, we can then show that

$$P(\rho < T) \leq \frac{e^{hT}}{v_R}\left[V(x_0, r_0) + hT + C_3(R)(\Delta^{\frac{1}{2}} + o(\Delta))\right]. \quad (4.37)$$

Step 3. Let $\tau = \rho \wedge \theta$. In the same way as Theorem 4.1 was proved we can show that

$$\mathbb{E}\left[\sup_{0 \leq t \leq \tau \wedge T} |X(t) - x(t)|^2\right] \leq C_4(R)(\Delta + o(\Delta)). \quad (4.38)$$

Now, let $\varepsilon, \delta \in (0, 1)$ be arbitrarily small. Set

$$\bar{\Omega} = \{\omega : \sup_{0 \leq t \leq T} |X(t) - x(t)|^2 \geq \delta\}$$

Using (4.38), we compute

$$\delta \mathbb{P}(\bar{\Omega} \cap \{\tau \geq T\}) = \delta \mathbb{E}\left[I_{\{\tau \geq T\}} I_{\bar{\Omega}}\right]$$
$$\leq \mathbb{E}\left[I_{\{\tau \geq T\}} \sup_{0 \leq t \leq \tau \wedge T} |X(t) - x(t)|^2\right]$$
$$\leq \mathbb{E}\left[\sup_{0 \leq t \leq \tau \wedge T} |X(t) - x(t)|^2\right]$$
$$\leq C_4(R)(\Delta + o(\Delta)).$$

This, together with (4.33) and (4.37), yields that

$$\mathbb{P}(\bar{\Omega}) \leq \mathbb{P}(\bar{\Omega} \cap \{\tau \geq T\}) + \mathbb{P}(\tau < T)$$
$$\leq \mathbb{P}(\bar{\Omega} \cap \{\tau \geq T\}) + \mathbb{P}(\theta < T) + \mathbb{P}(\rho < T)$$
$$\leq \frac{C_4(R)}{\delta}(\Delta + o(\Delta)) + \frac{2e^{hT}}{v_R}[V(x_0, r_0) + hT]$$
$$+ \frac{e^{hT}}{v_R} C_3(R)(\Delta^{\frac{1}{2}} + o(\Delta)).$$

Recalling that $v_R \to \infty$ as $R \to \infty$, we can choose R sufficiently large for

$$\frac{2e^{hT}}{v_R}[V(x_0, r_0) + hT] < \frac{\varepsilon}{2},$$

and then choose Δ sufficiently small for

$$\frac{C_4(R)}{\delta}(\Delta + o(\Delta)) + \frac{e^{hT}}{v_R} C_3(R)(\Delta^{\frac{1}{2}} + o(\Delta)) < \frac{\varepsilon}{2}$$

to obtain

$$\mathbb{P}(\bar{\Omega}) = \mathbb{P}\left(\sup_{0 \leq t \leq T} |X(t) - x(t)|^2 \geq \delta\right) < \varepsilon.$$

This proves the assertion (4.31). □

4.3 Caratheodory's Approximations

In this section, we will introduce another useful approximation procedure, known as the Caratheodory approximation method. We shall consider the same n-dimensional SDE (4.1) as in the previous section.

Let us begin with the definition of the Caratheodory approximate solutions. For a given step size $\Delta > 0$, define $X_\Delta(t) = x_0$ for $-\Delta \leq t \leq 0$ and

$$X_\Delta(t) = x_0 + \int_0^t f(X_\Delta(s-\Delta), r(s))ds + \int_0^t g(X_\Delta(s-\Delta), r(s))dB(s) \tag{4.39}$$

for $0 < t \leq T$. Note that for $0 \leq t \leq \Delta$, $X_\Delta(t)$ can be computed by

$$X_\Delta(t) = x_0 + \int_0^t f(x_0, r(s))ds + \int_0^t g(x_0, r(s))dB(s);$$

then for $\Delta < t \leq 2\Delta$,

$$X_\Delta(t) = X_\Delta(\Delta) + \int_\Delta^t f(X_\Delta(s-\Delta), r(s))ds$$
$$+ \int_\Delta^t g(X_\Delta(s-\Delta), r(s))dB(s)$$

and so on. In other words, $X_\Delta(t)$ can be computed step-by-step on the intervals $[0, \Delta]$, $(\Delta, 2\Delta], \cdots$. We need to prepare two lemmas in order to establish the main results.

Lemma 4.2 *Assume that f and g obey the linear growth condition (4.6). Then for any $p \geq 2$ there is a constant H, which is dependent on p, T, K, x_0 but independent of Δ, such that*

$$\mathbb{E}\left[\sup_{0 \leq t \leq T} |X_\Delta(t)|^p\right] \leq H. \qquad (4.40)$$

The proof is similar to that of Lemma 4.1 so is left to the reader as an exercise.

Lemma 4.3 *Let the linear growth condition (4.6) hold. Let $p \geq 2$ and $\Delta > 0$. Then for all $0 \leq s < t \leq T$ with $t - s \leq 1$,*

$$\mathbb{E}|X_\Delta(t) - X_\Delta(s)|^p \leq H_1(t-s)^{p/2}, \qquad (4.41)$$

where H_1 is a constant which is dependent on p, T, K, x_0 but independent of Δ.

Proof. Note that

$$X_\Delta(t) - X_\Delta(s) = \int_s^t f(X_\Delta(u-\Delta), r(u))du + \int_s^t g(X_\Delta(u-\Delta), r(u))dB(u).$$

Using the Hölder inequality and Theorem 2.11, we compute

$$\mathbb{E}|X_\Delta(t) - X_\Delta(s)|^p$$
$$\leq 2^{p-1}\mathbb{E}\left|\int_s^t f(X_\Delta(u-\Delta), r(u))du\right|^p$$
$$+ 2^{p-1}\mathbb{E}\left|\int_s^t g(X_\Delta(u-\Delta), r(u))dB(u)\right|^2$$
$$\leq 2^{p-1}(t-s)^{p-1}\mathbb{E}\int_s^t |f(X_\Delta(u-\Delta), r(u))|^p du$$
$$+ 2^{p-1}[p(p-1)/2]^{p/2}(t-s)^{(p-2)/2}\mathbb{E}\int_s^t |g(X_\Delta(u-\Delta), r(u))|^p du.$$

Making use of the linear growth condition (4.6) and then applying Lemma 4.2, we further compute

$$\mathbb{E}|X_\Delta(t) - X_\Delta(s)|^p \le C(t-s)^{(p-2)/2}\int_s^t (1 + \mathbb{E}|X_\Delta(u-\Delta)|^p)du$$

$$\le C(1+H)(t-s)^{p/2},$$

where C is a constant dependent on p and K. The assertion (4.41) hence follows by setting $H_1 = C(1+H)$. □

We can now state one of the main results in this section.

Theorem 4.6 *Let the global Lipschitz condition (4.10) hold and $p \ge 2$. Let $x(t)$ be the unique solution of equation (4.1). Then, for any $\Delta \in (0,1)$,*

$$\mathbb{E}\Big(\sup_{0 \le t \le T} |X_\Delta(t) - x(t)|^p\Big) \le C\Delta^{p/2}, \qquad (4.42)$$

where C is a positive constant independent of Δ.

Proof. It is easy to show from (4.1) and (4.39) that for any $t \in [0,T]$,

$$\mathbb{E}\Big(\sup_{0 \le s \le t} |X_\Delta(s) - x(s)|^p\Big)$$

$$\le (2T)^{p-1}\mathbb{E}\int_0^t |f(X_\Delta(s-\Delta), r(s)) - f(x(s), r(s))|^p ds$$

$$\le 2^{p-1}\mathbb{E}\Big(\sup_{0 \le t_1 \le t}\Big|\int_0^{t_1}[g(X_\Delta(s-\Delta), r(s)) - g(x(s), r(s))]dB(s)\Big|^p\Big).$$

But, by Theorem 2.13, we compute

$$\mathbb{E}\Big(\sup_{0 \le t_1 \le t}\Big|\int_0^{t_1}[g(X_\Delta(s-\Delta), r(s)) - g(x(s), r(s))]dB(s)\Big|^p\Big)$$

$$\le C_p \mathbb{E}\Big(\int_0^t |g(X_\Delta(s-\Delta), r(s)) - g(x(s), r(s))|^2 ds\Big)^{p/2}$$

$$\le C_p T^{(p-2)/2}\mathbb{E}\int_0^t |g(X_\Delta(s-\Delta), r(s)) - g(x(s), r(s))|^p ds,$$

where C_p is a constant dependent on only p as defined in Theorem 2.13. Combining these two inequalities and then using the global Lipschitz condition (4.10), we hence obtain that

$$\mathbb{E}\Big(\sup_{0 \le s \le t} |X_\Delta(s) - x(s)|^p\Big) \le c_1 \mathbb{E}\int_0^t |X_\Delta(s-\Delta) - x(s))|^p ds,$$

where $c_1 = \bar{K}^{p/2}\big[(2T)^{p-1} + 2^{p-1}C_p T^{(p-2)/2}\big]$. In view of Lemma 4.3, we observe that

$$\mathbb{E}|X_\Delta(s-\Delta) - X_\Delta(s)|^p \leq H_1 \Delta^{p/2} \quad \text{if } s \in [\Delta, T]$$

while

$$\mathbb{E}|X_\Delta(s-\Delta) - X_\Delta(s)|^p = \mathbb{E}|X_\Delta(0) - X_\Delta(s)|^p \leq H_1 \Delta^{p/2} \quad \text{if } s \in [0, \Delta].$$

We further then compute

$$\mathbb{E}\Big(\sup_{0 \leq s \leq t} |X_\Delta(s) - x(s)|^p\Big)$$

$$\leq c_1 2^{p-1} \int_0^t \big[\mathbb{E}|X_\Delta(s-\Delta) - X_\Delta(s)|^p + \mathbb{E}|X_\Delta(s) - x(s))|^p\big] ds$$

$$\leq c_1 2^{p-1} T H_1 \Delta^{p/2} + c_1 2^{p-1} \int_0^t \mathbb{E}\Big(\sup_{0 \leq u \leq s} |X_\Delta(u) - x(u)|^p\Big) ds.$$

Applying the Gronwall inequality yields

$$\mathbb{E}\Big(\sup_{0 \leq s \leq T} |X_\Delta(s) - x(s)|^p\Big) \leq c_1 2^{p-1} T H_1 \Delta^{p/2} e^{c_1 2^{p-1} T}.$$

The assertion (4.42) follows by setting $C = c_1 2^{p-1} T H_1 e^{c_1 2^{p-1} T}$. □

In practice, given any $\varepsilon > 0$, one can choose $\Delta \leq (\varepsilon/C)^{2/p}$ and then compute $X_\Delta(t)$ over the intervals $[0, \Delta], (\Delta, 2\Delta], \cdots$, step by step. Theorem 4.6 guarantees that this $X_\Delta(t)$ is closed enough to the accurate solution $x(t)$ in the sense

$$\mathbb{E}\Big(\sup_{0 \leq t \leq T} |X_\Delta(t) - x(t)|^p\Big) < \varepsilon.$$

Let us now replace the global Lipschitz condition (4.10) with a more general condition and show that the Caratheodory approximate solutions $X_\Delta(t)$ still converge to the true solution $x(t)$.

Theorem 4.7 *Let $p \geq 2$. Assume that there exists a continuous increasing concave function $\kappa : \mathbb{R}_+ \to \mathbb{R}_+$ such that*

$$\int_0^1 \frac{du}{\kappa(u)} = \infty, \qquad (4.43)$$

and for all $x, y \in \mathbb{R}^n$ and $i \in \mathbb{S}$,

$$|f(x,i) - f(y,i)|^p \vee |g(x,i) - g(y,i)|^p \leq \kappa(|x-y|^p). \qquad (4.44)$$

Then the SDE (4.1) has a unique solution $x(t)$. Moreover, the Caratheodory approximate solutions $X_\Delta(t)$ converge to $x(t)$ in the sense that

$$\lim_{\Delta \to 0} \mathbb{E}\left(\sup_{0 \le t \le T} |X_\Delta(t) - x(t)|^p \right) = 0. \tag{4.45}$$

Proof. We leave the proof of existence and uniqueness of the solution to the SDE (4.1) to the reader as an exercise. We shall also only show the assertion (4.45) for $p = 2$ but leave the more general case of $p > 2$ to the reader as another exercise.

First of all, let us show that the conditions imposed imply the linear growth condition (4.6). In fact, given that $\kappa(\cdot)$ is concave and increasing, there must exist a positive number a such that

$$\kappa(u) \le a(1 + u) \qquad \text{on } u \ge 0.$$

Besides, let $b = \max_{i \in \mathbb{S}}(|f(0,i)|^2 \vee |g(0,i)|^2) < \infty$. Then, for any $(x, i) \in \mathbb{R}^n \times \mathbb{S}$,

$$|f(x,i)|^2 \vee |g(x,i)|^2$$
$$\le 2(|f(0,t)|^2 \vee |g(0,t)|^2) + 2(|f(x,t) - f(0,t)|^2 \vee |g(x,t) - g(0,t)|^2)$$
$$\le 2b + 2\kappa(|x|^2) \le 2b + 2a(1 + |x|^2) \le 2(a+b)(1 + |x|^2),$$

which gives

$$|f(x,i)| \vee |g(x,i)| \le \sqrt{2(a+b)}(1 + |x|).$$

That is, the linear growth condition (4.6) is fulfilled with $K = \sqrt{2(a+b)}$. Thus, Lemmas 4.2 and 4.3 hold under the conditions imposed.

By the Hölder inequality and the Doob martingale inequality, we can show from (4.1) and (4.39) that for any $t \in [0, T]$,

$$\mathbb{E}\left(\sup_{0 \le s \le t} |X_\Delta(s) - x(s)|^2 \right)$$
$$\le 2T\mathbb{E} \int_0^t |f(X_\Delta(s-\Delta), r(s)) - f(x(s), r(s))|^2 ds$$
$$\le 8\mathbb{E} \int_0^t |g(X_\Delta(s-\Delta), r(s)) - g(x(s), r(s))|^2 ds.$$

By condition (4.44) we can then show that

$$\mathbb{E}\Big(\sup_{0\le s\le t} |X_\Delta(s) - x(s)|^2\Big)$$
$$\le 4(T+4)\int_0^t \Big[\mathbb{E}\kappa(|X_\Delta(s-\Delta) - X_\Delta(s)|^2) + \mathbb{E}\kappa(|X_\Delta(s) - x(s)|^2)\Big]ds.$$

Since κ is concave, by the Jensen inequality,

$$\mathbb{E}\kappa(|X_\Delta(s-\Delta) - X_\Delta(s)|^2) \le \kappa(\mathbb{E}|X_\Delta(s-\Delta) - X_\Delta(s)|^2)$$

and similarly for $\mathbb{E}\kappa(|X_\Delta(s) - x(s)|^2)$, whence

$$\mathbb{E}\Big(\sup_{0\le s\le t} |X_\Delta(s) - x(s)|^2\Big)$$
$$\le 4(T+4)\int_0^t \Big[\kappa(\mathbb{E}|X_\Delta(s-\Delta) - X_\Delta(s)|^2) + \kappa(\mathbb{E}|X_\Delta(s) - x(s)|^2)\Big]ds$$
$$\le 4T(T+4)\kappa(H_1\Delta) + 4(T+4)\int_0^t \kappa\Big(\mathbb{E}\Big[\sup_{0\le u\le s} |X_\Delta(u) - x(u)|^2\Big]\Big)ds.$$

Applying the Bihari inequality (i.e. Theorem 2.3), we obtain that

$$\mathbb{E}\Big(\sup_{0\le s\le T} |X_\Delta(s) - x(s)|^2\Big)$$
$$\le G^{-1}\Big(G\big(4T(T+4)\kappa(H_1\Delta)\big) + 4T(T+4)\Big), \qquad (4.46)$$

where

$$G(u) = \int_1^u \frac{ds}{\kappa(s)} \quad \text{on } u > 0$$

and $G^{-1}(\cdot)$ is the inverse function of G. By condition (4.43) we observe that

$$\lim_{u\downarrow 0} G(u) = -\infty \quad \text{and} \quad \lim_{u\to -\infty} G^{-1}(u) = 0.$$

Hence, for any $\varepsilon > 0$, there is a $\delta < 0$ such that

$$G^{-1}(u) < \varepsilon \quad \forall u \le \delta.$$

On the other hand, condition (4.43) implies that $\kappa(0) = 0$ whence there is a $\Delta^* > 0$ such that

$$G\big(4T(T+4)\kappa(H_1\Delta)\big) + 4T(T+4) \le \delta \quad \forall \Delta \le \Delta^*.$$

Therefore, we obtain from (4.46) that

$$\mathbb{E}\left(\sup_{0\leq s\leq T} |X_\Delta(s) - x(s)|^2\right) < \varepsilon \quad \forall \Delta \leq \Delta^*.$$

This means

$$\lim_{\Delta \to 0} \mathbb{E}\left(\sup_{0\leq s\leq T} |X_\Delta(s) - x(s)|^2\right) = 0$$

as required. □

Let us further replace the global Lipschitz condition (4.10) with the local Lipschitz condition (4.23). As the Euler–Maruyama method, the following theorem shows that the Caratheodory approximate solutions will converge to the true solution under the local Lipschitz condition and the bounded pth moment condition ($p > 2$).

Theorem 4.8 *Let the local Lipschitz condition (4.23) hold. Assume also that for some $p > 2$ there is a constant H, such that*

$$\mathbb{E}\left[\sup_{0\leq t\leq T} |x(t)|^p\right] \vee \mathbb{E}\left[\sup_{0\leq t\leq T} |X_\Delta(t)|^p\right] \leq H \quad (4.47)$$

for all sufficiently small Δ, where H is independent of Δ. Then the Caratheodory approximate solution converges to the exact solution of equation (4.1) in the sense that

$$\lim_{\Delta \to 0} \mathbb{E}\left[\sup_{0\leq t\leq T} |X_\Delta(t) - x(t)|^2\right] = 0.$$

Proof. Choose a sufficiently small step size Δ and a sufficiently large integer R. Define the stopping times

$$\tau_R = T \wedge \inf\{t \in [0, T] : |X_\Delta(t)| \geq R\},$$
$$\rho_R = T \wedge \inf\{t \in [0, T] : |x(t)| \geq R\},$$
$$\theta_R = \tau_R \wedge \rho_R.$$

Let

$$e(t) = X_\Delta(t) - x(t).$$

Using the Young inequality (2.3) we can show that for any $\delta > 0$,

$$\mathbb{E}\left[\sup_{0\le t\le T}|e(t)|^2\right]$$
$$\le \mathbb{E}\left[\sup_{0\le t\le T}|e(t\wedge\theta_R)|^2 I_{\{\theta_R>T\}}\right] + \frac{2\delta}{p}\mathbb{E}\left[\sup_{0\le t\le T}|e(t)|^p\right]$$
$$+ \frac{1-2/p}{\delta^{2/(p-2)}}\mathbb{P}(\tau_R\le T \text{ or } \rho_R\le T). \tag{4.48}$$

By (4.47),

$$\mathbb{P}(\tau_R\le T) = \mathbb{E}\left[I_{\{\tau_R\le T\}}\frac{|X_\Delta(\tau_R)|^p}{R^p}\right] \le \frac{1}{R^p}\mathbb{E}\left[\sup_{0\le t\le T}|X_\Delta(t)|^p\right] \le \frac{H}{R^p}.$$

A similar result can be derived for ρ_R, so that

$$\mathbb{P}(\tau_R\le T \text{ or } \rho_R\le T) \le \frac{2H}{R^p}.$$

Note also from (4.47) that

$$\mathbb{E}\left[\sup_{0\le t\le T}|e(t)|^p\right] \le 2^{p-1}\left(\mathbb{E}\left[\sup_{0\le t\le T}|X_\Delta(t)|^p\right] + \mathbb{E}\left[\sup_{0\le t\le T}|x(t)|^p\right]\right) \le 2^p H.$$

Substituting these bounds into (4.48) gives

$$\mathbb{E}\left[\sup_{0\le t\le T}|e(t)|^2\right] \le \mathbb{E}\left[\sup_{0\le t\le T}|X_\Delta(t\wedge\theta_R) - x(t\wedge\theta_R)|^2\right]$$
$$+ \frac{2^{p+1}\delta H}{p} + \frac{2(p-2)H}{p\delta^{2/(p-2)}R^p}. \tag{4.49}$$

In the similar way as Theorem 4.6 was proved, we can show that

$$\mathbb{E}\left[\sup_{0\le t\le T}|X_\Delta(t\wedge\theta_R) - x(t\wedge\theta_R)|^2\right] \le C_R\Delta, \tag{4.50}$$

where C_R is a constant independent of Δ. Substituting this into (4.49) gives

$$\mathbb{E}\left[\sup_{0\le t\le T}|e(t)|^2\right] \le C_R\Delta + \frac{2^{p+1}\delta H}{p} + \frac{2(p-2)H}{p\delta^{2/(p-2)}R^p}. \tag{4.51}$$

Now, given any $\varepsilon > 0$, we can choose δ so that

$$\frac{2^{p+1}\delta H}{p} < \frac{\epsilon}{3},$$

then choose R sufficiently large for

$$\frac{2(p-2)H}{p\delta^{2/(p-2)}R^p} < \frac{\epsilon}{3},$$

and finally choose Δ sufficiently small for

$$C_R \Delta + o(\Delta) < \frac{\epsilon}{3},$$

so that, in (4.51),

$$\mathbb{E}\left[\sup_{0 \le t \le T} |e(t)|^2\right] < \epsilon$$

as required. $\square$

The proof above shows that the reason we require $p > 2$ is to have (4.51) so that we can control the error

$$\mathbb{E}\left[\sup_{0 \le t \le T} |X_\Delta(t) - x(t)|^2\right] < \varepsilon$$

for any prescribed ε by choosing the step size Δ sufficiently small.

We also observe from Lemmas 4.1 and 4.2 that the linear growth condition (4.6) guarantees the pth moment bounded condition (4.47). So the following useful corollary is immediate.

Corollary 4.9 *Let the coefficients f and g of equation (4.1) obey the local Lipschitz condition (4.23) and the linear growth condition (4.6). Then the Caratheodory approximate solution converges to the exact solution of equation (4.1) in the sense that*

$$\lim_{\Delta \to 0} \mathbb{E}\left[\sup_{0 \le t \le T} |X_\Delta(t) - x(t)|^2\right] = 0.$$

4.4 Split-Step Backward Euler Scheme

In this section we impose further assumptions on the SDE (4.1). Theorem 4.2 requires bounds on the pth moment of the exact and numerical solution, namely condition (4.30) that is sometimes difficult to verify in practice for the EM method discussed in Section 4.2 and indeed may fail to hold. In this section we introduce an implicit version of the EM method, known as the backward Euler (BE) method, for which moment bounds, and hence a convergence result, can be obtained.

We make the following assumptions on the SDE.

Assumption 4.10 For each $i \in \mathbb{S}$, the functions $f(\cdot, i)$ and $g(\cdot, i)$ in the SDE (4.1) are C^1 and there exist constants $\mu, c > 0$, independent of i, such that

$$\langle x - y, f(x,i) - f(y,i)\rangle \le \mu |x-y|^2 \quad \forall x, y \in \mathbb{R}^n, \tag{4.52}$$

$$|g(x,i) - g(y,i)|^2 \le c|x-y|^2 \quad \forall x, y \in \mathbb{R}^n. \tag{4.53}$$

Note that we work with the case $\mu > 0$. In the deterministic setting there is a lot of attention paid to the contractive case $\mu < 0$. This case is of less interest here because, for most diffusion processes, contractivity is destroyed. Hence $\mu > 0$ is a natural assumption.

It follows from Assumption 4.10 that

$$\langle f(x,i), x\rangle = \langle f(x,i) - f(0,i), x\rangle + \langle f(0,i), x\rangle$$

$$\le \mu |x|^2 + |f(0,i)||x| \le \tfrac{1}{2}|f(0,i)|^2 + (\mu + \tfrac{1}{2})|x|^2$$

and

$$|g(x,i)|^2 \le 2|g(0,i)|^2 + 2|g(x) - g(0)|^2 \le 2|g(0,i)|^2 + 2c|x|^2.$$

This gives

$$\langle f(x,i), x\rangle \vee |g(x,i)|^2 \le K(1 + |x|^2) \quad \forall (x,i) \in \mathbb{R}^n \times \mathbb{S}, \tag{4.54}$$

where

$$K := \max\{\tfrac{1}{2}|f(0,i)|^2 \vee 2|g(0,i)|^2 : i \in \mathbb{S}\} \vee (\mu + \tfrac{1}{2}) \vee 2c. \tag{4.55}$$

The inequality (4.54) will prove very useful in what follows. We note from Theorem 3.17 that $f(\cdot, i)$, $g(\cdot, i) \in C^1$ for each $i \in \mathbb{S}$ and (4.54) ensure the existence of a unique solution to the SDE (4.1).

The inequality (4.52) in Assumption 4.10, which is known as a *one-sided Lipschitz condition*, has been exploited successfully in the deterministic numerical analysis literature [Burrage and Butcher (1979); Butcher (1975); Dekker and Verwer (1984); Stuart and Humphries (1996)] and in the case of SDEs has been used in [Higham et al. (2002); Hu (1996); Schurz (1997)]. The condition (4.52) is closely related to the *monotone* condition (3.34). Any f of the form $f(x,i) = -x^p + a_i x$, where the integer $p \ge 3$ is odd, satisfies (4.52) and further examples can be found in [Stuart and Humphries (1996)].

It is also very useful to observe from Theorem 3.24 that under Assumption 4.10 the solution of the SDE (4.1) has bounded pth moment for each $p > 2$. More precisely, for each $p > 2$, there is $C = C(p,T) > 0$ such that the solution of (4.1) obeys

$$\mathbb{E}\left[\sup_{0\le t\le T}|x(t)|^p\right] \le C(1+|x_0|^p). \qquad (4.56)$$

We now consider the split-step Backward Euler (SSBE) method. Given a step size $\Delta > 0$, the SSBE approximate solution is defined by taking $X_0 = x_0$ and, generally,

$$X_k^\star = X_k + f(X_k^\star, r_k^\Delta)\Delta, \qquad (4.57)$$

$$X_{k+1} = X_k^\star + g(X_k^\star, r_k^\Delta)\Delta B_k, \qquad (4.58)$$

where $\Delta B_k = B((k+1)\Delta) - B(k\Delta)$ and r_k^Δ is the same as defined in Section 4.2. We state our convergence theorem here and then give a sequence of results that lead to a proof.

Theorem 4.11 *Consider the split-step Backward Euler method (4.57)–(4.58) applied to the SDE (4.1) under Assumption 4.10. There exists a continuous-time extension $X(t)$ of the numerical solution (so that $X(k\Delta) = X_k$) for which*

$$\lim_{\Delta\to 0}\mathbb{E}\left[\sup_{0\le t\le T}|X(t)-x(t)|^2\right] = 0.$$

Note that (4.57) is an implicit equation that must be solved in order to obtain the intermediate approximation $X_k^\star$. Having obtained $X_k^\star$, adding the appropriate stochastic increment $g(X_k^\star)\Delta B_k$ produces the next approximation X_{k+1} in (4.58). The SSBE method reduces to the deterministic Backward Euler method [Dekker and Verwer (1984); Hairer and Wanner (1996)] when $f(x,i) = f(x)$ (i.e. independent of the Markov chain) and $g \equiv 0$ and x_0 is non-random. The method is studied in [Higham et al. (2002)] for SDEs without Markovian switching. The method is also effective for inheriting ergodicity; for related reasons it is effective here in enabling the derivation of moment bounds. Another stochastic extension of the deterministic Backward Euler method will be considered in the next section.

Our proof of Theorem 4.11 relies on showing that SSBE has two key properties under Assumption 4.10: (a) it may be regarded as EM applied to a modified SDE of a similar form, and (b) it produces solutions with all moments bounded. The first property is established in the next lemma and corollary.

Lemma 4.4 *Let Assumption 4.10 hold and suppose* $\Delta \in (0, \Delta_c), \Delta_c < 1/(2K)$, *where* K *is defined in (4.55). Fix any* $i \in \mathbb{S}$. *Given* $d \in \mathbb{R}^n$ *the implicit equation*

$$c = d + \Delta f(c, i) \tag{4.59}$$

has a unique solution c. *If we define the functions* $F_\Delta(\cdot, i)$, $f_\Delta(\cdot, i)$ *and* $g_\Delta(\cdot, i)$ *by*

$$F_\Delta(d, i) = c, \quad f_\Delta(d, i) = f(F_\Delta(d, i), i), \quad g_\Delta(d, i) = g(F_\Delta(d, i), i), \tag{4.60}$$

then $F_\Delta(\cdot, i), f_\Delta(\cdot, i), g_\Delta(\cdot, i) \in C^1$, $g_\Delta(\cdot, i) \to g(\cdot, i)$ *and* $f_\Delta(\cdot, i) \to f(\cdot, i)$ *as* $\Delta \to 0$ *in* C^1 *uniformly on compact sets and for any* $a, b \in \mathbb{R}^n$,

$$|f_\Delta(a, i)| \leq \frac{|f(a, i)|}{1 - \Delta \mu}, \tag{4.61}$$

$$|F_\Delta(a, i) - F_\Delta(b, i)|^2 \leq \frac{1}{1 - 2\Delta \mu}|a - b|^2, \tag{4.62}$$

$$\langle a - b, f_\Delta(a, i) - f_\Delta(b, i) \rangle \leq \frac{\mu}{1 - 2\mu \Delta}|a - b|^2. \tag{4.63}$$

Further, $g_\Delta(\cdot, i)$ *is globally Lipschitz and there exist* $K' > 0$ *such that*

$$\langle f_\Delta(a, i), a \rangle \vee |g_\Delta(a, i)|^2 \leq K'(1 + |a|^2) \quad \forall a \in \mathbb{R}^n. \tag{4.64}$$

Proof. It is quite standard to show the existence and uniqueness for (4.59) via a contraction mapping theorem, which also establishes the C^1 smoothness of $f_\Delta(\cdot, i)$ and $F_\Delta(\cdot, i)$ and the convergence property of $f_\Delta(\cdot, i)$, see [Dekker and Verwer (1984); Stuart and Humphries (1996)]. The smoothness and convergence properties of $g_\Delta(\cdot, i)$ follow from $g_\Delta(\cdot, i) = g(F_\Delta(\cdot, i), i)$. We leave these standard proofs to the reader as an exercise.

To obtain the bound (4.61) *etc*, we use the homotopy method. Fix any $i \in \mathbb{S}$ and write $f(\cdot, i) = f(\cdot)$. Let $h = h(\tau)$ obey

$$h = d + \Delta f(h) + (\tau - 1)\Delta f(d), \tag{4.65}$$

where τ is our homotopy parameter. For $\tau = 1$, h solves the equation (4.59). For $\tau = 0$ we have

$$h - d = \Delta \left(f(h) - f(d) \right)$$

and so, using Assumption 4.10,
$$|h-d|^2 = \Delta \langle h-d, f(h) - f(d) \rangle \le \Delta \mu |h-d|^2.$$
Note that $\beta > \mu$ so that $2\Delta\mu < 1$. It follows that $h = d$ is the unique solution to (4.65) when $\tau = 0$. Differentiating (4.65) with respect to τ gives
$$\dot{h} = \Delta \frac{\partial f}{\partial x}(h)\dot{h} + \Delta f(d).$$
So
$$|\dot{h}|^2 - \Delta \langle \dot{h}, \frac{\partial f}{\partial x}(h)\dot{h} \rangle = \Delta \langle \dot{h}, f(d) \rangle. \tag{4.66}$$
Setting $x - y = \epsilon u$ in (4.52) and letting $\epsilon \to 0$, we see that
$$\langle u, \frac{\partial f}{\partial x}(y)u \rangle \le \mu |u|^2, \quad \text{for any } u, y \in \mathbb{R}^n.$$
Hence, in (4.66),
$$|\dot{h}|^2 - \Delta \mu |\dot{h}|^2 \le \Delta |\dot{h}|\,|f(d)|.$$
So
$$|\dot{h}| \le \Delta \frac{|f(d)|}{1 - \Delta\mu}.$$
It follows that $h(\tau)$ exists uniquely for all $\tau > 0$ and
$$|h(1) - d| = \left| \int_0^1 \dot{h}(s)\,ds \right| \le \Delta \frac{|f(d)|}{1 - \Delta\mu},$$
which establishes (4.61).

To obtain (4.62) we note that if $c^{(1)} = d^{(1)} + \Delta f(c^{(1)})$ and $c^{(2)} = d^{(2)} + \Delta f(c^{(2)})$ then
$$|c^{(1)} - c^{(2)}|^2 - \Delta \langle f(c^{(1)}) - f(c^{(2)}), c^{(1)} - c^{(2)} \rangle = \langle d^{(1)} - d^{(2)}, c^{(2)} - c^{(2)} \rangle$$
and so, using Assumption 4.10,
$$(1 - \Delta\mu)|c^{(1)} - c^{(2)}|^2 \le \frac{1}{2}|d^{(1)} - d^{(2)}|^2 + \frac{1}{2}|c^{(1)} - c^{(2)}|^2$$
which gives (4.62).

Next, note from the implicit definition (4.60) that $f_\Delta(a)$ is equivalent to $f(a + \Delta f_\Delta(a))$. Using (4.52) we thus have
$$\langle f_\Delta(a) - f_\Delta(b), a + \Delta f_\Delta(a) - b - \Delta f_\Delta(b) \rangle \le \mu |a + \Delta f_\Delta(a) - b - \Delta f_\Delta(b)|^2.$$

Hence,

$$\langle f_\Delta(a) - f_\Delta(b), a - b\rangle + \Delta|f_\Delta(a) - f_\Delta(b)|^2$$
$$\leq \mu|a - b|^2 + 2\mu\langle a - b, f_\Delta(a) - f_\Delta(b)\rangle\Delta$$
$$+ \mu\Delta^2|f_\Delta(a) - f_\Delta(b)|^2$$

and (4.63) follows.

The global Lipschitz property of g_Δ follows from (4.62).

Finally, we use $f_\Delta(a) = f(a + \Delta f_\Delta(a))$ and (4.54) to give

$$\langle f_\Delta(a), a + \Delta f_\Delta(a)\rangle \leq K(1 + |a + \Delta f(a)|^2).$$

Hence

$$(1 - 2K\Delta)\langle f_\Delta(a), a\rangle \leq K + K|a|^2 + [K\Delta^2 - \Delta]|f_\Delta(a)|^2 \leq K(1 + |a|^2).$$

Since g_Δ is globally Lipschitz, the inequality (4.64) follows. The proof is therefore complete. □

Corollary 4.12 *Let Assumption 4.10 hold and suppose $\Delta \in (0, \Delta_c), \Delta_c < 1/(2K)$, where K is defined in (4.54). Then SSBE applied to (4.1) is equivalent to EM applied to the modified SDE*

$$dx_\Delta(t) = f_\Delta(x_\Delta(t), r(t))dt + g_\Delta(x_\Delta(t), r(t))dB(t) \qquad (4.67)$$

on $0 \leq t \leq T$ with initial value $x_\Delta(0) = x_0$, where f_Δ, g_Δ are defined in Lemma 4.4.

Proof. Lemma 4.4 allows us to express the SSBE method (4.57)–(4.58) in the form

$$X_{k+1} = X_k + \Delta f_\Delta(X_k, r_k^\Delta) + g_\Delta(X_k, r_k^\Delta)\Delta B_k, \qquad (4.68)$$

and the result is then immediate. □

Next, we show that the solution of the modified SDE (4.67) has bounded moments and converges strongly to $x(t)$.

Lemma 4.5 *Under Assumption 4.10, for each $p > 2$, there is $C = C(p, T) > 0$ such that*

$$\mathbb{E}\left[\sup_{0 \leq t \leq T} |x_\Delta(t)|^p\right] \leq C(1 + |x_0|^p), \qquad (4.69)$$

provided Δ is sufficiently small. In addition

$$\lim_{\Delta \to 0} \mathbb{E}\left[\sup_{0 \le t \le T} |x(t) - x_\Delta(t)|^2\right] = 0. \qquad (4.70)$$

Proof. It follows from Lemma 4.4 that for sufficiently small Δ the functions f_Δ and g_Δ satisfy (4.54) with K replaced by $2K$. By Theorem 3.24, we obtain (4.69).

Now, to prove (4.70) we note from Lemma 4.4 that given $R > 0$ there is a function $K_R : (0, \infty) \to (0, \infty)$ such that $K_R(\Delta) \to 0$ as $\Delta \to 0$ and

$$|f_\Delta(u, i) - f(u, i)|^2 \vee |g_\Delta(u, i) - g(u, i)|^2 \le K_R(\Delta) \qquad (4.71)$$

for $\forall (u, i) \in \mathbb{R}^n \times \mathbb{S}$ with $|u| \le R$, provided Δ is sufficiently small. Also, since $f(\cdot, i), g(\cdot, i) \in C^1$, there is a constant H_R such that

$$|f(u, i) - f(v, i)|^2 \vee |g(u, i) - g(v, i)|^2 \le H_R |u - v|^2 \qquad (4.72)$$

for all $i \in \mathbb{S}$ and those $u, v \in \mathbb{R}^n$ with $|u| \vee |v| \le R$. From (4.56) and (4.69) we have

$$\mathbb{E}\left[\sup_{0 \le t \le T} |x(t)|^p\right] \vee \mathbb{E}\left[\sup_{0 \le t \le T} |x_\Delta(t)|^p\right] \le H := C(1 + |y_0|^p). \qquad (4.73)$$

The remainder of the proof follows in a similar manner to that of Theorem 4.2. Define

$$\tau_R = \inf\{t \ge 0 : |x(t)| \ge R\}, \quad \rho_R = \inf\{t \ge 0 : |x_\Delta(t)| \ge R\}$$

and $\theta_R = \tau_R \wedge \rho_R$. For any $\delta > 0$, in the same way that (4.27) was obtained, we have

$$\mathbb{E}\left[\sup_{0 \le t \le T} |y(t) - y_\Delta(t)|^2\right] \le \mathbb{E}\left[\sup_{0 \le t \le T} |y(t \wedge \theta_R) - y_\Delta(t \wedge \theta_R)|^2\right]$$
$$+ \frac{2^{p+1}\delta H}{p} + \frac{(p-2)2K}{p\delta^{2/(p-2)} R^p}. \qquad (4.74)$$

To bound the first term on the right-hand side of (4.74), we observe that

$$x(t \wedge \theta_R) - x_\Delta(t \wedge \theta_R)$$
$$= \int_0^{t \wedge \theta_R} [f(x(s), r(s)) - f(x_\Delta(s), r(s))$$
$$\qquad + f(x_\Delta(s), r(s)) - f_\Delta(x_\Delta(s), r(s))] ds$$
$$+ \int_0^{t \wedge \theta_R} [g(x(s), r(s)) - g(x_\Delta(s), r(s))$$
$$\qquad + g(x_\Delta(s), r(s)) - g_\Delta(x_\Delta(s), r(s))] dB(s).$$

Using (4.71), (4.72), Cauchy–Schwartz and the Doob martingale inequality, we have that for any $\tau \leq T$,

$$\mathbb{E}\left[\sup_{0 \leq t \leq \tau} |x(t \wedge \theta_R) - x_\Delta(t \wedge \theta_R)|^2\right]$$
$$\leq 4H_R(T+4) \int_0^\tau \mathbb{E}\left[\sup_{0 \leq t \leq s} |x(t \wedge \theta_R) - x_\Delta(t \wedge \theta_R)|^2\right] ds$$
$$+ 4T(T+4)K_R(\Delta).$$

So the Gronwall inequality yields

$$\mathbb{E}\left[\sup_{0 \leq t \leq T} |x(t \wedge \theta_R) - x_\Delta(t \wedge \theta_R)|^2\right] \leq 4T(T+4)K_R(\Delta)e^{4H_R(T+4)T}.$$

Inserting this into (4.74) gives

$$\mathbb{E}\left[\sup_{0 \leq t \leq T} |x(t) - x_\Delta(t)|^2\right] \leq 4T(T+4)K_R(\Delta)e^{4H_R(T+4)T}$$
$$+ \frac{2^{p+1}\delta K}{p} + \frac{(p-2)2K}{p\delta^{2/(p-2)}R^p}. \qquad (4.75)$$

The final step of the proof follows that of Theorem 4.2. □

Now we show that the special structure of SSBE makes it possible for us to bound all moments of the numerical solution, under Assumption 4.10. We deal first with the discrete approximation and then with a continuous-time extension.

Lemma 4.6 *Suppose Assumption 4.10 holds and let $\Delta \leq \Delta_c < 1/(2K)$, where K is defined in (4.54). Then for each $p \geq 2$ there exists a $C =*

$C(p,T) > 0$ (independent of Δ) such that for the SSBE method (4.57)–(4.58)

$$\mathbb{E}\left(\sup_{k\Delta \in [0,T]} |X_k|^{2p}\right) \leq C.$$

Proof. In the following we assume that u and v are positive integers such that $u\Delta \leq v\Delta \leq T$. From (4.54) and (4.57) we have

$$|X_k^\star|^2 = \langle X_k, X_k^\star \rangle + \Delta \langle f(X_k^\star, r_k^\Delta), X_k^\star \rangle$$
$$\leq \tfrac{1}{2}|X_k|^2 + \tfrac{1}{2}|X_k^\star|^2 + \Delta K(1 + |X_k^\star|^2).$$

Thus

$$|X_k^\star|^2 \leq \frac{|X_k|^2 + 2K\Delta}{1 - 2K\Delta}. \tag{4.76}$$

From (4.58) and (4.76) we have

$$|X_{k+1}|^2 \leq |X_k|^2 + \frac{2K\Delta}{1 - 2K\Delta}|X_k|^2 + \frac{2K\Delta}{1 - 2K\Delta}$$
$$+ 2\langle X_k^\star, g(X_k^\star, r_k^\Delta)\Delta B_k \rangle + |g(X_k^\star, r_k^\Delta)\Delta B_k|^2.$$

Summing, and using the notation $H = (1 - 2K\Delta)^{-1}$, we obtain

$$|X_u|^2 \leq |X_0|^2 + 2K\Delta H \sum_{j=0}^{u-1} |X_j|^2 + 2K\Delta u H$$
$$+ 2\sum_{j=0}^{u-1} \langle X_j^\star, g(X_j^\star, r_j^\Delta)\Delta B_j \rangle + \sum_{j=0}^{u-1} |g(X_j^\star, r_j^\Delta)\Delta B_j|^2.$$

Raising both sides to the power p we have

$$\frac{1}{5^{p-1}}|X_u|^{2p}$$
$$\leq |X_0|^{2p} + (2K\Delta H)^p \left(\sum_{j=0}^{u-1} |X_j|^2\right)^p + (2KTH)^p$$
$$+ 2^p \left|\sum_{j=0}^{u-1} \langle X_j^\star, g(X_j^\star, r_j^\Delta)\Delta B_j \rangle\right|^p + \left(\sum_{j=0}^{u-1} |g(X_j^\star, r_j^\Delta)\Delta B_j|^2\right)^p$$

$$\leq |X_0|^{2p} + (2KH)^p T^{p-1} \Delta \sum_{j=0}^{u-1} |X_j|^{2p} + (2KTH)^p$$

$$+ 2^p \left| \sum_{j=0}^{u-1} \langle X_j^\star, g(X_j^\star, r_j^\Delta) \Delta B_j \rangle \right|^p + u^{p-1} \sum_{j=0}^{u-1} |g(X_j^\star, r_j^\Delta) \Delta B_j|^{2p}. \quad (4.77)$$

Now,

$$\mathbb{E}\left[\sup_{0 \leq u \leq v} \sum_{j=0}^{u-1} |X_j|^{2p} \right] = \sum_{j=0}^{v-1} \mathbb{E}|X_j|^{2p}. \quad (4.78)$$

Also, letting $C = C(p,T)$ be a constant that may change line by line,

$$\mathbb{E}\left[\sup_{0 \leq u \leq v} \sum_{j=0}^{u-1} |g(X_j^\star, r_j^\Delta) \Delta B_j|^{2p} \right]$$

$$= \mathbb{E} \sum_{j=0}^{v-1} |g(X_j^\star, r_j^\Delta) \Delta B_j|^{2p}$$

$$\leq \sum_{j=0}^{v-1} \mathbb{E}|g(X_j^\star, r_j^\Delta)|^{2p} \mathbb{E}|\Delta B_j|^{2p}$$

$$\leq C\Delta^p \sum_{j=0}^{v-1} \mathbb{E}[K + K|X_j^\star|^2]^p$$

$$\leq C\Delta^p \sum_{j=0}^{v-1} \mathbb{E}[1 + |X_j^\star|^{2p}]$$

$$\leq C\Delta^{p-1} + C\Delta^p \sum_{j=0}^{v-1} \mathbb{E}[|X_j|^2 + 2K\Delta]^p$$

$$\leq C\Delta^{p-1} + C\Delta^p \sum_{j=0}^{v-1} \mathbb{E}|X_j|^{2p}, \quad (4.79)$$

where we have used (4.54) and (4.76). Finally, using the Burkholder–Davis–

Gundy inequality (i.e. Theorem 2.13),

$$\mathbb{E}\left[\sup_{0\le u\le v}\left|\sum_{j=0}^{u-1}\langle X_j^\star, g(X_j^\star, r_j^\Delta)\rangle \Delta B_j\rangle\right|^p\right]$$

$$\le C\mathbb{E}\left[\sum_{j=0}^{v-1}|X_j^\star|^2|g(X_j^\star, r_j^\Delta)|^2\Delta\right]^{p/2}$$

$$\le C(\Delta)^{p/2}v^{p/2-1}\mathbb{E}\sum_{j=0}^{v-1}|X_j^\star|^p(K+K|X_j^\star|^2)^{p/2}$$

$$\le C\Delta\sum_{j=0}^{v-1}[1+\mathbb{E}|X_j^\star|^{2p}]$$

$$\le C\Delta\sum_{j=0}^{v-1}[1+\mathbb{E}(2K\Delta+|X_j|^2)^p]$$

$$\le C+C\Delta\sum_{j=0}^{v-1}\mathbb{E}|X_j|^{2p}. \tag{4.80}$$

Combining (4.77)–(4.80) we obtain

$$\mathbb{E}\left[\sup_{0\le u\le v}|X_u|^{2p}\right]\le C+C\Delta\sum_{j=0}^{v-1}\mathbb{E}|X_j|^{2p}$$

$$\le C+C\Delta\sum_{j=0}^{v-1}\mathbb{E}\left[\sup_{0\le u\le j}\mathbb{E}|X_u|^{2p}\right].$$

Using the discrete-type Gronwall inequality (i.e. Theorem 2.5) and noting that $v\Delta \le T$, we obtain

$$\mathbb{E}\left[\sup_{0\le u\le v}|X_u|^{2p}\right]\le Ce^{C\Delta v}\le Ce^{CT}$$

and the desired result follows. □

Corollary 4.13 *Suppose Assumption 4.10 holds and let $\Delta \in (0, \Delta_c), \Delta_c < 1/(2K)$, where K is defined in (4.55). Let $p \ge 2$. Then there exists a continuous-time extension $X(t)$ of the SSBE solution $\{X_k\}$*

and a constant $C = C(p,T) > 0$ *(independent of Δ) such that*

$$\mathbb{E}\left(\sup_{0 \leq t \leq T} |X(t)|^{2p}\right) \leq C.$$

Proof. We know that SSBE can be regarded as EM applied to the modified SDE (4.67). Hence, we may define $X(t)$ using (4.4) and (4.5) with f, g replaced by f_Δ, g_Δ. By definition we have, for $t_k = k\Delta$,

$$X(t_k + s) = X_k + s f_\Delta(X_k, r_k^\Delta) + g_\Delta(X_k, r_k^\Delta) \Delta B_k(s), \quad s \in [0, \Delta),$$

where

$$\Delta B_k(s) := B(t_k + s) - B(t_k).$$

But $X_k^\star = X_k + \Delta f_\Delta(X_k, r_k^\Delta)$ and so, for $a = s/\Delta$, we have

$$X(t_k + s) = a X_k^\star + (1-a) X_k + g_\Delta(X_k) \Delta B_k(s), \quad s \in [0, \Delta).$$

Since $\Delta \leq \Delta_c < 1/(2K)$, it follows from (4.76) that

$$|X(t_k + s)|^2 \leq C[1 + |X_k|^2 + |g_\Delta(X_k, r_k^\Delta) \Delta B_k(s)|^2].$$

Thus

$$\sup_{0 \leq t \leq T} |X(t)|^{2p}$$
$$\leq \sup_{0 \leq k\Delta \leq T} \sup_{0 \leq s \leq \Delta} |X(t_k + s)|^{2p}$$
$$\leq \sup_{0 \leq k\Delta \leq T} \sup_{0 \leq s \leq \Delta} C[1 + |X_k|^{2p} + |g_\Delta(X_k, r_k^\Delta) \Delta B_k(s)|^{2p}]$$
$$\leq C[1 + \sup_{0 \leq k\Delta \leq T} |X_k|^{2p} + \sup_{0 \leq s \leq \Delta} \sum_{j=0}^{u} |g_\Delta(X_j, r_j^\Delta) \Delta B_j(s)|^{2p}], \quad (4.81)$$

where $0 \leq u\Delta \leq T$. Now, using Doob's martingale inequality and (4.64)

$$\mathbb{E} \sup_{0 \leq s \leq \Delta} |g_\Delta(X_j, r_j^\Delta) \Delta B_j(s)|^{2p} \leq C \mathbb{E} |g_\Delta(X_j, r_j^\Delta) \Delta B_j(\Delta)|^{2p}$$
$$\leq C \mathbb{E} |g_\Delta(X_j, r_j^\Delta)|^{2p} \mathbb{E} |\Delta B_j(\Delta)|^{2p}$$
$$\leq C \Delta^p [1 + \mathbb{E} |X_j|^{2p}]$$
$$\leq C \Delta, \quad (4.82)$$

where C is a universal constant, independent of Δ. Since $u\Delta \leq T$, combining Lemma 4.6, (4.81) and (4.82) gives the desired result. $\square$

We can finally prove the main result of this section, Theorem 4.11.

Proof of Theorem 4.11. The proof now follows from an application of the triangle inequality: the SSBE method has solution close to the solution of an SDE with modified vector fields and the solution of this SDE in turn is close to that of the original SDE. More precisely, we may use Corollary 4.13 to define $X(t)$ and bound $\mathbb{E}(\sup_{0 \leq t \leq T} |X(t)|^p)$ while use Lemma 4.5 to bound $\mathbb{E}(\sup_{0 \leq t \leq T} |x_\Delta(t)|^p)$. We also know from Lemma 4.4 that f_Δ and g_Δ are uniformly locally Lipschitz for small Δ. Hence, we may follow the proof of Theorem 4.2 to give

$$\lim_{\Delta \to 0} \mathbb{E}\left[\sup_{0 \leq t \leq T} |X(t) - x_\Delta(t)|^2\right] = 0.$$

Combining this with (4.70) in Lemma 4.5 via the triangle inequality gives the result. $\square$

4.5 Backward Euler Scheme

The SSBE method (4.57)–(4.58) is a stochastic extension of the deterministic Backward Euler method. Another, perhaps more natural, extension of Backward Euler is given by $Z_0 = x_0$ and

$$Z_{k+1} = Z_k + \Delta f(Z_{k+1}, r_k^\Delta) + g(Z_k, r_k^\Delta)\Delta B_k. \tag{4.83}$$

Indeed, this implicit method has appeared frequently in the literature—it is a member of the *family of implicit Euler schemes* [Kloeden and Platen (1992), Sec. 12.2] and is sometimes called the semi-implicit Euler method. We will refer to the method (4.83) simply as the Backward Euler (BE) method for (4.1).

The BE method (4.83) requires an implicit equation to be solved. Under Assumption 4.10, the homotopy argument in the proof of Lemma 4.4 shows that for $2\mu\Delta < 1$ a unique solution exists with probability one. The next lemma points out a useful connection between BE and SSBE.

Lemma 4.7 *Let $\{X_k\}$ and $\{Z_k\}$ denote the SSBE and BE solutions, given by (4.57)–(4.58) and (4.83), respectively. Under Assumption 4.10, if $X_0 = Z_0 - \Delta f(Z_0, r_0)$ then*

$$Z_k = X_k + \Delta f_\Delta(X_k, r_k^\Delta), \quad \forall k \geq 0. \tag{4.84}$$

Proof. Let $Q_k^\star = Z_k$ and $Q_k = Z_k - \Delta f(Z_k, r_k^\Delta)$, where $\{Z_k\}$ is the BE solution (4.83). Then

$$Q_k^\star = Q_k + \Delta f(Q_k^\star, r_k^\Delta)$$

and, using (4.83),

$$Q_{k+1} = Z_{k+1} - \Delta f(Z_{k+1}, r_k^\Delta) = Q_k^\star + g(Q_k^\star, r_k^\Delta)\Delta B_k.$$

Hence, $\{Q_k\}$ is precisely the SSBE solution. This gives $X_k = Z_k - \Delta f(Z_k, r_k^\Delta)$. The relation (4.84) then follows from Lemma 4.4. □

Lemma 4.7 shows that the BE solution can be regarded as an $O(\Delta)$ perturbation of the SSBE solution. We may use this relation between BE and SSBE in order to obtain a convergence result for BE via Theorem 4.11. We first deal with the perturbation to the initial data.

Lemma 4.8 *Under Assumption 4.10, if $x(t)$ and $z(t)$ are solutions of the SDE (4.1) with initial conditions such that*

$$\mathbb{E}|x(0)|^2 \vee \mathbb{E}|z(0)|^2 \leq \infty,$$

then, for some constant M,

$$\mathbb{E}\left[\sup_{0 \leq t \leq T} |x(t) - z(t)|^2\right] \leq M\mathbb{E}|x(0) - z(0)|^2.$$

Proof. Letting $e(t) := x(t) - z(t)$ and applying the Itô formula to $|e(t)|^2$, we have

$$|e(t)|^2 = |e(0)|^2 + \int_0^t \Big[2\langle e(s), f(x(s), r(s)) - f(z(s), r(s))\rangle$$
$$+ |g(x(s), r(s)) - g(z(s), r(s))|^2\Big]ds$$
$$+ \int_0^t 2\langle e(s), [g(x(s), r(s)) - g(z(s), r(s))]dB(s)\rangle.$$

Using Assumption 4.10 we obtain that, for $0 \leq t_1 \leq T$,

$$\mathbb{E}\Big(\sup_{0 \leq t \leq t_1} |e(t)|^2\Big) \leq \mathbb{E}|e(0)|^2 + (2\mu + c)\int_0^{t_1} \mathbb{E}|e(s)|^2 ds + J(t_1), \quad (4.85)$$

where

$$J(t_1) = \mathbb{E}\Big(\sup_{0 \leq t \leq t_1} \int_0^t 2\langle e(s), [g(x(s), r(s)) - g(z(s), r(s))]dB(s)\rangle\Big).$$

But, by the Burkholder–Davis–Gundy inequality and Exercise 2.5, we derive that

$$J(t_1) \leq 3\mathbb{E}\bigg(\int_0^{t_1} 4|e(s)|^2|g(x(s),r(s)) - g(z(s),r(s))|^2 ds\bigg)^{\frac{1}{2}}$$

$$\leq 6\mathbb{E}\bigg(\sup_{0\leq s\leq t_1}|e(s)|^2 \int_0^{t_1}|e(s)|^2 ds\bigg)^{\frac{1}{2}}$$

$$\leq \tfrac{1}{2}\mathbb{E}\bigg(\sup_{0\leq s\leq t_1}|e(s)|^2\bigg) + 18\int_0^{t_1}\mathbb{E}|e(s)|^2 ds.$$

Substituting this into (4.85) yields

$$\mathbb{E}\bigg(\sup_{0\leq t\leq t_1}|e(t)|^2\bigg) \leq 2\mathbb{E}|e(0)|^2 + 2(2\mu+c+18)\int_0^{t_1}\mathbb{E}|e(s)|^2 ds.$$

The Gronwall inequality then shows

$$\mathbb{E}\bigg(\sup_{0\leq t\leq T}|e(t)|^2\bigg) \leq 2e^{2(2\mu+c+18)T}\mathbb{E}|e(0)|^2.$$

The required assertion follows by setting $M = 2e^{2(2\mu+c+18)T}$. □

Theorem 4.14 *Consider the Backward Euler method (4.83) applied to the SDE (4.1) under Assumption 4.10. Assume moreover that f obeys the following polynomial bounded*

$$|f(x,i)| \leq K_1(1+|x|^q) \quad \forall (x,i) \in \mathbb{R}^n \times \mathbb{S} \tag{4.86}$$

for some constants $K_1 > 0$ and $q \geq 1$. Then there exists a continuous-time extension $Z(t)$ of the numerical solution (so that $Z(k\Delta) = Z_k$) for which

$$\lim_{\Delta \to 0} \mathbb{E}\bigg[\sup_{0\leq t\leq T}|Z(t) - x(t)|^2\bigg] = 0 \tag{4.87}$$

Proof. Let $X(t)$ denote the continuous-time extension to SSBE defined in Theorem 4.11, with initial data $X(0) = x_0 - \Delta f(x_0, r_0)$. Also, let $Z(t) = X(t) + \Delta f_\Delta(X(t), r(t))$, so that, from Lemma 4.7, $Z(t)$ is a continuous-time extension to the BE solution with $Z(0) = x_0$. We let $\widehat{x}_\Delta(t)$ denote the solution to (4.1) with initial data $x_\Delta(0) = x_0 - \Delta f(x_0, r_0)$.

From Lemma 4.8 we have

$$\mathbb{E}\bigg[\sup_{0\leq t\leq T}|x(t) - \widehat{x}_\Delta(t)|^2\bigg] \leq M\Delta^2|f(x_0,r_0)|^2. \tag{4.88}$$

Also, the SSBE convergence result in Theorem 4.11 shows that

$$\lim_{\Delta \to 0} \mathbb{E}\left[\sup_{0 \le t \le T} |\hat{x}_\Delta(t) - X(t)|^2\right] = 0. \tag{4.89}$$

Further, by Lemma 4.4 as well as condition (4.86), we derive that

$$\mathbb{E}\left[\sup_{0 \le t \le T} |X(t) - Z(t)|^2\right] \le \Delta^2 \mathbb{E}\left[\sup_{0 \le t \le T} |f_\Delta(X(t), r(t))|^2\right]$$
$$\le \frac{\Delta^2}{(1-\Delta\mu)^2} \mathbb{E}\left[\sup_{0 \le t \le T} |f(X(t), r(t))|^2\right]$$
$$\le \frac{\Delta^2 K_1^2}{(1-\Delta\mu)^2}\left(1 + \mathbb{E}\left[\sup_{0 \le t \le T} |X(t)|^{2q}\right]\right). \tag{4.90}$$

Combining (4.88)–(4.90) and using Corollary 4.13 we obtain immediately the required assertion (4.87). □

4.6 Stochastic Theta Method

Another useful numerical scheme is the stochastic theta (ST) method which is defined by setting $Y_0 = x_0$ and forming

$$Y_{k+1} = Y_k + [(1-\theta)f(Y_k, r_k^\Delta) + \theta f(Y_{k+1}, r_k^\Delta)]\Delta + g(Y_k, r_k^\Delta)\Delta B_k, \tag{4.91}$$

where $\theta \in [0,1]$ is a free parameter that is specified *a priori*. Clearly, this method reduces to the EM method (4.3) if $\theta = 0$ while it becomes the BE method (4.83) if $\theta = 1$. But it is the combination of the EM and BE when $\theta \in (0,1)$, and in this case, (4.91) represents a nonlinear system that is to be solved for Y_{k+1}.

We introduce the continuous approximation

$$Y(t) = x_0 + \int_0^t [(1-\theta)f(Y_1(s), \bar{r}(s)) + \theta f(Y_2(s), \bar{r}(s))]ds$$
$$+ \int_0^t g(Y_1(s), \bar{r}(s))dB(s), \tag{4.92}$$

where

$$Y_1(t) = Y_k, \quad Y_2(t) = Y_{k+1} \quad \text{and} \quad \bar{r}(t) = r_k^\Delta \quad \text{for } t \in [k\Delta, (k+1)\Delta).$$

We impose the global Lipschitz condition: There is a constant $\bar{K} > 0$ such that

$$|f(x,i) - f(y,i)| \vee |g(x,i) - g(y,i)| \le \bar{K}|x-y| \qquad (4.93)$$

for all $x, y \in \mathbb{R}^n$ and $i \in \mathbb{S}$. As showed in Section 4.2.1, this condition implies the linear growth condition

$$|f(x,i)| \vee |g(x,i)| \le K(1+|x|), \quad \forall (x,i) \in \mathbb{R}^n \times \mathbb{S} \qquad (4.94)$$

with $K = \bar{K} \vee \max\{|f(0,i)| \vee |g(0,i)| : i \in \mathbb{S}\}$.

Our first lemma concerns the existence of solution to the implicit equation (4.91).

Lemma 4.9 *Under the global Lipschitz condition (4.93), if $\bar{K}\theta\Delta < 1$, then equation (4.91) can be solved uniquely for Y_{k+1} given Y_k, with probability 1.*

Proof. Writing (4.91) as $Y_{k+1} = F(Y_{k+1})$, we have, using (4.93),

$$|F(u) - F(v)| = |\theta f(u, r_k^\Delta)\Delta - \theta f(v, r_k^\Delta)\Delta| \le \bar{K}\theta\Delta|u-v|, \quad \forall u, v \in \mathbb{R}^n.$$

The required result hence follows from the classical Banach contraction mapping theorem. $\square$

Our next two lemmas concerns the pth moment of the ST approximate solution. It shows that under condition (4.94) any pth moment, especially the 2nd moment, is finite.

Lemma 4.10 *Suppose (4.94) holds and let $\Delta < \min\{1, 1/(6K)\}$. Then there exists a $C > 0$ (independent of Δ) such that for the ST method (4.91)*

$$\sup_{k\Delta \in [0,T]} \mathbb{E}|Y_k|^2 \le C.$$

Proof. It follows from (4.91) that

$$\mathbb{E}|Y_{k+1}|^2$$
$$= \mathbb{E}|Y_k|^2 + 2\mathbb{E}\langle Y_k^T[(1-\theta)f(Y_k, r_k^\Delta) + \theta f(Y_{k+1}, r_k^\Delta)]\Delta\rangle$$
$$+ \mathbb{E}\big|[(1-\theta)f(Y_k, r_k^\Delta) + \theta f(Y_{k+1}, r_k^\Delta)]\Delta + g(Y_k, r_k^\Delta)\Delta B_k\big|^2$$
$$\leq \mathbb{E}|Y_k|^2 + 2\Delta\mathbb{E}\big[|Y_k|(|f(Y_k, r_k^\Delta)| + |f(Y_{k+1}, r_k^\Delta)|)\big]$$
$$+ 3\mathbb{E}\Big(|f(Y_k, r_k^\Delta)|^2\Delta^2 + |f(Y_{k+1}, r_k^\Delta)|^2\Delta^2 + |g(Y_k, r_k^\Delta)|^2|\Delta B_k|^2\Big)$$
$$\leq \mathbb{E}|Y_k|^2 + 2K\Delta\mathbb{E}\big[|Y_k|(2 + |Y_k| + |Y_{k+1}|)\big]$$
$$+ 6K^2\mathbb{E}\Big([2 + |Y_k|^2 + |Y_{k+1}|^2]\Delta^2 + (1 + |Y_k|^2)\Delta\Big)$$
$$\leq \mathbb{E}|Y_k|^2 + 2K\Delta\mathbb{E}\big[1 + 2.5|Y_k|^2 + 0.5|Y_{k+1}|^2)\big]$$
$$+ K\Delta\mathbb{E}[2 + |Y_k|^2 + |Y_{k+1}|^2] + 6K^2\Delta(1 + \mathbb{E}|Y_k|^2)$$
$$\leq \mathbb{E}|Y_k|^2 + 2K(2 + 3K)\Delta + 6K(1 + K)\Delta\mathbb{E}|Y_k|^2 + 2K\Delta\mathbb{E}|Y_{k+1}|^2.$$

Let u be any positive integer such that $u\Delta \leq T$. Summing the inequality above for k from 0 to $u-1$, we obtain

$$\mathbb{E}|Y_u|^2$$
$$\leq \mathbb{E}|Y_0|^2 + 2K(2+3K)T + 6K(1+K)\Delta\sum_{k=1}^{u-1}\mathbb{E}|Y_k|^2 + 2K\Delta\sum_{k=1}^{u-1}\mathbb{E}|Y_{k+1}|^2$$
$$\leq |x_0|^2 + 2K(2+3K)T + 2K(4+3K)\Delta\sum_{k=1}^{u-1}\mathbb{E}|Y_k|^2 + 2K\Delta\mathbb{E}|Y_u|^2.$$

Noting that $2K\Delta \leq 1/3$, we have

$$\mathbb{E}|Y_u|^2 \leq \frac{3}{2}\Big[|x_0|^2 + 2K(2+3K)T + 2K(4+3K)\Delta\sum_{k=1}^{u-1}\mathbb{E}|Y_k|^2\Big].$$

Using the discrete-type Gronwall inequality (i.e. Theorem 2.5) and recalling that $M\Delta \leq T$, we obtain

$$\mathbb{E}|Y_u|^2 \leq \frac{3}{2}[|x_0|^2 + 2K(2+3K)T]e^{3K(4+3K)T}$$

and the desired result follows. □

Lemma 4.11 *Suppose (4.94) holds and let $\Delta < \min\{1, 1/(6K)\}$. Then there exists a $H > 0$ (independent of Δ) such that the continuous ST ap-*

proximate solution (4.92) has the property that

$$\mathbb{E}\left[\sup_{0\le t\le T}|Y(t)|^p\right]\le H. \tag{4.95}$$

Proof. It follows from (4.92) that

$$|Y(t)|^2 \le 4|x_0|^2 + 4T\int_0^t \left[|f(Y_1(s),\bar{r}(s))|^2 + |f(Y_2(s),\bar{r}(s))|^2\right]ds$$
$$+ 4\left|\int_0^t g(Y_1(s),\bar{r}(s))dB(s)\right|^2.$$

Using the Doob martingale inequality and condition (4.94) we get

$$\mathbb{E}\left[\sup_{0\le t\le T}|Y(t)|^p\right]$$
$$\le 4|x_0|^2 + 4T\mathbb{E}\int_0^T 2K^2(2+|Y_1(s)|^2+|Y_2(s)|^2)ds$$
$$+ 14\int_0^T 2K^2(1+|Y_1(s)|^2)ds$$
$$\le 4|x_0|^2 + 16T^2K^2 + 28TK^2 + (8TK^2+28K^2)\int_0^T E|Y_1(s)|^2$$
$$+ 8TK^2\int_0^T \mathbb{E}|Y_2(s)|^2 ds$$
$$\le 4|x_0|^2 + 4TK^2(4T+7) + 4TK^2(4T+7)\left[\sup_{k\Delta\in[0,T]}\mathbb{E}|Y_k|^2\right].$$

Applying Lemma 4.10 gives the assertion (4.95). □

Theorem 4.15 *Under the global Lipschitz condition (4.93),*

$$\mathbb{E}\left[\sup_{0\le t\le T}|Y(t)-x(t)|^2\right]\le C\Delta + o(\Delta), \tag{4.96}$$

where C is a positive constant independent of Δ.

Proof. By the Hölder inequality and the Doob martingale inequality, it is not difficult to show that for $0 \leq t \leq T$,

$$\mathbb{E}\left(\sup_{0\leq s\leq t} |Y(s) - x(s)|^2\right)$$
$$\leq 3T(1-\theta)^2 \mathbb{E}\int_0^t |f(Y_1(s), \bar{r}(s)) - f(x(s), r(s))|^2 ds$$
$$+ 3T\theta^2 \mathbb{E}\int_0^t |f(Y_2(s), \bar{r}(s)) - f(x(s), r(s))|^2 ds$$
$$+ 12\mathbb{E}\int_0^t |g(Y_1(s), \bar{r}(s)) - g(x(s), r(s))|^2 ds. \quad (4.97)$$

In the remainder of the proof C is a positive constant independent of Δ which may change line by line. In the same way in Theorem 4.1 was proved, we can show (Lemma 4.11 is needed here) that

$$\mathbb{E}\int_0^t |f(Y_j(s), \bar{r}(s)) - f(x(s), r(s))|^2 ds$$
$$\leq 2\bar{K}^2 \int_0^t \mathbb{E}|Y_j(s) - x(s)|^2 ds + C\Delta + o(\Delta), \quad j = 1, 2, \quad (4.98)$$

and

$$\mathbb{E}\int_0^t |g(Y_1(s), \bar{r}(s)) - g(x(s), r(s))|^2 ds$$
$$\leq 2\bar{K}^2 \int_0^t \mathbb{E}|Y_1(s) - x(s)|^2 ds + C\Delta + o(\Delta). \quad (4.99)$$

Substituting (4.98) and (4.99) into (4.97) yields that

$$\mathbb{E}\left(\sup_{0\leq s\leq t} |Y(s) - x(s)|^2\right)$$
$$\leq C \int_0^t \left(\mathbb{E}|Y_1(s) - x(s)|^2 + \mathbb{E}|Y_2(s) - x(s)|^2\right) ds + C\Delta + o(\Delta). \quad (4.100)$$

Note, for $j = 1, 2$,

$$\mathbb{E}|Y_j(s) - x(s)|^2 \leq 2\mathbb{E}|Y(s) - x(s)|^2 + 2\mathbb{E}|Y(s) - Y_j(s)|^2. \quad (4.101)$$

But, it can be showed (as an exercise) by Lemma 4.11 that

$$\mathbb{E}|Y_j(s) - Y(s)|^2 \leq C\Delta, \quad j = 1, 2. \quad (4.102)$$

Putting (4.101) and (4.102) into (4.100) we see that

$$\mathbb{E}\Big(\sup_{0\le s\le t}|Y(s)-x(s)|^2\Big) \le C\int_0^t \mathbb{E}|Y(s)-x(s)|^2 ds + C(\Delta+o(\Delta))$$

$$\le C\int_0^t \mathbb{E}\Big(\sup_{0\le r\le s}|Y(r)-x(r)|^2\Big)ds + C\Delta + o(\Delta)$$

and the required result (4.96) follows from the Gronwall inequality. □

4.7 Exercises

4.1 Recall the statement of Lemma 4.1 and show that

$$\mathbb{E}\Big[\sup_{0\le t\le T}|x(t)|^p\Big] \le H,$$

where H is the same as defined in the proof of Lemma 4.1.

4.2 Show (4.28).

4.3 Prove Theorem 4.4. (Hint: Using the same notations as defined in the proof of Theorem 4.2, show

$$\lim_{R\to\infty}\mathbb{E}\Big[\sup_{0\le t\le T}|e(t)|^2 I_{\{\tau_R\le T \text{ or } \rho_R\le T\}}\Big] = 0$$

and then show Theorem 4.4 in a similar way as Theorem 4.2 was proved.)

4.4 Prove Lemma 4.2.

4.5 Prove that under conditions (4.43) and (4.44), the SDE (4.1) has a unique solution.

4.6 Prove the assertion (4.45) of Theorem 4.7 for general $p\ge 2$.

4.7 Show (4.48).

4.8 Show (4.50).

4.9 By referring to [Dekker and Verwer (1984); Stuart and Humphries (1996)], show the existence and uniqueness for (4.59) via a contraction mapping theorem, and establish the C^1 smoothness of $f_\Delta(\cdot,i)$ and $F_\Delta(\cdot,i)$ and the convergence property of $f_\Delta(\cdot,i)$ as stated in Lemma 4.4.

4.10 Prove (4.98) and (4.99).

4.11 Prove (4.102).

Chapter 5

Boundedness and Stability

5.1 Introduction

In this chapter we will discuss the boundedness and stability of solutions to SDEs with Markovian switching. The main technique will be the method of Lyapunov functions known as Lyapunov's second method.

During the past century Lyapunov's second method has gained increasing significance and has given decisive impetus for modern development of stability theory of dynamic systems. A manifest advantage of this method is that it does not require the knowledge of solutions of equations and thus has exhibited a great power in applications. On the other hand, since Itô introduced his stochastic calculus more than 50 years ago, the theory of stochastic differential equations has been developed very quickly. Naturally, Lyapunov's second method has been developed to deal with stochastic stability by many authors. There are several books available expounding the main ideas of Lyapunov's second method for stochastic differential equations e.g. [Arnold (1974); Arnold(1998); Friedman (1975); Has'minskii (1980); Kolmanovskii and Myshkis (1992); Kolmanovskii and Nosov (1986); Kushner 1967); Ladde and Lakshmikantham (1980); Mao (1991); Mao (1994); Mao (1997)]. It is now well recognised that the method of Lyapunov functions can be utilized to investigate various qualitative and quantitative properties of stochastic differential equations. The original ideas have been refined, generalised and extended in several directions. Many new concepts have been introduced. Especially, some new ideas and approaches which might provide an exciting prospect of further advancement are still in the initial stages of investigation.

In recent years, Lyapunov's second method has been developed to investigate the stability and boundedness of solutions to SDEs with Markovian switching by many authors e.g. [Basak et al. (1996); Ji and Chizeck (1990); Lewin (1986); Mao (1999a); Mao (2001a); Mao (2002a); Mao and Shaikhet (2000); Mariton (1990); Shaikhet (1996); Yuan and Mao (2003a)]. In this chapter we will systematically present Lyapunov's second method in the study of stability and boundedness.

Let us consider a nonlinear SDE with Markovian switching

$$dx(t) = f(x(t), t, r(t))dt + g(x(t), t, r(t))dB(t), \quad t \geq t_0 (\geq 0) \quad (5.1)$$

with initial values $x(t_0) = x_0$ and $r(t_0) = r_0$, where

$$f : \mathbb{R}^n \times \mathbb{R}_+ \times \mathbb{S} \to \mathbb{R}^n \text{ and } g : \mathbb{R}^n \times \mathbb{R}_+ \times \mathbb{S} \to \mathbb{R}^{n \times m}.$$

As before, $B(t)$, $t \geq 0$, is an m-dimensional Brownian motion while $r(t)$, $t \geq 0$, is a right-continuous Markov chain taking values in a finite state space $\mathbb{S}$ with the generator $\Gamma = (\gamma_{ij})_{N \times N}$ given by

$$\mathbb{P}\{r(t+\Delta) = j | r(t) = i\} = \begin{cases} \gamma_{ij}\Delta + o(\Delta) & \text{if } i \neq j, \\ 1 + \gamma_{ii}\Delta + o(\Delta) & \text{if } i = j, \end{cases}$$

where $\Delta > 0$. Here $\gamma_{ij} > 0$ is the transition rate from i to j if $i \neq j$ while

$$\gamma_{ii} = -\sum_{j \neq i} \gamma_{ij}.$$

We always assume that the Markov chain $r(\cdot)$ is independent of the Brownian motion $B(\cdot)$. Assume that both f and g are sufficiently smooth (see Chapter 3) so that equation (5.1) has a unique global solution, which is denoted by $x(t; t_0, x_0, r_0)$. For convenience, we will let the initial values x_0 and r_0 be non-random, namely $x_0 \in \mathbb{R}^n$ and $r_0 \in \mathbb{S}$, but the theory developed in this chapter can be generalised without any difficulty to cope with the case of random initial values.

For any open subset G of $\mathbb{R}^n$, let $C^{2,1}(G \times \mathbb{R}_+ \times \mathbb{S}; \mathbb{R}_+)$ denote the family of all non-negative functions $V(x, t, i)$ on $G \times \mathbb{R}_+ \times \mathbb{S}$ which are continuously twice differentiable in x and once in t. If $V \in C^{2,1}(G \times \mathbb{R}_+ \times \mathbb{S}; \mathbb{R}_+)$, define an operator LV from $G \times \mathbb{R}_+ \times \mathbb{S}$ to $\mathbb{R}$ by

$$\begin{aligned} LV(x,t,i) = & V_t(x,t,i) + V_x(x,t,i)f(x,t,i) \\ & + \frac{1}{2}\text{trace}[g^T(x,t,i)V_{xx}(x,t,i)g(x,t,i)] \\ & + \sum_{j=1}^{N} \gamma_{ij} V(x,t,j), \end{aligned} \quad (5.2)$$

where

$$V_t(x,t,i) = \frac{\partial V(x,t,i)}{\partial t},$$

$$V_x(x,t,i) = \left(\frac{\partial V(x,t,i)}{\partial x_1}, \ldots, \frac{\partial V(x,t,i)}{\partial x_n}\right),$$

$$V_{xx}(x,t,i) = \left(\frac{\partial^2 V(x,t,i)}{\partial x_i \partial x_j}\right)_{n\times n}.$$

We shall also need some new notations. Let $\mathcal{K}$ denote the family of all continuous increasing functions $\kappa : \mathbb{R}_+ \to \mathbb{R}_+$ such that $\kappa(0) = 0$ while $\kappa(u) > 0$ for $u > 0$. If $\kappa \in \mathcal{K}$, its inverse function is denoted by κ^{-1} with domain $[0, \kappa(\infty))$, where $\kappa(\infty) = \lim_{u\to\infty} \kappa(u)$. Let $\mathcal{K}_\infty$ denote the family of all functions $\kappa \in \mathcal{K}$ with property $\kappa(\infty) = \infty$. Let $\mathcal{K}_V$ denote the family of all convex functions $\kappa \in \mathcal{K}$ while $\mathcal{K}_\wedge$ denote the family of all concave functions $\kappa \in \mathcal{K}$. It is easy to observe that if $\kappa \in \mathcal{K}_V$, then $\kappa(u) \to \infty$ monotonically as $u \to \infty$. In other words, $\mathcal{K}_V \subset \mathcal{K}_\infty$.

5.2 Asymptotic Boundedness

Let us begin with the definition of asymptotic boundedness of pth moment.

Definition 5.1 Let $p > 0$. Equation (5.1) is said to be *asymptotically bounded in pth moment* if there is a positive constant H such that

$$\limsup_{t\to\infty} \mathbb{E}|x(t;t_0,x_0,r_0)|^p \leq H \tag{5.3}$$

for all $(t_0,x_0,r_0) \in \mathbb{R}_+ \times \mathbb{R}^n \times \mathbb{S}$. When $p = 2$ we say equation (5.1) is *asymptotically bounded in mean square*.

The following theorem gives a criterion on the asymptotic boundedness of pth moment in terms of a Lyapunov function.

Theorem 5.2 *Assume that there exist functions $V \in C^{2,1}(\mathbb{R}^n \times \mathbb{R}_+ \times \mathbb{S}; \mathbb{R}_+)$ and $\kappa \in \mathcal{K}_V$ and positive numbers p, β, λ such that*

$$\kappa(|x|^p) \leq V(x,t,i) \tag{5.4}$$

and

$$LV(x,t,i) \leq -\lambda V(x,t,i) + \beta \tag{5.5}$$

for all $(x, t, i) \in \mathbb{R}^n \times \mathbb{R}_+ \times \mathbb{S}$. Then

$$\limsup_{t \to \infty} \mathbb{E}|x(t; t_0, x_0, r_0)|^p \leq \kappa^{-1}\left(\frac{\beta}{\lambda}\right) \tag{5.6}$$

for all $(t_0, x_0, r_0) \in \mathbb{R}_+ \times \mathbb{R}^n \times \mathbb{S}$. That is, equation (5.1) is asymptotically bounded in pth moment.

Proof. Fix any $(t_0, x_0, r_0) \in \mathbb{R}_+ \times \mathbb{R}^n \times \mathbb{S}$ and write $x(t; t_0, x_0, r_0) = x(t)$ for simplicity. Let k be a positive integer. Define the stopping time

$$\rho_k = \inf\{t \geq t_0 : |x(t)| \geq k\}.$$

Clearly, $\rho_k \to \infty$ almost surely as $k \to \infty$. Let $t_k = \rho_k \wedge t$ for any $t \geq t_0$. The generalised Itô formula shows that

$$\mathbb{E}\left[e^{\lambda t_k} V(x(t_k), t_k, r(t_k))\right] = V(x_0, t_0, r_0) + \mathbb{E}\int_{t_0}^{t_k} e^{\lambda s} LV(x(s), s, r(s)) ds$$

$$+ \lambda \mathbb{E}\int_{t_0}^{t_k} e^{\lambda s} V(x(s), s, r(s)) ds.$$

By conditions (5.4) and (5.5),

$$\mathbb{E}\left[e^{\lambda t_k} \kappa(|x(t_k)|^p)\right] \leq V(x_0, t_0, r_0) + \beta \int_0^t e^{\lambda s} ds = V(x_0, t_0, r_0) + \frac{\beta}{\lambda}\left[e^{\lambda t} - 1\right].$$

Letting $k \to \infty$ gives

$$\mathbb{E}\left[e^{\lambda t} \kappa(|x(t)|^p)\right] \leq V(x_0, t_0, r_0) + \frac{\beta}{\lambda}\left[e^{\lambda t} - 1\right].$$

By the well-known Jensen inequality,

$$\kappa(\mathbb{E}|x(t)|^p) \leq \frac{\beta}{\lambda} + e^{-\lambda t} V(x_0, t_0, r_0)$$

so

$$\mathbb{E}|x(t)|^p \leq \kappa^{-1}\left(\frac{\beta}{\lambda} + e^{-\lambda t} V(x_0, t_0, r_0)\right).$$

Letting $t \to \infty$ yields the assertion (5.6). $\square$

The following result gives a new criterion on asymptotic boundedness of pth moment where the conditions are described in terms of an M-matrix.

Theorem 5.3 Let $p \geq 2, \alpha > 0$ and β_i ($i \in \mathbb{S}$) be constants. Assume that for all $(x, t, i) \in \mathbb{R}^n \times \mathbb{R}_+ \times \mathbb{S}$,

$$x^T f(x,t,i) + \frac{p-1}{2}|g(x,t,i)|^2 \leq \beta_i |x|^2 + \alpha. \tag{5.7}$$

If

$$\mathcal{A} := -\text{diag}(p\beta_1, \cdots, p\beta_N) - \Gamma \tag{5.8}$$

is a nonsingular M-matrix, then equation (5.1) is asymptotically bounded in pth moment.

Proof. By Theorem 2.10 there is a vector $\vec{q} = (q_1, \cdots, q_N)^T \gg 0$ such that

$$\vec{\lambda} = (\lambda_1, \cdots, \lambda_N)^T := \mathcal{A}\vec{q} \gg 0. \tag{5.9}$$

Define the function $V : \mathbb{R}^n \times \mathbb{R}_+ \times \mathbb{S} \to \mathbb{R}_+$ by

$$V(x,t,i) = q_i |x|^p.$$

By (5.7) and (5.9) we compute the operator LV from $\mathbb{R}^n \times \mathbb{R}_+ \times \mathbb{S}$ to $\mathbb{R}$ as follows:

$$LV(x,t,i) = pq_i|x|^{p-2}x^T f(x,t,i) + \tfrac{1}{2}pq_i|x|^{p-2}|g(x,t,i)|^2$$

$$+ \tfrac{1}{2}p(p-2)q_i|x|^{p-4}|x^T g(x,t,i)|^2 + \sum_{j=1}^N \gamma_{ij} q_j |x|^p$$

$$\leq pq_i|x|^{p-2}\left[x^T f(x,t,i) + \frac{p-1}{2}|g(x,t,i)|^2\right] + \sum_{j=1}^N \gamma_{ij} q_j |x|^p$$

$$\leq pq_i|x|^{p-2}(\beta_i|x|^2 + \alpha) + \sum_{j=1}^N \gamma_{ij} q_j |x|^p$$

$$\leq \left(p\beta_i q_i + \sum_{i=1}^N \gamma_{ij} q_j\right)|x|^p + \alpha p q_i |x|^{p-2}$$

$$= -\lambda_i |x|^p + \alpha p q_i |x|^{p-2}. \tag{5.10}$$

By the elementary inequality

$$a^c b^{1-c} \leq ac + b(1-c) \quad \forall a, b \geq 0, \ c \in [0,1]$$

we compute

$$\alpha p q_i |x|^{p-2} = \alpha p q_i \left(\frac{2}{\lambda_i}\right)^{(p-2)/p} \left[\frac{\lambda_i}{2}|x|^p\right]^{(p-2)/p}$$

$$= \left[(\alpha p q_i)^{p/2}\left(\frac{2}{\lambda_i}\right)^{(p-2)/2}\right]^{2/p} \left[\frac{\lambda_i}{2}|x|^p\right]^{(p-2)/p}$$

$$\leq \frac{2}{p}(\alpha p q_i)^{p/2}\left(\frac{2}{\lambda_i}\right)^{(p-2)/2} + \frac{\lambda_i(p-2)}{2p}|x|^p.$$

Substituting this into (5.10) gives

$$LV(x,t,i) \leq -\frac{\lambda_i(p+2)}{2p}|x|^p + \frac{2}{p}(\alpha p q_i)^{p/2}\left(\frac{2}{\lambda_i}\right)^{(p-2)/2}.$$

If we set

$$\lambda = \min_{i \in \mathbb{S}} \left[\frac{\lambda_i(p+2)}{2pq_i}\right], \quad \beta = \max_{i \in \mathbb{S}} \left[\frac{2}{p}(\alpha p q_i)^{p/2}\left(\frac{2}{\lambda_i}\right)^{(p-2)/2}\right],$$

then

$$LV(x,t,i) \leq -\lambda V(x,t,i) + \beta.$$

Moreover, define $\kappa \in \mathcal{K}_V$ by $\kappa(u) = (\min_{i \in \mathbb{S}} q_i)u$ for $u \geq 0$. Then all the conditions of Theorem 5.2 are satisfied so the theorem shows

$$\limsup_{t \to \infty} \mathbb{E}|x(t;t_0,x_0,r_0)|^p \leq \frac{\beta}{\lambda \min_{i \in \mathbb{S}} q_i}$$

for all $(t_0, x_0, r_0) \in \mathbb{R}_+ \times \mathbb{R}^n \times \mathbb{S}$. In other words, equation (5.1) is asymptotically bounded in pth moment. $\square$

Example 5.4 Let $B(t)$ be a scalar Brownian motion. Let α and σ be constants. Consider the Ornstein–Uhlenbeck process

$$dx(t) = \alpha x(t)dt + \sigma dB(t), \quad t \geq t_0. \tag{5.11}$$

Given initial value $x(t_0) = x_0 \in \mathbb{R}^n$, we compute by the Itô formula that

$$e^{-2\alpha t}\mathbb{E}|x(t)|^2 = e^{-2\alpha t_0}|x_0|^2 + \sigma^2 \int_{t_0}^{t} e^{-2\alpha s} ds$$

$$= e^{-2\alpha t_0}|x_0|^2 + \frac{\sigma^2}{2\alpha}\left[e^{-2\alpha t_0} - e^{-2\alpha t}\right].$$

That is
$$\mathbb{E}|x(t)|^2 = |x_0|^2 e^{2\alpha(t-t_0)} + \frac{\sigma^2}{2\alpha}\left[e^{2\alpha(t-t_0)} - 1\right].$$
So
$$\lim_{t\to\infty} \mathbb{E}|x(t)|^2 = \begin{cases} \frac{\sigma^2}{2|\alpha|} & \text{if } \alpha < 0, \\ \infty & \text{if } \alpha > 0. \end{cases}$$

In other words, equation (5.11) is asymptotically bounded in mean square if $\alpha < 0$ but it is not if $\alpha > 0$.

Now let $r(t)$ be a right-continuous Markov chain taking values in $\mathbb{S} = \{1, 2\}$ with generator
$$\Gamma = (\gamma_{ij})_{2\times 2} = \begin{pmatrix} -4 & 4 \\ \gamma & -\gamma \end{pmatrix},$$
where $\gamma > 0$. Assume that $B(t)$ and $r(t)$ are independent. Consider a one-dimensional SDE with Markovian switching
$$dx(t) = \alpha(r(t))x(t)dt + \sigma dB(t) \tag{5.12}$$
on $t \geq t_0$, where σ is a constant, $\alpha(1) = 1$ and $\alpha(2) = -1/2$. This system can be regarded as the result of two equations
$$dx(t) = x(t)dt + \sigma dB(t) \tag{5.13}$$
and
$$dx(t) = -\frac{1}{2}x(t)dt + \sigma dB(t) \tag{5.14}$$
switching from one to the other according to the law of the Markov chain. From the property of the Ornstein–Uhlenbeck process (5.11) we observe that equation (5.13) is not asymptotically bounded in mean square although equation (5.14) is. However, we shall see that due to the Markovian switching the overall system (5.12) will be asymptotically bounded in mean square for certain γ. In fact, for
$$p = 2, \quad f(x,t,i) = a(i)x, \quad g(x,t,i) = \sigma$$
it is easy to see that condition (5.7) holds with
$$\beta_1 = 1, \quad \beta_2 = -\frac{1}{2}, \quad \alpha = \sigma^2.$$

So the matrix defined by (5.8) becomes

$$\mathcal{A} = -\mathrm{diag}(2,-1) - \Gamma = \begin{pmatrix} 2 & -4 \\ -\gamma & 1+\gamma \end{pmatrix}.$$

Since $\gamma > 0$, this is an M-matrix if and only if

$$2(1+\gamma) - 4\gamma > 0, \text{ namely } \gamma < 1.$$

By Theorem 5.3 we can therefore conclude that equation (5.12) is asymptotically bounded in mean square if $\gamma \in (0,1)$.

Example 5.5 Let $B(t)$ be a scalar Brownian motion. Let $r(t)$ be a right-continuous Markov chain taking values in $\mathbb{S} = \{1,2\}$ with generator

$$\Gamma = \begin{pmatrix} -\gamma_{12} & \gamma_{12} \\ \gamma_{21} & -\gamma_{21} \end{pmatrix},$$

where $\gamma_{12} > 0$ and $\gamma_{21} > 0$. Assume that $B(t)$ and $r(t)$ are independent. Consider a one-dimensional non-linear SDE with Markovian switching

$$dx(t) = f(x(t), r(t))dt + g(x(t), r(t))dB(t) \tag{5.15}$$

on $t \geq t_0$, where

$$f(x,i) = \begin{cases} x - x^3 + 2 & \text{if } i=1, \\ x - 1 & \text{if } i=2, \end{cases} \quad g(x,i) = \begin{cases} x^2 + 2 & \text{if } i=1, \\ x+1 & \text{if } i=2. \end{cases}$$

For $p \in [2,3)$, compute

$$xf(x,1) + \frac{p-1}{2}|g(x,1)|^2 = x^2 - x^4 + 2x + \frac{p-1}{2}(x^4 + 4x^2 + 4)$$

$$\leq -\frac{3-p}{2}x^4 + 6x^2 + 5.$$

Using the elementary inequality

$$-x^4 \leq -\delta x^2 + \delta^2 \quad \forall \delta > 0, \ x \in \mathbb{R},$$

we then have

$$xf(x,1) + \frac{p-1}{2}|g(x,1)|^2 \leq -\left[\frac{(3-p)\delta}{2} - 6\right]x^2 + \frac{(3-p)\delta^2}{2} + 5.$$

Also compute

$$xf(x,2) + \frac{p-1}{2}|g(x,2)|^2 = x^2 - x + \frac{p-1}{2}(x^2 + 2x + 1)$$
$$= \frac{p+1}{2}x^2 + (p-2)x + \frac{p-1}{2}.$$

But

$$(p-2)x = 2\sqrt{\frac{3-p}{2}} \; x \; \frac{p-2}{\sqrt{2(3-p)}} \leq \frac{3-p}{2}x^2 + \frac{(p-2)^2}{2(3-p)}.$$

So

$$xf(x,2) + \frac{p-1}{2}|g(x,2)|^2 \leq 2x^2 + \frac{(p-2)^2}{2(3-p)} + \frac{p-1}{2}.$$

We therefore obtain that

$$xf(x,i) + \frac{p-1}{2}|g(x,i)|^2 \leq \beta_i x^2 + \alpha,$$

where

$$\beta_i = \begin{cases} -\left[\frac{(3-p)\delta}{2} - 6\right] & \text{if } i = 1, \\ 2 & \text{if } i = 2, \end{cases}$$

and

$$\alpha = \max\left\{\frac{(3-p)\delta^2}{2} + 5, \frac{(p-2)^2}{2(3-p)} + \frac{p-1}{2}\right\}.$$

So the matrix defined by (5.8) becomes

$$\mathcal{A} = -\text{diag}(p\beta_1, p\beta_2) - \Gamma = \begin{pmatrix} p(0.5(3-p)\delta - 6) + \gamma_{12} & -\gamma_{12} \\ -\gamma_{21} & -2p + \gamma_{21} \end{pmatrix}.$$

If we let $\gamma_{21} \geq 6$, then $-2p + \gamma_{21} > 0$ whenever $p \in [2,3)$. Moreover, we can choose δ sufficiently large for

$$p(0.5(3-p)\delta - 6) + \gamma_{12} > 0$$

and

$$[p(0.5(3-p)\delta - 6) + \gamma_{12}](-2p + \gamma_{21}) > \gamma_{12}\gamma_{21},$$

namely for $\mathcal{A}$ to be an M-matrix. By Theorem 5.3 we can therefore conclude that equation (5.15) is asymptotically bounded in pth moment ($p \in [2,3)$) as long as $\gamma_{21} \geq 6$.

5.3 Exponential Stability

In this section we shall discuss the exponential stability of equation (5.1). Here, the key idea is to show not only that some quantity decays with time, but also that the decay is exponential. Applying this idea to the pth moment of the solution gives *moment exponential stability*. Similarly, asking for exponential decay for almost all sample paths leads to *almost sure exponential stability*. Such properties are highly-valued in many application areas. For this purpose we impose the following assumption.

Assumption 5.6 Assume that for each $k = 1, 2, \cdots$, there is an $h_k > 0$ such that

$$|f(x,t,i)| \vee |g(x,t,i)| \leq h_k |x|$$

for all $0 \leq t \leq k$, $i \in \mathbb{S}$ and those $x \in \mathbb{R}^n$ with $|x| \leq k$.

We first note that Assumption 5.6 implies that

$$f(0,t,i) \equiv 0 \quad \text{and} \quad g(0,t,i) \equiv 0.$$

It is therefore easy to observe that the solution of equation (5.1) will remain to be zero if it starts from zero, namely $x(t; t_0, 0, r_0) \equiv 0$. In other words, zero is an *equilibrium* or *stationary state*. This solution $x(t; t_0, 0, r_0) \equiv 0$ is often called a *trivial solution*.

However it is not so obvious to observe from Assumption 5.6 that any solution of equation (5.1) starting from a non-zero state will remain to be non-zero. In the sequel we will need this non-zero property so we prove it now.

Lemma 5.1 *Under Assumption 5.6,*

$$\mathbb{P}\{x(t; t_0, x_0, r_0) \neq 0 \text{ on } t \geq t_0\} = 1$$

for all $(t_0, x_0, r_0) \in \mathbb{R}_+ \times (\mathbb{R}^n - \{0\}) \times \mathbb{S}$. That is, almost all the sample paths of any solution of equation (5.1) starting from a non-zero state will never be zero.

Proof. If the Lemma were false, there would exist some $t_0 \geq 0$, $x_0 \neq 0$ and $r_0 \in \mathbb{S}$ such that

$$\mathbb{P}\{\tau < \infty\} > 0,$$

where τ is the first time of zero of the corresponding solution, namely

$$\tau = \inf\{t \geq t_0 : x(t; t_0, x_0, r_0) = 0\}.$$

Let us write $x(t; t_0, x_0, r_0) = x(t)$ for simplicity. Hence we can find an integer $k > t_0 \vee (1 + |x_0|)$ sufficiently large for $\mathbb{P}(G) > 0$, where

$$G = \{\tau \leq k \text{ and } |x(t)| \leq k - 1 \text{ for all } t_0 \leq t \leq \tau\}.$$

On the other hand, by Assumption 5.6, there exists a positive constant h_k such that

$$|f(x, t, i)| \vee |g(x, t, i)| \leq h_k |x| \quad \text{if } |x| \leq k, \ t_0 \leq t \leq k, \ i \in \mathbb{S}.$$

Let $V(x, t, i) = |x|^{-1}$. Then, for $0 < |x| \leq k$, $t_0 \leq t \leq k$ and $i \in \mathbb{S}$,

$$LV(x, t, i) = -|x|^{-3} x^T f(x, t, i) + \frac{1}{2}\left(-|x|^{-3}|g(x, t, i)|^2 + 3|x|^{-5}|x^T g(x, t, i)|^2\right)$$
$$\leq |x|^{-2}|f(x, t, i)| + |x|^{-3}|g(x, t, i)|^2$$
$$\leq h_k(1 + h_k)|x|^{-1}.$$

For any $\varepsilon \in (0, |x_0|)$, define a stopping time

$$\tau_\varepsilon = \inf\{t \geq t_0 : |x(t)| \notin (\varepsilon, k)\}.$$

By the Itô formula,

$$\mathbb{E}\left[e^{-h_k(1+h_k)(\tau_\varepsilon \wedge k)}|x(\tau_\varepsilon \wedge T)|^{-1}\right] = |x_0|^{-1}e^{-h_k(1+h_k)t_0}$$
$$+ \mathbb{E}\int_{t_0}^{\tau_\varepsilon \wedge k} e^{-h_k(1+h_k)s}\left[-h_k(1+h_k)|x(s)|^{-1} + LV(x(s), s, r(s))\right]ds$$
$$\leq |x_0|^{-1}e^{-h_k(1+h_k)t_0}.$$

Note that for $\omega \in G$ we have $\tau_\varepsilon \leq k$ and $|x(\tau_\varepsilon)| = \varepsilon$. We then see from the above inequality that

$$\mathbb{E}\left[e^{-h_k(1+h_k)k}\varepsilon^{-1}I_G\right] \leq |x_0|^{-1}e^{-h_k(1+h_k)t_0}.$$

That is

$$\mathbb{P}(G) \leq \varepsilon|x_0|^{-1}e^{h_k(1+h_k)(k-t_0)}.$$

Letting $\varepsilon \to 0$ yields $\mathbb{P}(G) = 0$, which is in contradiction with $\mathbb{P}(G) > 0$. The proof is complete. □

We shall now begin to establish the general theory of both pth moment and almost sure exponential stability for equation (5.1). Let us first give the definition.

Definition 5.7 For $p > 0$, the trivial solution of equation (5.1) or, simply, equation (5.1) is said to be *pth moment exponentially stable* if the pth moment Lyapunov exponent

$$\limsup_{t \to \infty} \frac{1}{t} \log(\mathbb{E}|x(t; t_0, x_0, r_0)|^p) < 0$$

for all $(t_0, x_0, r_0) \in \mathbb{R}_+ \times \mathbb{R}^n \times \mathbb{S}$. When $p = 2$, it is said to be *exponentially stable in mean square*. Moreover, it is said to be *almost surely exponentially stable* if the sample Lyapunov exponent

$$\limsup_{t \to \infty} \frac{1}{t} \log(|x(t; t_0, x_0, r_0)|) < 0 \quad a.s.$$

for all $(t_0, x_0, r_0) \in \mathbb{R}_+ \times \mathbb{R}^n \times \mathbb{S}$.

The stability criteria to be established will be described in terms of Lyapunov functions. Lemma 5.1 shows that $x(t; t_0, x_0, r_0)$ will never reach zero whenever $x_0 \ne 0$. So in what follows we will only need a $C^{2,1}$-function $V(x, t, i)$ defined on $\mathbb{R}_0^n \times \mathbb{R}_+ \times \mathbb{S}$, where $\mathbb{R}_0^n = \mathbb{R}^n - \{0\}$, namely $V \in C^{2,1}(\mathbb{R}_0^n \times \mathbb{R}_+ \times \mathbb{S}; \mathbb{R}_+)$.

Theorem 5.8 *Let Assumption 5.6 hold. Let p, λ, c_1, c_2 be positive numbers. Assume that there exists a function $V \in C^{2,1}(\mathbb{R}_0^n \times \mathbb{R}_+ \times \mathbb{S}; \mathbb{R}_+)$ such that*

$$c_1|x|^p \le V(x, t, i) \le c_2|x|^p \tag{5.16}$$

and

$$LV(x, t, i) \le -\lambda|x|^p \tag{5.17}$$

for all $(x, t, i) \in \mathbb{R}_0^n \times \mathbb{R}_+ \times \mathbb{S}$. Then

$$\limsup_{t \to \infty} \frac{1}{t} \log(\mathbb{E}|x(t; t_0, x_0, r_0)|^p) \le -\frac{\lambda}{c_2} \tag{5.18}$$

for all $(t_0, x_0, r_0) \in \mathbb{R}_+ \times \mathbb{R}^n \times \mathbb{S}$. In other words, the trivial solution of equation (5.1) is pth moment exponentially stable and the pth moment Lyapunov exponent is not greater than $-\lambda/c_2$.

Proof. Clearly (5.18) holds if $x_0 = 0$. So we only need to show (5.18) for $x_0 \ne 0$. Fix such x_0 and $t_0 \geq 0$, $r_0 \in \mathbb{S}$ arbitrarily and write $x(t; t_0, x_0, r_0) = x(t)$. For each integer $k \geq 1$, define a stopping time

$$\tau_k = \inf\{t \geq t_0 : |x(t)| \geq k\}.$$

Obviously $\tau_k \to \infty$ almost surely as $k \to \infty$. Noting that $0 < |x(t)| \leq k$ if $t_0 \leq t \leq \tau_k$, we can apply the generalised Itô formula to derive that for any $t \geq t_0$,

$$\mathbb{E}\left[e^{(\lambda/c_2)(t\wedge\tau_k)} V(x(t\wedge\tau_k), t\wedge\tau_k, r(t\wedge\tau_k))\right]$$
$$= V(x_0, t_0, r_0) e^{(\lambda/c_2) t_0}$$
$$+ \mathbb{E}\int_{t_0}^{t\wedge\tau_k} e^{(\lambda/c_2)s} \left[(\lambda/c_2) V(x(s), s, r(s)) + LV(x(s), s, r(s))\right] ds$$
$$\leq c_2 |x_0|^p e^{(\lambda/c_2) t_0} + \mathbb{E}\int_0^{t\wedge\tau_k} e^{(\lambda/c_2)s} \left[(\lambda/c_2) c_2 |x(s)|^p - \lambda |x(s)|^p\right] ds$$
$$\leq c_2 |x_0|^p e^{(\lambda/c_2) t_0}.$$

Consequently

$$c_1 \mathbb{E}\left[e^{(\lambda/c_2)(t\wedge\tau_k)} |x(t\wedge\tau_k)|^p\right] \leq c_2 |x_0|^p e^{(\lambda/c_2) t_0}.$$

Letting $k \to \infty$ gives

$$\mathbb{E}|x(t)|^p \leq \frac{c_2}{c_1} |x_0|^p e^{-(\lambda/c_2)(t-t_0)}$$

and the required assertion (5.18) follows. □

The following theorem shows that if Assumption 5.6 is slightly strengthened, then the pth moment exponential stability implies the almost sure exponential stability.

Theorem 5.9 *Assume that there is a positive constant K such that*

$$|f(x, t, i)| \vee |g(x, t, i)| \leq K|x| \quad \text{for all } (x, t, i) \in \mathbb{R}^n \times \mathbb{R}_+ \times \mathbb{S}. \tag{5.19}$$

Let $p > 0$ and $\lambda > 0$. If, for $(t_0, x_0, r_0) \in \mathbb{R}_+ \times \mathbb{R}^n \times \mathbb{S}$,

$$\limsup_{t\to\infty} \frac{1}{t} \log(\mathbb{E}|x(t; t_0, x_0, r_0)|^p) \leq -\lambda \tag{5.20}$$

then

$$\limsup_{t\to\infty} \frac{1}{t} \log(|x(t; t_0, x_0, r_0)|) \leq -\frac{\lambda}{p} \quad a.s. \tag{5.21}$$

In other words, under (5.19) the pth moment exponential stability implies the almost sure exponential stability.

Proof. For the sake of simplicity we only prove the case of $t_0 = 0$, but it can be done in the same way when $t_0 > 0$. Fix any $x_0 \in \mathbb{R}^n$ and $r_0 \in \mathbb{S}$. Write $x(t; 0, x_0, r_0) = x(t)$. Let $\varepsilon \in (0, \gamma/2)$ be arbitrary. By (5.20), there is a positive constant M such that

$$\mathbb{E}|x(t)|^p \leq M e^{-(\lambda-\varepsilon)t} \quad \text{on } t \geq 0. \tag{5.22}$$

Let $\delta > 0$ be sufficiently small for

$$(3K)^p(\delta^p + C_p \delta^{p/2}) < \frac{1}{2}, \tag{5.23}$$

where C_p is the constant given by the well-known Burkholder–Davis–Gundy inequality (i.e. Theorem 2.13). Let $k = 1, 2, \cdots$. Noting that for any $a, b, c \geq 0$,

$$(a+b+c)^p \leq [3(a \vee b \vee c)]^p = 3^p(a^p \vee b^p \vee c^p) \leq 3^p(a^p + b^p + c^p),$$

we have that

$$\mathbb{E}\left[\sup_{(k-1)\delta \leq t \leq k\delta} |x(t)|^p\right]$$
$$\leq 3^p \mathbb{E}|x((k-1)\delta)|^p + 3^p \mathbb{E}\left(\int_{(k-1)\delta}^{k\delta} |f(x(s), s, r(s))|ds\right)^p$$
$$+ 3^p \mathbb{E}\left[\sup_{(k-1)\delta \leq t \leq k\delta}\left|\int_{(k-1)\delta}^{t} g(x(s), s, r(s))dB(s)\right|^p\right]. \tag{5.24}$$

By (5.22),

$$\mathbb{E}|x((k-1)\delta)|^p \leq M e^{-(\lambda-\varepsilon)(k-1)\delta}. \tag{5.25}$$

Compute that

$$\mathbb{E}\left(\int_{(k-1)\delta}^{k\delta} |f(x(s), s, r(s))|ds\right)^p$$
$$\leq \mathbb{E}\left(\delta \sup_{(k-1)\delta \leq s \leq k\delta} |f(x(s), s, r(s))|\right)^p$$
$$\leq (K\delta)^p \mathbb{E}\left[\sup_{(k-1)\delta \leq s \leq k\delta} |x(s)|^p\right]. \tag{5.26}$$

Compute also that

$$\mathbb{E}\left[\sup_{(k-1)\delta \leq t \leq k\delta}\left|\int_{(k-1)\delta}^{t} g(x(s), s, r(s))dB(s)\right|^{p}\right]$$

$$\leq C_p \mathbb{E}\left(\int_{(k-1)\delta}^{k\delta} |g(x(s), s, r(s))|^2 ds\right)^{\frac{p}{2}}$$

$$\leq C_p \mathbb{E}\left(\delta \sup_{(k-1)\delta \leq s \leq k\delta} |g(x(s), s, r(s))|^2\right)^{\frac{p}{2}}$$

$$\leq C_p K^p \delta^{p/2} \mathbb{E}\left[\sup_{(k-1)\delta \leq s \leq k\delta} |x(s)|^p\right]. \qquad (5.27)$$

Substituting (5.25)–(5.27) into (5.24) yields that

$$\mathbb{E}\left[\sup_{(k-1)\delta \leq t \leq k\delta} |x(t)|^p\right] \leq M 3^p e^{-(\lambda-\varepsilon)(k-1)\delta}$$

$$+ (3K)^p (\delta^p + C_p \delta^{p/2}) \mathbb{E}\left[\sup_{(k-1)\delta \leq t \leq k\delta} |x(t)|^p\right].$$

Making use of (5.23) we obtain that

$$\mathbb{E}\left[\sup_{(k-1)\delta \leq t \leq k\delta} |x(t)|^p\right] \leq 2M 3^p e^{-(\lambda-\varepsilon)(k-1)\delta}. \qquad (5.28)$$

Hence

$$\mathbb{P}\left\{\omega: \sup_{(k-1)\delta \leq t \leq k\delta} |x(t)| > e^{-(\lambda-2\varepsilon)(k-1)\delta/p}\right\} \leq 2M 3^p e^{-\varepsilon(k-1)\delta}$$

In view of the well-known Borel–Cantelli lemma, we see that for almost all $\omega \in \Omega$,

$$\sup_{(k-1)\delta \leq t \leq k\delta} |x(t)| \leq e^{-(\lambda-2\varepsilon)(k-1)\delta/p} \qquad (5.29)$$

holds for all but finitely many k. Hence there exists a $k_0(\omega)$, for all $\omega \in \Omega$ excluding a $\mathbb{P}$-null set, for which (5.29) holds whenever $k \geq k_0$. Consequently, for almost all $\omega \in \Omega$,

$$\frac{1}{t}\log(|x(t)|) \leq -\frac{(\lambda-2\varepsilon)(k-1)\delta}{pt} \leq -\frac{(\lambda-2\varepsilon)(k-1)}{pk}$$

if $(k-1)\delta \le t \le k\delta$ and $k \ge k_0$. Therefore

$$\limsup_{t\to\infty} \frac{1}{t}\log(|x(t)|) \le -\frac{\lambda-2\varepsilon}{p} \quad a.s.$$

and the required (5.21) follows by letting $\varepsilon \to 0$. □

In application we often use the functions of the form

$$V(x,t,i) = (x^T Q_i x)^{p/2} \tag{5.30}$$

for some symmetric positive-definite matrices Q_i. Note that, for $x \ne 0$,

$$V_x(x,t,i) = p(x^T Q_i x)^{(p-2)/2} x^T Q_i$$

and

$$V_{xx}(x,t,i) = p(x^T Q_i x)^{(p-2)/2} Q_i + \frac{p(p-2)}{2}(x^T Q_i x)^{(p-4)/2} Q_i x x^T Q_i.$$

Note also that

$$\text{trace}\big[g^T(x,t,i) Q_i x x^T Q_i g(x,t,i)\big] = x^T Q_i g(x,t,i) g^T(x,t,i) Q_i x$$

$$= |x^T Q_i g(x,t,i)|^2.$$

Hence

$$\begin{aligned}
LV(x,t,i) &= p(x^T Q_i x)^{(p-2)/2} x^T Q_i f(x,t,i) \\
&\quad + \frac{p}{2}(x^T Q_i x)^{(p-2)/2} \text{trace}\big[g^T(x,t,i) Q_i g(x,t,i)\big] \\
&\quad + \frac{p(p-2)}{2}(x^T Q_i x)^{(p-4)/2} |x^T Q_i g(x,t,i)|^2 \\
&\quad + \sum_{j=1}^{N} \gamma_{ij} (x^T Q_j x)^{p/2}.
\end{aligned} \tag{5.31}$$

We can now easily establish the following useful corollary.

Corollary 5.10 *Let Assumption 5.6 hold. Let $p > 0$ and $\lambda > 0$. Assume that there exist N symmetric positive-definite matrices Q_i such that for all*

$(x,t,i) \in \mathbb{R}_0^n \times \mathbb{R}_+ \times \mathbb{S}$,

$$px^T Q_i f(x,t,i) + \frac{p}{2}\text{trace}\big[g^T(x,t,i)Q_i g(x,t,i)\big]$$
$$+ \frac{p(p-2)}{2}(x^T Q_i x)^{-2}|x^T Q_i g(x,t,i)|^2$$
$$+ (x^T Q_i x)^{(2-p)/2} \sum_{j=1}^{N} \gamma_{ij}(x^T Q_j x)^{p/2} \leq -\lambda |x|^2. \qquad (5.32)$$

Then the trivial solution of equation (5.1) is pth moment exponentially stable. If, moreover, condition (5.19) is satisfied, then the trivial solution is also almost surely exponentially stable.

Proof. Let $V(x,t,i)$ be defined by (5.30). Clearly, for $(x,t,i) \in \mathbb{R}_0^n \times \mathbb{R}_+ \times \mathbb{S}$,

$$\left[\min_{1 \leq i \leq N} \lambda_{\min}(Q_i)\right]^{\frac{p}{2}} |x|^p \leq V(x,t,i) \leq \left[\max_{1 \leq i \leq N} \lambda_{\max}(Q_i)\right]^{\frac{p}{2}} |x|^p.$$

Moreover, by (5.31) and (5.32),

$$LV(x,t,i) \leq -\lambda |x|^2 (x^T Q_i x)^{(p-2)/2} \leq -\left[\min_{1 \leq i \leq N} \lambda_{\min}(Q_i)\right]^{\frac{p-2}{2}} \lambda |x|^p$$

if $p \geq 2$ while

$$LV(x,t,i) \leq -\left[\max_{1 \leq i \leq N} \lambda_{\max}(Q_i)\right]^{\frac{p-2}{2}} \lambda |x|^p$$

if $0 < p < 2$. The conclusions follow from Theorems 5.8 and 5.9. □

In the remainder of this section we shall use the theory of M-matrices to establish some sufficient criteria for the exponential stability. These criteria can be verified much more easily than the general results obtained above and prove to be very useful in practice. For this purpose, we impose the following assumption:

Assumption 5.11 Assume that there are constants $K > 0$, $\alpha_i \in \mathbb{R}$, $\sigma_i \geq 0$ and $\rho_i \geq 0$ $(i \in \mathbb{S})$ such that

$$|f(x,t,i)| \leq K|x|, \quad x^T f(x,t,i) \leq \alpha_i |x|^2,$$
$$|g(x,t,i)| \leq \rho_i |x|, \quad |x^T g(x,t,i)| \geq \sigma_i |x|^2,$$

for all $(x,t,i) \in \mathbb{R}^n \times \mathbb{R}_+ \times \mathbb{S}$.

Accordingly, for each $p \geq 0$, define an $N \times N$ matrix

$$\mathcal{A}(p) = \text{diag}(\theta_1(p), \cdots, \theta_N(p)) - \Gamma, \tag{5.33}$$

where

$$\theta_i(p) = \frac{p}{2}\Big[(2-p)\sigma_i^2 - \rho_i^2\Big] - p\alpha_i.$$

Theorem 5.12 *Let Assumption 5.11 hold and $0 < p < 2$. If $\mathcal{A}(p)$ is a nonsingular M-matrix, then the trivial solution of equation (5.1) is pth moment exponentially stable and it is also almost surely exponentially stable.*

Proof. By Theorem 2.10, there exists a vector $\beta = (\beta_1, \cdots, \beta_N)^T \gg 0$ such that

$$\mathcal{A}(p)\beta \gg 0.$$

Set $\mathcal{A}(p)\beta = \bar{\beta} = (\bar{\beta}_1, \cdots, \bar{\beta}_N)^T$. Then

$$\theta_i(p)\beta_i - \sum_{j=1}^{N}\gamma_{ij}\beta_j = \bar{\beta}_i > 0, \quad 1 \leq i \leq N. \tag{5.34}$$

Define $V(x,t,i) = \beta_i|x|^p$ for $(x,t,i) \in \mathbb{R}_0^n \times \mathbb{R}_+ \times \mathbb{S}$. Using (5.31) with $Q_i = \beta_i I$ (I = the identity matrix) and Assumption 5.11, we can derive that

$$LV(x,t,i) = p\beta_i|x|^{p-2}x^T f(x,t,i) + \frac{1}{2}p\beta_i|x|^{p-2}|g(x,t,i)|^2$$

$$- \frac{1}{2}p(2-p)\beta_i|x|^{p-4}|x^T g(x,t,i)|^2 + \sum_{j=1}^{N}\gamma_{ij}\beta_j|x|^p$$

$$\leq \left(\left[p\alpha_i + \frac{1}{2}p\rho_i^2 - \frac{1}{2}p(2-p)\sigma_i^2\right]\beta_i + \sum_{j=1}^{N}\gamma_{ij}\beta_j\right)|x|^p$$

$$= -\bar{\beta}_i|x|^p \leq -\lambda|x|^p,$$

where $\lambda = \min_{1 \leq i \leq N} \bar{\beta}_i > 0$ by (5.34). By Theorem 5.8, the trivial solution of equation (5.1) is pth moment exponentially stable while, by Theorem 5.9, the trivial solution is also almost surely exponentially stable. The proof is therefore complete. □

This theorem tells us that under Assumption 5.11, the trivial solution will be almost surely exponentially stable as long as we can find a $p \in (0,2)$ for $\mathcal{A}(p)$ to be a nonsingular M-matrix. We now look for the conditions which guarantee the existence of such p and hence the almost sure exponential stability. Let us present a useful lemma.

Lemma 5.2 *Let $\mathcal{A}(p)$ be defined by (5.33) for $p \geq 0$. Then*

$$\frac{d}{dp}\det\mathcal{A}(0) = \begin{vmatrix} \sigma_1^2 - \rho_1^2/2 - \alpha_1, & -\gamma_{12}, & \cdots, & -\gamma_{1N} \\ \sigma_2^2 - \rho_2^2/2 - \alpha_2, & -\gamma_{22}, & \cdots, & -\gamma_{2N} \\ \vdots & & & \vdots \\ \sigma_N^2 - \rho_N^2/2 - \alpha_N, & -\gamma_{N2}, & \cdots, & -\gamma_{NN} \end{vmatrix}.$$

Proof. It is easy to see that

$$\det\mathcal{A}(p) = \begin{vmatrix} \theta_1(p), & -\gamma_{12}, & \cdots, & -\gamma_{1N} \\ \theta_2(p), & \theta_2(p) - \gamma_{22}, & \cdots, & -\gamma_{2N} \\ \vdots & & & \vdots \\ \theta_{N-1}(p), & -\gamma_{N-1,2}, & \cdots, & -\gamma_{N-1,N} \\ \theta_N(p), & -\gamma_{N2}, & \cdots, & \theta_N(p) - \gamma_{NN} \end{vmatrix}$$

$$= \sum_{i=1}^{N} \theta_i(p) M_i(p),$$

where $M_i(p)$ is the corresponding minor of $\theta_i(p)$ in the first column. More precisely,

$$M_1(p) = (-1)^{1+1} \begin{vmatrix} \theta_2(p) - \gamma_{22}, & \cdots, & -\gamma_{2N} \\ \vdots & & \vdots \\ -\gamma_{N-1,2}, & \cdots, & -\gamma_{N-1,N} \\ -\gamma_{N2}, & \cdots, & \theta_N(p) - \gamma_{NN} \end{vmatrix},$$

$$\vdots$$

$$M_N(p) = (-1)^{N+1} \begin{vmatrix} -\gamma_{12}, & \cdots, & -\gamma_{1N} \\ \theta_2(p) - \gamma_{22}, & \cdots, & -\gamma_{2N} \\ \vdots & & \vdots \\ -\gamma_{N-1,2}, & \cdots, & -\gamma_{N-1,N} \end{vmatrix}.$$

Noting that

$$\theta_i(0) = 0 \quad \text{and} \quad \frac{d}{dp}\theta_i(0) = \sigma_i^2 - \rho_i^2/2 - \alpha_i,$$

we have

$$\frac{d}{dp}\det \mathcal{A}(0) = \sum_{i=1}^{N}(\sigma_i^2 - \rho_i^2/2 - \alpha_i)M_i(0),$$

but this is the required assertion. □

We shall also need a classical result.

Lemma 5.3 (Minkowski, 1907) *If $A = (a_{ij}) \in Z^{N \times N}$ has all of its row sums positive, that is*

$$\sum_{j=1}^{N} a_{ij} > 0 \quad \text{for all } 1 \leq i \leq N,$$

then $\det A > 0$.

We can now establish a very useful result on the almost sure exponential stability.

Theorem 5.13 *Let Assumption 5.11 hold. Assume*

$$\begin{vmatrix} \sigma_1^2 - \rho_1^2/2 - \alpha_1, & -\gamma_{12}, & \cdots, & -\gamma_{1N} \\ \sigma_2^2 - \rho_2^2/2 - \alpha_2, & -\gamma_{22}, & \cdots, & -\gamma_{2N} \\ & \vdots & & \vdots \\ \sigma_N^2 - \rho_N^2/2 - \alpha_N, & -\gamma_{N2}, & \cdots, & -\gamma_{NN} \end{vmatrix} > 0. \quad (5.35)$$

Assume also that for some $u \in \mathbb{S}$,

$$\gamma_{iu} > 0 \quad \text{for all } i \in \mathbb{S}, \ i \neq u. \quad (5.36)$$

Then the trivial solution of equation (5.1) is almost surely exponentially stable.

Proof. It is known that a determinant will not change its value by switching the ith row with the jth row and then switching the ith column with the jth column. It is also known that given a nonsingular M-matrix, if we switch the ith row with the jth row and then switch the ith column with the jth column, then the new matrix is still a nonsingular M-matrix. We may therefore assume $u = N$ without loss of generality, that is

$$\gamma_{iN} > 0 \quad \text{for all } 1 \leq i \leq N-1 \quad (5.37)$$

instead of (5.36). It is easy to see that $\det \mathcal{A}(0) = 0$. Hence by (5.35), (5.37) and Lemma 5.2, we can find a $p > 0$ sufficiently small for

$$\det \mathcal{A}(p) > 0 \tag{5.38}$$

and

$$\theta_i(p) = \frac{p}{2}\left[(2-p)\sigma_i^2 - \rho_i^2\right] - p\alpha_i > -\gamma_{iN}, \quad 1 \leq i \leq N-1. \tag{5.39}$$

For each $k = 1, 2, \cdots, N-1$, consider the leading principal sub-matrix

$$\mathcal{A}_k(p) := \begin{bmatrix} \theta_1(p) - \gamma_{11}, & -\gamma_{12}, & \cdots, & -\gamma_{1k} \\ -\gamma_{21}, & \theta_2(p) - \gamma_{22}, & \cdots, & -\gamma_{2k} \\ \vdots & & & \vdots \\ -\gamma_{k1}, & -\gamma_{k2}, & \cdots, & \theta_k(p) - \gamma_{kk} \end{bmatrix}$$

of $\mathcal{A}(p)$. Clearly $\mathcal{A}_k(p) \in Z^{k \times k}$. Moreover, by (5.39), each row of this sub-matrix has the sum

$$\theta_i(p) - \sum_{j=1}^{k} \gamma_{ij} \geq \theta_i(p) + \gamma_{iN} > 0.$$

By Lemma 5.3, $\det \mathcal{A}_k(p) > 0$. In other words, we have shown that all the leading principal minors of $\mathcal{A}(p)$ are positive. By Theorem 2.10, $\mathcal{A}(p)$ is a nonsingular M-matrix so the conclusion follows from Theorem 5.12. □

This result can be improved slightly if we take the value of $\sigma_i^2 - \rho_i^2/2 - \alpha_i$ into account. By reordering the states of the Markov chain if necessary, we may assume that for some integer $0 \leq v \leq N$

$$\sigma_i^2 - \frac{\rho_i^2}{2} - \alpha_i \begin{cases} > 0 & \text{if } 1 \leq i \leq v, \\ \leq 0 & \text{otherwise.} \end{cases} \tag{5.40}$$

Especially, all $\sigma_i^2 - \rho_i^2/2 - \alpha_i \leq 0$ if $v = 0$ but all $\sigma_i^2 - \rho_i^2/2 - \alpha_i > 0$ if $v = N$.

Theorem 5.14 *Let Assumption 5.11 hold.*

(i) *If (5.40) holds for $v = N$, then the trivial solution of equation (5.1) is almost surely exponentially stable.*
(ii) *If (5.35) is satisfied and (5.40) holds for some integer v, $0 \leq v < N$, and, moreover, for some u, $v < u \leq N$,*

$$\gamma_{iu} > 0 \quad \text{for all } v < i \leq N, \ i \neq u, \tag{5.41}$$

then the trivial solution of equation (5.1) is still almost surely exponentially stable.

Proof. (i) Note that for every $i \in \mathbb{S}$, $\theta_i(0) = 0$ and
$$\frac{d}{dp}\theta_i(0) = \sigma_i^2 - \frac{\rho_i^2}{2} - \alpha_i > 0$$
since (5.40) holds for $v = N$. We can then choose $p > 0$ so small that $\theta_i(p) > 0$ for all $1 \le i \le N$. Consequently, every row of $\mathcal{A}(p)$ has a positive sum. By Lemma 5.3, we see easily that all the leading principal minors of $\mathcal{A}(p)$ are positive so $\mathcal{A}(p)$ is a nonsingular M-matrix. By Theorem 5.12, the trivial solution of equation (5.1) is almost surely exponentially stable.

(ii) In the same spirit as in the proof of Theorem 5.13 we may assume without loss of generality that (5.41) holds for $u = N$, that is
$$\gamma_{iN} > 0 \quad \text{for all } v < i \le N - 1. \tag{5.42}$$
By the assumptions, we can find a $p > 0$ sufficiently small for
$$\det \mathcal{A}(p) > 0 \tag{5.43}$$
and
$$\theta_i(p) > \begin{cases} 0 & \text{if } 1 \le i \le v, \\ -\gamma_{iN} & \text{if } v < i \le N - 1. \end{cases} \tag{5.44}$$
For each $k = 1, 2, \cdots, N-1$, consider the leading principal sub-matrix $\mathcal{A}_k(p)$ of $\mathcal{A}(p)$ as defined in the proof of Theorem 5.19. If $k \le v$, then
$$\text{the sum of its } i\text{th row } \ge \theta_i(p) > 0$$
and hence $\det \mathcal{A}_k(p) > 0$ by Lemma 5.3. If $k > v$, then
$$\text{the sum of its } i\text{th row } \ge \begin{cases} \theta_i(p) > 0 & \text{if } 1 \le i \le v, \\ \theta_i(p) + \gamma_{iN} > 0 & \text{if } v < i \le k \end{cases}$$
and again $\det \mathcal{A}_k(p) > 0$. In other words, we have shown that all the leading principal minors of $\mathcal{A}(p)$ are positive. By Theorem 2.10, $\mathcal{A}(p)$ is a nonsingular M-matrix so the conclusion follows from Theorem 5.12. $\square$

Let us stress that in general we need to reorder the states of the Markov chain when we apply part (ii) of this theorem. It is also interesting to point out that if $\sigma_i^2 - \rho_i^2/2 - \alpha_i > 0$ for some i, then the trivial solution of equation
$$dx(t) = f(x(t), t, i)dt + g(x(t), t, i)dB(t)$$

is almost surely exponentially stable (see [Mao (1997), Section 4.3]). Hence part (i) of Theorem 5.14 tells us that if every individual equation of

$$dx(t) = f(x(t), t, i)dt + g(x(t), t, i)dB(t), \quad 1 \leq i \leq N \tag{5.45}$$

is almost surely exponentially stable (guaranteed by $\sigma_i^2 - \rho_i^2/2 - \alpha_i > 0$ for all $1 \leq i \leq N$), then as the result of Markovian switching, the overall behaviour, i.e. equation (5.1) remains stable. On the other hand, part (ii) of this theorem tells us a more interesting result that some individuals in (5.45) are stable while some may not, but as the result of Markovian switching, the overall behaviour, i.e. equation (5.1) may be stable. We shall illustrate this point through examples later.

Theorem 5.12 gives a criterion for the pth moment exponential stability in the case when $0 < p < 2$. To discuss the case when $p \geq 2$ we impose, instead of Assumption 5.11, the following one:

Assumption 5.15 For every $i \in \mathbb{S}$, there are two constants $\alpha_i \in \mathbb{R}$ and $\rho_i \geq 0$ such that

$$x^T f(x, t, i) \leq \alpha_i |x|^2 \quad \text{and} \quad |g(x, t, i)| \leq \rho_i |x|$$

for all $(x, t) \in \mathbb{R}^n \times \mathbb{R}_+$.

Accordingly, define an $N \times N$ matrix

$$\bar{\mathcal{A}}(p) = \text{diag}(\bar{\theta}_1(p), \cdots, \bar{\theta}_N(p)) - \Gamma, \tag{5.46}$$

where

$$\bar{\theta}_i(p) = -\frac{p(p-1)}{2}\rho_i^2 - p\alpha_i.$$

Theorem 5.16 *Let Assumption 5.15 hold and $p \geq 2$. If $\bar{\mathcal{A}}(p)$ is a nonsingular M-matrix, then the trivial solution of equation (5.1) is pth moment exponentially stable. If there is moreover a $K > 0$ such that*

$$|f(x, t, i)| \leq K|x| \quad \forall (x, t, i) \in \mathbb{R}^n \times \mathbb{R}_+ \times \mathbb{S},$$

then the trivial solution is also almost surely exponentially stable.

This theorem can be proved in the same way as in the proof of Theorem 5.12 and hence the details are left to the reader.

5.3.1 Nonlinear Jump Systems

When $g(x,t,i) \equiv 0$ equation (5.1) reduces to a special but important nonlinear jump system

$$\dot{x}(t) = f(x(t), t, r(t)). \tag{5.47}$$

A special case is the linear jump equation

$$\dot{x}(t) = A(r(t))x(t) \tag{5.48}$$

which has been discussed by several authors e.g. [Ji and Chizeck (1990); Mariton (1990)]. Using the theory established above we can easily obtain a number of useful results on the stability of equation (5.47).

Clearly, instead of Assumption 5.11 or 5.15, we now need only to impose a simple one: There are constants $K > 0$ and $\alpha_i \in \mathbb{R}$ ($i \in \mathbb{S}$) such that

$$|f(x,t,i)| \leq K|x|, \quad x^T f(x,t,i) \leq \alpha_i |x|^2 \tag{5.49}$$

for all $(x, t, i) \in \mathbb{R}^n \times \mathbb{R}_+ \times \mathbb{S}$. For $p \geq 0$, define an $N \times N$ matrix

$$\tilde{\mathcal{A}}(p) = \text{diag}(-p\alpha_1, \cdots, -p\alpha_N) - \Gamma. \tag{5.50}$$

Corollary 5.17 *Let (5.49) hold and $p > 0$. If $\tilde{\mathcal{A}}(p)$ is a nonsingular M-matrix, then the trivial solution of equation (5.47) is pth moment exponentially stable and it is also almost surely exponentially stable.*

This corollary follows from Theorems 5.12 and 5.16 directly. We also have the following result which follows from Theorem 5.14.

Corollary 5.18 *Let (5.49) hold.*

(i) *If $\alpha_i < 0$ for all $1 \leq i \leq N$, then the trivial solution of equation (5.47) is almost surely exponentially stable.*

(ii) *Assume that*

$$\begin{vmatrix} -\alpha_1, & -\gamma_{12}, & \cdots, & -\gamma_{1N} \\ -\alpha_2, & -\gamma_{22}, & \cdots, & -\gamma_{2N} \\ \vdots & & & \vdots \\ -\alpha_N, & -\gamma_{N2}, & \cdots, & -\gamma_{NN} \end{vmatrix} > 0. \tag{5.51}$$

Assume also that for some integer $0 \leq v < N$,

$$\alpha_i \begin{cases} < 0 & \text{if } 1 \leq i \leq v, \\ \geq 0 & \text{otherwise} \end{cases} \tag{5.52}$$

and, moreover, for some $v < u \leq N$,

$$\gamma_{iu} > 0 \quad \text{for all } v < i \leq N, \ i \neq u, \tag{5.53}$$

then the trivial solution of equation (5.47) is almost surely exponentially stable.

Again we need to stress that in general we need to reorder the states of the Markov chain when we apply part (ii) of this corollary.

It is very interesting to have another look at equation (5.1) from the perturbation point of view. That is, we may regard equation (5.1) as a stochastically perturbed system of equation (5.47). If equation (5.47) is stable and the stochastic perturbation is sufficiently small, we would expect the perturbed equation remains stable. The following corollary describes this situation precisely.

Corollary 5.19 *Let Assumption 5.15 hold and $p \geq 2$. If $\tilde{A}(p)$ defined by (5.50) is a nonsingular M-matrix and*

$$\frac{2}{p(p-1)} \begin{bmatrix} \rho_1^{-2} \\ \vdots \\ \rho_N^{-2} \end{bmatrix} \gg \tilde{A}^{-1}(p) \begin{bmatrix} 1 \\ \vdots \\ 1 \end{bmatrix}, \tag{5.54}$$

where we set $\rho_i^{-2} = \infty$ when $\rho_i = 0$, then the trivial solution of equation (5.1) is pth moment exponentially stable. If there is moreover a $K > 0$ such that

$$|f(x,t,i)| \leq K|x| \quad \forall (x,t,i) \in \mathbb{R}^n \times \mathbb{R}_+ \times \mathbb{S},$$

then the trivial solution is also almost surely exponentially stable.

Proof. By Theorem 2.10, $\tilde{A}^{-1}(p) \geq 0$ (namely, its every element is nonnegative). It is easy to see that every row of $\tilde{A}^{-1}(p)$ has at least one positive element. Hence

$$\beta = \begin{bmatrix} \beta_1 \\ \vdots \\ \beta_N \end{bmatrix} := \tilde{A}^{-1}(p) \begin{bmatrix} 1 \\ \vdots \\ 1 \end{bmatrix} \gg 0.$$

By (5.54), $\frac{2}{p(p-1)}\rho_i^{-2} > \beta_i$ or $1 > \frac{p(p-1)}{2}\rho_i^2 \beta_i$ for every $1 \leq i \leq N$. Hence

$$\tilde{A}(p)\beta = (1,\cdots,1)^T \gg \frac{p(p-1)}{2}\text{diag}(\rho_1^2,\cdots,\rho_N^2)\beta,$$

that is
$$\tilde{\mathcal{A}}(p)\beta - \frac{p(p-1)}{2}\text{diag}(\rho_1^2, \cdots, \rho_N^2)\beta \gg 0.$$

But, by definition (5.46),
$$\bar{\mathcal{A}}(p) = \tilde{\mathcal{A}}(p) - \frac{p(p-1)}{2}\text{diag}(\rho_1^2, \cdots, \rho_N^2).$$

We therefore have
$$\bar{\mathcal{A}}(p)\beta \gg 0.$$

By Theorem 2.10, $\bar{\mathcal{A}}(p)$ is a nonsingular M-matrix. Now the conclusions follow from Theorem 5.16. $\square$

When ρ_i's are known, it is better to apply Theorem 5.16. However, when ρ_i's are unknown, the above corollary gives a robust bound of stability for ρ_i's.

When $0 < p < 2$, we can also apply Theorem 5.12 to obtain a similar result assuming Assumption 5.11 hold and $\tilde{\mathcal{A}}(p)$ is a nonsingular M-matrix. The details are left to the reader.

5.3.2 Multi-Dimensional Linear Equations

Let us now consider linear stochastic differential equations with Markovian switching of the form

$$dx(t) = A(r(t))x(t)dt + \sum_{k=1}^{m} G_k(r(t))x(t)dB_k(t). \qquad (5.55)$$

Here $A(i)$ and $G_k(i)$ are all $n \times n$ matrices and we shall write $A(i) = A_i$ and $G_k(i) = G_{ki}$. Clearly, if we define

$$f(x,t,i) = A_i x \quad \text{and} \quad g(x,t,i) = (G_{1i}x, \cdots, G_{mi}x), \qquad (5.56)$$

then equation (5.55) can be written as equation (5.1). Moreover, we have

$$x^T f(x,t,i) = x^T A_i x = \frac{1}{2}x^T(A_i + A_i^T)x \leq \frac{1}{2}\lambda_{\max}(A_i + A_i^T)|x|^2$$

and

$$|f(x,t,i)| = |A_i x| \leq \|A_i\|\,|x| \leq \left(\max_{1 \leq i \leq N} \|A_i\|\right)|x|.$$

Compute

$$|x^T g(x,t,i)|^2 = |x^T(G_{1i}x,\cdots,G_{ki}x)|^2 = \sum_{k=1}^m (x^T G_{ki} x)^2$$

$$= \sum_{k=1}^m \left(\frac{1}{2}x^T(G_{ki}+G_{ki}^T)x\right)^2 \geq \frac{1}{4}\sum_{k=1}^m \lambda_{\min}^2(G_{ki}+G_{ki}^T)|x|^4.$$

Hence

$$|x^T g(x,t,i)| \geq \frac{1}{2}\sqrt{\sum_{k=1}^m \lambda_{\min}^2(G_{ki}+G_{ki}^T)}\,|x|^2.$$

Furthermore, we have

$$|g(x,t,i)| = \sqrt{\sum_{k=1}^m |G_{ki}x|^2} \leq \sqrt{\sum_{k=1}^m \|G_{ki}\|^2}\,|x|.$$

Summarising the aboves we see that with definition (5.56), Assumptions 5.11 and 5.15 are satisfied with

$$\begin{array}{ll} \alpha_i = \frac{1}{2}\lambda_{\max}(A_i+A_i^T), & \sigma_i = \frac{1}{2}\sqrt{\sum_{k=1}^m \lambda_{\min}^2(G_{ki}+G_{ki}^T)}, \\ \rho_i = \sqrt{\sum_{k=1}^m \|G_{ki}\|^2}, & K = \max_{1\leq i\leq N}\|A_i\|. \end{array} \quad (5.57)$$

Therefore, by the theorems obtained above, we have the following results immediately.

Corollary 5.20 *Let the parameters be specified by (5.57). The trivial solution of equation (5.55) is both pth moment and almost surely exponentially stable if one of the following statements is true:*

(i) $0 < p < 2$ and $\mathcal{A}(p)$ defined by (5.33) is a nonsingular M-matrix.
(ii) $p \geq 2$ and $\bar{\mathcal{A}}(p)$ defined by (5.46) is a nonsingular M-matrix.

Corollary 5.21 *Let the parameters be specified by (5.57). The trivial solution of equation (5.55) is almost surely exponentially stable if one of the following statements is true:*

(i) *(5.40) holds for $v = N$.*
(ii) *(5.35) and (5.36) hold.*
(iii) *(5.35) holds, (5.40) holds for some integer $0 \leq v < N$ and, moreover, for some $v < u \leq N$,*

$$\gamma_{iu} > 0 \quad \text{for all } v < i \leq N,\ i \neq u.$$

5.3.3 Scalar Linear Equations

Let us now turn to discuss a special but important scalar linear SDE with Markovian switching of the form

$$dx(t) = \mu(r(t))x(t)dt + \delta(r(t))x(t)dB(t) \qquad (5.58)$$

on $t \geq t_0$ with initial data $x(t_0) = x_0 \in \mathbb{R}$ and $r(t_0) = r_0 \in \mathbb{S}$. Here in this subsection, we let $B(t)$ be a scalar Brownian motion to avoid complicated notations but the theory holds for multi-dimensional Brownian motions. Moreover, μ and δ are mappings from $\mathbb{S} \to \mathbb{R}$ and we shall write $\mu(i) = \mu_i$ and $\delta(i) = \delta_i$.

Of course, this scalar linear equation is a special case of the SDE (5.55) so the results in the previous subsection, namely Corollaries 5.20 and 5.21 can be applied to it but the parameters defined by (5.57) become

$$\sigma_i = \mu_i, \quad \bar{\sigma}_i = \rho_i = |\delta_i|, \quad K = \max_{1 \leq i \leq N} |\mu_i|.$$

However, this equation has its explicit solution and, making use of it, we can obtain "if and only if" stability criteria. Let us establish the explicit solution first.

Theorem 5.22 *The linear SDE (5.58) has the explicit solution*

$$x(t) = x_0 \exp\left\{ \int_{t_0}^{t} [\mu(r(s)) - \tfrac{1}{2}\delta^2(r(s))]ds + \int_{t_0}^{t} \delta(r(s))dB(s) \right\}. \qquad (5.59)$$

Proof. The formula (5.59) holds obviously when $x_0 = 0$. For $x_0 < 0$, it is easy to see that $x(t; t_0, x_0, r_0) = -x(t; t_0, -x_0, r_0)$ we hence only need to show the formula for $x_0 > 0$. Write $x(t; t_0, x_0, r_0) = x(t)$. We observe from Lemma 5.1 that $x(t) > 0$ on $t \geq t_0$ with probability 1. We can then apply the Itô formula to $\log(x(t))$ to have

$$d[\log(x(t))] = [\mu(r(t)) - \tfrac{1}{2}\delta^2(r(t))]dt + \delta(r(t))dB(t),$$

namely

$$\log(x(t)) = \log(x_0) + \int_{t_0}^{t}[\mu(r(s)) - \tfrac{1}{2}\delta^2(r(s))]ds + \int_{t_0}^{t} \delta(r(s))dB(s).$$

This implies the desired formula (5.59). □

In order to obtain the "if and only if" criteria on exponential stability we assume, as a standing hypothesis in this section, that the Markov chain is *irreducible*. This is equivalent to the condition that for any $i, j \in \mathbb{S}$, one can find finite numbers $i_1, i_2, \cdots, i_k \in \mathbb{S}$ such that

$$\gamma_{i,i_1}\gamma_{i_1,i_2}\cdots\gamma_{i_k,j} > 0.$$

The algebraic interpretation of irreducibility is $\text{rank}(\Gamma) = N - 1$. Under this condition, the Markov chain has a unique stationary (probability) distribution $\pi = (\pi_1, \pi_2, \cdots, \pi_N) \in \mathbb{R}^{1 \times N}$ which can be determined by solving the following linear equation

$$\pi \Gamma = 0 \qquad (5.60)$$

subject to

$$\sum_{j=1}^{N} \pi_j = 1 \quad \text{and} \quad \pi_j > 0 \quad \forall j \in \mathbb{S}.$$

The following theorem gives the sufficient and necessary condition for the SDE (5.58) to be almost surely exponentially stable.

Theorem 5.23 *The sample Lypunov exponent of the SDE (5.58) is*

$$\lim_{t \to \infty} \frac{1}{t} \log(|x(t)|) = \sum_{j=1}^{N} \pi_j(\mu_j - \tfrac{1}{2}\delta_j^2) \quad a.s. \qquad (5.61)$$

(for $x_0 \neq 0$ of course). Hence the SDE (5.58) is almost surely exponentially stable if and only if

$$\sum_{j=1}^{N} \pi_j(\alpha_j - \tfrac{1}{2}\delta_j^2) < 0. \qquad (5.62)$$

Proof. For any $x_0 \neq 0$, it follows from (5.59) that

$$\log(|x(t)|) = \log(|x_0|) + \int_{t_0}^{t} [\mu(r(s)) - \tfrac{1}{2}\delta^2(r(s))]ds + \int_{t_0}^{t} \delta(r(s))dB(s). \qquad (5.63)$$

By Theorem 1.6,

$$\lim_{t \to \infty} \frac{1}{t} \int_{t_0}^{t} \delta(r(s))dB(s) = 0 \quad a.s.$$

while by the ergodic property of the Markov chain,

$$\lim_{t\to\infty} \frac{1}{t}\int_{t_0}^t [\mu(r(s)) - \tfrac{1}{2}\delta^2(r(s))]ds = \sum_{j=1}^N \pi_j(\alpha_j - \tfrac{1}{2}\delta_j^2) \quad a.s.$$

Dividing both sides of (5.63) by t and then letting $t \to \infty$ we hence obtain the assertion (5.61). □

The following theorem gives the sufficient and necessary condition for the SDE (5.58) to be pth moment exponentially stable. It should be pointed out that the proof for the pth moment exponential stability of a linear scalar SDE *without* Markovian switching is rather simple while the proof below for the SDE with Markovian switching becomes much more complicated.

Theorem 5.24 *The pth moment Lypunov exponent of the SDE (5.58) is*

$$\lim_{t\to\infty} \frac{1}{t} \log(\mathbb{E}|x(t)|^p) = \sum_{j=1}^N \pi_j p[\mu_j + \tfrac{1}{2}(p-1)\delta_j^2] \quad (5.64)$$

(for $x_0 \ne 0$ of course). Hence the SDE (5.58) is pth moment exponentially stable if and only if

$$\sum_{j=1}^N \pi_j[\alpha_j + \tfrac{1}{2}(p-1)\delta_j^2] < 0. \quad (5.65)$$

Proof. It is well known (see Chapter 1) that almost every sample path of the Markov chain $r(\cdot)$ is a right continuous step function with a finite number of sample jumps in any finite subinterval of $\mathbb{R}_+ := [t_0, \infty)$. Hence there is a sequence of finite stopping times $t_0 = \tau_0 < \tau_1 < \cdots < \tau_k \to \infty$ such that

$$r(t) = \sum_{k=0}^\infty r(\tau_k) I_{[\tau_k, \tau_{k+1})}(t), \quad t \ge t_0.$$

For any integer $z > 0$, it then follows from (5.59) that

$$|x(t \wedge \tau_z)|^p$$
$$= |x_0|^p \exp\Big\{ \int_{t_0}^{t\wedge\tau_z} [p\mu(r(s)) - \tfrac{1}{2}p\delta^2(r(s))]ds + \int_0^{t\wedge\tau_z} p\delta(r(s))dB(s) \Big\}$$

$$= \xi(t \wedge \tau_z) \exp\left\{ -\int_{t_0}^{t\wedge\tau_z} \tfrac{1}{2} p^2 \delta^2(r(s)) ds + \int_{t_0}^{t\wedge\tau_z} p\delta(r(s)) dB(s) \right\}$$

$$= \xi(t \wedge \tau_z) \prod_{k=0}^{z-1} \zeta_k,$$

where

$$\xi(t \wedge \tau_z) = |x_0|^p \exp\left\{ \int_{t_0}^{t\wedge\tau_z} [p\mu(r(s)) + \tfrac{1}{2} p(p-1)\delta^2(r(s))] ds \right\},$$

$$\zeta_k = \exp\left\{ -\tfrac{1}{2} p^2 \delta^2(r(t \wedge \tau_k))(t \wedge \tau_{k+1} - t \wedge \tau_k) \right.$$
$$\left. + p\delta(r(t \wedge \tau_k))[B(t \wedge \tau_{k+1}) - B(t \wedge \tau_k)] \right\}.$$

Let $\mathcal{G}_t = \sigma(\{r(u)\}_{u \geq t_0}, \{B(s)\}_{t_0 \leq s \leq t})$, namely the σ-algebra generated by $\{r(u)\}_{u \geq t_0}$ and $\{B(s)\}_{t_0 \leq s \leq t}$. Compute

$$\mathbb{E}|x(t \wedge \tau_z)|^p = \mathbb{E}\Big(\xi(t \wedge \tau_z) \prod_{k=0}^{z-1} \zeta_k\Big) = \mathbb{E}\Big\{\mathbb{E}\Big(\xi(t \wedge \tau_z) \prod_{k=0}^{z-1} \zeta_k \Big| \mathcal{G}_{t \vee \tau_{z-1}}\Big)\Big\}$$

$$= \mathbb{E}\Big\{\Big[\xi(t \wedge \tau_z) \prod_{k=0}^{z-2} \zeta_k\Big] \mathbb{E}\Big(\zeta_{z-1} \Big| \mathcal{G}_{t \vee \tau_{z-1}}\Big)\Big\}. \tag{5.66}$$

Define, for $i \in \mathbb{S}$,

$$\zeta_{z-1}(i) = \exp\left\{ -\tfrac{1}{2} p\delta_i^2 (t \wedge \tau_z - t \wedge \tau_{z-1}) + p\delta_i [B(t \wedge \tau_z) - B(t \wedge \tau_{z-1})] \right\}.$$

By Theorem 1.38 (i.e. the exponential martingale formula), we have

$$\mathbb{E}\zeta_{z-1}(i) = 1, \quad i \in \mathbb{S}.$$

Then

$$\mathbb{E}\Big(\zeta_{z-1} \Big| \mathcal{G}_{t\vee\tau_{z-1}}\Big) = \mathbb{E}\Big(\sum_{i\in\mathbb{S}} I_{\{r(t\wedge\tau_{z-1})=i\}} \zeta_{z-1}(i) \Big| \mathcal{G}_{t\vee\tau_{z-1}}\Big)$$

$$= \sum_{i\in\mathbb{S}} I_{\{r(t\wedge\tau_{z-1})=i\}} \mathbb{E}\Big(\zeta_{z-1}(i) \Big| \mathcal{G}_{t\vee\tau_{z-1}}\Big).$$

Noting that $t\wedge\tau_z - t\wedge\tau_{z-1}$ is $\mathcal{G}_{t\vee\tau_{z-1}}$-measurable while $B(t\wedge\tau_z) - B(t\wedge\tau_{z-1})$ is independent of $\mathcal{G}_{t\vee\tau_{z-1}}$, we have

$$\mathbb{E}\Big(\zeta_{z-1}(i) \Big| \mathcal{G}_{t\vee\tau_{z-1}}\Big) = \mathbb{E}\zeta_{z-1}(i) = 1,$$

whence
$$\mathbb{E}\left(\zeta_{z-1}\big|\mathcal{G}_{t\vee\tau_{z-1}}\right) = 1.$$

Substituting this into (5.66) yields

$$\mathbb{E}|x(t\wedge\tau_z)|^p = \mathbb{E}\left[\xi(t\wedge\tau_z)\prod_{k=0}^{z-2}\zeta_k\right]. \tag{5.67}$$

Repeating this procedure implies

$$\mathbb{E}|x(t\wedge\tau_z)|^p = \mathbb{E}\xi(t\wedge\tau_z).$$

Letting $z\to\infty$ we obtain

$$\mathbb{E}|x(t)|^p = \mathbb{E}\left\{|x_0|^p\exp\left[\int_{t_0}^t [p\mu(r(s)) + \tfrac{1}{2}p(p-1)\delta^2(r(s))]ds\right]\right\}. \tag{5.68}$$

Now, by the ergodic property of the Markov chain (see e.g. [Anderson (1991)]), we have

$$\lim_{t\to\infty}\frac{1}{t}\int_0^t [p\mu(r(s)) + \tfrac{1}{2}p(p-1)\delta^2(r(s))]ds$$
$$= \sum_{j\in\mathbb{S}}\pi_j(p\alpha_j + \tfrac{1}{2}p(p-1)\delta_j^2) := \gamma \quad a.s. \tag{5.69}$$

Let $\varepsilon > 0$ be arbitrary. It follows from (5.68) that

$$e^{-(\gamma-\varepsilon)t}\mathbb{E}|x(t)|^p$$
$$= \mathbb{E}\left\{|x_0|^2\exp\left[-(\gamma-\varepsilon)t + \int_0^t [p\mu(r(s)) + \tfrac{1}{2}p(p-1)\delta^2(r(s))]ds\right]\right\}.$$

By (5.69),

$$\lim_{t\to\infty}\exp\left[-(\gamma-\varepsilon)t + \int_0^t [p\mu(r(s)) + \tfrac{1}{2}p(p-1)\delta^2(r(s))]ds\right] = \infty \quad a.s.$$

Hence

$$\lim_{t\to\infty} e^{-(\gamma-\varepsilon)t}\mathbb{E}|x(t)|^p = \infty,$$

which implies

$$\mathbb{E}|x(t)|^p \geq e^{(\gamma-\varepsilon)t} \quad \text{for all sufficiently large } t,$$

whence
$$\liminf_{t\to\infty} \frac{1}{t}\log(\mathbb{E}|x(t)|^p) \geq \gamma - \varepsilon.$$

Similar, we can show
$$\limsup_{t\to\infty} \frac{1}{t}\log(\mathbb{E}|x(t)|^p) \leq \gamma + \varepsilon.$$

Since ε is arbitrary, we must have
$$\lim_{t\to\infty} \frac{1}{t}\log(\mathbb{E}|x(t)|^p) = \gamma,$$

which is the required assertion (5.64). □

5.3.4 *Examples*

Let us now discuss a number of examples to illustrate our theory.

Example 5.25 Let $B(t)$ be a scalar Brownian motion. Let $r(t)$ be a right-continuous Markov chain taking values in $\mathbb{S} = \{1, 2\}$ with the generator $\Gamma = (\gamma_{ij})_{2\times 2}$:
$$-\gamma_{11} = \gamma_{12} > 0, \quad -\gamma_{22} = \gamma_{21} > 0.$$

Of course $B(t)$ and $r(t)$ are assumed to be independent. Consider a one-dimensional stochastic differential equation with Markovian switching of the form
$$dx(t) = f(x(t), t, r(t))dt + b(r(t))x(t)dB(t) \tag{5.70}$$

on $t \geq 0$. Here
$$\begin{array}{ll} f(x,t,i) = \sin(x)\cos(2t) \text{ and } b(i) = 1 \text{ if } i = 1 \\ f(x,t,i) = 2x \qquad\qquad\quad \text{and } b(i) = 2 \text{ if } i = 2 \end{array} \tag{5.71}$$

for $(x, t, i) \in \mathbb{R} \times \mathbb{R}_+ \times \mathbb{S}$. Note
$$xf(x,t,i) \leq |x|^2 \text{ if } i = 1 \text{ but } \leq 2|x|^2 \text{ if } i = 2$$

and
$$|f(x,t,i)| \leq 2|x|.$$

Applying Theorem 5.13 with
$$\sigma_1 = \rho_1 = 2, \quad \sigma_2 = \rho_2 = 1, \quad \alpha_1 = 1, \quad \alpha_2 = 2$$

we can therefore conclude that the trivial solution of equation (5.70) is almost surely exponentially stable if

$$\begin{vmatrix} 2^2 - 2^2/2 - 1 & -\gamma_{12} \\ 1^2 - 1^2/2 - 2 & -\gamma_{22} \end{vmatrix} = \begin{vmatrix} 1 & -\gamma_{12} \\ -1.5 & \gamma_{21} \end{vmatrix} = \gamma_{21} - 1.5\gamma_{12} > 0, \tag{5.72}$$

that is $\gamma_{21} > 1.5\gamma_{12}$. To see what this example shows us, we regard equation (5.70) as the result of the following two equations

$$dx(t) = \sin(x(t))\cos(2t)dt + 2x(t)dB(t) \tag{5.73}$$

and

$$dx(t) = 2x(t)dt + x(t)dB(t) \tag{5.74}$$

switching from one to the other according to the movement of the Markov chain $r(t)$. We observe that equation (5.73) is almost surely exponentially stable since the Lyapunov exponent is not greater than $\lambda_1 = -1$ while equation (5.74) is almost surely exponentially unstable since the Lyapunov exponent is $\lambda_2 = 1.5$. However, as the result of Markovian switching, the overall behaviour, i.e. equation (5.70) will be almost surely exponentially stable as long as the transition rate γ_{21} from the unstable equation (5.74) to the stable equation (5.73) is greater than $\lambda_2/|\lambda_1| = 1.5$ time of the transition rate γ_{12} from the stable equation (5.73) to the unstable equation (5.74).

Example 5.26 Let $B(t)$ be a scalar Brownian motion. Let $r(t)$ be a right-continuous Markov chain taking values in $\mathbb{S} = \{1, 2, 3\}$ with the generator

$$\Gamma = \begin{bmatrix} -2 & 1 & 1 \\ 3 & -4 & 1 \\ 1 & 1 & -2 \end{bmatrix}.$$

Assume that $B(t)$ and $r(t)$ are independent. Consider a three-dimensional semilinear stochastic differential equation with Markovian switching of the form

$$dx(t) = A(r(t))x(t)dt + g(x(t), t, r(t))dB(t) \tag{5.75}$$

on $t \geq 0$. Here

$$A(1) = A_1 = \begin{bmatrix} -2 & -1 & -2 \\ 2 & -2 & 1 \\ 1 & -2 & -3 \end{bmatrix}, \quad A(2) = A_2 = \begin{bmatrix} 0.5 & 1 & 0.5 \\ -0.8 & 0.5 & 1 \\ -0.7 & -0.9 & 0.2 \end{bmatrix},$$

$$A(3) = A_3 = \begin{bmatrix} -0.5 & -0.9 & -1 \\ 1 & -0.6 & -0.7 \\ 0.8 & 1 & -1 \end{bmatrix}.$$

Moreover, $g : \mathbb{R}^3 \times \mathbb{R}_+ \times \mathbb{S} \to \mathbb{R}^3$ satisfying

$$|g(x,t,i)| \le \rho_i |x|, \quad (x,t) \in \mathbb{R}^3 \times \mathbb{R}_+$$

for each $i \in \mathbb{S}$, where $\rho_i > 0$. Assume we are required to find out if the trivial solution is 3rd moment exponentially stable. When ρ_i's are unknown, we apply Corollary 5.19 to obtain the bounds for them. So we compute $\alpha_i = \frac{1}{2}\lambda_{\max}(A_i + A_i^T)$:

$$\alpha_1 = -1.21925, \quad \alpha_2 = 0.60359, \quad \alpha_3 = -0.47534.$$

The matrix $\tilde{\mathcal{A}}(3)$ defined by (5.50) becomes

$$\tilde{\mathcal{A}}(3) = \text{diag}(3.65775, -1.81077, 1.42602) - \Gamma = \begin{bmatrix} 5.65775 & -1 & -1 \\ -3 & 2.18923 & -1 \\ -1 & -1 & 3.42602 \end{bmatrix}.$$

Compute

$$\tilde{\mathcal{A}}^{-1}(3) = \begin{bmatrix} 0.320056 & 0.217923 & 0.157027 \\ 0.555295 & 0.905147 & 0.426279 \\ 0.255501 & 0.327806 & 0.462142 \end{bmatrix}.$$

By Theorem 2.10, $\mathcal{A}(3)$ is a nonsingular M-matrix. Compute

$$\tilde{\mathcal{A}}^{-1}(3)(1,1,1)^T = (0.69501, 1.88672, 1.04545)^T.$$

In view of Corollary 5.19, we can conclude that if

$$\frac{1}{3}(\rho_1^{-2}, \rho_2^{-2}, \rho_3^{-2})^T \gg (0.69501, 1.88672, 1.04545)^T,$$

that is

$$\rho_1 < 0.69253, \quad \rho_2 < 0.42032, \quad \rho_3 < 0.56466, \tag{5.76}$$

then the trivial solution of equation (5.75) is 3rd moment exponentially stable and is also almost surely exponentially stable.

On the other hand, suppose we are given

$$\rho_1 = 0.7, \quad \rho_2 = 0.3, \quad \rho_3 = 0.5 \tag{5.77}$$

so (5.76) is not satisfied. However, we may still apply Theorem 5.16. Note that the matrix $\bar{A}(3)$ defined by (5.46) becomes

$$\bar{A}(3) = \text{diag}(2.18775, -2.08077, 0.67602) - \Gamma = \begin{bmatrix} 4.18775 & -1 & -1 \\ -3 & 1.91923 & -1 \\ -1 & -1 & 2.67602 \end{bmatrix}.$$

It is easy to verify that all the leading principal minors of $\bar{A}(3)$ are positive and hence $\bar{A}(3)$ is a nonsingular M-matrix. By Theorem 5.16, we see that under (5.77) the trivial solution of equation (5.75) is still both 3rd moment and almost surely exponentially stable.

Example 5.27 Consider equation (5.1) but we let $r(t)$ be a right-continuous Markov chain taking values in $\mathbb{S} = \{1, 2, 3, 4\}$ with the generator

$$\Gamma = \begin{bmatrix} -2 & 1 & 1 & 0 \\ 1 & -2 & 1 & 0 \\ 1 & 2 & -4 & 1 \\ 1 & 2 & 0 & -3 \end{bmatrix}.$$

We assume that Assumption 5.11 is satisfied with the parameters specified as follows:

$$\alpha_1 = -1, \; \alpha_2 = -2, \; \alpha_3 = 0, \quad \alpha_4 = 1,$$
$$\sigma_1 = 1, \quad \sigma_2 = 0.5, \; \sigma_3 = 1, \quad \sigma_4 = 0.5,$$
$$\rho_1 = 1.5, \; \rho_2 = 1, \quad \rho_3 = 1.5, \; \rho_4 = 1,$$

and $K > 0$. Compute

$$\sigma_i^2 - \frac{\rho_i^2}{2} - \alpha_i = \begin{cases} 0.875 & \text{if } i = 1, \\ 1.750 & \text{if } i = 2, \\ -0.125 & \text{if } i = 3, \\ -1.250 & \text{if } i = 4. \end{cases} \quad (5.78)$$

Verify (5.35)

$$\begin{vmatrix} 0.875 & -1 & -1 & 0 \\ 1.75 & 2 & -1 & 0 \\ -0.125 & -2 & 4 & -1 \\ -1.25 & -2 & 0 & 3 \end{vmatrix} = 44.125 > 0.$$

We also see from (5.78) that (5.40) holds for $v = 2$. Moreover, (5.41) holds for $u = 4$ since $\gamma_{34} = 1 > 0$. Hence, by Theorem 5.14, the trivial solution of equation (5.1) with the parameters specified as above is almost surely exponentially stable.

5.4 Moment and Almost Sure Asymptotic Stability

In the previous we have discussed the exponential stability in detail. However, there are many equations whose solutions will tend to zero asymptotically but may not exponentially. It is therefore very useful to obtain some criteria on the asymptotic stability, which is the aim of this section. Let us begin with the definition.

Definition 5.28 For $p > 0$, the trivial solution of equation (5.1) or, simply, equation (5.1) is said to be *asymptotically stable in pth moment* if

$$\lim_{t \to \infty} \mathbb{E}(|x(t; t_0, x_0, r_0)|^p) = 0$$

for all $(t_0, x_0, r_0) \in \mathbb{R}_+ \times \mathbb{R}^n \times \mathbb{S}$. When $p = 2$, it is said to be *asymptotically stable in mean square*. Moreover, it is said to be *almost surely asymptotically stable* or *asymptotically stable with probability 1* if

$$\lim_{t \to \infty} x(t; t_0, x_0, r_0) = 0 \quad a.s.$$

for all $(t_0, x_0, r_0) \in \mathbb{R}_+ \times \mathbb{R}^n \times \mathbb{S}$.

Let us also introduce some new notations which will be used in this section as well as the other sections below. Let $L^1(\mathbb{R}_+; \mathbb{R}_+)$ denote the family of functions $\gamma : \mathbb{R}_+ \to \mathbb{R}_+$ such that $\int_0^\infty \gamma(t)dt < \infty$. Denote by $D(\mathbb{R}_+; \mathbb{R}_+)$ the family of functions $\eta : \mathbb{R}_+ \to \mathbb{R}_+$ such that $\int_0^\infty \eta(t)dt = \infty$. Moreover, denote by $\Psi(\mathbb{R}_+; \mathbb{R}_+)$ the family of all continuous functions $\psi : \mathbb{R}_+ \to \mathbb{R}_+$ such that for any $\delta > 0$ and any increasing sequence $\{t_k\}_{k \geq 1}$

$$\sum_{k=1}^\infty \int_{t_k}^{t_k + \delta} \psi(t)dt = \infty.$$

There are many such functions. For example, if ψ is a uniformly continuous non-negative function on $\mathbb{R}_+$ with the property that for any $\delta > 0$ and any increasing sequence $\{t_k\}_{k \geq 1}$

$$\limsup_{k \to \infty} \left[\max_{t_k \leq t \leq t_k + \delta} \psi(t) \right] > 0,$$

then $\psi \in \Psi(\mathbb{R}_+; \mathbb{R}_+)$. In fact, since δ is arbitrary, it follows from the above that

$$\limsup_{k \to \infty} \left[\max_{t_k \leq t \leq t_k + \delta/2} \psi(t) \right] > 0.$$

One can then find a subsequence $\{\tau_{\bar{k}}\}$ such that $t_{\bar{k}} \leq \tau_{\bar{k}} \leq t_{\bar{k}} + \delta/2$ and

$$\varepsilon := \liminf_{\bar{k} \to \infty} \psi(\tau_{\bar{k}}) > 0.$$

Since ψ is uniformly continuous on $\mathbb{R}_+$, one can find a $\bar{\delta} \in (0, \delta/2)$ and an integer k_0 such that

$$\psi(t) \geq \frac{\varepsilon}{2} \quad \text{for all } \tau_{\bar{k}} \leq t \leq \tau_{\bar{k}} + \bar{\delta}, \ \bar{k} \geq k_0.$$

Therefore

$$\sum_{k=1}^{\infty} \int_{t_k}^{t_k+\delta} \psi(t) dt \geq \sum_{\bar{k}=k_0}^{\infty} \int_{\tau_{\bar{k}}}^{\tau_{\bar{k}}+\bar{\delta}} \frac{\varepsilon}{2} dt = \infty.$$

This means $\psi \in \Psi(\mathbb{R}_+; \mathbb{R}_+)$ as claimed. From this property we see that $|\sin t|$, $\cos^2(t)$ etc, belong to $\Psi(\mathbb{R}_+; \mathbb{R}_+)$.

Theorem 5.29 *Assume that there are functions $V \in C^{2,1}(\mathbb{R}^n \times \mathbb{R}_+ \times \mathbb{S}; \mathbb{R}_+)$ and $\gamma \in L^1(\mathbb{R}_+; \mathbb{R}_+)$, and constants $p, \alpha > 0$, such that*

$$LV(x, t, i) \leq \gamma(t) - \alpha |x|^p \tag{5.79}$$

for all $(x, t, i) \in \mathbb{R}^n \times \mathbb{R}_+ \times \mathbb{S}$. Moreover, assume that there is a constant $K > 0$ such that

$$|f(x, t, i)| + |g(x, t, i)| \leq K|x| \tag{5.80}$$

for all $(x, t, i) \in \mathbb{R}^n \times \mathbb{R}_+ \times \mathbb{S}$. Then equation (5.1) is not only asymptotically stable in pth moment but also almost surely asymptotically stable.

Proof. Fix any $(t_0, x_0, r_0) \in \mathbb{R}_+ \times \mathbb{R}^n \times \mathbb{S}$ and write $x(t; t_0, x_0, r_0) = x(t)$. Without loss of any generality we may let $t_0 = 0$; the case of $t_0 > 0$ can be proved in the same way. We divide the whole proof into three steps.

Step 1. By the generalised Itô formula and condition (5.79),

$$0 \leq V(x_0, 0, r_0) + \int_0^t \gamma(s) ds - \alpha \int_0^t \mathbb{E}|x(s)|^p ds.$$

Letting $t \to \infty$ yields

$$\int_0^\infty \mathbb{E}|x(t)|^p dt < \infty. \tag{5.81}$$

We now claim that if $\delta > 0$ is sufficiently small for

$$(3K)^p[\delta^p + C_p \delta^{p/2}] < \frac{2}{3}, \tag{5.82}$$

then

$$\int_0^\infty \mathbb{E}\left(\sup_{t \leq s \leq t+\delta} |x(s)|^p\right) dt < \infty, \tag{5.83}$$

where C_p is the constant given by the Burkholder–David–Gundy inequality (i.e. Theorem 2.13). Note that for any $a, b, c \geq 0$,

$$(a+b+c)^p \leq [3(a \vee b \vee c)]^p \leq 3^p(a^p + b^p + c^p).$$

It follows from equation (5.1) that for any $t \geq 0$,

$$\mathbb{E}\left(\sup_{t \leq s \leq t+\delta} |x(s)|^p\right)$$

$$\leq 3^p \mathbb{E}|x(t)|^p + 3^p \mathbb{E}\left(\int_t^{t+\delta} |f(x(s), s, r(s))| ds\right)^p$$

$$+ 3^p \mathbb{E}\left(\sup_{t \leq s \leq t+\delta} \left|\int_t^s g(x(u), u, r(u)) dB(u)\right|^p\right). \tag{5.84}$$

By condition (5.80), compute

$$\mathbb{E}\left(\int_t^{t+\delta} |f(x(s), s, r(s))| ds\right)^p \leq (\delta K)^p \mathbb{E}\left(\sup_{t \leq s \leq t+\delta} |x(s)|^p\right). \tag{5.85}$$

Also, by the Burkholder–David–Gundy inequality,

$$\mathbb{E}\left(\sup_{t \leq s \leq t+\delta} \left|\int_t^s g(x(u), u, r(u)) dB(u)\right|^p\right)$$

$$\leq C_p \mathbb{E}\left(\int_t^{t+\delta} |g(x(s), s, r(s))|^2 ds\right)^{p/2}$$

$$\leq C_p \mathbb{E}\left(\delta K^2 \sup_{t \leq s \leq t+\delta} |x(s)|^2\right)^{p/2}$$

$$\leq C_p \delta^{p/2} K^p \mathbb{E}\left(\sup_{t \leq s \leq t+\delta} |x(s)|^p\right). \tag{5.86}$$

Substituting (5.85) and (5.86) into (5.84) and making use of (5.82) we obtain that

$$\mathbb{E}\left(\sup_{t\leq s\leq t+\delta}|x(s)|^p\right) \leq 3^p\mathbb{E}|x(t)|^p + \frac{2}{3}\mathbb{E}\left(\sup_{t\leq s\leq t+\delta}|x(s)|^p\right).$$

Hence

$$\mathbb{E}\left(\sup_{t\leq s\leq t+\delta}|x(s)|^p\right) \leq 3^{p+1}\mathbb{E}|x(t)|^p.$$

This and (5.81) imply (5.83) as claimed.

Step 2. Let us now prove the asymptotic stability in pth moment. Let $k = 1, 2, \cdots$. Note that

$$\mathbb{E}\left(\sup_{k\delta/2\leq s\leq (k+1)\delta/2}|x(s)|^p\right) \leq \mathbb{E}\left(\sup_{t\leq s\leq t+\delta}|x(s)|^p\right)$$

whenever $(k-1)\delta/2 \leq t \leq k\delta/2$. Thus

$$\mathbb{E}\left(\sup_{k\delta/2\leq s\leq (k+1)\delta/2}|x(s)|^p\right) \leq \frac{2}{\delta}\int_{(k-1)\delta/2}^{k\delta/2}\mathbb{E}\left(\sup_{t\leq s\leq t+\delta}|x(s)|^p\right)dt.$$

Therefore, by (5.83),

$$\sum_{k=1}^{\infty}\mathbb{E}\left(\sup_{k\delta/2\leq s\leq (k+1)\delta/2}|x(s)|^p\right)$$
$$\leq \frac{2}{\delta}\sum_{k=1}^{\infty}\int_{(k-1)\delta/2}^{k\delta/2}\mathbb{E}\left(\sup_{t\leq s\leq t+\delta}|x(s)|^p\right)dt$$
$$= \frac{2}{\delta}\int_0^{\infty}\mathbb{E}\left(\sup_{t\leq s\leq t+\delta}|x(s)|^p\right)dt < \infty. \quad (5.87)$$

We must then have

$$\lim_{k\to\infty}\mathbb{E}\left(\sup_{k\delta/2\leq s\leq (k+1)\delta/2}|x(s)|^p\right) = 0$$

which implies

$$\lim_{s\to\infty}\mathbb{E}(|x(s)|^p) = 0.$$

In other words, equation (5.1) is asymptotically stable in pth moment.

Step 3. For any integer M, it follows from (5.87) that

$$\mathbb{E}\left[\sum_{k=1}^{M}\left(\sup_{k\delta/2\leq s\leq (k+1)\delta/2}|x(s)|^p\right)\right] \leq \frac{2}{\delta}\int_0^\infty \mathbb{E}\left(\sup_{t\leq s\leq t+\delta}|x(s)|^p\right)dt < \infty.$$

Letting $M \to \infty$ yields

$$\mathbb{E}\left[\sum_{k=1}^{\infty}\left(\sup_{k\delta/2\leq s\leq (k+1)\delta/2}|x(s)|^p\right)\right] \leq \frac{2}{\delta}\int_0^\infty \mathbb{E}\left(\sup_{t\leq s\leq t+\delta}|x(s)|^p\right)dt < \infty.$$

Consequently

$$\sum_{k=1}^{\infty}\left(\sup_{k\delta/2\leq s\leq (k+1)\delta/2}|x(s)|^p\right) < \infty \quad a.s.$$

whence

$$\lim_{k\to\infty}\left(\sup_{k\delta/2\leq s\leq (k+1)\delta/2}|x(s)|^p\right) = 0 \quad a.s.$$

We therefore must have

$$\lim_{t\to\infty} x(t) = 0 \quad a.s.$$

that is, equation (5.1) is asymptotically stable with probability 1. □

Theorem 5.30 *Assume that there are functions $V \in C^{2,1}(\mathbb{R}^n \times \mathbb{R}_+ \times \mathbb{S}; \mathbb{R}_+)$, $\gamma \in L^1(\mathbb{R}_+; \mathbb{R}_+)$, $\kappa_1, \kappa_2 \in \mathcal{K}_V$, $\psi \in \Psi(\mathbb{R}_+; \mathbb{R}_+)$, and a constant $p \geq 2$, such that*

$$\kappa_1(|x|^p) \leq V(x,t,i) \tag{5.88}$$

and

$$LV(x,t,i) \leq \gamma(t) - \psi(t)\kappa_2(|x|^p) \tag{5.89}$$

for all $(x,t,i) \in \mathbb{R}^n \times \mathbb{R}_+ \times \mathbb{S}$. Assume moreover there is an $h > 0$ such that

$$|f(x,t,i)| + |g(x,t,i)| \leq h(1+|x|) \tag{5.90}$$

for all $(x,t,i) \in \mathbb{R}^n \times \mathbb{R}_+ \times \mathbb{S}$. Then equation (5.1) is asymptotically stable in pth moment.

To prove this theorem let us present an interesting lemma.

Lemma 5.4 *Let (5.90) hold. Let $p \geq 2$ and $x(t)$ be a solution of equation (5.1). If the pth moment of the solution is bounded, say*

$$\mathbb{E}|x(t)|^p \leq K \quad \forall t \geq 0,$$

then

$$|\mathbb{E}|x(t)|^p - \mathbb{E}|x(s)|^p| \leq hp(1 + 2h(p-1))(1 + K)(t - s)$$

for all $0 \leq s < t < \infty$. In other words, the pth moment $\mathbb{E}|x(t)|^p$ of the solution is uniformly continuous on the entire $t \in \mathbb{R}_+$.

Proof. By the Itô formula,

$$\mathbb{E}|x(t)|^p - \mathbb{E}|x(s)|^p$$
$$= \mathbb{E}\int_s^t \bigg[p|x(u)|^{p-2}x^T(u)f(x(u), u, r(u))$$
$$+ \frac{p}{2}|x(u)|^{p-2}|g(x(u), u, r(u))|^2$$
$$+ \frac{p(p-2)}{2}|x(u)|^{p-4}|x^T(u)g(x(u), u, r(u))|^2\bigg]du.$$

Hence

$$|\mathbb{E}|x(t)|^p - \mathbb{E}|x(s)|^p|$$
$$\leq \mathbb{E}\int_s^t \bigg[p|x(u)|^{p-1}|f(x(u), u, r(u))|$$
$$+ \frac{p(p-1)}{2}|x(u)|^{p-2}|g(x(u), u, r(u))|^2\bigg]du.$$

Using (5.90) we obtain that

$$|\mathbb{E}|x(t)|^p - \mathbb{E}|x(s)|^p|$$
$$\leq \mathbb{E}\int_s^t \bigg[hp|x(u)|^{p-1}[1 + |x(u)|]$$
$$+ \frac{3h^2p(p-1)}{2}|x(u)|^{p-2}[1 + |x(u)|^2]\bigg]du.$$

Note

$$|x(u)|^{p-1} \leq 1 + |x(u)|^p, \quad |x(u)|^{p-2} \leq 1 + |x(u)|^p.$$

Therefore

$$|\mathbb{E}|x(t)|^p - \mathbb{E}|x(s)|^p| \leq hp(1 + 2h(p-1)) \int_s^t [1 + \mathbb{E}|x(u)|^p]du$$
$$\leq hp(1 + 2h(p-1))(1 + K)(t-s)$$

as required. □

We can now prove Theorem 5.30

Proof of Theorem 5.30. Fix any $(t_0, x_0, r_0) \in \mathbb{R}_+ \times \mathbb{R}^n \times \mathbb{S}$ and write $x(t; t_0, x_0, r_0) = x(t)$. It is easy to show by the Itô formula and the Jensen inequality that

$$\sup_{t_0 \leq t < \infty} \kappa_1(\mathbb{E}|x(t)|^p) < \infty \tag{5.91}$$

and

$$\int_{t_0}^\infty \psi(s)\kappa_2(\mathbb{E}|x(s)|^p)ds < \infty. \tag{5.92}$$

Recalling the fact that the $\mathcal{K}_\vee$-function κ_1 has the property that $\kappa_1(v) \to \infty$ as $v \to \infty$, we see that $\mathbb{E}|x(t)|^p$ must be bounded, say

$$\mathbb{E}|x(t)|^p \leq K \quad \forall t \geq t_0$$

for some $K > 0$. By Lemma 5.4,

$$|\mathbb{E}|x(t)|^p - \mathbb{E}|x(s)|^p| \leq C(t-s), \quad t_0 \leq s < t < \infty, \tag{5.93}$$

where $C = hp(1 + 2h(p-1))(1 + K)$. We now claim

$$\lim_{t \to \infty} \mathbb{E}|x(t)|^p = 0. \tag{5.94}$$

If this is not true, then there is some $\varepsilon > 0$ and a sequence $\{t_k\}_{k \geq 1}$ satisfying $t_0 \leq t_k < t_k + 1 \leq t_{k+1}$ such that

$$\mathbb{E}|x(t_k)|^p \geq \varepsilon, \quad k \geq 1. \tag{5.95}$$

Let $\delta = 1 \wedge [\varepsilon/2C]$. Then, for $t_k \leq s \leq t_k + \delta$, it follows from (5.93) and (5.95) that

$$\mathbb{E}|x(s)|^p \geq \mathbb{E}|x(t_k)|^p - |\mathbb{E}|x(s)|^p - \mathbb{E}|x(t_k)|^p|$$
$$\geq \varepsilon - C(s - t_k) \geq \varepsilon - C\delta \geq \frac{\varepsilon}{2}.$$

Consequently

$$\int_{t_0}^{\infty} \psi(s)\kappa_2(\mathbb{E}|x(s)|^p)ds \geq \sum_{k=1}^{\infty} \int_{t_k}^{t_k+\delta} \psi(s)\kappa_2(\mathbb{E}|x(s)|^p)ds$$

$$\geq \frac{\varepsilon}{2} \sum_{k=1}^{\infty} \int_{t_k}^{t_k+\delta} \psi(s)ds = \infty$$

due to the assumption of $\psi \in \Psi(\mathbb{R}_+; \mathbb{R}_+)$. But this is in contradiction with (5.92) so (5.94) must hold and this completes the proof. $\square$

Theorem 5.31 *Assume that there are functions $V \in C^{2,1}(\mathbb{R}^n \times \mathbb{R}_+ \times \mathbb{S}; \mathbb{R}_+)$, $\gamma \in L^1(\mathbb{R}_+; \mathbb{R}_+)$, $\kappa_1, \kappa_3 \in \mathcal{K}_V$, $\kappa_2 \in \mathcal{K}_\wedge$, $\eta \in D(\mathbb{R}_+; \mathbb{R}_+)$, and a constant $p > 0$, such that*

$$\kappa_1(|x|^p) \leq V(x,t,i) \leq \kappa_2(|x|^p) \tag{5.96}$$

and

$$LV(x,t,i) \leq \gamma(t) - \eta(t)\kappa_3(|x|^p) \tag{5.97}$$

for all $(x,t,i) \in \mathbb{R}^n \times \mathbb{R}_+ \times \mathbb{S}$. Then equation (5.1) is asymptotically stable in pth moment.

Proof. We prove the case of $t_0 = 0$ but it can be done similarly when $t_0 > 0$. Fix any $(x_0, r_0) \in \mathbb{R}^n \times \mathbb{S}$ and write $x(t; 0, x_0, r_0) = x(t)$. By the generalised Itô formula,

$$0 \leq \mathbb{E}V(x(t), t, r(t))$$
$$= C_1 + \mathbb{E}\int_0^t LV(x(s), s, r(s))ds$$
$$= C_1 + \int_0^t \gamma(s)ds - \int_0^t \eta(s)\mathbb{E}\kappa_3(|x(s)|^p)]ds - \mathbb{E}\int_0^t \zeta(s)ds, \tag{5.98}$$

where $C_1 = V(x_0, 0, r_0)$ and

$$\zeta(s) = \gamma(s) - \eta(s)\kappa_3(|x(s)|^p) - LV(x(s), s, r(s))$$

which is, by condition (5.97), a non-negative process. It follows easily that

$$\int_0^{\infty} \eta(s)\mathbb{E}\kappa_3(|x(s)|^p)] < \infty. \tag{5.99}$$

Moreover, $\mathbb{E}\int_0^t \zeta(s)ds$ is nondecreasing so has a limit as $t \to \infty$ and the limit must be finite by (5.98). Letting $t \to \infty$ at both sides of (5.98) yields that

$$0 \leq \lim_{t\to\infty} \mathbb{E}V(x(t), t, r(t)) < \infty. \tag{5.100}$$

We now claim

$$\lim_{t\to\infty} \mathbb{E}V(x(t), t, r(t)) = 0. \tag{5.101}$$

If this is not true, then

$$\rho := \lim_{t\to\infty} \mathbb{E}V(x(t), t, r(t)) > 0$$

whence there is some $T > 0$ such that

$$\mathbb{E}V(x(t), t, r(t)) \geq \frac{\rho}{2}, \quad t \geq T.$$

By condition (5.96) and the Jensen inequality

$$\kappa_2(\mathbb{E}|x(t)|^p) \geq \frac{\rho}{2}, \quad t \geq T.$$

This shows that there is some $\rho_1 > 0$ such that

$$\mathbb{E}|x(t)|^p \geq \rho_1, \quad t \geq T.$$

We now compute

$$\int_T^\infty \eta(t)\mathbb{E}\kappa_3(|x(t)|^p)dt \geq \int_T^\infty \eta(t)\kappa_3(\mathbb{E}|x(t)|^p)dt$$

$$\geq \kappa_3(\rho_1)\int_T^\infty \eta(t)dt = \infty,$$

but this is in contradiction with (5.99) so (5.101) must hold. It is then easy to show from (5.101) and condition (5.96) that

$$\lim_{t\to\infty} \kappa_1(\mathbb{E}|x(t)|^p) = 0$$

whence

$$\lim_{t\to\infty} \mathbb{E}|x(t)|^p = 0$$

as required. □

Let us now discuss a number of examples to illustrate our theory. In the following examples, $B(t)$ will always be a one-dimensional Brownian motion. We will also let the dimension of the equations and the state space of the Markov chains be relatively small in order to facilitate the calculations, thereby enabling the theory of this section to be clearly illustrated.

Example 5.32 Consider a one-dimensional stochastic differential equation

$$dx(t) = f(x(t), t, r(t))dt + g(x(t), t, r(t))dB(t), \qquad (5.102)$$

where $r(t)$ is a Markov chain in the state space $\mathbb{S} = \{1, 2\}$ with the generator Γ. Let us give different coefficients f and g and generator Γ to illustrate different situations covered by Theorems 5.29–5.31.

Case 1. Let

$$f(x, t, i) = \begin{cases} -x & \text{if } i = 1, \\ \sin x/(1+t) & \text{if } i = 2, \end{cases}$$

and

$$g(x, t, i) = \begin{cases} \sin x/(1+t) & \text{if } i = 1, \\ x & \text{if } i = 2, \end{cases}$$

while

$$\Gamma = \begin{bmatrix} -1 & 1 \\ 1 & -1 \end{bmatrix}.$$

Define

$$V(x, t, i) = \begin{cases} x^2 & \text{if } i = 1, \\ 2.5x^2 & \text{if } i = 2. \end{cases}$$

Compute

$$LV(x, t, 1) \leq (1+t)^{-2} - 0.5x^2,$$

and

$$LV(x, t, 2) = 2x \sin x/(1+t) + x^2 - 1.5x^2$$
$$\leq 4(1+t)^{-1} - 0.25x^2.$$

So

$$LV(x, t, i) \leq 4(1+t)^{-1} - 0.25x^2.$$

Applying Theorem 5.29 we can conclude that equation (5.102) is not only asymptotically stable in mean square but also with probability 1.

Case 2. Let
$$f(x,t,i) = \begin{cases} (-1-\sin^2 t)x & \text{if } i=1, \\ (1-\sin^2 t)x & \text{if } i=2, \end{cases}$$

and
$$g(x,t,i) = \begin{cases} \sin t\sqrt{2|\sin(x^2)|} & \text{if } i=1, \\ \sin x/(1+t) & \text{if } i=2, \end{cases}$$

while
$$\Gamma = \begin{bmatrix} -1 & 1 \\ 3 & -3 \end{bmatrix}.$$

Define
$$V(x,t,i) = \begin{cases} x^2 & \text{if } i=1, \\ 3x^2 & \text{if } i=2. \end{cases}$$

Compute
$$LV(x,t,1) = -2\sin^2 t(x^2 - |\sin(x^2)|),$$

and
$$LV(x,t,2) = 6x^2(1-\sin^2 t) + \frac{3\sin^2 x}{(1+t)^2} - 6x^2$$
$$\leq \frac{3}{(1+t)^2} - 6x^2 \sin^2 t.$$

So
$$LV(x,t,i) \leq \frac{3}{(1+t)^2} - 2\sin^2 t(x^2 - |\sin(x^2)|).$$

Applying Theorem 5.30 with $p=2$,
$$\gamma(t) = \frac{3}{(1+t)^2}, \quad \psi(t) = 2\sin^2 t$$

$$\kappa_1(v) = v, \quad k_2(v) = \begin{cases} v - |\sin v| & \text{if } 0 \leq v \leq \pi/2, \\ v - 1 & \text{if } v > \pi/2, \end{cases}$$

we can conclude that equation (5.102) is asymptotically stable in mean square.

Case 3. Let
$$f(x,t,i) = \begin{cases} -(3+t)x/(1+t) & \text{if } i = 1, \\ (-1+3t)x/4(1+t) & \text{if } i = 2, \end{cases}$$

and
$$g(x,t,i) = \begin{cases} 2\sqrt{|\sin x^2|/(1+t)} & \text{if } i = 1, \\ \sqrt{(x^2+|\sin x^2|)/(1+t)} & \text{if } i = 2, \end{cases}$$

while
$$\Gamma = \begin{bmatrix} -2 & 2 \\ 3 & -3 \end{bmatrix}.$$

Define
$$V(x,t,i) = \begin{cases} x^2 & \text{if } i = 1, \\ 2x^2 & \text{if } i = 2. \end{cases}$$

Compute
$$LV(x,t,1) = -\frac{4}{1+t}(x^2 - |\sin x^2|),$$

and
$$LV(x,t,2) = -\frac{2}{1+t}(x^2 - |\sin x^2|).$$

So
$$LV(x,y,t,i) \leq -\frac{2}{1+t}(x^2 - |\sin x^2|).$$

Applying Theorem 5.31 with $p = 2$, $\gamma(t) = 0$,
$$\eta(t) = \frac{2}{1+t}, \quad \kappa_1(v) = v, \quad \kappa_2(v) = 2v,$$

while
$$k_3(v) = \begin{cases} v - |\sin v| & \text{if } 0 \leq v \leq \pi/2, \\ v - 1 & \text{if } v > \pi/2, \end{cases}$$

we can conclude that equation (5.102) is asymptotically stable in mean square.

Example 5.33 We give one more example to illustrate the 3rd moment stability. Let $r(t)$ be a Markov chain taking values in $\mathbb{S} = \{1,2\}$ with generator

$$\Gamma = \begin{bmatrix} -1 & 1 \\ 1 & -1 \end{bmatrix}.$$

Consider a one-dimensional stochastic differential equation

$$dx(t) = \alpha(r(t))x(t)dt + \sigma(x(t), t, r(t))dB(t) \quad (5.103)$$

on $t \geq 0$, where

$$\alpha(1) = \frac{1}{4}, \quad \alpha(2) = -3,$$

while $\sigma : \mathbb{R} \times \mathbb{R}_+ \times \mathbb{S} \to \mathbb{R}$ satisfying

$$|\sigma(x,t,1)|^2 \leq e^{-t} + \frac{|x|^2}{64}, \quad |\sigma(x,t,2)|^2 \leq e^{-t} + \frac{|x|^2}{2}.$$

Define

$$V(x,t,i) = \begin{cases} 8|x|^3 & \text{if } i = 1, \\ |x|^3 & \text{if } i = 2. \end{cases}$$

Compute

$$LV(x,t,1) \leq -|x|^3 + 24|x|(e^{-t} + |x|^2/64)$$

Using the elementary inequality, for any $\varepsilon > 0$,

$$|x|e^{-t} = (\varepsilon|x|^3)^{1/3}(\varepsilon^{-1/2}e^{-3t/2})^{2/3}$$

$$\leq \varepsilon|x|^3/3 + 2\varepsilon^{-1/2}e^{-3t/2}/3.$$

we can then show that

$$LV(x,t,1) \leq C_1 e^{-3t/2} - 0.5|x|^3.$$

for some $C_1 > 0$. Similarly we can show that

$$LV(x,t,2) \leq C_2 e^{-3t/2} - 0.25|x|^3$$

for some $C_2 > 0$. So

$$LV(x,t,i) \leq (C_1 \vee C_2)e^{-3t/2} - 0.25|x|^3.$$

By Theorem 3.1 we see that the trivial solution of equation (5.103) is asymptotically stable in 3rd moment.

5.5 Stability in Probability

In the previous sections we discuss the property that the solutions will converge to the equilibrium state 0 either with probability 1 or in pth moment. Sometimes in practice it is a little bit too strong to require the probability of the convergence be 1 but it would be satisfactory if the probability is large, say 95% or 99%. We sometimes also require the solution starting near to the equilibrium state will remain close to the equilibrium state with large probability. These are the concept of stability in probability.

In this section we shall again let $f(0, t, i) \equiv 0$ and $g(0, t, i) \equiv 0$ so equation (5.1) has the trivial solution $x(t; t_0, 0, i) = 0$ or the equilibrium state 0. We shall also use the notations $\mathbb{S}_\delta = \{x \in \mathbb{R}^n : |x| < \delta\}$ and $\bar{\mathbb{S}}_\delta = \{x \in \mathbb{R}^n : |x| \leq \delta\}$.

Definition 5.34   (i) The trivial solution of equation (5.1) is said to be *stochastically stable* or *stable in probability* if for every triple of $\varepsilon \in (0,1)$, $\rho > 0$ and $t_0 \geq 0$, there exists a $\delta = \delta(\varepsilon, \rho, t_0) > 0$ such that

$$\mathbb{P}\{|x(t; t_0, x_0, i)| < \rho \text{ for all } t \geq t_0\} \geq 1 - \varepsilon$$

for any $(x_0, i) \in \mathbb{S}_\delta \times \mathbb{S}$.

(ii) The trivial solution is said to be *stochastically asymptotically stable* or *asymptotically stable in probability* if it is stochastically stable and, moreover, for every pair of $\varepsilon \in (0,1)$ and $t_0 \geq 0$, there exists a $\delta_0 = \delta_0(\varepsilon, t_0) > 0$ such that

$$\mathbb{P}\{\lim_{t \to \infty} x(t; t_0, x_0) = 0\} \geq 1 - \varepsilon$$

whenever $(x_0, i) \in \mathbb{S}_{\delta_0} \times \mathbb{S}$.

(iii) The trivial solution is said to be *stochastically asymptotically stable in the large* if it is stochastically stable and, moreover,

$$\mathbb{P}\{\lim_{t \to \infty} x(t; t_0, x_0, i) = 0\} = 1 \quad \forall (t_0, x_0, i) \in \mathbb{R}_+ \times \mathbb{R}^n \times \mathbb{S}.$$

Theorem 5.35   *Let $h > 0$. Assume that there are two functions $V \in C^{2,1}(\mathbb{S}_h \times \mathbb{R}^n \times \mathbb{S}; \mathbb{R}_+)$ and $\mu \in \mathcal{K}$ such that $V(0, t, i) \equiv 0$,*

$$V(x, t, i) = \mu(|x|) \quad \text{and} \quad LV(x, t, i) \leq 0 \qquad (5.104)$$

for all $(x,t,i) \in \mathbb{S}_h \times \mathbb{R}_+ \times \mathbb{S}$. Then the trivial solution of equation (5.1) is stochastically stable.

Proof. Let $\varepsilon \in (0,1)$, $\rho > 0$ and $t_0 \geq 0$ be arbitrary. Without loss of generality we may assume that $\rho < h$. By the continuity of function V and the fact $V(0, t_0, i) = 0$, we can find a $\delta = \delta(\varepsilon, \rho, t_0) > 0$ such that

$$\frac{1}{\varepsilon} \sup_{(x,i) \in \mathbb{S}_\delta \times \mathbb{S}} V(x, t_0, i) \leq \mu(\rho). \tag{5.105}$$

It is easy to see that $\delta < \rho$. Now fix any $(x_0, i) \in \mathbb{S}_\delta \times \mathbb{S}$ and write $x(t; t_0, x_0, i) = x(t)$. Let τ be the first exit time of $x(t)$ from $\mathbb{S}_\rho$, that is

$$\tau = \inf\{t \geq t_0 : x(t) \notin \mathbb{S}_\rho\}.$$

By the generalised Itô formula, for any $t \geq t_0$,

$$\mathbb{E}V(x(\tau \wedge t), \tau \wedge t, r(\tau \wedge t)) = V(x_0, t_0, i) + \mathbb{E} \int_{t_0}^{\tau \wedge t} LV(x(s), s, r(s)) ds.$$

Making use of (5.104) yields

$$\mathbb{E}V(x(\tau \wedge t), \tau \wedge t, r(\tau \wedge t)) \leq V(x_0, t_0, i). \tag{5.106}$$

Note that $|x(\tau \wedge t)| = |x(\tau)| = r$ if $\tau \leq t$. Hence, by (5.104),

$$\mathbb{E}V(x(\tau \wedge t), \tau \wedge t, r(\tau \wedge t)) \geq \mathbb{E}\left[I_{\{\tau \leq t\}} V(x(\tau), \tau, r(\tau))\right] \geq \mu(r) \mathbb{P}\{\tau \leq t\}.$$

This, together with (5.106) and (5.105), implies

$$\mathbb{P}\{\tau \leq t\} \leq \varepsilon.$$

Letting $t \to \infty$ we get $\mathbb{P}\{\tau < \infty\} \leq \varepsilon$, that is

$$\mathbb{P}\{|x(t)| < \rho \text{ for all } t \geq t_0\} \geq 1 - \varepsilon$$

as required. □

Theorem 5.36 *Let $h > 0$. Assume that there are functions $V \in C^{2,1}(\mathbb{S}_h \times \mathbb{R}^n \times \mathbb{S}; \mathbb{R}_+)$ and $\mu_1, \mu_2, \mu_3 \in \mathcal{K}$ such that*

$$\mu_1(|x|) \leq V(x, t, i) \leq \mu_2(|x|) \quad \text{and} \quad LV(x, t, i) \leq -\mu_3(|x|) \tag{5.107}$$

for all $(x, t, i) \in \mathbb{S}_h \times \mathbb{R}_+ \times \mathbb{S}$. Then the trivial solution of equation (5.1) is stochastically asymptotically stable.

Proof. We know from Theorem 5.35 that the trivial solution is stochastically stable. So we only need to show that for any $\varepsilon \in (0,1)$ and $t_0 \geq 0$, there is a $\delta_0 = \delta_0(\varepsilon, t_0) > 0$ such that

$$\mathbb{P}\{\lim_{t\to\infty} x(t; t_0, x_0, i) = 0\} \geq 1 - \varepsilon \tag{5.108}$$

whenever $(x_0, i) \in \mathbb{S}_{\delta_0} \times \mathbb{S}$. Fix $\varepsilon \in (0,1)$ and $t_0 \geq 0$ arbitrarily. By Theorem 5.35, there is a $\delta_0 = \delta_0(\varepsilon, t_0) > 0$ such that

$$\mathbb{P}\{|x(t; t_0, x_0, i)| < h/2\} \geq 1 - \frac{\varepsilon}{4} \tag{5.109}$$

whenever $(x_0, i) \in \mathbb{S}_{\delta_0} \times \mathbb{S}$. Fix any $(x_0, i) \in \mathbb{S}_{\delta_0} \times \mathbb{S}$ and write $x(t; t_0, x_0, i) = x(t)$. Let $0 < \beta < |x_0|$ be arbitrary, and choose $\alpha \in (0, \beta)$ sufficiently small for

$$\frac{\mu_2(\alpha)}{\mu_1(\beta)} \leq \frac{\varepsilon}{4}. \tag{5.110}$$

Define the stopping times

$$\tau_\alpha = \inf\{t \geq t_0 : |x(t)| \leq \alpha\}$$

and

$$\tau_h = \inf\{t \geq t_0 : |x(t)| \geq h/2\}.$$

By the generalised Itô formula and (5.107), we can derive that for any $t \geq t_0$,

$$0 \leq \mathbb{E}V(x(\tau_\alpha \wedge \tau_h \wedge t), \tau_\alpha \wedge \tau_h \wedge t, r(\tau_\alpha \wedge \tau_h \wedge t))$$
$$= V(x_0, t_0, i) + \mathbb{E}\int_{t_0}^{\tau_\alpha \wedge \tau_h \wedge t} LV(x(s), s, r(s))ds$$
$$\leq V(x_0, t_0, i) - \mu_3(\alpha)\mathbb{E}(\tau_\alpha \wedge \tau_h \wedge t - t_0).$$

Consequently

$$(t - t_0)\mathbb{P}\{\tau_\alpha \wedge \tau_h \geq t\} \leq \mathbb{E}(\tau_\alpha \wedge \tau_h \wedge t - t_0) \leq \frac{V(x_0, t_0, i)}{\mu_3(\alpha)}.$$

This implies immediately that

$$\mathbb{P}\{\tau_\alpha \wedge \tau_h < \infty\} = 1.$$

But, by (5.109), $\mathbb{P}\{\tau_h < \infty\} \leq \varepsilon/4$. Hence

$$1 = \mathbb{P}\{\tau_\alpha \wedge \tau_h < \infty\} \leq \mathbb{P}\{\tau_\alpha < \infty\} + \mathbb{P}\{\tau_h < \infty\} \leq \mathbb{P}\{\tau_\alpha < \infty\} + \frac{\varepsilon}{4},$$

which yields
$$\mathbb{P}\{\tau_\alpha < \infty\} \geq 1 - \frac{\varepsilon}{4}. \tag{5.111}$$

Choose θ sufficiently large for
$$\mathbb{P}\{\tau_\alpha < \theta\} \geq 1 - \frac{\varepsilon}{2}.$$

Then
$$\mathbb{P}\{\tau_\alpha < \tau_h \wedge \theta\} \geq \mathbb{P}(\{\tau_\alpha < \theta\} \cap \{\tau_h = \infty\})$$
$$\geq \mathbb{P}\{\tau_\alpha < \theta\} - \mathbb{P}\{\tau_h < \infty\} \geq 1 - \frac{3\varepsilon}{4}. \tag{5.112}$$

Now, define two stopping times
$$\sigma = \begin{cases} \tau_\alpha & \text{if } \tau_\alpha < \tau_h \wedge \theta, \\ \infty & \text{otherwise} \end{cases}$$

and
$$\tau_\beta = \inf\{t > \sigma : |x(t)| \geq \beta\}.$$

We can then show by the generalised Itô formula that for any $t \geq \theta$,
$$\mathbb{E}V(x(\tau_\beta \wedge t), \tau_\beta \wedge t, r(\tau_\beta \wedge t)) \leq \mathbb{E}V(x(\sigma \wedge t), \sigma \wedge t, r(\sigma \wedge t)).$$

Noting that $V(x(\tau_\beta \wedge t), \tau_\beta \wedge t, r(\tau_\beta \wedge t)) = V(x(\sigma \wedge t), \sigma \wedge t, r(s \wedge t)) = V(x(t), t)$ on $\omega \in \{\tau_\alpha \geq \tau_h \wedge \theta\}$, we get
$$\mathbb{E}\Big[I_{\{\tau_\alpha < \tau_h \wedge \theta\}} V(x(\tau_\beta \wedge t), \tau_\beta \wedge t, r(\tau_\beta \wedge t))\Big]$$
$$\leq \mathbb{E}\Big[I_{\{\tau_\alpha < \tau_h \wedge \theta\}} \mathbb{E}V(x(\tau_\alpha), \tau_\alpha, r(\tau_\alpha))\Big].$$

Using (5.107) and the fact $\{\tau_\beta \leq t\} \subset \{\tau_\alpha < \tau_h \wedge \theta\}$ we further obtain
$$\mu_1(\beta)\mathbb{P}\{\tau_\beta \leq t\} \leq \mu_2(\alpha).$$

This, together with (5.110), yields
$$\mathbb{P}\{\tau_\beta \leq t\} \leq \frac{\varepsilon}{4}.$$

Letting $t \to \infty$ we have
$$\mathbb{P}\{\tau_\beta < \infty\} \leq \frac{\varepsilon}{4}.$$

It then follows, using (5.112) as well, that

$$\mathbb{P}\{\sigma < \infty \text{ and } \tau_\beta = \infty\} \geq \mathbb{P}\{\tau_\alpha < \tau_h \wedge \theta\} - \mathbb{P}\{\tau_\beta < \infty\} \geq 1 - \varepsilon.$$

But this means that

$$\mathbb{P}\{\omega : \limsup_{t \to \infty} |x(t)| \leq \beta\} \geq 1 - \varepsilon.$$

Since β is arbitrary, we must have

$$\mathbb{P}\{\omega : \lim_{t \to \infty} x(t) = 0\} \geq 1 - \varepsilon$$

as required. □

Theorem 5.37 *Assume that there are functions $V \in C^{2,1}(\mathbb{R}^n \times \mathbb{R}^n \times \mathbb{S}; \mathbb{R}_+)$, $\mu_1, \mu_2 \in \mathcal{K}_\infty$ and $\mu_3 \in \mathcal{K}$ such that*

$$\mu_1(|x|) \leq V(x,t,i) \leq \mu_2(|x|) \quad \text{and} \quad LV(x,t,i) \leq -\mu_3(|x|) \quad (5.113)$$

for all $(x,t,i) \in \mathbb{R}^n \times \mathbb{R}_+ \times \mathbb{S}$. Then the trivial solution of equation (5.1) is stochastically asymptotically stable in the large

Proof. By Theorem 5.35, the trivial solution of equation is stochastically stable. So we only need to show that

$$\mathbb{P}\{\lim_{t \to \infty} x(t; t_0, x_0, i) = 0\} = 1 \quad (5.114)$$

for all $(t_0, x_0, i) \in \mathbb{R}_+ \times \mathbb{R}^n \times \mathbb{S}$. Fix any such (t_0, x_0, i) and write $x(t; t_0, x_0, i) = x(t)$ again. Let $\varepsilon \in (0,1)$ be arbitrary. By (5.113) and $\kappa_1 \in \mathcal{K}_\infty$, we can find an $h > |x_0|$ sufficiently large for

$$\inf_{|x| \geq h, t \geq 0, j \in \mathbb{S}} V(x,t,j) \geq \frac{4V(x_0, t_0, i)}{\varepsilon}. \quad (5.115)$$

Define the stopping time

$$\tau_h = \inf\{t \geq t_0 : |x(t)| \geq h\}.$$

By the generalised Itô formula, we can show that for any $t \geq t_0$,

$$\mathbb{E}V(x(\tau_h \wedge t), \tau_h \wedge t, r(\tau_h \wedge t)) \leq V(x_0, t_0, i). \quad (5.116)$$

But, by (5.115), we see that

$$\mathbb{E}V(x(\tau_h \wedge t), \tau_h \wedge t, r(\tau_h \wedge t)) \geq \frac{4V(x_0, t_0, i)}{\varepsilon} \mathbb{P}\{\tau_h \leq t\}.$$

It then follows from (5.116) that
$$\mathbb{P}\{\tau_h \leq t\} \leq \frac{\varepsilon}{4}.$$

Letting $t \to \infty$ gives $\mathbb{P}\{\tau_h < \infty\} \leq \varepsilon/4$. That is

$$\mathbb{P}\{|x(t)| \leq h \text{ for all } t \geq t_0\} \geq 1 - \frac{\varepsilon}{4}. \tag{5.117}$$

From here, we can show in the same way as in the proof of Theorem 5.36 that
$$\mathbb{P}\{\lim_{t \to \infty} x(t) = 0\} \geq 1 - \varepsilon.$$

Since ε is arbitrary, the required (5.114) must hold. □

Let us discuss an example to illustrate the theory.

Example 5.38 Assume that the coefficients f and g of equation (5.1) have the expansions
$$\begin{aligned} f(x,t,i) &= F(t,i)x + o(|x|), \\ g(x,t,i) &= (G_1(t,i)x, \cdots, G_m(t,i)x) + o(|x|) \end{aligned} \tag{5.118}$$

in a neighbourhood of the origin $x = 0$ uniformly with respect to $t \geq 0$, where $F(t,i)$, $G_k(t,i)$ are all bounded Borel-measurable $n \times n$-matrix-valued functions. Assume that there are symmetric positive-definite matrices Q_i ($1 \leq i \leq N$) such that the symmetric matrix

$$Q_i F(t,i) + F^T(t,i) Q_i + \sum_{k=1}^{m} G_k^T(t,i) Q_i G_k(t,i) + \sum_{j=1}^{N} \gamma_{ij} Q_j$$

is negative-definite uniformly in $t \in \mathbb{R}_+$ for every $i \in \mathbb{S}$, that is

$$\lambda_{\max}\left(Q_i F(t,i) + F^T(t,i) Q_i + \sum_{k=1}^{m} G_k^T(t,i) Q_i G_k(t,i) + \sum_{j=1}^{N} \gamma_{ij} Q_j\right)$$
$$\leq -\lambda_i < 0 \tag{5.119}$$

for all $t \geq 0$. Now, define $V(x,t,i) = x^T Q_i x$. It is obvious that

$$\min_{i \in \mathbb{S}} \lambda_{\min}(Q_i)|x|^2 \leq V(x,t,i) \leq \max_{i \in \mathbb{S}} \lambda_{\max}(Q_i)|x|^2.$$

Moreover, with $\lambda := \min_{i \in \mathbb{S}} \lambda_i > 0$,

$$LV(x,t) = x^T \Big(Q_i F(t,i) + F^T(t,i) Q_i + \sum_{k=1}^m G_k^T(t,i) Q_i G_k(t,i)$$
$$+ \sum_{j=1}^N \gamma_{ij} Q_j \Big) x + o(|x|^2)$$
$$\leq -\lambda |x|^2 + o(|x|^2).$$

Hence $LV(x,t,i)$ is negative-definite in a sufficiently small neighbourhood of $x = 0$ for $t \geq 0$ and $i \in \mathbb{S}$. By Theorem 5.36 we therefore conclude that under (5.118) and (5.119), the trivial solution of equation (5.1) is stochastically asymptotically stable.

5.6 Asymptotic Stability in Distribution

In the previous sections we have discussed whether a solution of a given stochastic differential equation will converge to 0 (the equilibrium state) almost surely or in pth moment or in probability. However the solution will sometimes not converge to 0 but do converge to a random variable in distribution. This is the concept of asymptotic stability in distribution. A simple example is the Ornstein–Uhlenbeck process

$$dX(t) = \alpha X(t) dt + \sigma dB(t), \quad t \geq 0.$$

Given any initial value $X(0) = x \in \mathbb{R}$, it has the unique solution

$$X(t) = e^{\alpha t} x + \sigma \int_0^t e^{\alpha(t-s)} dB(s).$$

It is easy to observe that when $\alpha < 0$, the distribution of the solution $X(t)$ converges to the normal distribution $N(0, \sigma^2/2|\alpha|)$ as $t \to \infty$ for arbitrary initial value x.

To avoid the notations becoming too complicated we only consider an autonomous stochastic differential equation with Markovian switching of the form

$$dX(t) = f(X(t), r(t)) dt + g(X(t), r(t)) dB(t) \qquad (5.120)$$

on $t \geq 0$ with initial value $X(0) = x \in \mathbb{R}^n$. Here we use the capital letter X to denote the solution instead of the little letter x, which is now used to denote the initial value. Moreover, the coefficients

$$f : \mathbb{R}^n \times \mathbb{S} \to \mathbb{R}^n \text{ and } g : \mathbb{R}^n \times \mathbb{S} \to \mathbb{R}^{n \times m}$$

are assumed to satisfy the following assumption:

Assumption 5.39 Both f and g satisfy the local Lipschitz condition and the linear growth condition. That is, for each $k = 1, 2, \ldots$, there is an $h_k > 0$ such that

$$|f(x, i) - f(y, i)| + |g(x, i) - g(y, i)| \leq h_k |x - y|$$

for all $i \in \mathbb{S}$ and those $x, y \in \mathbb{R}^n$ with $|x| \vee |y| \leq k$; and there is moreover an $h > 0$ such that

$$|f(x, i)| + |g(x, i)| \leq h(1 + |x|)$$

for all $x \in \mathbb{R}^n$ and $i \in \mathbb{S}$.

We observe from Chapter 3 that under Assumption 5.39, equation (5.120) has a unique continuous solution $X(t)$ on $t \geq 0$. Accordingly, we let $C^2(\mathbb{R}^n \times \mathbb{S}; \mathbb{R}_+)$ denote the family of all non-negative functions $V(x, i)$ on $\mathbb{R}^n \times \mathbb{S}$ which are continuously twice differentiable in x. If $V \in C^2(\mathbb{R}^n \times \mathbb{S}; \mathbb{R}_+)$, define an operator LV from $\mathbb{R}^n \times \mathbb{S}$ to $\mathbb{R}$ by

$$LV(x, i) = V_x(x, i) f(x, i) + \frac{1}{2} \text{trace}[g^T(x, i) V_{xx}(x, i) g(x, i)]$$
$$+ \sum_{j=1}^{N} \gamma_{ij} V(x, j), \qquad (5.121)$$

where

$$V_x(x, i) = \left(\frac{\partial V(x, i)}{\partial x_1}, \ldots, \frac{\partial V(x, t, i)}{\partial x_n} \right), \quad V_{xx}(x, i) = \left(\frac{\partial^2 V(x, i)}{\partial x_i \partial x_j} \right)_{n \times n}.$$

Let $y(t)$ denote the $\mathbb{R}^n \times \mathbb{S}$-valued process $(X(t), r(t))$. Then $y(t)$ is a time homogeneous Markov process. Let $p(t, x, i, dy \times \{j\})$ denote the transition probability of the process $y(t)$. Let $P(t, x, i, A \times D)$ denote the probability of event $\{y(t) \in A \times D\}$ given initial condition $y(0) = (x, i)$, this is

$$P(t, x, i, A \times D) = \sum_{j \in D} \int_A p(t, x, i, dy \times \{j\}).$$

Definition 5.40 The process $y(t)$ is said to be *asymptotically stable in distribution* if there exists a probability measure $\pi(\cdot \times \cdot)$ on $\mathbb{R}^n \times \mathbb{S}$ such that the transition probability $p(t, x, i, dy \times \{j\})$ of $y(t)$ converges weakly to $\pi(dy \times \{j\})$ as $t \to \infty$ for every $(x, i) \in \mathbb{R}^n \times \mathbb{S}$. Equation (5.120) is said to be *asymptotically stable in distribution* if $y(t)$ is asymptotically stable in distribution.

To highlight the initial values, we let $r_i(t)$ be the Markov chain starting from state $i \in \mathbb{S}$ at $t = 0$ and denote by $X^{x,i}(t)$ the solution of equation (5.120) with initial conditions $X(0) = x \in \mathbb{R}^n$ and $r(0) = i$. To establish the main result of this section we will need two more assumptions.

Assumption 5.41 For some $p > 0$, the solutions of equation (5.120) have the property that

$$\sup_{0 \le t < \infty} \mathbb{E}|X^{x,i}(t)|^p < \infty \quad \forall (x, i) \in \mathbb{R}^n \times \mathbb{S}. \tag{5.122}$$

Assumption 5.42 For some $p > 0$, the solutions of equation (5.120) have the property that

$$\lim_{t \to \infty} \mathbb{E}|X^{x,i}(t) - X^{y,i}(t)|^p = 0 \text{ uniformly in } (x, y, i) \in K \times K \times \mathbb{S}. \tag{5.123}$$

for any compact subset K of $\mathbb{R}^n$.

By the well-known Chebyshev inequality, it is easy to observe that Assumption 5.41 guarantees that for any $(x, i) \in \mathbb{R}^n \times \mathbb{S}$, the family of transition probabilities $\{p(t, x, i, dy \times \{j\}) : t \ge 0\}$ is tight. That is, for any $\varepsilon > 0$ there is a compact subset $K = K(\varepsilon, x, i)$ of $\mathbb{R}^n$ such that

$$P(t, x, i, K \times \mathbb{S}) \ge 1 - \varepsilon \quad \forall t \ge 0. \tag{5.124}$$

We can now state the main result of this section.

Theorem 5.43 *Under Assumptions 5.39, 5.41 and 5.42, equation (5.120) is asymptotically stable in distribution.*

To prove this theorem we need to introduce more notations. Let $\mathcal{P}(\mathbb{R}^n \times \mathbb{S})$ denote all probability measures on $\mathbb{R}^n \times \mathbb{S}$. For $P_1, P_2 \in \mathcal{P}(\mathbb{R}^n \times \mathbb{S})$ define metric $d_\mathbb{L}$ as follows:

$$d_\mathbb{L}(P_1, P_2) = \sup_{f \in \mathbb{L}} \left| \sum_{i=1}^N \int_{\mathbb{R}^n} f(x, i) P_1(dx, i) - \sum_{i=1}^N \int_{\mathbb{R}^n} f(x, i) P_2(dx, i) \right|,$$

where

$$\mathbb{L} = \{f : \mathbb{R}^n \times \mathbb{S} \to \mathbb{R} :$$
$$|f(x, i) - f(y, j)| \le |x - y| + |i - j| \text{ and } |f(\cdot, \cdot)| \le 1\}.$$

Let us now present three lemmas.

Lemma 5.5 *Under Assumption 5.39, for every $p > 0$ and any compact subset K of $\mathbb{R}^n$,*

$$\sup_{(x,i)\in K\times \mathbb{S}} \mathbb{E}\left[\sup_{0\leq s\leq t} |X^{x,i}(s)|^p\right] < \infty \quad \forall t \geq 0. \tag{5.125}$$

The proof is left to the reader as an exercise.

Lemma 5.6 *Let Assumptions 5.39 and 5.42 hold. Then, for any compact subset K of $\mathbb{R}^n$,*

$$\lim_{t\to\infty} d_{\mathbb{L}}(p(t,x,i,\cdot\times\cdot), p(t,y,j,\cdot\times\cdot)) = 0 \tag{5.126}$$

uniformly in $x, y \in K$ and $i, j \in \mathbb{S}$.

Proof. For any pair of $i, j \in \mathbb{S}$, define the stopping time

$$\beta_{ij} = \inf\{t \geq 0 : r_i(t) = r_j(t)\}. \tag{5.127}$$

Due to the ergodicity of the Markov chain, $\beta_{ij} < \infty$ a.s. (see [Anderson (1991)]). So, for any $\varepsilon > 0$, there exists a positive number T such that

$$\mathbb{P}\{\beta_{ij} \leq T\} > 1 - \frac{\varepsilon}{8} \quad \forall i, j \in \mathbb{S}. \tag{5.128}$$

For such T, by Lemma 5.5, there is a sufficiently large $R > 0$ for

$$\mathbb{P}(\Omega_{x,i}) > 1 - \frac{\varepsilon}{16} \quad \forall (x,i) \in K \times S, \tag{5.129}$$

where $\Omega_{x,i} = \{|X^{x,i}(t)| \leq R \ \forall t \in [0,T]\}$.

Now, fix any $x, y \in K$ and $i, j \in \mathbb{S}$. Set $\Omega_1 = \Omega_{x,i} \cap \Omega_{y,j}$. For any $f \in \mathbb{L}$ and $t \geq T$, compute

$$|\mathbb{E}f(X^{x,i}(t), r_i(t)) - \mathbb{E}f(X^{y,j}(t), r_j(t))|$$
$$\leq 2P\{\beta_{ij} > T\} + \mathbb{E}\Big(I_{\{\beta_{ij}\leq T\}}|f(X^{x,i}(t), r_i(t)) - f(X^{y,j}(t), r_j(t))|\Big)$$

$$\le \frac{\varepsilon}{4} + \mathbb{E}\Big[I_{\{\beta_{ij}\le T\}}\mathbb{E}\big(|f(X^{x,i}(t),r_i(t)) - f(X^{y,j}(t),r_j(t))|\ \big|\ \mathcal{F}_{\beta_{ij}}\big)\Big]$$

$$\le \frac{\varepsilon}{4} + \mathbb{E}\Big[I_{\{\beta_{ij}\le T\}}\mathbb{E}|f(X^{u,k}(t-\beta_{ij}),r_k(t-\beta_{ij}))$$

$$\qquad\qquad - f(X^{v,k}(t-\beta_{ij}),r_k(t-\beta_{ij}))|\Big]$$

$$\le \frac{\varepsilon}{4} + \mathbb{E}\Big[I_{\{\beta_{ij}\le T\}}\mathbb{E}\big(2\wedge |X^{u,k}(t-\beta_{ij}) - X^{v,k}(t-\beta_{ij})|\big)\Big]$$

$$\le \frac{\varepsilon}{4} + 2\mathbb{P}(\Omega - \Omega_1)$$

$$\quad + \mathbb{E}\Big[I_{\Omega_1\cap\{\beta_{ij}\le T\}}\mathbb{E}\big(2\wedge |X^{u,k}(t-\beta_{ij}) - X^{v,k}(t-\beta_{ij})|\big)\Big], \qquad (5.130)$$

where $u = X^{x,i}(\beta_{ij})$, $v = X^{y,j}(\beta_{ij})$ and $k = r_i(\beta_{ij}) = r_j(\beta_{ij})$. However, if $p \ge 1$, by Assumption 5.42, there is a $T_1 > 0$ such that

$$\mathbb{E}|X^{x,i}(t) - X^{y,i}(t)|^p < \left(\frac{\varepsilon}{2}\right)^p \quad \forall t \ge T_1$$

whenever $|x|\vee|y|\le R$ and $i \in \mathbb{S}$, and hence

$$\mathbb{E}\big(2\wedge |X^{x,i}(t) - X^{y,i}(t)|\big) \le \big(\mathbb{E}|X^{x,i}(t) - X^{y,i}(t)|^p\big)^{1/p} \le \frac{\varepsilon}{2}.$$

But, if $p \in (0,1)$, then, by Assumption 5.42 again, there is a $T_1 > 0$ such that

$$\mathbb{E}|X^{x,i}(t) - X^{y,i}(t)|^p < \frac{\varepsilon}{8} \quad \forall t \ge T_1$$

whenever $|x|\vee|y|\le R$ and $i \in \mathbb{S}$, and hence

$$\mathbb{E}\big(2\wedge |X^{x,i}(t) - X^{y,i}(t)|\big)$$

$$\le 2\mathbb{P}\{|X^{x,i}(t) - X^{y,i}(t)|\ge 2\} + \mathbb{E}\big(I_{\{|X^{x,i}(t)-X^{y,i}(t)|<2\}}|X^{x,i}(t) - X^{y,i}(t)|\big)$$

$$\le 2^{1-p}\mathbb{E}|X^{x,i}(t) - X^{y,i}(t)|^p + \mathbb{E}\big(2^{1-p}|X^{x,i}(t) - X^{y,i}(t)|^p\big)$$

$$\le 2^{2-p}\mathbb{E}|X^{x,i}(t) - X^{y,i}(t)|^p \le \frac{\varepsilon}{2}.$$

In other words, for any $p > 0$, there is a $T_1 > 0$ such that

$$\mathbb{E}\big(2\wedge |X^{x,i}(t) - X^{y,i}(t)|\big) < \varepsilon/2 \quad \forall t \ge T_1 \qquad (5.131)$$

whenever $|x|\vee|y|\le R$ and $i \in \mathbb{S}$. Note that given $\omega \in \Omega_1 \cap \{\beta_{ij}\le T\}$, $|u|\vee|v|\le R$. So, by (5.131),

$$\mathbb{E}\big(2\wedge |X^{u,k}(t-\beta_{ij}) - X^{v,k}(t-\beta_{ij})|\big) < \varepsilon/2 \quad \forall t \ge T+T_1. \qquad (5.132)$$

It therefore follows from (5.129), (5.130) and (5.132) that

$$|\mathbb{E}f(X^{x,i}(t), r_i(t)) - \mathbb{E}f(X^{y,j}(t), r_j(t))| \leq \frac{\varepsilon}{4} + \frac{\varepsilon}{4} + \frac{\varepsilon}{2} = \varepsilon \quad \forall t \geq T + T_1.$$

Since f etc, are arbitrary, we must have that

$$\sup_{f \in \mathbb{L}} |\mathbb{E}f(X^{x,i}(t), r_i(t)) - \mathbb{E}f(X^{y,j}(t), r_j(t))| \leq \varepsilon \quad \forall t \geq T + T_1,$$

namely

$$d_{\mathbb{L}}(p(t, x, i, \cdot \times \cdot), p(t, y, j, \cdot \times \cdot)) \leq \varepsilon \quad \forall t \geq T + T_1$$

for all $x, y \in K$ and $i, j \in \mathbb{S}$. The proof is therefore complete. □

Lemma 5.7 *Let Assumptions 5.39, 5.41 and 5.42 hold. Then for any $(x, i) \in \mathbb{R}^n \times \mathbb{S}$, $\{p(t, x, i, \cdot \times \cdot) : t \geq 0\}$ is Cauchy in the space $\mathcal{P}(\mathbb{R}^n \times \mathbb{S})$ with metric $d_{\mathbb{L}}$.*

Proof. Fix any $(x, i) \in \mathbb{R}^n \times \mathbb{S}$. We need to show that for any $\varepsilon > 0$, there is a $T > 0$ such that

$$d_{\mathbb{L}}(p(t+s, x, i, \cdot \times \cdot), p(t, x, i, \cdot \times \cdot)) \leq \varepsilon \quad \forall t \geq T, \ s > 0.$$

This is equivalent to

$$\sup_{f \in \mathbb{L}} |\mathbb{E}f(X^{x,i}(t+s), r_i(t+s)) - \mathbb{E}f(X^{x,i}(t), r_i(t))| \leq \varepsilon \quad (5.133)$$

for any $t \geq T$ and $s > 0$. Now, for any $f \in \mathbb{L}$ and $t, s > 0$, compute

$$|\mathbb{E}f(X^{x,i}(t+s), r_i(t+s)) - \mathbb{E}f(X^{x,i}(t), r_i(t))|$$
$$= |\mathbb{E}[\mathbb{E}(f(X^{x,i}(t+s), r_i(t+s))|\mathcal{F}_s)] - \mathbb{E}f(X^{x,i}(t), r_i(t))|$$
$$= |\sum_{l=1}^{N} \int_{\mathbb{R}^n} \mathbb{E}f(X^{z,l}(t), r_l(t))p(s, x, i, dz \times \{l\}) - \mathbb{E}f(X^{y,j}(t), r_i(t))|$$
$$\leq \sum_{l=1}^{N} \int_{\mathbb{R}^n} |\mathbb{E}f(X^{z,l}(t), r_l(t)) - \mathbb{E}f(X^{x,i}(t), r_i(t))|\, p(s, x, i, dz \times \{l\})$$
$$\leq 2p(s, x, i, \bar{\mathbb{S}}_R^c \times \mathbb{S})$$
$$+ \sum_{l=1}^{N} \int_{\bar{\mathbb{S}}_R} |\mathbb{E}f(X^{z,l}(t), r_l(t)) - \mathbb{E}f(X^{x,i}(t), r_i(t))|$$
$$\times p(s, x, i, dz \times \{l\}), \quad (5.134)$$

where $\bar{\mathbb{S}}_R = \{x \in \mathbb{R}^n : |x| \leq R\}$ and $\bar{\mathbb{S}}_R^c = \mathbb{R}^n - \bar{\mathbb{S}}_R$. By Assumption 5.41 or (5.124), there is a positive number R sufficiently large for

$$p(s, x, i, \bar{\mathbb{S}}_R^c \times \mathbb{S}) < \frac{\varepsilon}{4} \quad \forall s \geq 0. \tag{5.135}$$

On the other hand, by Lemma 5.6, there is a $T > 0$ such that

$$\sup_{f \in \mathbb{L}} |\mathbb{E}f(X^{z,l}(t), r_l(t)) - \mathbb{E}f(X^{x,i}(t), r_i(t))| < \frac{\varepsilon}{2} \quad \forall t \geq T \tag{5.136}$$

whenever $(z, l) \in \bar{\mathbb{S}}_R \times \mathbb{S}$. Substituting (5.135) and (5.136) into (5.134) yields

$$|\mathbb{E}f(X^{x,i}(t+s), r_i(t+s)) - \mathbb{E}f(X^{x,i}(t), r_i(t))| < \varepsilon \quad \forall t \geq T, \; s > 0.$$

Since f is arbitrary, the desired inequality (5.133) must hold. □

We can now easily prove the main result Theorem 5.43.

Proof of Theorem 5.43. By definition, we need to show that there exists a probability measure $\pi(\cdot \times \cdot) \in \mathcal{P}(\mathbb{R}^n \times \mathbb{S})$ such that for any $(x, i) \in \mathbb{R}^n \times \mathbb{S}$, the transition probabilities $\{p(t, x, i, \cdot \times \cdot) : t \geq 0\}$ converge weakly to $\pi(\cdot \times \cdot)$. Recalling the well-known fact that the weak convergence of probability measures is a metric concept (see [Ikeda and Watanabe (1981), Proposition 2.5]), we therefore need to show that for any $(x, i) \in \mathbb{R}^n \times \mathbb{S}$,

$$\lim_{t \to \infty} d_\mathbb{L}(p(t, x, i, \cdot \times \cdot), \pi(\cdot \times \cdot)) = 0. \tag{5.137}$$

By Lemma 5.7, $\{p(t, 0, 1, \cdot \times \cdot) : t \geq 0\}$ is Cauchy in the space $\mathcal{P}(\mathbb{R}^n \times \mathbb{S})$ with metric $d_\mathbb{L}$. So there is a unique $\pi(\cdot \times \cdot) \in \mathcal{P}(\mathbb{R}^n \times \mathbb{S})$ such that

$$\lim_{t \to \infty} d_\mathbb{L}(p(t, 0, 1, \cdot \times \cdot), \pi(\cdot \times \cdot)) = 0.$$

Now, for any $(x, i) \in \mathbb{R}^n \times \mathbb{S}$, by Lemma 5.6,

$$\lim_{t \to \infty} d_\mathbb{L}(p(t, x, i, \cdot \times \cdot), \pi(\cdot \times \cdot))$$
$$\leq \lim_{t \to \infty} \Big[d_\mathbb{L}(p(t, 0, 1, \cdot \times \cdot), \pi(\cdot \times \cdot)) + d_\mathbb{L}(p(t, x, i, \cdot \times \cdot), p(t, 0, 1, \cdot \times \cdot)) \Big]$$
$$= 0$$

as required. □

Theorem 5.43 depends on Assumptions 5.39, 5.41 and 5.42. We observe that Assumption 5.39 is usually imposed to guarantee the existence and uniqueness of the solution while Assumption 5.41 is concerned with the moment boundedness of the solution, which has been discussed in Section

6.2. It is therefore necessary to establish sufficient criteria for Assumption 5.42 to be satisfied so that Theorem 5.43 is applicable. We observe that Assumption 5.42 means that two solutions starting from different initial values will converge to each other in pth moment. This is somehow similar to the asymptotic stability in pth moment discussed in Section 6.4, but we clearly need to consider the difference between two solutions of equation (5.120) starting from different initial values, namely

$$X^{x,i}(t) - X^{y,i}(t)$$
$$= x - y + \int_0^t [f(X^{x,i}(s), r_i(s)) - f(X^{y,i}, r_i(s))]ds$$
$$+ \int_0^t [g(X^{x,i}(s), r_i(s)) - g(X^{y,i}, r_i(s))]dB(s). \quad (5.138)$$

Moreover, Assumption 5.42 requires the convergence of (5.123) be uniformly in initial values, which makes it significantly different from what we discussed in Section 6.4.

To establish sufficient criteria for Assumption 5.42 to be satisfied, we need to introduce a new operator. For a given function $U \in C^2(\mathbb{R}^n \times \mathbb{S}; \mathbb{R}_+)$, we define an operator $\mathcal{L}U : \mathbb{R}^n \times \mathbb{R}^n \times \mathbb{S} \to \mathbb{R}$ associated with equation (5.138) by

$$\mathcal{L}U(x,y,i)$$
$$= \sum_{j=1}^N \gamma_{ij} U(x-y, j) + U_x(x-y, i)[f(x,i) - f(y,i)]$$
$$+ \frac{1}{2}\text{trace}\Big([g(x,i) - g(y,i)]^T U_{xx}(x-y,i)[g(x,i) - g(y,i)]\Big). \quad (5.139)$$

Please note the difference between this operator and the other LV defined by (5.121).

Lemma 5.8 *Let Assumptions 5.39 and 5.41 hold and $p \geq 2$. Assume that there exist functions $U \in C^2(\mathbb{R}^n \times \mathbb{S}; \mathbb{R}_+)$ and $\kappa_1, \kappa_2 \in \mathcal{K}_\mathcal{V}$ such that*

$$U(0,i) = 0 \quad \forall i \in \mathbb{S}, \quad (5.140)$$
$$\kappa_1(|x|^p) \leq U(x,i) \quad \forall (x,i) \in \mathbb{R}^n, \quad (5.141)$$
$$\mathcal{L}U(x,y,i) \leq -\kappa_2(|x-y|^p) \quad \forall (x,y,i) \in \mathbb{R}^n \times \mathbb{R}^n \times \mathbb{S}. \quad (5.142)$$

Then Assumption 5.42 is satisfied.

Proof. We divide the whole proof into three steps.

Step 1. Let $x, y \in \mathbb{R}^n$ and $i \in \mathbb{S}$. Let k be a positive number and define the stopping time

$$\tau_k = \inf\{t > 0 : |X^{x,i}(t) - X^{y,i}(t)| \geq k\}.$$

Let $t_k = \tau_k \wedge t$. By the generalised Itô formula

$$\mathbb{E}U(X^{x,i}(t_k) - X^{y,i}(t_k), r_i(t_k))$$

$$= U(x - y, i) + \mathbb{E}\int_0^{t_k} \mathcal{L}U(X^{x,i}(s), X^{y,i}(s), r_i(s))ds.$$

Using (5.141) and (5.142) and then letting $k \to \infty$ produce

$$\mathbb{E}\kappa_1(|X^{x,i}(t) - X^{y,i}(t)|^p)$$

$$\leq U(x - y, i) - \mathbb{E}\int_0^t \kappa_2(|X^{x,i}(s) - X^{y,i}(s)|^p)ds.$$

By the well-known Jensen inequality we obtain that

$$\kappa_1(\mathbb{E}|X^{x,i}(t) - X^{y,i}(t)|^p) \leq U(x - y, i) \quad \forall t \geq 0 \tag{5.143}$$

and

$$\int_0^\infty \kappa_2(\mathbb{E}|X^{x,i}(t) - X^{y,i}(t)|^p)dt \leq U(x - y, i) < \infty. \tag{5.144}$$

Step 2. We now claim that

$$\lim_{t \to \infty} \mathbb{E}|X^{x,i}(t) - X^{y,i}(t)|^p = 0. \tag{5.145}$$

If this is not true, then

$$\limsup_{t \to \infty} \mathbb{E}|X^{x,i}(t) - X^{y,i}(t)|^p > 3^{p-1}\varepsilon$$

for some $\varepsilon > 0$. Thus there exists a sequence $\{t_k\}_{k \geq 1}$ with $t_{k+1} > t_k + 1$ such that

$$\mathbb{E}|X^{x,i}(t_k) - X^{y,i}(t_k)|^p \geq 3^{p-1}\varepsilon \quad \forall k \geq 1. \tag{5.146}$$

For $t_k \leq t \leq t_k + 1$, it is easy to see from (5.138) that

$$\mathbb{E}|X^{x,i}(t) - X^{y,i}(t)|^p \geq 3^{1-p}\mathbb{E}|X^{x,i}(t_k) - X^{y,i}(t_k)|^p$$
$$- \mathbb{E}|\int_{t_k}^{t} [f(X^{x,i}(s), r_i(s)) - f(X^{y,i}(s), r_i(s))]ds|^p$$
$$- \mathbb{E}|\int_{t_k}^{t} [g(X^{x,i}(s), r_i(s)) - g(X^{y,i}(s), r_i(s))]dB(s)|^p. \quad (5.147)$$

However, by the Hölder inequality and Assumption 5.39, we derive that

$$\mathbb{E}|\int_{t_k}^{t} [f(X^{x,i}(s), r_i(s)) - f(X^{y,i}(s), r_i(s))]ds|^p$$
$$\leq \mathbb{E}\int_{t_k}^{t} |f(X^{x,i}(s), r_i(s)) - f(X^{y,i}(s), r_i(s))|^p ds$$
$$\leq c_1 \int_{t_k}^{t} \left[1 + \mathbb{E}|X^{x,i}(s)|^p + \mathbb{E}|X^{y,i}(s)|^p\right] ds, \quad (5.148)$$

where c_1 is a constant dependent of only p and h (described in Assumption 5.39). Moreover, by the Burkholder–Davis–Gundy inequality *etc*, we compute

$$\mathbb{E}|\int_{t_k}^{t} [g(X^{x,i}(s), r_i(s)) - g(X^{y,i}(s), r_i(s))]dB(s)|^p$$
$$\leq C_p \mathbb{E}\left[\int_{t_k}^{t} |g(X^{x,i}(s), r_i(s)) - g(X^{y,i}(s), r_i(s))|^2 ds\right]^{p/2}$$
$$\leq C_p E \int_{t_k}^{t} |g(X^{x,i}(s), r_i(s)) - g(X^{y,i}(s), r_i(s))|^p ds$$
$$\leq c_2 \int_{t_k}^{t} \left[1 + \mathbb{E}|X^{x,i}(s)|^p + \mathbb{E}|X^{y,i}(s)|^p\right] ds, \quad (5.149)$$

where C_p is a constant described by the Burkholder–Davis–Gundy inequality and c_2 is a constant dependent of p and h only. By Assumption 5.41 we therefore see from (5.148) and (5.149) that there is $\delta \in (0,1)$ such that

$$\mathbb{E}|\int_{t_k}^{t} [f(X^{x,i}(s), r_i(s)) - f(X^{y,i}(s), r_i(s))]ds|^p$$
$$+ \mathbb{E}|\int_{t_k}^{t} [g(X^{x,i}(s), r_i(s)) - g(X^{y,i}(s), r_i(s))]dB(s)|^p$$
$$\leq \frac{\varepsilon}{2} \quad \forall t \in [t_k, t_k + \delta], \; k \geq 1.$$

It therefore follows from (5.147) and (5.146) that

$$\mathbb{E}|X^{x,i}(t) - X^{y,i}(t)|^p \geq \frac{\varepsilon}{2} \quad \forall t \in [t_k, t_k + \delta], \ k \geq 1. \tag{5.150}$$

Consequently

$$\int_0^\infty \kappa_2(\mathbb{E}|X^{x,i}(t) - X^{y,i}(t)|^p)dt \geq \sum_{k=1}^\infty \int_{t_k}^{t_k+\delta} \kappa_2(\varepsilon/2)dt = \infty$$

but this contradicts with (5.144) so (5.145) must hold.

Step 3. We can now show Assumption 5.42, namely (5.123). Let $\varepsilon > 0$ be arbitrary. It is easy to observe from (5.143) that there is a $\delta > 0$ such that

$$\mathbb{E}|X^{x,i}(t) - X^{y,i}(t)|^p < \frac{\varepsilon}{3^p} \quad \forall t \geq 0 \tag{5.151}$$

provided $x, y \in \mathbb{R}^n$ with $|x - y| < \delta$.

Now, given any any compact subset K of $\mathbb{R}^n$, we can find finite vectors $x_1, ..., x_u \in K$ such that $\cup_{k=1}^u B(x_k, \delta) \supseteq K$, where $B(x_k, \delta) = \{x \in \mathbb{R}^n : |x - x_k| < \delta\}$. From Step 2 we observe that there is a $T > 0$ such that

$$\mathbb{E}|X^{x_k,i}(t) - X^{x_l,i}(t)|^p < \frac{\varepsilon}{3^p} \quad \forall t \geq T, \ 1 \leq k, l \leq u \text{ and } i \in \mathbb{S}. \tag{5.152}$$

Consequently, for any $(x, y, i) \in K \times K \times \mathbb{S}$, find x_l, x_k for $|x - x_l| < \delta$ and $|y - x_k| < \delta$. It then follows from (5.151) and (5.152) that

$$\mathbb{E}|X^{x,i}(t) - X^{y,i}(t)|^p \leq 3^{p-1}\Big(\mathbb{E}|X^{x,i}(t) - X^{x_l,i}(t)|^p$$
$$+ \mathbb{E}|X^{y,i}(t) - X^{x_k,i}(t)|^p + \mathbb{E}|X^{x_k,i}(t) - X^{x_l,i}(t)|^p\Big)$$
$$< \varepsilon \quad \forall t \geq T.$$

This proves the required assertion. □

The proof of Lemma 5.8 was rather technical. However, we shall now demonstrate that if condition (5.142) is replaced by (5.154) below, then Assumption 5.42 can be shown much more easily even without Assumptions 5.39 and 5.41. Besides, the following lemma does not require $p \geq 2$.

Lemma 5.9 *Assume that there exist functions* $U \in C^2(\mathbb{R}^n \times \mathbb{S}; \mathbb{R}_+)$, $\kappa_1 \in \mathcal{K}_V$ *and constants* $\lambda, p > 0$ *such that*

$$\kappa_1(|x|^p) \leq U(x, i) \quad \forall (x, i) \in \mathbb{R}^n, \tag{5.153}$$
$$\mathcal{L}U(x, y, i) \leq -\lambda U(x - y, i) \quad \forall (x, y, i) \in \mathbb{R}^n \times \mathbb{R}^n \times \mathbb{S}. \tag{5.154}$$

Then Assumption 5.42 is satisfied.

Proof. Let K be any compact subset of $\mathbb{R}^n$. For any $(x, y, i) \in K \times K \times \mathbb{S}$, by Ito's formula and condition (5.154), we can show that

$$e^{\lambda t}\mathbb{E}U(X^{x,i}(s) - X^{y,i}(s), r_i(t)) \leq U(x - y, i) \qquad (5.155)$$

Let

$$C_K = \sup_{(x,y,i) \in K \times K \times \mathbb{S}} U(x - y, i) < \infty.$$

Then, by condition (5.153) and the Jensen inequality,

$$\kappa_1(\mathbb{E}|X^{x,i}(t) - X^{y,i}(t)|^p) \leq C_K e^{-\lambda t},$$

which implies

$$\mathbb{E}|X^{x,i}(t) - X^{y,i}(t)|^p \leq \kappa_1^{-1}(C_K e^{-\lambda t}).$$

Noting that $\lim_{v \to 0} \kappa_1^{-1}(v) = 0$, we obtain that

$$\lim_{t \to \infty} \mathbb{E}|X^{x,i}(t) - X^{y,i}(t)|^p = 0 \text{ uniformly in } (x, y, i) \in K \times K \times \mathbb{S}$$

as desired. $\square$

The following lemma gives one more criterion for Assumption 5.42 that does not require Assumptions 5.39 and 5.41 and $p \geq 2$ either.

Lemma 5.10 *Let $p > 0$. Assume that there exist functions $U \in C^2(\mathbb{R}^n \times \mathbb{S}; \mathbb{R}_+)$, $\kappa_1, \kappa_2 \in \mathcal{K}_\vee$ and $\varphi \in \mathcal{K}_\wedge$ such that*

$$\kappa_1(|x|^p) \leq U(x, i) \leq \varphi(|x|^p) \quad \forall (x, i) \in \mathbb{R}^n, \qquad (5.156)$$

$$\mathcal{L}U(x, y, i) \leq -\kappa_2(|x - y|^p) \quad \forall (x, y, i) \in \mathbb{R}^n \times \mathbb{R}^n \times \mathbb{S}. \qquad (5.157)$$

Then Assumption 5.42 is satisfied.

Proof. Let K be any compact subset of $\mathbb{R}^n$ and set

$$C_K = \sup_{(x,y,i) \in K \times K \times \mathbb{S}} U(x - y, i) < \infty.$$

Fix any $(x, y, i) \in K \times K \times \mathbb{S}$. It is easy to show that

$$0 \leq C_K - \int_0^t \kappa_2(\mathbb{E}|X^{x,i}(s) - X^{y,i}(s)|^p)ds \quad \forall t \geq 0. \qquad (5.158)$$

For any $\varepsilon > 0$ choose $\delta > 0$ sufficiently small for

$$\varphi(u) \leq \kappa_1(\varepsilon) \quad \forall u \in [0, \delta]. \qquad (5.159)$$

Let $T = C_K/\kappa_2(\delta)$. It is easy to see from (5.158) that there must be a $t_1 \leq T$ such that
$$\mathbb{E}|X^{x,i}(t_1) - X^{y,i}(t_1)|^p \leq \delta.$$
Now, for any $t \geq T$, it is easy to show that
$$\mathbb{E}U(X^{x,i}(t) - X^{y,i}(t), r_i(t)) \leq \mathbb{E}U(X^{x,i}(t_1) - X^{y,i}(t_1), r_i(t_1)).$$
By (5.156) and the Jensen inequality,
$$\kappa_1(\mathbb{E}|X^{x,i}(t) - X^{y,i}(t)|^p) \leq \varphi(\mathbb{E}|X^{x,i}(t_1) - X^{y,i}(t_1)|^p) \leq \kappa_1(\varepsilon).$$
This, together with (5.159), yields
$$\mathbb{E}|X^{x,i}(t) - X^{y,i}(t)|^p \leq \varepsilon \quad \forall t \geq T.$$
Since T is independent of (x, y, i), we must have
$$\lim_{t \to \infty} \mathbb{E}|X^{x,i}(t) - X^{y,i}(t)|^p = 0 \text{ uniformly in } (x, y, i) \in K \times K \times \mathbb{S}$$
as required. $\square$

Let us now use the theory obtained above to establish a new criterion in terms of M-matrices, which can be verified easily in applications. For the general theory of M-matrices please see Chapter 2.

Theorem 5.44 *Let Assumption 5.39 hold. Assume that for all $x, y \in \mathbb{R}^n$ and $i \in \mathbb{S}$,*
$$x^T f(x, i) \leq \beta_i |x|^2 + \alpha, \tag{5.160}$$
$$(x - y)^T(f(x, i) - f(y, i)) \leq \beta_i |x - y|^2, \tag{5.161}$$
$$|g(x, i)|^2 \leq \delta_i |x|^2 + \alpha, \tag{5.162}$$
$$|g(x, i) - g(y, i)|^2 \leq \delta_i |x - y|^2, \tag{5.163}$$
where α, β_i and δ_i are constants. If
$$\mathcal{A} := -\mathrm{diag}(2\beta_1 + \delta_1, \cdots, 2\beta_N + \delta_N) - \Gamma \tag{5.164}$$
is a nonsingular M-matrix, then equation (5.120) is asymptotically stable in distribution.

Proof. By Theorem 2.10, there is a vector $\vec{q} = (q_1, \cdots, q_N)^T \gg 0$ such that
$$\vec{\lambda} = (\lambda_1, \cdots, \lambda_N)^T := \mathcal{A}\vec{q} \gg 0.$$

Set $\lambda = \min_{1\le i\le N} \lambda_i > 0$ and $q = \max_{1\le i\le N} q_i > 0$. Define functions $V, U : \mathbb{R}^n \times \mathbb{S} \to \mathbb{R}_+$ by

$$V(x,i) = U(x,i) = q_i|x|^2.$$

By (5.160)–(5.163) we compute the operator LV from $\mathbb{R}^n \times \mathbb{S}$ to $\mathbb{R}$ as follows:

$$LV(x,i) = 2q_i x^T f(x,i) + q_i|g(x,i)|^2 + \sum_{j=1}^{N} \gamma_{ij} q_j |x|^2$$

$$\le \left(2\beta_i q_i + \delta_i q_i + \sum_{i=1}^{N} \gamma_{ij} q_j\right)|x|^2 + 3q_i \alpha$$

$$= -\lambda_i |x|^2 + 3q_i \alpha$$

$$\le -\lambda |x|^2 + 3q\alpha.$$

By Theorem 5.2 we see that Assumption 5.41 is satisfied. Moreover, compute the operator $\mathcal{L}U$ from $\mathbb{R}^n \times \mathbb{R}^n \times \mathbb{S}$ to $\mathbb{R}$:

$$\mathcal{L}U(x,y,i) = 2q_i(x-y)^T (f(x,i) - f(y,i))$$

$$+ q_i|g(x,i) - g(y,i)|^2 + \sum_{j=1}^{N} \gamma_{ij} q_j |x-y|^2$$

$$\le \left(2\beta_i q_i + \delta_i q_i + \sum_{i=1}^{N} \gamma_{ij} q_j\right)|x-y|^2$$

$$\le -\lambda |x-y|^2.$$

By Lemma 5.10, Assumption 5.42 is satisfied too so the conclusion follows from Theorem 5.43. □

Example 5.45 Let $B(t)$ be a scalar Brownian motion. Let α and σ be constants. Consider the Ornstein–Uhlenbeck process

$$dX(t) = \alpha X(t)dt + \sigma dB(t), \quad t \ge 0. \tag{5.165}$$

Given any initial value $X(0) = x \in \mathbb{R}$, it has the unique solution

$$X(t) = e^{\alpha t} x + \sigma \int_0^t e^{\alpha(t-s)} dB(s). \tag{5.166}$$

It is easy to observe that when $\alpha < 0$, the distribution of the solution $X(t)$ converges to the normal distribution $N(0, \sigma^2/2|\alpha|)$ as $t \to \infty$ for arbitrary x, but when $\alpha \ge 0$, the distribution will not converge. In other words, equation (5.165) is asymptotically stable in distribution if $\alpha < 0$ but it is not if $\alpha \geq 0$.

Now let $r(t)$ be a right-continuous Markov chain taking values in $\mathbb{S} = \{1,2\}$ with generator

$$\Gamma = (\gamma_{ij})_{2\times 2} = \begin{bmatrix} -4 & 4 \\ \gamma & -\gamma \end{bmatrix},$$

where $\gamma > 0$. Assume that $B(t)$ and $r(t)$ are independent. Consider the one-dimensional stochastic differential equation with Markovian switching

$$dX(t) = \alpha(r(t))X(t)dt + \sigma dB(t) \tag{5.167}$$

on $t \geq 0$, where σ is a constant, $\alpha(1) = 1$ and $\alpha(2) = -1/2$. This system can be regarded as the result of two equations

$$dX(t) = X(t)dt + \sigma dB(t) \tag{5.168}$$

and

$$dX(t) = -\frac{1}{2}X(t)dt + \sigma dB(t) \tag{5.169}$$

switching from one to the other according to the law of the Markov chain. From the property of the Ornstein–Uhlenbeck process (5.165) we observe that equation (5.168) is not asymptotically stable in distribution through equation (5.169) is. However, we shall see that due to the Markovian switching the overall system (5.167) will be asymptotically stable in distribution. In fact, with obvious definitions of f and g, it is easy to see conditions (5.160)–(5.163) hold with

$$\beta_1 = 1, \quad \beta_2 = -\frac{1}{2}, \quad \delta_1 = \delta_2 = 0, \quad \alpha = \sigma^2.$$

So the matrix defined by (5.164) becomes

$$\mathcal{A} = -\text{diag}(2,-1) - \Gamma = \begin{bmatrix} 2 & -4 \\ -\gamma & 1+\gamma \end{bmatrix}.$$

Since $\gamma > 0$, this is a nonsingular M-matrix if and only if

$$2(1+\gamma) - 4\gamma > 0, \text{ namely } \gamma < 1.$$

By Theorem 5.44 we can therefore conclude that equation (5.167) is asymptotically stable in distribution if $\gamma \in (0,1)$.

Example 5.46 More generally, let us consider an n-dimensional linear SDE with Markovian switching

$$dX(t) = [A(r(t))X(t) + a(r(t))]dt$$
$$+ \sum_{k=1}^{m}[G_k(r(t))X(t) + g_k(r(t))]dB_k(t). \qquad (5.170)$$

Here $A(i)$ and $G_k(i)$ are all $n \times n$ matrices while $a(i)$ and $g_k(i)$ are $\mathbb{R}^n$-vectors. We shall write

$$A(i) = A_i, \quad G_k(i) = G_{ki}, \quad a(i) = a_i, \quad g_k(i) = g_{ki}.$$

Clearly, if we define

$$f(x, i) = A_i x + a_i$$

and

$$g(x, i) = (G_{1i}x + g_{ki}, \cdots, G_{mi}x + g_{mi}),$$

then equation (5.170) can be written as equation (5.120). Compute

$$x^T f(x, i) = x^T(A_i x + a_i) = \frac{1}{2}x^T(A_i + A_i^T)x + x^T a_i$$
$$\leq \frac{1}{2}\lambda_{\max}(A_i + A_i^T)|x|^2 + |a_i||x|$$

and

$$|g(x,i)|^2 = \sum_{k=1}^{m}|G_{ki}x + g_{ki}|^2$$
$$\leq \sum_{k=1}^{m}\Big(\|G_{ki}\|^2|x|^2 + 2\|G_{ki}\|\|g_{ki}\||x| + |g_{ki}|^2\Big).$$

So for any $\varepsilon > 0$ we can find a constant $\alpha_\varepsilon > 0$ such that

$$x^T f(x,i) \leq \Big(\frac{1}{2}\lambda_{\max}(A_i + A_i^T) + \varepsilon\Big)|x|^2 + \alpha_\varepsilon$$

and

$$|g(x,i)|^2 \leq \Big(\sum_{k=1}^{m}\|G_{ki}\|^2 + \varepsilon\Big)|x|^2 + \alpha_\varepsilon.$$

We also have
$$(x-y)^T(f(x,i) - f(y,i)) = (x-y)^T A_i (x-y)$$
$$\leq \frac{1}{2}\lambda_{\max}(A_i + A_i^T)|x-y|^2$$

and
$$|g(x,i) - g(y,i)|^2 \leq \Big(\sum_{k=1}^m \|G_{ki}\|^2\Big)|x-y|^2.$$

Consequently, inequalities (5.160)–(5.163) hold with
$$\alpha = \alpha_\varepsilon, \quad \beta_i = \frac{1}{2}\lambda_{\max}(A_i + A_i^T) + \varepsilon, \quad \delta_i = \sum_{k=1}^m \|G_{ki}\|^2 + \varepsilon.$$

If
$$-\mathrm{diag}\Big(\lambda_{\max}(A_1 + A_1^T) + \sum_{k=1}^m \|G_{k1}\|^2, \cdots,$$
$$\lambda_{\max}(A_N + A_N^T) + \sum_{k=1}^m \|G_{kN}\|^2\Big) - \Gamma \quad (5.171)$$

is a nonsingular M-matrix, then we can find a sufficiently small $\varepsilon > 0$ for
$$-\mathrm{diag}(2\beta_1 + \delta_1, \cdots, 2\beta_N + \delta_N) - \Gamma$$

remains to be a nonsingular M-matrix. By Theorem 5.44 we can therefore conclude that the linear SDE (5.170) is asymptotically stable in distribution if matrix (5.171) is a nonsingular M-matrix.

5.7 Exercises

5.1 Let $p > 0$ and Q be a symmetric positive-definite $n \times n$-matrix. Define
$$V(x) = (x^T Q x)^{p/2}, \quad x \in \mathbb{R}^n - \{0\}.$$

Show
$$V_x(x) = p(x^T Q x)^{(p-2)/2} x^T Q$$

and
$$V_{xx}(x) = p(x^T Q x)^{(p-2)/2} Q + \frac{p(p-2)}{2}(x^T Q x)^{(p-4)/2} Q x x^T Q.$$

5.2 For $0 < p < 2$, form a similar statement as Corollary 5.19 and then prove it by applying Theorem 5.12.

5.3 Prove Theorem 5.16.

5.4 Prove Lemma 5.5.

5.5 Let $x(t)$ be a solution of equation (5.73). Show
$$\limsup_{t\to\infty} \frac{1}{t}\log(|x(t)|) \leq -1 \quad a.s.$$

5.6 Let $x(t)$ be a solution of equation (5.74) with initial value $x(0) \neq 0$. Show
$$\lim_{t\to\infty} \frac{1}{t}\log(|x(t)|) = 1 \quad a.s.$$

5.7 Under the conditions of Theorem 5.30, show (5.91) and (5.81).

5.8 Let $B(t)$, $t \geq 0$, be a scalar Brownian motion and let $x(t)$ be the Ornstein–Uhlenbeck process:
$$dx(t) = -x(t)dt + dB(t), \quad t \geq 0.$$
Show the probability distribution of $x(t)$ will converge to the standard normal distribution $N(0,1)$.

5.9 Let $B(t)$ be a scalar Brownian motion. Let $r(t)$ be a right-continuous Markov chain taking values in $\mathbb{S} = \{1,2\}$ with generator
$$\Gamma = \begin{pmatrix} -\gamma_{12} & \gamma_{12} \\ \gamma_{21} & -\gamma_{21} \end{pmatrix},$$
where $\gamma_{12} > 0$ and $\gamma_{21} > 0$. Consider a scalar SDE
$$dx(t) = f(x(t), r(t))dt + g(x(t), r(t))dB(t),$$
where, for $(x,i) \in \mathbb{R} \times \mathbb{S}$,
$$f(x,i) = \begin{cases} -x^3 & \text{if } i = 1, \\ x & \text{if } i = 2, \end{cases} \quad g(x,i) = \begin{cases} 1-x & \text{if } i = 1, \\ x & \text{if } i = 2. \end{cases}$$
Show that for any $p \geq 2$, if $\gamma_{21} > \frac{1}{2}p(p+1)$, then the SDE is asymptotically bounded in pth moment.

5.10 Consider the SDE (5.75) in Example 5.26. Establish an upper bound for each ρ_i ($1 \leq i \leq 3$) so that the trivial solution of the SDE (5.75) is exponentially stable in mean square.

228   SDEs with Markovian Switching

5.11 Let $B(t)$ be a scalar Brownian motion. Let $r(t)$ be a right-continuous Markov chain taking values in $\mathbb{S} = \{1,2\}$ with generator

$$\Gamma = \begin{pmatrix} -1 & 1 \\ 2 & -2 \end{pmatrix}.$$

Consider a scalar linear SDE

$$dx(t) = \mu(r(t))x(t)dt + \delta(r(t))x(t)dB(t),$$

where μ, $\delta : \mathbb{S} \to \mathbb{R}$ are given by

$$\mu(1) = 1, \quad \mu(2) = 2, \quad \delta(1) = 2, \quad \delta(2) = 1.$$

Show that the linear SDE is almost surely exponentially stable. Moreover, give an upper bound p^* for p such that if $0 < p < p^*$ then the linear SDE is exponentially stable in pth moment.

Chapter 6

Numerical Methods for Asymptotic Properties

6.1 Introduction

We discussed the asymptotic properties of SDEs with Markovian switching in Chapter 5 while we introduced several numerical methods to approximate the solutions of SDEs in a finite time interval in Chapter 4. In this chapter we will employ numerical methods to study asymptotic properties, namely stability and stationary distribution.

Let us explain our motivations. Suppose we are required to find out whether a given stochastic differential equation with Markovian switching is exponentially stable in mean square. Although Chapter 5 provides us with several possible techniques to address the stability problem including the Lyapunov method, we may fail to solve the problem due to, for example, the absence of an appropriate Lyapunov function. In this case, we may carry out careful numerical simulations using a numerical method with a 'small' step size Δ. Two key questions are then:

(Q1) *If the SDE is exponentially stable in mean square, will the numerical method be exponentially stable in mean square for sufficiently small Δ?*

(Q2) *If the numerical method is exponentially stable in mean square for small Δ, can we infer that the underlying SDE is exponentially stable in mean square?*

These questions deal with an asymptotic ($t \to \infty$) property and hence they cannot be answered directly by applying the finite-time convergence results established in Chapter 4. Results that answer (Q1) and (Q2) for SDEs without Markovian switching can be found in [Higham (2000); Higham et al. (2003); Saito and Mitsui (1996); Schurz (1997)] but we will develop the theory to cope with Markvian switching. A similar approach will also be developed to study the stationary distribution of SDEs with Markovian switching.

6.2 Euler–Maruyama's Method and Exponential Stability

Consider an n-dimensional Itô SDE with Markovian switching

$$dx(t) = f(x(t), r(t))dt + g(x(t), r(t))dB(t) \quad (6.1)$$

on $t \geq 0$ with initial data $x(0) = x_0 \in L^2_{\mathcal{F}_0}(\Omega; \mathbb{R}^n)$ and $r(0) = r_0 \in L_{\mathcal{F}_0}(\Omega; \mathbb{S})$. Here $L_{\mathcal{F}_0}(\Omega; \mathbb{S})$ denotes the space of all $\mathcal{F}_0$-measurable $\mathbb{S}$-valued random variables, $B(t)$ and $r(t)$ are the same as before while

$$f : \mathbb{R}^n \times \mathbb{S} \to \mathbb{R}^n \quad \text{and} \quad g : \mathbb{R}^n \times \mathbb{S} \to \mathbb{R}^{n \times m}$$

are assumed to be sufficiently smooth for the existence and uniqueness of the solution. We also assume that

$$f(0, i) = 0 \quad \text{and} \quad g(0, i) = 0 \quad \forall i \in \mathbb{S} \quad (6.2)$$

so equation (6.1) admits the trivial solution $x(t) \equiv 0$. In this section we consider exponential stability in mean square of the trivial solution, which was defined in Chapter 5, but we give an alternative definition as follows.

Definition 6.1 The SDE (6.1) is said to be *exponentially stable in mean square* if there is a pair of positive constants λ and M such that, for all initial data $x_0 \in L^2_{\mathcal{F}_0}(\Omega; \mathbb{R}^n)$ and $r(0) = r_0 \in L_{\mathcal{F}_0}(\Omega; \mathbb{S})$,

$$\mathbb{E}|x(t)|^2 \leq M\mathbb{E}|x_0|^2 e^{-\lambda t}, \quad \forall t \geq 0. \quad (6.3)$$

We refer to λ as a *rate constant* and M as a *growth constant*.

Given a step size $\Delta > 0$, let $X(t)$ be the continuous Euler–Maruyama (EM) approximate solution to the SDE (6.1). This was defined in Chapter 4 but we recall here its detailed definition: Let $t_k = k\Delta$ for $k \geq 0$ and $r_k^\Delta = r(t_k)$. Compute the discrete approximations $X_k \approx x(t_k)$ by setting $X_0 = x_0$, $r_0^\Delta = r_0$ and forming

$$X_{k+1} = X_k + f(X_k, r_k^\Delta)\Delta + g(X_k, r_k^\Delta)\Delta B_k, \quad (6.4)$$

where $\Delta B_k = B(t_{k+1}) - B(t_k)$. Let

$$\bar{X}(t) = X_k, \quad \bar{r}(t) = r_k^\Delta \quad \text{for } t \in [t_k, t_{k+1}) \quad (6.5)$$

and define the continuous EM approximate solution

$$X(t) = x_0 + \int_0^t f(\bar{X}(s), \bar{r}(s))ds + \int_0^t g(\bar{X}(s), \bar{r}(s))dB(s). \tag{6.6}$$

Note that $X(t_k) = \bar{X}(t_k) = X_k$, that is $X(t)$ and $\bar{X}(t)$ coincide with the discrete solution at the grid-points.

Following Definition 6.1, we now define exponential stability in mean square for the EM continuous time approximation $X(t)$.

Definition 6.2 For a given step size $\Delta > 0$, the EM method is said to be *exponentially stable in mean square* on the SDE (6.1) if there is a pair of positive constants γ and H such that with initial data $x_0 \in L^2_{\mathcal{F}_0}(\Omega; \mathbb{R}^n)$ and $r_0 \in L_{\mathcal{F}_0}(\Omega; \mathbb{S})$

$$\mathbb{E}|X(t)|^2 \leq H\mathbb{E}|x_0|^2 e^{-\gamma t}, \quad \forall t \geq 0. \tag{6.7}$$

We wish to know whether the numerical method shares exponential mean-square stability with the SDE. First of all, let us impose the global Lipschitz condition

Assumption 6.3 There is a positive constant K such that

$$|f(x,i) - f(y,i)|^2 \vee |g(x,i) - g(y,i)|^2 \leq K|x-y|^2 \tag{6.8}$$

for all $(x, y, i) \in \mathbb{R}^n \times \mathbb{R}^n \times \mathbb{S}$.

Recalling (6.2) we observe from this assumption that the following linear growth condition

$$|f(x,i)|^2 \vee |g(x,i)|^2 \leq K|x|^2 \tag{6.9}$$

holds for all $(x, i) \in \mathbb{R}^n \times \mathbb{S}$.

Let us now present a key lemma.

Lemma 6.1 *Under (6.2) and Assumption 6.3, for all sufficiently small $\Delta < 1$ the continuous EM approximate solution (6.6) with initial data x_0 and r_0 satisfies, for any $T > 0$,*

$$\sup_{0 \leq t \leq T} \mathbb{E}|X(t)|^2 \leq B_{x_0,T}, \tag{6.10}$$

where $B_{x_0,T} = 3\mathbb{E}|x_0|^2 e^{3(T+1)KT}$, and

$$\sup_{0 \leq t \leq T} \mathbb{E}|X(t) - x(t)|^2 \leq \left(\sup_{0 \leq t \leq T} \mathbb{E}|X(t)|^2 \right) C_T \Delta \tag{6.11}$$

where C_T depends on T but not on x_0, r_0 and Δ.

Proof. It can be shown (the details are left to the reader as an exercise) from (6.6) and (6.9) that, for $0 \leq t \leq T$,

$$\mathbb{E}|X(t)|^2 \leq 3\mathbb{E}|x_0|^2 + 3(T+1)K \int_0^t \mathbb{E}|\bar{X}(s)|^2 ds. \qquad (6.12)$$

Since the right-hand side term is non-decreasing in t, we have

$$\sup_{0 \leq t \leq t_1} \mathbb{E}|X(t)|^2 \leq 3\mathbb{E}|x_0|^2 + 3(T+1)K \int_0^{t_1} \mathbb{E}|\bar{X}(s)|^2 ds$$

$$\leq 3\mathbb{E}|x_0|^2 + 3(T+1)K \int_0^{t_1} \Big(\sup_{0 \leq t \leq s} \mathbb{E}|X(t)|^2 \Big) ds.$$

for any $t_1 \in [0, T]$. The Gronwall inequality hence yields

$$\sup_{0 \leq t \leq T} \mathbb{E}|X(t)|^2 \leq 3\mathbb{E}|x_0|^2 e^{3(T+1)KT},$$

which is the required assertion (6.10).

To show the assertion (6.11), we derive from (6.1) and (6.6) that, for $0 \leq t \leq T$,

$$\mathbb{E}|X(t) - x(t)|^2$$

$$\leq 2T\mathbb{E} \int_0^t |f(\bar{X}(s), \bar{r}(s)) - f(x(s), r(s))|^2 ds$$

$$+ 2\mathbb{E} \int_0^t |g(\bar{X}(s), \bar{r}(s)) - g(x(s), r(s))|^2 ds$$

$$\leq 4K(T+1) \int_0^t \mathbb{E}|\bar{X}(s) - x(s)|^2 ds$$

$$+ 4T\mathbb{E} \int_0^t |f(\bar{X}(s), \bar{r}(s)) - f(\bar{X}(s), r(s))|^2$$

$$+ 4\mathbb{E} \int_0^t |g(\bar{X}(s), \bar{r}(s)) - g(\bar{X}(s), r(s))|^2 ds. \qquad (6.13)$$

By using (6.9), it can be shown in the same way as Theorem 4.1 was proved that

$$\mathbb{E}\int_0^t |f(\bar{X}(s), \bar{r}(s)) - f(\bar{X}(s), r(s))|^2$$
$$+ \mathbb{E}\int_0^t |g(\bar{X}(s), \bar{r}(s)) - g(\bar{X}(s), r(s))|^2 ds$$
$$\leq c_1 \Delta \left(\sup_{0 \leq t \leq T} \mathbb{E}|X(t)|^2 \right), \tag{6.14}$$

where c_1 and the following c_2 are both positive constants dependent on T but independent of x_0, r_0 and Δ. Moreover, note

$$\mathbb{E}|\bar{X}(s) - x(s)|^2 \leq 2\mathbb{E}|\bar{X}(s) - X(s)|^2 + 2\mathbb{E}|X(s) - x(s)|^2. \tag{6.15}$$

Let $k = k(s)$ be the integer part of s/Δ so $k\Delta \leq s < (k+1)\Delta$. It then follows from (6.6) that

$$X(s) - \bar{X}(s) = X(s) - X_k$$
$$= f(X_k, r_k^\Delta)(s - k\Delta) + g(X_k, r_k^\Delta)(B(s) - B(k\Delta)).$$

Thus

$$\mathbb{E}|X(s) - \bar{X}(s)|^2 \leq 2(\Delta^2 + \Delta)K\mathbb{E}|X_k|^2 \leq 4K\Delta \left(\sup_{0 \leq t \leq T} \mathbb{E}|X(t)|^2 \right). \tag{6.16}$$

Combining (6.13)–(6.16) together yields

$$\mathbb{E}|X(t) - x(t)|^2 \leq c_2 \int_0^t \mathbb{E}|X(s) - x(s)|^2 ds$$
$$+ c_2 \Delta \left(\sup_{0 \leq t \leq T} \mathbb{E}|X(t)|^2 \right).$$

The Gronwall inequality hence implies that, for any $t \in [0, T]$,

$$\mathbb{E}|X(t) - x(t)|^2 \leq \left(\sup_{0 \leq t \leq T} \mathbb{E}|X(t)|^2 \right) c_2 \Delta e^{c_2 T}.$$

which is the required assertion (6.11) by setting $C_T = c_2 e^{c_2 T}$. □

Lemma 6.1 emphasises the dependence of C upon T as this is important in the subsequent analysis. We remark that (6.11) says that the method has strong finite-time convergence order of at least $\frac{1}{2}$, with a 'squared error constant' that is linearly proportional to $\sup_{0 \leq t \leq T} \mathbb{E}|X(t)|^2$.

It is useful to observe from Corollary 3.22 or Exercise 6.2 below that under Assumption 6.3 the solution of equation (6.1) satisfies

$$\sup_{0\leq t\leq T} \mathbb{E}|x(t)|^2 < \infty, \quad \forall T > 0. \tag{6.17}$$

The following lemma gives a positive answer to question (Q1) from section 6.1.

Lemma 6.2 *Let (6.2) and Assumption 6.3 hold. Assume that the SDE (6.1) is exponentially stable in mean square and satisfies (6.3). Then there exists a $\Delta^\star > 0$ such that for every $0 < \Delta \leq \Delta^\star$ the EM method is exponentially stable in mean square on the SDE (6.1) with rate constant $\gamma = \frac{1}{2}\lambda$ and growth constant $H = 2Me^{\frac{1}{2}\lambda(1+(4\log M)/\lambda)}$.*

Proof. Choose $T = 1 + (4\log M)/\lambda$, so that

$$Me^{-\lambda T} \leq e^{-\frac{3}{4}\lambda T}. \tag{6.18}$$

Now, for any $\alpha > 0$,

$$\mathbb{E}|X(t)|^2 \leq (1+\alpha)\mathbb{E}|X(t) - x(t)|^2 + (1+1/\alpha)\mathbb{E}|x(t)|^2. \tag{6.19}$$

$$\sup_{0\leq t\leq 2T} \mathbb{E}|X(t)|^2 \leq (1+\alpha) \sup_{0\leq t\leq 2T} \mathbb{E}|X(t)|^2 C_{2T}\Delta + (1+1/\alpha)M\mathbb{E}|x_0|^2.$$

Taking Δ sufficiently small, this rearranges to

$$\sup_{0\leq t\leq 2T} \mathbb{E}|X(t)|^2 \leq \frac{(1+1/\alpha)M\mathbb{E}|x_0|^2}{1 - (1+\alpha)C_{2T}\Delta}. \tag{6.20}$$

Now, taking the supremum over $[T, 2T]$ in (6.19), using Lemma 6.1 and the bound (6.20), and also the stability condition (6.3), gives

$$\sup_{T\leq t\leq 2T} \mathbb{E}|X(t)|^2 \leq \frac{(1+\alpha)(1+1/\alpha)M\mathbb{E}|x_0|^2}{1 - (1+\alpha)C_{2T}\Delta} C_{2T}\Delta$$
$$+ (1+1/\alpha)M\mathbb{E}|x_0|^2 e^{-\lambda T}.$$

We write this as

$$\sup_{T\leq t\leq 2T} \mathbb{E}|X(t)|^2 \leq R(\Delta)\mathbb{E}|x_0|^2, \tag{6.21}$$

where

$$R(\Delta) := \frac{(1+\alpha)(1+1/\alpha)}{1 - (1+\alpha)C_{2T}\Delta} C_{2T}\Delta M + (1+1/\alpha)Me^{-\lambda T}.$$

Putting $\alpha = 1/\sqrt{\Delta}$ and using (6.18) we see that for sufficiently small Δ

$$R(\Delta) \leq 2\sqrt{\Delta}C_{2T}M + (1+\sqrt{\Delta})e^{-\frac{3}{4}\lambda T}.$$

The right-hand side of this inequality is equal to $e^{-\frac{3}{4}\lambda T}$ when $\Delta = 0$ and increases monotonically with Δ. Hence, by taking Δ sufficiently small we may ensure that

$$R(\Delta) \leq e^{-\frac{1}{2}\lambda T}. \tag{6.22}$$

In (6.21) this gives

$$\sup_{T \leq t \leq 2T} \mathbb{E}|X(t)|^2 \leq e^{-\frac{1}{2}\lambda T}\mathbb{E}|x_0|^2 \leq e^{-\frac{1}{2}\lambda T} \sup_{0 \leq t \leq T} \mathbb{E}|X(t)|^2.$$

Now, let $\hat{x}(t)$ be the solution to the SDE (6.1) for $t \in [T, \infty)$ with the initial data $\hat{x}(T) = X(T)$ and $r(T) = \bar{r}(T)$ at $t = T$. Since the SDE (6.1) is time-homogeneous, we may shift (6.3) to obtain

$$\mathbb{E}|\hat{x}(t)|^2 \leq M\mathbb{E}|X(T)|^2 e^{-\lambda(t-T)}, \quad \forall t \geq T. \tag{6.23}$$

Similarly, Lemma 6.1 gives

$$\sup_{T \leq t \leq 3T} \mathbb{E}|X(t) - \hat{x}(t)|^2 \leq \left(\sup_{0 \leq t \leq T} \mathbb{E}|X(t)|^2 \right) C_{2T}\Delta, \quad \forall t \geq T. \tag{6.24}$$

Copying the previous analysis, we have

$$\mathbb{E}|X(t)|^2 \leq (1+\alpha)\mathbb{E}|X(t) - \hat{x}(t)|^2 + (1+1/\alpha)\mathbb{E}|\hat{x}(t)|^2. \tag{6.25}$$

Taking the supremum over $[T, 3T]$ and using (6.23) and (6.24), we obtain

$$\sup_{T \leq t \leq 3T} \mathbb{E}|X(t)|^2 \leq (1+\alpha) \sup_{T \leq t \leq 3T} \mathbb{E}|X(t)|^2 C_{2T}\Delta + (1+1/\alpha)M\mathbb{E}|X(T)|^2.$$

This gives

$$\sup_{T \leq t \leq 3T} \mathbb{E}|X(t)|^2 \leq \frac{(1+1/\alpha)M\mathbb{E}|X(T)|^2}{1 - (1+\alpha)C_{2T}\Delta}.$$

Now, taking the supremum over $[2T, 3T]$ in (6.25), in place of (6.21) we arrive at

$$\sup_{2T \leq t \leq 3T} \mathbb{E}|X(t)|^2 \leq R(\Delta)\mathbb{E}|X(T)|^2.$$

Continuing this approach and using (6.22) gives

$$\sup_{(i+1)T \leq t \leq (i+2)T} \mathbb{E}|X(t)|^2 \leq e^{-\frac{1}{2}\lambda T}\mathbb{E}|X(iT)|^2, \quad \text{for } i \geq 0. \tag{6.26}$$

From (6.26) we see that

$$\sup_{(i+1)T\leq t\leq (i+2)T} \mathbb{E}|X(t)|^2 \leq e^{-\frac{1}{2}\lambda T}e^{-\frac{1}{2}\lambda T} \sup_{(i-1)T\leq t\leq iT} \mathbb{E}|X(t)|^2$$

$$\vdots$$

$$\leq e^{-\frac{1}{2}\lambda T(i+1)} \sup_{0\leq t\leq T} \mathbb{E}|X(t)|^2. \qquad (6.27)$$

Now, using $\alpha = 1/\sqrt{\Delta}$ in (6.20), for sufficiently small Δ we see that

$$\sup_{0\leq t\leq T} \mathbb{E}|X(t)|^2 \leq 2M\mathbb{E}|x_0|^2. \qquad (6.28)$$

It follows from (6.27) and (6.28) that

$$\sup_{(i+1)T\leq t\leq (i+2)T} \mathbb{E}|X(t)|^2 \leq e^{-\frac{1}{2}\lambda T(i+1)} 2M\mathbb{E}|x_0|^2$$

$$= 2Me^{\frac{1}{2}\lambda T}\mathbb{E}|x_0|^2 e^{-\frac{1}{2}\lambda T(i+2)}.$$

Hence, the numerical method is exponentially stable in mean square with $\gamma = \frac{1}{2}\lambda$ and $H = 2Me^{\frac{1}{2}\lambda T}$. $\qquad \square$

The next lemma gives a positive answer to question (Q2) from section 6.1.

Lemma 6.3 *Let (6.2) and Assumption 6.3 hold. Assume that for a step size $\Delta > 0$, the numerical method is exponentially stable in mean square with rate constant γ and growth constant H. If Δ satisfies*

$$C_{2T}e^{\gamma T}(\Delta + \sqrt{\Delta}) + 1 + \sqrt{\Delta} \leq e^{\frac{1}{4}\gamma T} \quad \text{and} \quad C_T\Delta \leq 1, \qquad (6.29)$$

where $T := 1 + (4\log H)/\gamma$, then the SDE (6.1) is exponentially stable in mean square with rate constant $\lambda = \frac{1}{2}\gamma$ and growth constant $M = 2He^{\frac{1}{2}\gamma T}$.

Proof. First, note that

$$e^{-\frac{3}{4}\gamma T}H \leq e^{-\frac{1}{2}\gamma T}. \qquad (6.30)$$

For any $\alpha > 0$ we have

$$\mathbb{E}|x(t)|^2 \leq (1+\alpha)\mathbb{E}|X(t) - x(t)|^2 + (1 + 1/\alpha)\mathbb{E}|X(t)|^2. \qquad (6.31)$$

Using Lemma 6.1 and (6.7) in (6.31), we obtain

$$\sup_{T\leq t\leq 2T} \mathbb{E}|x(t)|^2$$
$$\leq (1+\alpha) \sup_{T\leq t\leq 2T} \mathbb{E}|X(t) - x(t)|^2 + (1+1/\alpha) \sup_{T\leq t\leq 2T} \mathbb{E}|X(t)|^2$$
$$\leq (1+\alpha)C_{2T}\Delta \sup_{0\leq t\leq 2T} \mathbb{E}|X(t)|^2 + (1+1/\alpha) \sup_{T\leq t\leq 2T} \mathbb{E}|X(t)|^2$$
$$\leq (1+\alpha)C_{2T}\Delta H\mathbb{E}|x_0|^2 + (1+1/\alpha)H\mathbb{E}|x_0|^2 e^{-\gamma T}$$
$$\leq \left[(1+\alpha)C_{2T}\Delta e^{\gamma T} + (1+1/\alpha)\right] H\mathbb{E}|x_0|^2 e^{-\gamma T}. \qquad (6.32)$$

Setting $\alpha = 1/\sqrt{\Delta}$ gives

$$\sup_{T\leq t\leq 2T} \mathbb{E}|x(t)|^2 \leq \left[C_{2T}e^{\gamma T}(\Delta + \sqrt{\Delta}) + 1 + \sqrt{\Delta}\right] H\mathbb{E}|x_0|^2 e^{-\gamma T}. \qquad (6.33)$$

Using (6.29) and (6.30), we then have

$$\sup_{T\leq t\leq 2T} \mathbb{E}|x(t)|^2 \leq e^{-\frac{3}{4}\gamma T} H\mathbb{E}|x_0|^2 \leq e^{-\frac{1}{2}\gamma T}\mathbb{E}|x_0|^2$$
$$\leq e^{-\frac{1}{2}\gamma T} \sup_{0\leq t\leq T} \mathbb{E}|x(t)|^2. \qquad (6.34)$$

Now let $\hat{X}(t)$ for $t \in [T, \infty)$ denote the continuous EM approximation that arises from applying the EM method with initial data $x(T)$ and $r(T)$ at time $t = T$. Then using similar arguments to those that produced (6.32) and (6.33), we have

$$\sup_{2T\leq t\leq 3T} \mathbb{E}|x(t)|^2$$
$$\leq (1+\alpha) \sup_{2T\leq t\leq 3T} \mathbb{E}|\hat{X}(t) - x(t)|^2 + (1+1/\alpha) \sup_{2T\leq t\leq 3T} \mathbb{E}|\hat{X}(t)|^2$$
$$\leq (1+\alpha)C_{2T}\Delta \sup_{T\leq t\leq 3T} \mathbb{E}|\hat{X}(t)|^2 + (1+1/\alpha) \sup_{2T\leq t\leq 3T} \mathbb{E}|\hat{X}(t)|^2$$
$$\leq (1+\alpha)C_{2T}\Delta H\mathbb{E}|x(T)|^2 + (1+1/\alpha)H\mathbb{E}|x(T)|^2 e^{-\gamma T}$$
$$\leq \left[(1+\alpha)C_{2T}\Delta e^{\gamma T} + (1+1/\alpha)\right] H\mathbb{E}|x(T)|^2 e^{-\gamma T}$$
$$\leq e^{-\frac{3}{4}\gamma T} H\mathbb{E}|x(T)|^2$$
$$\leq e^{-\frac{1}{2}\gamma T}\mathbb{E}|x(T)|^2$$
$$\leq e^{-\frac{1}{2}\gamma T} \sup_{T\leq t\leq 2T} \mathbb{E}|x(t)|^2.$$

Generally, this approach may be used to show that
$$\sup_{iT\leq t\leq (i+1)T} \mathbb{E}|x(t)|^2 \leq e^{-\frac{1}{2}\gamma T} \sup_{(i-1)T\leq t\leq iT} \mathbb{E}|x(t)|^2, \quad i\geq 1.$$

Hence,
$$\sup_{iT\leq t\leq (i+1)T} \mathbb{E}|x(t)|^2 \leq e^{-\frac{1}{2}\gamma iT} \sup_{0\leq t\leq T} \mathbb{E}|x(t)|^2. \tag{6.35}$$

Now
$$\sup_{0\leq t\leq T} \mathbb{E}|x(t)|^2 \leq \sup_{0\leq t\leq T} \mathbb{E}|X(t)-x(t)|^2 + \sup_{0\leq t\leq T} \mathbb{E}|X(t)|^2$$
$$\leq (C_T\Delta + 1)H\mathbb{E}|x_0|^2$$
$$\leq 2H\mathbb{E}|x_0|^2,$$

using (6.29). In (6.35) this gives
$$\sup_{iT\leq t\leq (i+1)T} \mathbb{E}|x(t)|^2 \leq e^{-\frac{1}{2}\gamma(i+1)T}e^{\frac{1}{2}\gamma T}2H\mathbb{E}|x_0|^2,$$

which proves the required result. □

Lemmas 6.2 and 6.3 lead to the following important necessary and sufficient theorem.

Theorem 6.4 *Under (6.2) and Assumption 6.3, the SDE (6.1) is exponentially stable in mean square if and only if there exists a $\Delta > 0$ such that the EM method is exponentially stable in mean square with rate constant γ, growth constant H, step size Δ and global error constant C_T for $T := 1 + (4\log H)/\gamma$ satisfying (6.29).*

Proof. The 'if' part of the theorem follows from Lemma 6.3 directly. To prove the 'only if' part, suppose the SDE is exponentially stable in mean square with rate constant λ and growth constant M. Lemma 6.2 shows that there is a $\Delta^* > 0$ such that for any step size $0 < \Delta \leq \Delta^*$, the EM method is exponentially stable in mean square with rate constant $\gamma = \frac{1}{2}\lambda$ and growth constant $H = 2Me^{\frac{1}{2}\lambda(1+(4\log M)/\lambda)}$. Noting that both of these constants are independent of Δ, it follows that we may reduce Δ if necessary until (6.29) becomes satisfied. □

We emphasise that Theorem 6.4 is an 'if and only if' result, which shows that, under Assumption 6.3 and for sufficiently small Δ, exponential stability of the EM method is equivalent to exponential stability of the SDE. Thus it is feasible to investigate exponential stability of the SDE from careful numerical simulations.

6.3 Euler–Maruyama's Method and Lyapunov Exponents

In Lemmas 6.2 and 6.3, we found new rate constants that were within a factor $\frac{1}{2}$ of the given ones. If we are interested only in asymptotic decay rates then it is useful to adopt the definition given in Chapter 5 that eliminate the growth constant completely. To be more precise, we recall the definition below.

Definition 6.5 Equation (6.1) is said to have *2nd moment Lyapunov exponent bounded by* $-\lambda < 0$ if, with any initial data $x_0 \in L^2_{\mathcal{F}_0}(\Omega; \mathbb{R}^n)$ and $r_0 \in L_{\mathcal{F}_0}(\Omega; \mathbb{S})$,

$$\limsup_{t \to \infty} \frac{1}{t} \log \left(\mathbb{E}|x(t)|^2 \right) \leq -\lambda. \tag{6.36}$$

Similarly we may define the 2nd moment Lyapunov exponent bound for the EM method.

Definition 6.6 For a given step size $\Delta > 0$, the EM method is said to have *2nd moment Lyapunov exponent bounded by* $-\gamma < 0$ on the SDE (6.1) if, with initial data $x_0 \in L^2_{\mathcal{F}_0}(\Omega; \mathbb{R}^n)$ and $r_0 \in L_{\mathcal{F}_0}(\Omega; \mathbb{S})$,

$$\limsup_{t \to \infty} \frac{1}{t} \log \left(\mathbb{E}|X(t)|^2 \right) \leq -\gamma. \tag{6.37}$$

We note that the λ appearing as a rate constant in Definition 6.1 is equivalent to the λ appearing in the 2nd moment Lyapunov exponent bound in Definition 6.5, and similarly for γ in Definitions 6.2 and 6.6. Theorem 6.7 below shows that by taking Δ sufficiently small, we can make the 2nd moment Lyapunov exponent bounds for the SDE and numerical method arbitrarily close.

Theorem 6.7 *Let (6.2) and Assumption 6.3 hold. If the SDE (6.1) is exponentially stable in mean square with rate constant* λ, *then given any* $\epsilon \in (0, \lambda)$ *there exists a* $\Delta^* > 0$ *such that for all* $0 < \Delta \leq \Delta^*$ *the EM method has 2nd moment Lyapunov exponent bounded by* $-\lambda + \epsilon$. *Conversely, if the EM method is exponentially stable in mean square for sufficiently small step size* Δ *with fixed values of* γ *and* H, *then the SDE has 2nd moment Lyapunov exponent bounded by* $-\gamma$.

Proof. This proof is similar to the proofs of Lemmas 6.2 and 6.3. Suppose the SDE (6.1) has rate constant λ and growth constant M. Given ϵ, choose

$T = 1 + (2\log M)/\epsilon$, so that

$$Me^{-\lambda T} \leq e^{-(\lambda-\frac{1}{2}\epsilon)T}. \tag{6.38}$$

Now, as in the proof of Lemma 6.2, the inequality (6.21) holds. (Note that T, and hence the constant C_{2T}, depend upon ϵ.) Using (6.38), we have, for sufficiently small Δ,

$$R(\Delta) \leq 2\sqrt{\Delta}C_{2T}M + (1+\sqrt{\Delta})e^{-(\lambda-\frac{1}{2}\epsilon)T},$$

and hence there exists a $\Delta^\star$ such that for all $\Delta \leq \Delta^\star$

$$\sup_{T \leq t \leq 2T} \mathbb{E}|X(t)|^2 \leq e^{-(\lambda-\epsilon)T}\mathbb{E}|x_0|^2 \leq e^{-(\lambda-\epsilon)T} \sup_{0 \leq t \leq T} \mathbb{E}|X(t)|^2.$$

Continuing as in the proof of Lemma 6.2 we find that for each $\Delta \leq \Delta^\star$ the numerical method is exponentially stable in mean square with $\gamma = \lambda - \epsilon$ and $H = 2Me^{(\lambda-\epsilon)T}$. The first part of the theorem then follows. (Note, however, that T depends upon ϵ, and hence we cannot conclude that $H = H(\epsilon)$ is uniformly bounded.)

To prove the converse, for any $\varepsilon \in (0, \gamma)$, we may choose $T = 1 + (2\log H)/\epsilon$ so that

$$He^{-\gamma T} \leq e^{-(\gamma-\frac{1}{2}\epsilon)T},$$

and then in place of (6.34) we have

$$\sup_{T \leq t \leq 2T} \mathbb{E}|x(t)|^2 \leq e^{-(\gamma-\epsilon)T} \sup_{0 \leq t \leq T} \mathbb{E}|x(t)|^2.$$

Continuing in this way, we find that the SDE is exponentially stable in mean square with $\lambda = \gamma - \epsilon$ and $M = 2He^{(\gamma-\epsilon)T}$. This means that the SDE has 2nd moment Lyapunov exponent bounded by $-\gamma + \varepsilon$. Since the ε is arbitrary, the result then follows. (Note that, as for the first part of the proof, T depends upon ϵ, and hence we cannot conclude that $M = M(\epsilon)$ is uniformly bounded.) □

In the proof of Theorem 6.7 we found it necessary to have T increasing with ϵ in order to control the intermediate growth allowed by a growth factor greater than unity. In the special case where the growth factor equals unity we have the following stronger result.

Theorem 6.8 *Assume that (6.2) and Assumption 6.3 hold. If the SDE (6.1) is exponentially stable in mean square with rate constant λ and growth constant $M = 1$, then given any $\epsilon \in (0, \lambda)$ there exists a $\Delta^\star > 0$ such that* for all $0 < \Delta \leq \Delta^*$ the numerical method is exponentially stable in mean square with rate constant $-\lambda + \epsilon$ and growth constant $2e^\lambda$. Conversely, if the numerical method is exponentially stable in mean square for sufficiently small step size Δ with fixed values of γ and $H = 1$, then the SDE is exponentially stable in mean square with rate constant $-\gamma$ and growth constant $2e^\gamma$.

Proof. The result can be proved in a similar manner to Theorem 6.7, using $T = 1$. □

6.4 Generalised Results and Stochastic Theta Method

It is important to observe that the proofs of Lemmas 6.2 and 6.3 as well as the theorems in the previous two sections are not only based on the assertions of Lemma 6.1 rather than the global Lipschitz Assumption 6.3 but also independent of the EM method used. In other words, the results presented there will hold for any numerical method applied to the SDE (6.1) as long as the corresponding continuous approximate solution $X(t)$ obeys properties (6.10) and (6.11). This observation leads to the more general treatment below.

We suppose that a numerical method is available which, given a step size $\Delta > 0$, computes discrete approximations $X_k \approx x(k\Delta)$, with $X_0 = x_0$. We also suppose that there is a well-defined interpolation process that extends the discrete approximation $\{X_k\}_{k \geq 1}$ to a continuous-time approximation $\{X(t)\}_{t \in \mathbb{R}_+}$, with $X(k\Delta) = X_k$. Such a process could be obtained by the EM method as seen in the previous sections but will be illustrated for the class of stochastic theta methods in this section. Moreover, we assume that this continuous approximation obeys the following natural finite-time convergence condition.

Assumption 6.9 For all sufficiently small Δ the continuous approximation $X(t)$ satisfies, for any $T > 0$,

$$\sup_{0 \leq t \leq T} \mathbb{E}|X(t)|^2 < B_{x_0,T}, \tag{6.39}$$

where $B_{x_0,T}$ depends on x_0 and T, but not upon Δ and r_0, and

$$\sup_{0 \leq t \leq T} \mathbb{E}|X(t) - x(t)|^2 \leq \left(\sup_{0 \leq t \leq T} \mathbb{E}|X(t)|^2\right) C_T \Delta \tag{6.40}$$

where C_T depends on T but not on x_0, r_0 and Δ.

It is useful to remark that Assumption 6.9 guarantees the finite 2nd moment of the true solution of the SDE (6.1), namely
$$\sup_{0 \leq t \leq T} \mathbb{E}|x(t)|^2 < \infty, \quad \forall T \geq 0.$$

The exponential stability in mean square of the numerical method can be defined in the same way as for the EM method. In fact, just change "the EM method" in Definition 6.2 by "the numerical method". According to the observation pointed out in the beginning of this section, we have the following general result.

Theorem 6.10 *Under Assumption 6.9, the assertions of Theorems 6.4, 6.7 and 6.8 hold for the numerical method.*

Let us now strengthen the bound (6.40) in Assumption 6.9 by forcing the 'squared error constant' to be linearly proportional to $\mathbb{E}|x_0|^2$, rather than $\sup_{0 \leq t \leq T} \mathbb{E}|X(t)|^2$. The motivation for this is twofold: (a) the proofs of the two key lemmas (i.e. Lemmas 6.2 and 6.3) become simpler and more symmetric, and (b) the stronger bound (6.41) can be established for the case of the stochastic theta method on globally Lipschitz SDEs (to be done in the last part of this section). However, the constant $\bar{C}_T$ in (6.41) arising from that analysis is, in general, much larger than the C_T in (6.40). Hence, the restriction on Δ in Theorem 6.4 is typically much less stringent than that in Theorem 6.12.

Assumption 6.11 For sufficiently small Δ the numerical method applied to (6.1) with initial data x_0 and r_0 satisfies, for any $T > 0$,
$$\sup_{0 \leq t \leq T} \mathbb{E}|X(t)|^2 < B_{x_0,T},$$
where $B_{x_0,T}$ depends on x_0 and T, but not upon Δ and r_0, and
$$\sup_{0 \leq t \leq T} \mathbb{E}|X(t) - x(t)|^2 \leq \bar{C}_T \Delta \mathbb{E}|x_0|^2, \tag{6.41}$$
where $\bar{C}_T$ depends on T but not on x_0, r_0 and Δ.

Lemma 6.4 *Assume that the SDE (6.1) is exponentially stable in mean square and satisfies (6.3), and that Assumption 6.11 holds. Let $T := 1 + (4\log M)/\lambda$. Choose $\Delta^* > 0$ such that for all $0 < \Delta \leq \Delta^*$*
$$\left(\Delta + \sqrt{\Delta}\right)\bar{C}_T + \left(\sqrt{\Delta} + 1\right)M \leq 2M \tag{6.42}$$

and

$$\left(\Delta + \sqrt{\Delta}\right) \bar{C}_{2T} + \left(\sqrt{\Delta} + 1\right) e^{-\frac{3}{4}\lambda T} \leq e^{-\frac{1}{2}\lambda T}. \tag{6.43}$$

Then for all $0 < \Delta \leq \Delta^*$ the numerical method is exponentially stable in mean square with rate constant $\gamma = \frac{1}{2}\lambda$ and growth constant $H = 2Me^{\frac{1}{2}\lambda T}$.

Proof. Starting with (6.19), choosing $\alpha = 1/\sqrt{\Delta}$ and using Assumption 6.11 and condition (6.42) we have

$$\sup_{0 \leq t \leq T} \mathbb{E}|X(t)|^2 \leq \left[(\Delta + \sqrt{\Delta})\bar{C}_T + (\sqrt{\Delta} + 1)M\right] \mathbb{E}|x_0|^2$$

$$\leq 2M\mathbb{E}|x_0|^2. \tag{6.44}$$

Now, let $\hat{y}^{[i]}(t)$ be the solution to the SDE (6.1) for $t \in [iT, \infty)$ with the initial data $\hat{y}^{[i]}(iT) = X(iT)$ and $r(iT) = \bar{r}(iT)$. Then, using (6.19),

$$\sup_{(i+1)T \leq t \leq (i+2)T} \mathbb{E}|X(t)|^2 \leq (1+\alpha) \sup_{iT \leq t \leq (i+2)T} \mathbb{E}|X(t) - \hat{y}^{[i]}(t)|^2$$

$$+ (1 + \frac{1}{\alpha}) \sup_{iT \leq t \leq (i+2)T} \mathbb{E}|\hat{y}^{[i]}(t)|^2.$$

Choosing $\alpha = 1/\sqrt{\Delta}$, using Assumption 6.11 and condition (6.43), and noting that $Me^{-\lambda T} \leq e^{-\frac{3}{4}\lambda T}$, we find that

$$\sup_{(i+1)T \leq t \leq (i+2)T} \mathbb{E}|X(t)|^2$$

$$\leq (\Delta + \sqrt{\Delta})\bar{C}_{2T}\mathbb{E}|X(iT)|^2 + (\sqrt{\Delta} + 1)Me^{-\lambda T}\mathbb{E}|X(iT)|^2 \tag{6.45}$$

$$\leq \left[(\Delta + \sqrt{\Delta})\bar{C}_{2T} + (\sqrt{\Delta} + 1)e^{-\frac{3}{4}\lambda T}\right] \sup_{iT \leq t \leq (i+1)T} \mathbb{E}|X(t)|^2$$

$$\leq e^{-\frac{1}{2}\lambda T} \sup_{iT \leq t \leq (i+1)T} \mathbb{E}|X(t)|^2. \tag{6.46}$$

Combining (6.44) and (6.46) we deduce that

$$\sup_{nT \leq t \leq (n+1)T} \mathbb{E}|X(t)|^2 \leq e^{-\frac{1}{2}\lambda nT} \sup_{0 \leq t \leq T} \mathbb{E}|X(t)|^2$$

$$\leq 2Me^{\frac{1}{2}\lambda T} e^{-\frac{1}{2}\lambda(n+1)T} \mathbb{E}|x_0|^2,$$

and the result follows. □

The following is proved by techniques almost identical to those used in the preceding lemma so the details are left to the reader as an exercise.

Lemma 6.5 *Assume that the numerical method is exponentially stable in mean square with rate constant γ and growth constant H for some step size $\Delta > 0$ and Assumption 6.11 holds. Let $T := 1 + (4\log H)/\gamma$. Then if*

$$\left(\Delta + \sqrt{\Delta}\right)\bar{C}_T + \left(\sqrt{\Delta} + 1\right)H \leq 2H, \tag{6.47}$$

and

$$\left(\Delta + \sqrt{\Delta}\right)\bar{C}_{2T} + \left(\sqrt{\Delta} + 1\right)e^{-\frac{3}{4}\gamma T} \leq e^{-\frac{1}{2}\gamma T}, \tag{6.48}$$

the SDE (6.1) is exponentially stable in mean square with rate constant $\lambda = \frac{1}{2}\gamma$ and growth constant $M = 2He^{\frac{1}{2}\gamma T}$.

Just as Lemmas 6.2 and 6.3 combined to give Theorem 6.4, the next theorem follows from Lemmas 6.4 and 6.5.

Theorem 6.12 *Suppose that a numerical method satisfies Assumption 6.11. Then the SDE (6.1) is exponentially stable in mean square if and only if there exists a $\Delta > 0$ such that the numerical method is exponentially stable in mean square with rate constant γ, growth constant H, step size Δ and global error constant $\bar{C}_T$ for $T := 1 + (4\log H)/\gamma$ satisfying (6.47)–(6.48).*

The results above hold for any numerical method that obeys Assumption 6.9 or 6.11. The question is: what type of numerical method under what conditions will obey these assumptions? We have seen in Section 6.2 that under the global Lipschitz condition the EM method obeys Assumption 6.9. In the remainder of this section we will show that the stochastic theta method will obey both Assumptions 6.9 and 6.11 again under the global Lipschitz condition.

Let us now focus here on the class of stochastic theta methods, which was defined in Chapter 4 and recalled as follows: Set $X_0 = x_0$ and form, for $k = 0, 1, 2, \cdots$,

$$X_{k+1} = X_k + [(1-\theta)f(X_k, r_k^\Delta) + \theta f(X_{k+1}, r_k^\Delta)]\Delta + g(X_k, r_k^\Delta)\Delta B_k, \tag{6.49}$$

where $\theta \in [0, 1]$ is a free parameter that is specified *a priori*. As explained in Chapter 4, with the choice $\theta = 0$, (6.49) is the widely-used EM method. In this case, (6.49) is an explicit equation that defines X_{k+1}. However, (6.49) generally represents a nonlinear system that is to be solved for X_{k+1} given X_k. By Lemma 4.9, we note that under the global Lipschitz Assumption 6.3, if Δ is sufficiently small for $\Delta\theta\sqrt{K} < 1$, then equation (6.49) can be solved uniquely for X_{k+1} given X_k, with probability 1. In the remainder of this section, we always let Δ be so sufficiently small that the stochastic theta method is well defined.

Let us also recall the continuous approximation

$$X(t) = x_0 + \int_0^t [(1-\theta)f(z_1(s), \bar{r}(s)) + \theta f(z_2(s), \bar{r}(s))]ds$$
$$+ \int_0^t g(z_1(s), \bar{r}(s))dB(s), \tag{6.50}$$

where

$$z_1(t) = x_k, \ z_2(t) = x_{k+1} \text{ and } \bar{r}(t) = r_k^\Delta \text{ for } t \in [k\Delta, (k+1)\Delta).$$

Note that $X(k\Delta) = X_k$, and hence $X(t)$ is an interpolant to the discrete stochastic theta method solution. We also note that $z_1(k\Delta) = z_2((k-1)\Delta) = X_k$. Let us first show that the stochastic theta method obeys (6.39).

Lemma 6.6 *Under (6.2) and Assumption 6.3, for all sufficiently small $\Delta < 1/(3 + 3K)$, the continuous approximation $X(t)$ defined by (6.50) satisfies*

$$\sup_{0 \le t \le T} \mathbb{E}|X(t)|^2 \le \alpha_T \mathbb{E}|x_0|^2, \quad \forall T \ge 0, \tag{6.51}$$

where $\alpha_T = 3 + 9K(T+1)e^{1.5(4+5K)(T+1)}$.

Proof. It follows from (6.49) that

$$\mathbb{E}|X_{k+1}|^2$$
$$= \mathbb{E}|X_k|^2 + 2\mathbb{E}\Big(X_k^T[(1-\theta)f(X_k, r_k^\Delta) + \theta f(X_{k+1}, r_k^\Delta)]\Delta\Big)$$
$$+ \mathbb{E}\big|[(1-\theta)f(X_k, r_k^\Delta) + \theta f(X_{k+1}, r_k^\Delta)]\Delta + g(X_k, r_k^\Delta)\Delta B_k\big|^2$$
$$\le \mathbb{E}|X_k|^2 + \Delta\mathbb{E}\big[2|X_k|^2 + |f(X_k, r_k^\Delta)|^2 + |f(X_{k+1}, r_k^\Delta)|^2\big)\big]$$
$$+ 3\mathbb{E}\Big(|f(X_k, r_k^\Delta)|^2\Delta^2 + |f(X_{k+1}, r_k^\Delta)|^2\Delta^2 + |g(X_k, r_k^\Delta)|^2|\Delta B_k|^2\Big).$$

Using (6.9), which follows from (6.2) and Assumption 6.3, we further derive that

$$\mathbb{E}|X_{k+1}|^2 \le \mathbb{E}|X_k|^2 + \Delta\mathbb{E}\big[2|X_k|^2 + K|X_k|^2 + K|X_{k+1}|^2\big)\big]$$
$$+ 3K\mathbb{E}\Big(|X_k|^2\Delta^2 + |X_{k+1}|^2\Delta^2 + |X_k|^2\Delta\Big)$$
$$\le \mathbb{E}|X_k|^2 + (3+4K)\Delta\mathbb{E}|X_k|^2 + (1+K)\Delta\mathbb{E}|X_{k+1}|^2,$$

where we have noted that $3K\Delta < 1$. Let M be any positive integer such that $M \leq [T/\Delta] + 1$, where $[T/\Delta]$ is the integer part of T/Δ. Summing the inequality above for k from 0 to $M-1$, we obtain

$$\mathbb{E}|X_M|^2$$
$$\leq \mathbb{E}|X_0|^2 + (3+4K)\Delta \sum_{k=0}^{M-1} \mathbb{E}|X_k|^2 + (1+K)\Delta \sum_{k=0}^{M-1} \mathbb{E}|X_{k+1}|^2$$
$$\leq \mathbb{E}|x_0|^2 + (4+5K)\Delta \sum_{k=0}^{M-1} \mathbb{E}|X_k|^2 + (1+K)\Delta \mathbb{E}|X_M|^2.$$

Noting that $(1+K)\Delta \leq 1/3$, we have

$$\mathbb{E}|X_M|^2 \leq 1.5\mathbb{E}|x_0|^2 + 1.5(4+5K)\Delta \sum_{k=0}^{M-1} \mathbb{E}|X_k|^2.$$

Using the discrete-type Gronwall inequality (i.e. Theorem 2.5) and recalling that $M\Delta \leq T+1$, we obtain

$$\mathbb{E}|X_M|^2 \leq 1.5\mathbb{E}|x_0|^2 e^{1.5(4+5K)\Delta M} \leq \beta_T \mathbb{E}|x_0|^2,$$

where $\beta_T = 1.5 e^{1.5(4+5K)(T+1)}$. Recalling the definitions of $z_1(t)$ and $z_2(t)$ we see

$$\sup_{0 \leq t \leq T} \mathbb{E}|z_j(t)|^2 \leq \beta_T \mathbb{E}|x_0|^2, \quad j=1,2. \tag{6.52}$$

It can be shown (the details are left to the reader as an exercise) from (6.50) and (6.9) that, for $0 \leq t \leq T$,

$$\mathbb{E}|X(t)|^2 \leq 3\mathbb{E}|x_0|^2 + 3K(T+1) \int_0^t [\mathbb{E}|z_1(s)|^2 + \mathbb{E}|z_2(s)|^2] ds. \tag{6.53}$$

By (6.52) we have

$$\mathbb{E}|X(t)|^2 \leq [3 + 6K\beta_T(T+1)]\mathbb{E}|x_0|^2, \quad \forall t \in [0,T]$$

which is the required assertion (6.51). $\square$

Lemma 6.7 *Under (6.2) and Assumption 6.3, for all sufficiently small $\Delta < 1/(2K)$, the continuous approximation $X(t)$ defined by (6.50) satisfies*

$$\sup_{0 \leq t \leq T} \left\{ \mathbb{E}|X(t) - z_1(t)|^2 \vee \mathbb{E}|X(t) - z_2(t)|^2 \right\}$$
$$\leq 2(K+1)\Delta \sup_{0 \leq t \leq T} \mathbb{E}|X(t)|^2 \tag{6.54}$$

for all $T > 0$.

Proof. Given any $0 \leq t \leq T$, let $k = [T/\Delta]$, the integer part of T/Δ, so $k\Delta \leq t < (k+1)\Delta$. It follows from (6.50) that

$$X(t) - z_1(t) = \left[(1-\theta)f(X_k, r_k^\Delta) + \theta f(X_{k+1}, r_k^\Delta)\right](t - k\Delta) + g(X_k, r_k^\Delta)[B(t) - B(k\Delta)], \qquad (6.55)$$

and

$$z_2(t) - X(t) = \left[(1-\theta)f(X_k, r_k^\Delta) + \theta f(X_{k+1}, r_k^\Delta)\right]((k+1)\Delta - t) + g(X_k, r_k^\Delta)[B((k+1)\Delta) - B(t)]. \qquad (6.56)$$

By Exercise 6.5 and condition (6.9), we compute from (6.55) that

$$\mathbb{E}|X(t) - z_1(t)|^2 \leq 2\Delta^2 K(\mathbb{E}|X_k|^2 + \mathbb{E}|X_{k+1}|^2) + 2\Delta K \mathbb{E}|X_k|^2$$
$$\leq (1 + 2K)\Delta \mathbb{E}|X_k|^2 + \Delta \mathbb{E}|X_{k+1}|^2$$
$$\leq 2(K+1)\Delta \sup_{0 \leq t \leq T} \mathbb{E}|X(t)|^2,$$

where we have used the condition that $2\Delta K < 1$. Similarly, we can show the same upper bound for $\mathbb{E}|z_2(t) - X(t)|^2$ and hence the assertion (6.54) follows. □

Lemma 6.8 *Under (6.2) and Assumption 6.3, for sufficiently small Δ the stochastic theta method solution (6.50) satisfies*

$$\sup_{0 \leq t \leq T} \mathbb{E}|X(t) - x(t)|^2 \leq \left(\sup_{0 \leq t \leq T} \mathbb{E}|X(t)|^2\right) C_T \Delta, \quad \forall T > 0, \qquad (6.57)$$

where C_T is a positive constant dependent on T but independent of x_0, r_0 and Δ.

Proof. It follows from (6.1) and (6.50) that for any $0 \leq t \leq T$,

$$X(t) - x(t) = \int_0^t \Big((1-\theta)[f(z_1(s), \bar{r}(s)) - f(x(s), r(s))]$$
$$+ \theta[f(z_2(s), \bar{r}(s)) - f(x(s), r(s))]\Big) ds$$
$$+ \int_0^t \Big(g(z_1(s), \bar{r}(s)) - g(x(s), r(s))\Big) dB(s).$$

We hence compute

$$\mathbb{E}|X(t) - x(t)|^2 \leq 2T\mathbb{E}\int_0^t \Big(|f(z_1(s), \bar{r}(s)) - f(x(s), r(s))|^2$$
$$+ |f(z_2(s), \bar{r}(s)) - f(x(s), r(s))|^2\Big)ds$$
$$+ 2\mathbb{E}\int_0^t |g(z_1(s), \bar{r}(s)) - g(x(s), r(s))|^2 ds$$
$$\leq 4TK\mathbb{E}\int_0^t \Big(|z_1(s) - x(s)|^2 + |z_2(s) - x(s)|^2\Big)ds$$
$$+ 4K\mathbb{E}\int_0^t |z_1(s) - x(s)|^2 ds + J(T)$$
$$\leq J(T) + 8K(2T+1)\int_0^t \mathbb{E}|X(s) - x(s)|^2 ds$$
$$+ 8K(T+1)\int_0^t \Big(\mathbb{E}|X(s) - z_1(s)|^2 + \mathbb{E}|X(s) - z_2(s)|^2\Big)ds$$
$$\leq J(T) + 8K(2T+1)\int_0^t \mathbb{E}|X(s) - x(s)|^2 ds$$
$$+ 32K(K+1)(T+1)T\Delta \sup_{0\leq t\leq T}\mathbb{E}|X(t)|^2,$$

where

$$J(T) := 4T\mathbb{E}\int_0^T \Big(|f(z_1(s), \bar{r}(s)) - f(z_1(s), r(s))|^2$$
$$+ |f(z_2(s), \bar{r}(s)) - f(z_2(s), r(s))|^2\Big)ds$$
$$+ 4\mathbb{E}\int_0^T |g(z_1(s), \bar{r}(s)) - g(z_1(s), r(s))|^2 ds.$$

But, by using (6.9), it can be shown in the same way as Theorem 4.1 was proved that

$$J(T) \leq c_1 \Delta \Big(\sup_{0\leq t\leq T}\mathbb{E}|X(t)|^2\Big),$$

where c_1 and the following c_2 are both positive constants dependent on T

but independent of x_0, r_0 and Δ. Hence

$$\mathbb{E}|X(t) - x(t)|^2 \leq c_2 \Delta \left(\sup_{0 \leq t \leq T} \mathbb{E}|X(t)|^2\right)$$
$$+ 8K(2T+1) \int_0^t \mathbb{E}|X(s) - x(s)|^2 ds.$$

An application of the Gronwall inequality gives a bound of the form

$$\mathbb{E}|X(t) - x(t)|^2 \leq \left(\sup_{0 \leq t \leq T} \mathbb{E}|X(t)|^2\right) C_T \Delta.$$

Since this holds for any $t \in [0, T]$, the assertion (6.57) must hold. □

Combining Lemmas 6.6–6.8 we can conclude the following result.

Theorem 6.13 *Under (6.2) and Assumption 6.3, for all sufficiently small Δ, the continuous approximation $X(t)$ defined by (6.50) obeys Assumption 6.9. Moreover, it obeys Assumption 6.11 with*

$$\bar{C}_T = C_T\left[3 + 9K(T+1)e^{1.5(4+5K)(T+1)}\right].$$

6.5 Asymptotic Stability in Distribution of the EM Method: Constant Step Size

Let us now proceed to investigate the stability in distribution of the EM method.

6.5.1 Stability in Distribution of the EM Method

We will use the EM method as in Section 6.2 but to hight the step size Δ, we set $X_0^\Delta = x$, $r_0^\Delta = i$ and form

$$X_{k+1}^\Delta = X_k^\Delta + f(X_k^\Delta, r_k^\Delta)\Delta + g(X_k^\Delta, r_k^\Delta)\Delta B_k, \tag{6.58}$$

where ΔB_k and r_k^Δ are the same as before.

Lemma 6.9 *Let $Y_k^\Delta = (X_k^\Delta, r_k^\Delta)$. Then*

$$\mathbb{P}(Y_{k+1}^\Delta \in A \times \{j\} | Y_k^\Delta = (x, i)) = \mathbb{P}(Y_1^\Delta \in A \times \{j\} | Y_0^\Delta = (x, i)) \tag{6.59}$$

for any Borel set $A \subset \mathbb{R}^n$ and $j \in \mathbb{S}$.

Proof. If $X_k^\Delta = x, r_k^\Delta = i$ and $X_0^\Delta = x, r_0^\Delta = i$, by (6.58) we have

$$X_{k+1}^\Delta = x + f(x,i)\Delta + g(x,i)\Delta B_k, \qquad (6.60)$$

and

$$X_1^\Delta = x + f(x,i)\Delta + g(x,i)\Delta B_1. \qquad (6.61)$$

Since ΔB_k and ΔB_1 are identical in probability law, comparing (6.60) with (6.61), we know that $(X_{k+1}^\Delta, r_{k+1}^\Delta)$ and (X_1^Δ, r_1^Δ) are identical in probability law under $X_k^\Delta = x, r_k^\Delta = i$ and $X_0^\Delta = x, r_0^\Delta = i$. In consequence,

$$\mathbb{P}(Y_{k+1}^\Delta \in A \times \{j\} | Y_k^\Delta = (x,i)) = \mathbb{P}(Y_1^\Delta \in A \times \{j\} | Y_0^\Delta = (x,i))$$

as required. $\square$

For any Borel set $A \subset \mathbb{R}^n$ and $x \in \mathbb{R}^n$, $i, j \in \mathbb{S}$, define

$$P^\Delta((x,i), A \times \{j\}) := \mathbb{P}(Y_1^\Delta \in A \times \{j\} | Y_0^\Delta = (x,i)),$$
$$P_k^\Delta((x,i), A \times \{j\}) := \mathbb{P}(Y_k^\Delta \in A \times \{j\} | Y_0^\Delta = (x,i)).$$

Theorem 6.14 $\{Y_k^\Delta\}_{k\geq 0}$ *is a homogeneous Markov process with transition probability kernel* $P^\Delta((x,i), A \times \{j\})$.

Proof. The homogeneous property follows from Lemma 6.9 so we only need to prove the Markov property. We write $X_k^\Delta = X_k$ and $r_k^\Delta = r_k$ simply. For $k \geq 0$, $x \in \mathbb{R}^n$ and $i \in \mathbb{S}$, define

$$\zeta_{k+1}^i = i + r_{k+1} - r_k$$

and

$$\xi_{k+1}^{x,i} = x + f(x,i)\Delta_{k+1} + g(x,i)\Delta w_{k+1}.$$

By (6.58) we know that $X_{k+1} = \xi_{k+1}^{X_k, r_k}$ and $r_{k+1} = \zeta_{k+1}^{r_k}$. Let $\mathcal{G}_{t_{k+1}} = \sigma\{w(t_{k+1}) - w(t_k)\}$. Clearly, $\mathcal{G}_{t_{k+1}}$ is independent of $\mathcal{F}_{t_k}$. Moreover, $\xi_{k+1}^{x,i}$ depends completely on the increment $B(t_{k+1}) - B(t_k)$ so is $\mathcal{G}_{t_{k+1}}$-measurable. Hence, $\xi_{k+1}^{x,i}$ is independent of $\mathcal{F}_{t_k}$. Noting that $\xi_{k+1}^{X_k, r_k}$ and $\zeta_{k+1}^{r_k}$ are conditional independent given (X_k, r_k) and of course given $\mathcal{F}_{t_k}$ we can apply Lemma 3.2 with $\bar{h}((x,i), \omega) = I_A(\xi_{k+1}^{x,i})$ and $\bar{h}(i, \omega) = I_{\{j\}}(r_{k+1}^i)$, respectively, to compute that

$$\mathbb{P}(Y_{k+1} \in A \times \{j\}|\mathcal{F}_{t_k}) = \mathbb{E}\left(I_{A \times \{j\}}(Y_{k+1})|\mathcal{F}_{t_k}\right)$$
$$= \mathbb{E}\left(I_{A \times \{j\}}(\xi_{k+1}^{X_k,r_k}, \zeta_{k+1}^{r_k})|\mathcal{F}_{t_k}\right)$$
$$= \mathbb{E}\left(I_A(\xi_{k+1}^{X_k,r_k})|\mathcal{F}_{t_k}\right) \mathbb{E}\left(I_{\{j\}}(\zeta_{k+1}^{r_k})|\mathcal{F}_{t_k}\right)$$
$$= \mathbb{E}\left(I_A(\xi_{k+1}^{x,i})\right)\Big|_{x=X_k, i=r_k} \mathbb{E}\left(I_{\{j\}}(\zeta_{k+1}^i)\right)\Big|_{i=r_k}$$
$$= \mathbb{P}(\xi_{k+1}^{x,i} \in A)\Big|_{x=X_k, i=r_k} \mathbb{P}(\zeta_{k+1}^i = j)\Big|_{i=r_k}$$
$$= \mathbb{P}((\xi_{k+1}^{x,i}, \zeta_{k+1}^i) \in A \times \{j\})\Big|_{x=X_k, i=r_k}$$
$$= \mathbb{P}(Y_{k+1} \in A \times \{j\}|Y_k),$$

which is the required Markov property. $\square$

We therefore see that $P^\Delta(\cdot, \cdot)$ defined above is the one-step transition probability measure while $P_k^\Delta(\cdot, \cdot)$ is the k-step transition probability measure. In what follows we will need another property of the Markov chain which is now described as a lemma.

Lemma 6.10 *There is a pair of positive numbers δ_0 and M such that*

$$\mathbb{P}\{r(t+\delta) \neq r(t)\} \leq M\delta \quad \text{for any } t \geq 0, \ 0 \leq \delta \leq \delta_0.$$

Proof. Compute

$$\mathbb{P}\{r(t+\delta) \neq r(t)\} = \sum_{i \in \mathbb{S}} \mathbb{P}\{r(t) = i\} \mathbb{P}\{r(t+\delta) \neq i | r(t) = i\}$$
$$= \sum_{i \in \mathbb{S}} \mathbb{P}\{r(t) = i\}(-\gamma_{ii}\delta + o(\delta))$$
$$\leq \left(\max_{i \in \mathbb{S}}(-\gamma_{ii})\right)\delta + o(\delta).$$

The assertion follows immediately. $\square$

In Section 5.6 we discussed the asymptotic stability in distribution for the true solution. Our aim here is to establish a sufficient criterion on asymptotic stability in distribution for the EM solution $\{Y_k^\Delta\}_{k \geq 0}$. To highlight the initial values, sometimes we will use notation $\{Y_k^{(x,i),\Delta}\}_{k \geq 0}$ instead of $\{Y_k^\Delta\}_{k \geq 0}$. By the concept of weak convergence as introduced in Section 5.6, let us now define the asymptotic stable in distribution for $\{Y_k^\Delta\}_{k \geq 0}$. The notations $\mathcal{P}(\mathbb{R}^n \times S)$ and $\mathbb{L}$ etc, were defined in Section 5.6.

Definition 6.15 For a given step size $\Delta > 0$, $\{Y_k^{(x,i),\Delta}\}_{k\geq 0}$ is said to be asymptotically stable in distribution if there exists a probability measure $\pi^\Delta(\cdot \times \cdot) \in \mathcal{P}(\mathbb{R}^n \times \mathbb{S})$ such that the k-step transition probability measure $P_k^\Delta((x,i), dy \times \{j\})$ converges weakly to $\pi^\Delta(dy \times \{j\})$ as $k \to \infty$ for every $(x,i) \in \mathbb{R}^n \times \mathbb{S}$, that is

$$\lim_{k\to\infty} \left(\sup_{F\in\mathbb{L}} |\mathbb{E}F(Y_k^{(x,i),\Delta}) - \mathbb{E}_{\pi^\Delta} F| \right) = 0,$$

where

$$\mathbb{E}_{\pi^\Delta} F = \sum_{i=1}^N \int_{\mathbb{R}^n} F(y,i) \pi^\Delta(dy, i).$$

Obviously the asymptotic stability in distribution of $\{Y_k^{(x,i),\Delta}\}_{k\geq 0}$ implies the existence of a unique invariant or stationary distribution for $\{Y_k^{(x,i),\Delta}\}_{k\geq 0}$. To show the stability in distribution we impose the following assumptions.

Assumption 6.16 For any $\varepsilon > 0$ and $(x,i) \in \mathbb{R}^n \times \mathbb{S}$, there exists a constant $R = R(\varepsilon, x, i) > 0$ such that

$$\mathbb{P}\{|X_k^{(x,i),\Delta}| \geq R\} < \varepsilon, \quad \forall k \geq 0. \tag{6.62}$$

Assumption 6.17 For any $\varepsilon > 0$ and any compact subset K of $\mathbb{R}^n$, there exists a positive integer $T = T(\varepsilon, K)$ such that

$$\mathbb{P}\{|X_k^{(x,i),\Delta} - X_k^{(z,i),\Delta}| < \varepsilon\} \geq 1 - \varepsilon, \quad \forall k \geq T \tag{6.63}$$

whenever $(x, z, i) \in K \times K \times \mathbb{S}$.

Assumption 6.18 For any $\varepsilon > 0$, $\bar{k} \geq 1$ and any compact subset K of $\mathbb{R}^n$, there exists a $R = R(\varepsilon, \bar{k}, K) > 0$ such that

$$\mathbb{P}\left\{ \sup_{0\leq k \leq \bar{k}} |X_k^{(x,i),\Delta}| \leq R \right\} > 1 - \varepsilon, \quad \forall (x,i) \in K \times \mathbb{S}. \tag{6.64}$$

We observe that Assumption 6.16 guarantees that for any $(x,i) \in \mathbb{R}^n \times \mathbb{S}$, the family of transition probability measures $\{P_k^\Delta((x,i), dy \times \{j\}) : k \geq 0\}$ is tight. That is, for any $\varepsilon > 0$ there is a compact subset $U = U(\varepsilon, x, i)$ of $\mathbb{R}^n$ such that

$$P_k^\Delta((x,i), U \times \mathbb{S}) \geq 1 - \varepsilon, \quad \forall k \geq 0. \tag{6.65}$$

In the next subsection we will discuss the conditions under which Assumptions 6.16–6.18 hold, but let us now state one of our main results in this section.

Theorem 6.19 *Under Assumptions 6.16–6.18, $\{Y_k^{(x,i),\Delta}\}_{k\geq 0}$ is asymptotically stable in distribution.*

Let us present three lemmas in order to prove the theorem.

Lemma 6.11 *Let Assumptions 6.17 and 6.18 hold. Then, for any compact subset K of $\mathbb{R}^n$,*

$$\lim_{k\to\infty} d_{\mathbb{L}}(P_k^{\Delta}((x,i),\cdot\times\cdot), P_k^{\Delta}((z,j),\cdot\times\cdot)) = 0 \tag{6.66}$$

uniformly in $x, z \in K$ and $i, j \in \mathbb{S}$.

Proof. For any pair of $i, j \in \mathbb{S}$, define the stopping time

$$\beta_{ij} = \inf\{k \geq 0 : r_k^{i,\Delta} = r_k^{j,\Delta}\}. \tag{6.67}$$

Recall that $r_k^{i,\Delta}$ is the Markov chain starting from state $i \in \mathbb{S}$ at $k = 0$ and due to the ergodicity of the Markov chain, $\beta_{ij} < \infty$ a.s. (see e.g. [Anderson (1991)]). So, for any $\varepsilon > 0$, there exists a positive integer number T such that

$$\mathbb{P}\{\beta_{ij} \leq T\} > 1 - \frac{\varepsilon}{8} \quad \text{for any } i, j \in \mathbb{S}. \tag{6.68}$$

For this T, by Assumption 6.18, there is a sufficiently large $R > 0$ for

$$\mathbb{P}(\Omega_{x,i}) > 1 - \frac{\varepsilon}{16} \quad \text{for any } (x,i) \in K \times \mathbb{S}, \tag{6.69}$$

where $\Omega_{x,i} = \{|X_k^{(x,i),\Delta}| \leq R \text{ for all } 0 \leq k \leq T\}$.

Now, fix any $x, z \in K$ and $i, j \in \mathbb{S}$ and set $\Omega_1 = \Omega_{x,i} \cap \Omega_{z,j}$. For any $F \in \mathbb{L}$ and $k \geq T$, compute

$$|\mathbb{E}F(X_k^{(x,i),\Delta}, r_k^{i,\Delta}) - \mathbb{E}F(X_k^{(z,j),\Delta}, r_k^{j,\Delta})|$$

$$\leq 2\mathbb{P}\{\beta_{ij} > T\} + \mathbb{E}\left(I_{\{\beta_{ij}\leq T\}}|F(X_k^{(x,i),\Delta}, r_k^{i,\Delta}) - F(X_k^{(z,j),\Delta}, r_k^{j,\Delta})|\right)$$

$$\leq \frac{\varepsilon}{4} + \mathbb{E}\left[I_{\{\beta_{ij}\leq T\}}\mathbb{E}\left(|F(X_k^{(x,i),\Delta}, r_k^{i,\Delta}) - F(X_k^{(z,j),\Delta}, r_k^{j,\Delta})| \,\Big|\, \mathcal{F}_{\beta_{ij}}\right)\right]$$

$$\leq \frac{\varepsilon}{4} + \mathbb{E}\left[I_{\{\beta_{ij}\leq T\}}\mathbb{E}|F(X_{k-\beta_{ij}}^{(u,l),\Delta}, r_{k-\beta_{ij}}^{l,\Delta}) - F(X_{k-\beta_{ij}}^{(v,l),\Delta}, r_{k-\beta_{ij}}^{l,\Delta})|\right]$$

$$\leq \frac{\varepsilon}{4} + \mathbb{E}\left[I_{\{\beta_{ij}\leq T\}}\mathbb{E}\left(2 \wedge |X_{k-\beta_{ij}}^{(u,l),\Delta} - X_{k-\beta_{ij}}^{(v,l),\Delta}|\right)\right]$$

$$\leq \frac{\varepsilon}{4} + 2\mathbb{P}(\Omega - \Omega_1) + \mathbb{E}\left[I_{\Omega_1\cap\{\beta_{ij}\leq T\}}\mathbb{E}\left(2 \wedge |X_{k-\beta_{ij}}^{(u,l),\Delta} - X_{k-\beta_{ij}}^{(v,l),\Delta}|\right)\right], \tag{6.70}$$

where $u = X^{(x,i),\Delta}_{\beta_{ij}}, v = X^{(z,j),\Delta}_{\beta_{ij}}$ and $l = r^{i,\Delta}_{\beta_{ij}} = r^{j,\Delta}_{\beta_{ij}}$. Note that for any $\omega \in \Omega_1 \cap \{\beta_{ij} \leq T\}$, we have $|u| \vee |v| \leq R$. So, by Assumption 6.17, there exists a positive integer T_1 such that

$$\mathbb{E}\left(2 \wedge |X^{(u,l),\Delta}_{k-\beta_{ij}} - X^{(v,l),\Delta}_{k-\beta_{ij}}|\right) < \varepsilon/2, \quad \forall k \geq T + T_1. \tag{6.71}$$

It therefore follows from (6.69), (6.70) and (6.71) that

$$|\mathbb{E}F(X^{(x,i),\Delta}_k, r^{i,\Delta}_k) - \mathbb{E}F(X^{(z,j),\Delta}_k, r^{j,\Delta}_k)| \leq \varepsilon, \quad \forall k \geq T + T_1.$$

Since F etc, are arbitrary, we must have that

$$\sup_{F \in \mathbb{L}} |\mathbb{E}F(X^{(x,i),\Delta}_k, r^{i,\Delta}_k) - \mathbb{E}F(X^{(z,j),\Delta}_k, r^{j,\Delta}_k)| \leq \varepsilon, \quad \forall k \geq T + T_1,$$

namely

$$d_\mathbb{L}(P^\Delta_k((x,i),\cdot \times \cdot), P^\Delta_k((z,j),\cdot \times \cdot)) \leq \varepsilon, \quad \forall k \geq T + T_1$$

for all $x, z \in K$ and $i, j \in \mathbb{S}$. The proof is complete. $\square$

Lemma 6.12 *Under Assumptions 6.16–6.18, for any $(x,i) \in \mathbb{R}^n \times \mathbb{S}$, $\{P^\Delta_k((x,i),\cdot \times \cdot)\}_{k \geq 1}$ is Cauchy in the space $\mathcal{P}(\mathbb{R}^n \times \mathbb{S})$ with metric $d_\mathbb{L}$.*

Proof. Fix any $(x,i) \in \mathbb{R}^n \times \mathbb{S}$. We need to show that for any $\varepsilon > 0$, there is a $T > 0$ such that

$$d_\mathbb{L}(P^\Delta_{k+u}((x,i),\cdot \times \cdot), P^\Delta_k((x,i),\cdot \times \cdot)) \leq \varepsilon$$

for any $k \geq T$ and $u > 0$. This is equivalent to

$$\sup_{F \in \mathbb{L}} |\mathbb{E}F(X^{(x,i),\Delta}_{k+u}, r^{i,\Delta}_{k+u}) - \mathbb{E}F(X^{(x,i),\Delta}_k, r^{i,\Delta}_k)| \leq \varepsilon \tag{6.72}$$

for any $k \geq T$ and $u > 0$. For any $F \in \mathbb{L}$ and $k, u > 0$, compute

$$|\mathbb{E}F(X^{(x,i),\Delta}_{k+u}, r^{i,\Delta}_{k+u}) - \mathbb{E}F(X^{(x,i),\Delta}_k, r^{i,\Delta}_k)|$$
$$= |\mathbb{E}[\mathbb{E}(F(X^{(x,i),\Delta}_{k+u}, r^{i,\Delta}_{k+u})|\mathcal{F}_{t_u})] - \mathbb{E}F(X^{(x,i),\Delta}_k, r^{i,\Delta}_k)|$$
$$= |\sum_{l=1}^N \int_{\mathbb{R}^n} \mathbb{E}F(X^{(z,l),\Delta}_k, r^{l,\Delta}_k) P^\Delta_u((x,i), dz \times \{l\}) - \mathbb{E}F(X^{(x,i),\Delta}_k, r^{i,\Delta}_k)|$$

$$\leq \sum_{l=1}^{N} \int_{\mathbb{R}^n} |\mathbb{E}F(X_k^{(z,l),\Delta}, r_k^{l,\Delta}) - \mathbb{E}F(X_k^{(x,i),\Delta}, r_k^{i,\Delta})| \, P_u^{\Delta}((x,i), dz \times \{l\})$$

$$\leq 2 P_u^{\Delta}((x,i), \bar{\mathbb{S}}_R^c \times \mathbb{S})$$

$$+ \sum_{l=1}^{N} \int_{\bar{\mathbb{S}}_R} |\mathbb{E}F(X_k^{(z,l),\Delta}, r_k^{l,\Delta}) - \mathbb{E}F(X_k^{(x,i),\Delta}, r_k^{i,\Delta})| \, P_u^{\Delta}((x,i), dz \times \{l\}),$$

(6.73)

where $\bar{\mathbb{S}}_R = \{x \in \mathbb{R}^n : |x| \leq R\}$ and $\bar{\mathbb{S}}_R^c = \mathbb{R}^n - \bar{\mathbb{S}}_R$. By Assumption 6.16 (or (6.65)), there is a positive number R sufficiently large for

$$P_u^{\Delta}((x,i), \bar{\mathbb{S}}_R^c \times \mathbb{S}) < \frac{\varepsilon}{4}, \quad \forall u > 0. \tag{6.74}$$

On the other hand, by Lemma 6.11, there is a $T > 0$ such that

$$\sup_{f \in \mathbb{L}} |\mathbb{E}F(X_k^{(z,l),\Delta}, r_k^{l,\Delta}) - \mathbb{E}F(X_k^{(x,i),\Delta}, r_k^{i,\Delta})| < \frac{\varepsilon}{2}, \quad \forall k \geq T \tag{6.75}$$

whenever $(z,l) \in \bar{\mathbb{S}}_R \times \mathbb{S}$. Substituting (6.74) and (6.75) into (6.73) yields

$$|\mathbb{E}F(X_{k+u}^{(x,i),\Delta}, r_{k+u}^{i,\Delta}) - \mathbb{E}F(X_k^{(x,i),\Delta}, r_k^{i,\Delta})| < \varepsilon, \quad \forall k \geq T, \, u > 0.$$

Since F is arbitrary, the desired inequality (6.72) must hold. □

We can now easily prove our main result Theorem 6.19.

Proof of Theorem 6.19. By definition, we need to show that there exists a unique probability measure $\pi^{\Delta}(\cdot \times \cdot) \in \mathcal{P}(\mathbb{R}^n \times \mathbb{S})$ such that for any $(x,i) \in \mathbb{R}^n \times \mathbb{S}$, the transition probability measures $\{P_k^{\Delta}((x,i), \cdot \times \cdot) : k \geq 0\}$ will converge weakly to $\pi^{\Delta}(\cdot \times \cdot)$. In other words, we need to show that for any $(x,i) \in \mathbb{R}^n \times \mathbb{S}$,

$$\lim_{k \to \infty} d_{\mathbb{L}}(P_k^{\Delta}((x,i), \cdot \times \cdot), \pi^{\Delta}(\cdot \times \cdot)) = 0. \tag{6.76}$$

By Lemma 6.12, $\{P_k^{\Delta}((0,1), \cdot \times \cdot) : k \geq 0\}$ is Cauchy in the space $\mathcal{P}(\mathbb{R}^n \times \mathbb{S})$ with metric $d_{\mathbb{L}}$. So there is a unique $\pi^{\Delta}(\cdot \times \cdot) \in \mathcal{P}(\mathbb{R}^n \times \mathbb{S})$ such that

$$\lim_{k \to \infty} d_{\mathbb{L}}(P_k^{\Delta}((0,1), \cdot \times \cdot), \pi^{\Delta}(\cdot \times \cdot)) = 0.$$

Now, for any $(x, i) \in \mathbb{R}^n \times \mathbb{S}$, by Lemma 6.11,

$$\lim_{k \to \infty} d_{\mathbb{L}}(P_k^\Delta((x,i), \cdot \times \cdot), \pi^\Delta(\cdot \times \cdot))$$
$$\leq \lim_{k \to \infty} \Big[d_{\mathbb{L}}(P_k^\Delta((0,1), \cdot \times \cdot), \pi^\Delta(\cdot \times \cdot))$$
$$\qquad + d_{\mathbb{L}}(P_k^\Delta((x,i), \cdot \times \cdot), P_k^\Delta((0,1), \cdot \times \cdot)) \Big]$$
$$= 0$$

as required. □

6.5.2 Sufficient Criteria for Assumptions 6.16–6.18

Theorem 6.19 depends on Assumptions 6.16–6.18. It is therefore necessary to establish sufficient criteria for these assumptions so that Theorem 6.19 is applicable. In this section we impose a standing hypothesis.

Assumption 6.20 Both f and g are globally Lipschitz continuous. That is, there exists a constant $H > 0$ such that

$$|f(u,i) - f(v,i)|^2 \vee |g(u,i) - g(v,i)|^2 \leq H|u - v|^2$$

for all $u, v \in \mathbb{R}^n$ and $i \in \mathbb{S}$.

It is easy to see from this assumption that

$$|f(u,i)|^2 \vee |g(u,i)|^2 \leq \rho(1 + |u|^2) \qquad (6.77)$$

with $\rho = 2 \max_{i \in \mathbb{S}} (H \vee |f(0,i)|^2 \vee |g(0,i)|^2)$.

Lemma 6.13 *Under Assumption 6.20, the EM solution (6.58) has the property that*

$$\mathbb{E}\Big(\sup_{0 \leq k \leq \bar{k}} |X_k^{(x,i),\Delta}|^2 \Big) \leq (1 + 3|x|^2) e^{3\rho \bar{k} \Delta (4 + \bar{k}\Delta)}, \quad \forall \bar{k} \geq 1. \qquad (6.78)$$

In particular, Assumption 6.18 holds.

This lemma can be proved in a similar way as Lemma 4.1 was proved so the details are left to the reader as an exercise.

Lemma 6.14 *Let Assumption 6.20 hold. Assume that there exists a pair of positive constants μ_1 and α as well as N symmetric positive-definite* matrices Q_i such that for all $(x,i) \in \mathbb{R}^n \times \mathbb{S}$

$$2x^T Q_i f(x,i) + \operatorname{trace}[g^T(x,i)Q_i g(x,i)] + \sum_{j=1}^{N} \gamma_{ij} x^T Q_j x \leq -\mu_1 |x|^2 + \alpha. \tag{6.79}$$

Let $\Delta < 1 \wedge \delta_0$ be sufficiently small such that $16\Delta\rho < 1$ and

$$32q\rho\Delta(H+M)\left(\frac{4q}{\mu_1} + 1 + \Delta^{-1/2}\right) + q\rho\sqrt{\Delta} \leq 0.5\mu_1, \tag{6.80}$$

where ρ is the linear growth condition, δ_0 and M are the numbers specified in Lemma 6.10, $q = \max\{\lambda_{\max}(Q_i) : i \in \mathbb{S}\}$ and $\tilde{q} = \min\{\lambda_{\min}(Q_i) : i \in \mathbb{S}\}$. Then

$$\mathbb{E}\left[|X_k^{(x,i),\Delta}|^2\right] \leq \frac{q}{\tilde{q}}\left[|x|^2 + 2 + \frac{4\alpha}{\mu_1}\right], \quad \forall k \geq 1. \tag{6.81}$$

In particular, Assumption 6.16 holds provided the step size Δ is sufficiently small.

Proof. Fix any $(x,i) \in \mathbb{R}^n \times \mathbb{S}$ and write $X_k^{(x,i),\Delta} = X_k$ and $r_k^{i,\Delta} = r_k$. Introduce the step processes

$$\bar{X}(t) = \sum_{k=0}^{\infty} X_k I_{[t_k,t_{k+1})}(t) \quad \text{and} \quad \bar{r}(t) = \sum_{k=0}^{\infty} r_k I_{[t_k,t_{k+1})}(t),$$

and define the continuous approximate solution

$$X(t) = X_0 + \int_0^t f(\bar{X}(s), \bar{r}(s))ds + \int_0^t g(\bar{X}(s), \bar{r}(s))dB(s). \tag{6.82}$$

Note that $\bar{X}(t_k) = X(t_k) = X_k$, that is $\bar{X}(t)$ and $X(t)$ coincide with the discrete solution at the gridpoints. For $t > 0$, there exists a k such that $t_k \leq t < t_{k+1}$. It follows from (6.82) that

$$X(t) = X_k + \int_{t_k}^t f(\bar{X}(s), \bar{r}(s))ds + \int_{t_k}^t g(\bar{X}(s), \bar{r}(s))dB(s).$$

By (6.77),

$$\mathbb{E}|X(t) - \bar{X}(t)|^2$$
$$\leq 2\mathbb{E}\left(|\int_{t_k}^t f(\bar{X}(s), \bar{r}(s))ds|^2 + |\int_{t_k}^t g(\bar{X}(s), \bar{r}(s))dB(s)|^2\right)$$
$$\leq 2(\Delta+1)\rho\mathbb{E}\int_{t_k}^t (1+|\bar{X}(s)|^2)ds \leq 4\Delta\rho(1+\mathbb{E}|\bar{X}(t)|^2). \qquad (6.83)$$

Let

$$V(x, i) = x^T Q_i x.$$

By the generalised Itô formula, we derive that, for any $\theta > 0$,

$$e^{\theta t}\mathbb{E}[X^T(t)Q_{r(t)}X(t)] = x^T Q_i x + \theta\mathbb{E}\int_0^t e^{\theta s} X^T(s)Q_{r(s)}X(s)ds$$
$$+ \mathbb{E}\int_0^t e^{\theta s}\Big\{2X^T(s)Q_{r(s)}f(\bar{X}(s), \bar{r}(s))$$
$$+ \text{trace}[g^T(\bar{X}(s), \bar{r}(s))Q_{r(s)}g(\bar{X}(s), \bar{r}(s))]$$
$$+ \sum_{j=1}^N \gamma_{r(s)j} X^T(s)Q_j X(s)\Big\}ds.$$

Using the elementary inequalities

$$\text{trace}[(A-D)^T Q_i (A-D)]$$
$$\leq (1+\sqrt{\Delta})\text{trace}[A^T Q_i A] + (1+\Delta^{-1/2})\text{trace}[D^T Q_i D],$$

for any $A, D \in \mathbb{R}^{n \times m}$, and

$$2x^T Q_i y \leq \theta q |x|^2 + \theta^{-1} q |y|^2, \quad \forall x, y \in \mathbb{R}^n$$

as well as condition (6.79), we can derive that

$$e^{\theta t}\mathbb{E}[X^T(t)Q_{r(t)}X(t)]$$
$$\leq q|x|^2 + \theta q \mathbb{E}\int_0^t e^{\theta s}|X(s)|^2 ds + \mathbb{E}\int_0^t e^{\theta s}\Big\{2X^T(s)Q_{r(s)}f(X(s), r(s))$$
$$+ (1+\sqrt{\Delta})\text{trace}[g^T(X(s), r(s))Q_{r(s)}g(X(s), r(s))]$$
$$+ \sum_{j=1}^N \gamma_{r(s)j} X^T(s)Q_j X(s)\Big\}ds$$

$$+ \mathbb{E} \int_0^t e^{\theta s} \Big\{ 2X^T(s) Q_{r(s)} [f(\bar{X}(s), \bar{r}(s)) - f(X(s), r(s))]$$
$$+ (1 + \Delta^{-1/2}) \text{trace} \Big([(g^T(\bar{X}(s), \bar{r}(s)) - g^T(X(s), r(s))]$$
$$Q_{r(s)} [g(\bar{X}(s), \bar{r}(s)) - g(X(s), r(s))] \Big) \Big\} ds$$
$$\leq q|x|^2 + \mathbb{E} \int_0^t e^{\theta s} \Big[\alpha + q\rho\sqrt{\Delta} + (2\theta q + q\rho\sqrt{\Delta} - \mu_1) |X(s)|^2 \Big] ds + J,$$
(6.84)

where
$$J := q\mathbb{E} \int_0^t e^{\theta s} \Big\{ \theta^{-1} |f(\bar{X}(s), \bar{r}(s)) - f(X(s), r(s))|^2$$
$$+ (1 + \Delta^{-1/2}) |g(\bar{X}(s), \bar{r}(s)) - g(X(s), r(s))|^2 \Big\} ds.$$

But by Assumption 6.20, we have
$$|f(\bar{X}(s), \bar{r}(s)) - f(X(s), r(s))|^2$$
$$\leq 2|f(\bar{X}(s), \bar{r}(s)) - f(\bar{X}(s), r(s))|^2 + 2|f(\bar{X}(s), r(s)) - f(X(s), r(s))|^2$$
$$\leq 8\rho(1 + |\bar{X}(s)|^2) I_{\{r(s) \neq \bar{r}(s)\}} + 2H|\bar{X}(s) - X(s)|^2.$$

Similarly, we have
$$|g(\bar{X}(s), \bar{r}(s)) - g(X(s), r(s))|^2$$
$$\leq 8\rho(1 + |\bar{X}(s)|^2) I_{\{r(s) \neq \bar{r}(s)\}} + 2H|\bar{X}(s) - X(s)|^2.$$

Moreover, let k be an integer such that $k\Delta \leq s < (k+1)\Delta$ and compute
$$\mathbb{E}\Big[(1 + |\bar{X}(s)|^2) I_{\{r(s) \neq \bar{r}(s)\}} \Big] = \mathbb{E}\Big(\mathbb{E}\Big[(1 + |X_k|^2) I_{\{r(s) \neq r_k\}} | \mathcal{F}_{t_k} \Big] \Big)$$
$$= \mathbb{E}\Big((1 + |X_k|^2) \mathbb{E}\Big[I_{\{r(s) \neq r_k\}} | \mathcal{F}_{t_k} \Big] \Big) = \mathbb{E}\Big((1 + |X_k|^2) \mathbb{P}\{r(s) \neq r_k\} \Big)$$
$$= \mathbb{E}\Big((1 + |\bar{X}(s)|^2) \mathbb{P}\{r(s) \neq \bar{r}(s)\} \Big). \quad (6.85)$$

Thus, by Lemma 6.10 and inequality (6.83), we derive that
$$J \leq q(\theta^{-1} + 1 + \Delta^{-1/2})$$
$$\times \int_0^t e^{\theta s} \Big\{ 8\rho \mathbb{E}\Big[(1 + |\bar{X}(s)|^2) \mathbb{P}\{r(s) \neq \bar{r}(s)\} \Big] + 2H\mathbb{E}|\bar{X}(s) - X(s)|^2 \Big\} ds$$
$$\leq 8q\rho\Delta(H + M)(\theta^{-1} + 1 + \Delta^{-1/2}) \int_0^t e^{\theta s} (1 + \mathbb{E}|\bar{X}(s)|^2) ds. \quad (6.86)$$

On the other hand, by inequality (6.83),

$$\mathbb{E}|\bar{X}(s)|^2 \leq 2\mathbb{E}|X(s)|^2 + 2\mathbb{E}|X(s) - \bar{X}(s)|^2$$
$$\leq 2\mathbb{E}|X(s)|^2 + 8\Delta\rho(1 + \mathbb{E}|\bar{X}(t)|^2).$$

Recalling that $16\Delta\rho \leq 1$ we obtain that

$$\mathbb{E}|\bar{X}(s)|^2 \leq 4\mathbb{E}|X(s)|^2 + 1.$$

It then follows from (6.86) that

$$J \leq 8q\rho\Delta(H+M)(\theta^{-1} + 1 + \Delta^{-1/2})\int_0^t e^{\theta s}(2 + 4\mathbb{E}|X(s)|^2)ds. \quad (6.87)$$

Substituting this into (6.84) gives

$$e^{\theta t}\mathbb{E}[X^T(t)Q_{r(t)}X(t)]$$
$$\leq q|x|^2 + \int_0^t e^{\theta s}\Big[\alpha + q\rho\sqrt{\Delta} + (2\theta q + q\rho\sqrt{\Delta} - \mu_1)|X(s)|^2\Big]ds$$
$$+ 8q\rho\Delta(H+M)(\theta^{-1} + 1 + \Delta^{-1/2})\int_0^t e^{\theta s}(2 + 4\mathbb{E}|X(s)|^2)ds. \quad (6.88)$$

Now, set $\theta = \mu_1/4q$. By (6.80),

$$32q\rho\Delta(H+M)(\theta^{-1} + 1 + \Delta^{-1/2}) + q\rho\sqrt{\Delta} \leq 0.5\mu_1.$$

We therefore obtain from (6.88) that

$$\tilde{q}e^{\theta t}\mathbb{E}|X(t)|^2 \leq q|x|^2 + \int_0^t e^{\theta s}(\alpha + 0.5\mu_1)ds \leq q|x|^2 + \frac{e^{\theta t}}{\theta}(\alpha + 0.5\mu_1).$$

That is,

$$\mathbb{E}|X(t)|^2 \leq \frac{q}{\tilde{q}}\Big[|x|^2 + \frac{4}{\mu_1}(\alpha + 0.5\mu_1)\Big] \leq \frac{q}{\tilde{q}}\Big[|x|^2 + 2 + \frac{4\alpha}{\mu_1}\Big]$$

and the assertion (6.81) follows. □

Lemma 6.15 *Let Assumption 6.20 hold. Assume that there exist N symmetric positive-definite matrices Q_i such that for all $(x, z, i) \in \mathbb{R}^n \times \mathbb{R}^n \times \mathbb{S}$,*

$$2(x-z)^T Q_i[f(x,i) - f(z,i)] + \text{trace}[(g(x,i) - g(z,i))^T Q_i(g(x,i) - g(z,i))]$$
$$+ \sum_{j=1}^N \gamma_{ij}(x-z)^T Q_j(x-z) \leq -\mu_2|x-z|^2. \quad (6.89)$$

If Δ is sufficiently small, then for any $(x, z, i) \in \mathbb{R}^n \times \mathbb{R}^n \times \mathbb{S}$,

$$\mathbb{E}|X_k^{(x,i),\Delta} - X_k^{(z,i),\Delta}|^2 \leq \frac{q}{\tilde{q}} e^{-(\mu_2/4q)k\Delta}|x-z|^2 \quad \forall k \geq 1, \tag{6.90}$$

where $\tilde{q}$ and q are the same as defined in Lemma 6.14. In particular, Assumption 6.17 holds as long as Δ is small enough.

Proof. Write $X_k^{(x,i),\Delta} = X_k^x$, $X_k^{(z,i),\Delta} = X_k^z$ and $r_k^{i,\Delta} = r_k$. Let $\bar{X}^x(t)$, $X^x(t)$ and $\bar{r}(t)$ be the same as defined in the proof of Lemma 6.14 and define $\bar{X}^z(t)$ and $X^z(t)$ similarly. For any $t > 0$, there exists an integer k such that $t_k \leq t < t_{k+1}$. By Assumption 6.20 and (6.82), it is easy to show that, if $\Delta < 1$,

$$\mathbb{E}|X^x(t) - X^z(t) - (\bar{X}^x(t) - \bar{X}^z(t))|^2$$
$$\leq 2\mathbb{E}\Big(\Big|\int_{t_k}^t [f(\bar{X}^x(s), \bar{r}(s)) - f(\bar{X}^z(s), \bar{r}(s))]ds\Big|^2$$
$$+ \Big|\int_{t_k}^t [g(\bar{X}^x(s), \bar{r}(s)) - g(\bar{X}^z(s), \bar{r}(s))]dB(s)\Big|^2\Big)$$
$$\leq 4\Delta H \mathbb{E}|\bar{X}^x(t) - \bar{X}^z(t)|^2. \tag{6.91}$$

Thus, if Δ is sufficiently small for $12\Delta H \leq 0.5$,

$$\mathbb{E}|X^x(t) - X^z(t)|^2$$
$$\leq 3\mathbb{E}|X^x(t) - X^z(t) - (\bar{X}^x(t) - \bar{X}^z(t))|^2 + 1.5|\bar{X}^x(t) - \bar{X}^z(t)|^2$$
$$\leq 2\mathbb{E}|\bar{X}^x(t) - \bar{X}^z(t)|^2. \tag{6.92}$$

Using (6.82) and the generalised Itô formula again we derive that, for any $\theta > 0$,

$$e^{\theta t}\mathbb{E}[(X^x(t) - X^z(t))^T Q_{r(t)}(X^x(t) - X^z(t))]$$
$$= (x-z)^T Q_i(x-z) + \int_0^t e^{\theta s}[J_1(s) + J_2(s) + J_3(s) + J_4(s)]ds, \tag{6.93}$$

where

$$J_1(s) := \theta \mathbb{E}\Big[(X^x(s) - X^z(s))^T Q_{r(s)}(X^x(s) - X^z(s))\Big],$$
$$J_2(s) := 2\mathbb{E}\Big[(X^x(s) - X^z(s))^T Q_{r(s)}(f(\bar{X}^x(s), \bar{r}(s)) - f(\bar{X}^z(s), \bar{r}(s)))\Big],$$

$$J_3(s) := \mathbb{E}\Big[\text{trace}[(g(\bar{X}^x(s),\bar{r}(s)) - g(\bar{X}^z(s),\bar{r}(s)))^T Q_{r(s)}$$
$$\times (g(\bar{X}^x(s),\bar{r}(s)) - g(\bar{X}^z(s),\bar{r}(s)))]\Big],$$
$$J_4(s) := \sum_{j=1}^{N} \mathbb{E}\Big[\gamma_{r(s)j}(X^x(t) - X^z(t))^T Q_j(X^x(t) - X^z(t))\Big].$$

By (6.92), it is easy to see that

$$J_1(s) \le 2\theta q \mathbb{E}|\bar{X}^x(t) - \bar{X}^z(t)|^2. \tag{6.94}$$

Write

$$J_2(s) = 2\mathbb{E}\Big[(\bar{X}^x(s) - \bar{X}^z(s))^T Q_{\bar{r}(s)}$$
$$\times (f(\bar{X}^x(s),\bar{r}(s)) - f(\bar{X}^z(s),\bar{r}(s)))\Big] + \bar{J}_2,$$

where

$$\bar{J}_2 = 2\mathbb{E}\Big[(X^x(s) - X^z(s) - (\bar{X}^x(s) - \bar{X}^z(s)))^T Q_{r(s)}$$
$$\times (f(\bar{X}^x(s),\bar{r}(s)) - f(\bar{X}^z(s),\bar{r}(s)))\Big]$$
$$+ 2\mathbb{E}\Big[(\bar{X}^x(s) - \bar{X}^z(s))^T (Q_{r(s)} - Q_{\bar{r}(s)})$$
$$\times (f(\bar{X}^x(s),\bar{r}(s)) - f(\bar{X}^z(s),\bar{r}(s)))\Big].$$

It is easy to show that

$$\bar{J}_2 \le \frac{q}{\sqrt{\Delta}}\mathbb{E}|X^x(s) - X^z(s) - (\bar{X}^x(s) - \bar{X}^z(s))|^2$$
$$+ qH\sqrt{\Delta}\mathbb{E}|\bar{X}^x(s) - \bar{X}^z(s)|^2$$
$$+ 4\bar{q}\sqrt{H}\mathbb{E}\Big[|\bar{X}^x(s) - \bar{X}^z(s)|^2 I_{\{r(s)\ne \bar{r}(s)\}}\Big],$$

where $\bar{q} = \max_{i\in\mathbb{S}}\|Q_i\|$. But, in the same way as (6.85) was done we can show that

$$\mathbb{E}\Big[|\bar{X}^x(s) - \bar{X}^z(s)|^2 I_{\{r(s)\ne\bar{r}(s)\}}\Big] = \mathbb{E}\Big[|\bar{X}^x(s) - \bar{X}^z(s)|^2 \mathbb{P}\{r(s)\ne\bar{r}(s)\}\Big].$$

By (6.91) and Lemma 6.10, we further compute that

$$\bar{J}_2 \le (5qH\sqrt{\Delta} + 4\bar{q}\sqrt{H}M\Delta)\mathbb{E}|\bar{X}^x(s) - \bar{X}^z(s)|^2$$
$$\le C_1\sqrt{\Delta}\mathbb{E}|\bar{X}^x(s) - \bar{X}^z(s)|^2,$$

where $C_1 = 5qH + 4\bar{q}\sqrt{H}M$. We therefore have that

$$J_2(s) \leq 2\mathbb{E}\Big[(\bar{X}^x(s) - \bar{X}^z(s))^T Q_{\bar{r}(s)}(f(\bar{X}^x(s),\bar{r}(s)) - f(\bar{X}^z(s),\bar{r}(s)))\Big]$$
$$+ C_1\sqrt{\Delta}\mathbb{E}|\bar{X}^x(s) - \bar{X}^z(s)|^2. \qquad (6.95)$$

Similarly, we can show that

$$J_3(s) \leq \mathbb{E}\Big[\text{trace}[(g(\bar{X}^x(s),\bar{r}(s)) - g(\bar{X}^z(s),\bar{r}(s)))^T Q_{\bar{r}(s)}$$
$$\times (g(\bar{X}^x(s),\bar{r}(s)) - g(\bar{X}^z(s),\bar{r}(s)))]\Big]$$
$$+ 2\bar{q}HM\Delta\mathbb{E}|\bar{X}^x(s) - \bar{X}^z(s)|^2. \qquad (6.96)$$

To compute J_4, we write

$$J_4(s) = \sum_{j=1}^{N} \mathbb{E}\Big[\gamma_{\bar{r}(s)j}(\bar{X}^x(t) - \bar{X}^z(t))^T Q_j(\bar{X}^x(t) - \bar{X}^z(t))\Big] + \bar{J}_4,$$

where

$$\bar{J}_4 = \sum_{j=1}^{N} \mathbb{E}\Big[\gamma_{r(s)j}(X^x(t) - X^z(t))^T Q_j[X^x(t) - X^z(t) - (\bar{X}^x(t) - \bar{X}^z(t))]\Big]$$
$$+ \sum_{j=1}^{N} \mathbb{E}\Big[\gamma_{r(s)j}[X^x(t) - X^z(t) - (\bar{X}^x(t) - \bar{X}^z(t))]^T Q_j(\bar{X}^x(t) - \bar{X}^z(t))\Big]$$
$$+ \sum_{j=1}^{N} \mathbb{E}\Big[(\gamma_{r(s)j} - \gamma_{\bar{r}(s)j})(\bar{X}^x(t) - \bar{X}^z(t))^T Q_j(\bar{X}^x(t) - \bar{X}^z(t))\Big].$$

In the same way as $\bar{J}_2$ was estimated we can show that

$$\bar{J}_4 \leq C_2\sqrt{\Delta}\mathbb{E}|\bar{X}^x(s) - \bar{X}^z(s)|^2,$$

where C_2 is a constant independent of Δ. We therefore obtain that

$$J_4(s) \leq \sum_{j=1}^{N} \mathbb{E}\Big[\gamma_{\bar{r}(s)j}(\bar{X}^x(t) - \bar{X}^z(t))^T Q_j(\bar{X}^x(t) - \bar{X}^z(t))\Big]$$
$$+ C_2\sqrt{\Delta}\mathbb{E}|\bar{X}^x(s) - \bar{X}^z(s)|^2. \qquad (6.97)$$

Substituting (6.94)–(6.97) into (6.93) and then making use of condition (6.89) we obtain that

$$e^{\theta t}\mathbb{E}[(X^x(t) - X^z(t))^T Q_{r(t)}(X^x(t) - X^z(t))]$$
$$\leq (x-z)^T Q_i(x-z)$$
$$+ \int_0^t e^{\theta s}[-\mu_2 + 2\theta q + C_3\sqrt{\Delta}]\mathbb{E}|\bar{X}^x(s) - \bar{X}^z(s)|^2 ds, \quad (6.98)$$

where C_3 is a constant independent of Δ. Now, choose $\theta = \mu_2/4q$. If Δ is sufficiently small for $C_3\sqrt{\Delta} < \mu_2/2$, it then follows from (6.98) that

$$\tilde{q} e^{(\mu_2/4q)t}\mathbb{E}|X^x(t) - X^z(t)|^2 \leq q|x-z|^2,$$

which implies the assertion (6.90) immediately. □

Corollary 6.21 *Let Assumption 6.20 and condition (6.89) hold. If Δ is sufficiently small (satisfying (6.80) and the condition of Lemma 6.15), then $\{Y_k^{(x,i),\Delta}\}_{k\geq 0}$ has a unique stationary distribution $\pi^\Delta(\cdot \times \cdot)$. Moreover, for any $\varepsilon > 0$ and $(x,i) \in \mathbb{R}^n \times \mathbb{S}$, there exists a $T > 0$ which is independent of Δ such that*

$$d_{\mathbb{L}}(P_k^\Delta((x,i), \cdot \times \cdot), \pi^\Delta(\cdot \times \cdot)) \leq \varepsilon \quad (6.99)$$

whenever $k\Delta \geq T$.

Proof. Since condition (6.89) implies (6.79), the existence of the unique stationary distribution $\pi^\Delta(\cdot \times \cdot)$ follows from Theorem 6.19 and Lemmas 6.13–6.15. To show (6.99) we compute, for any $F \in \mathbb{L}$ and $k, u > 0$, that

$$|\mathbb{E}F(X_{k+u}^{(x,i),\Delta}, r_{k+u}^{i,\Delta}) - \mathbb{E}F(X_k^{(x,i),\Delta}, r_k^{i,\Delta})|$$
$$\leq \sum_{l=1}^N \int_{\mathbb{R}^n} |\mathbb{E}F(X_k^{(z,l),\Delta}, r_k^{l,\Delta}) - \mathbb{E}F(X_k^{(x,i),\Delta}, r_k^{i,\Delta})| P_u^\Delta((x,i), dz \times \{l\})$$
$$\leq \sum_{l=1}^N \int_{\bar{\mathbb{S}}_R} |\mathbb{E}F(X_k^{(z,l),\Delta}, r_k^{l,\Delta}) - \mathbb{E}F(X_k^{(x,i),\Delta}, r_k^{i,\Delta})| P_u^\Delta((x,i), dz \times \{l\})$$
$$+ 2P_u^\Delta((x,i), \bar{\mathbb{S}}_R^c \times \mathbb{S}), \quad (6.100)$$

where R is a positive number, $\bar{\mathbb{S}}_R = \{x \in \mathbb{R}^n : |x| \leq R\}$ and $\bar{\mathbb{S}}_R^c = \mathbb{R}^n - \bar{\mathbb{S}}_R$. By Lemma 6.14, we can choose R, which is independent of Δ, sufficiently large for

$$P_u^\Delta((x,i), \bar{\mathbb{S}}_R^c \times \mathbb{S}) < \frac{\varepsilon}{2} \quad \text{for any } u \geq 1. \quad (6.101)$$

On the other hand, using (6.90) we can show, in the same way as Lemma 6.11 was proved, that there is a $T > 0$ such that

$$\sup_{f \in \mathbb{L}} |\mathbb{E}F(X_k^{(z,l),\Delta}, r_k^{l,\Delta}) - \mathbb{E}F(X_k^{(x,i),\Delta}, r_k^{i,\Delta})| < \frac{\varepsilon}{2}, \qquad (6.102)$$

whenever $(z,l) \in \bar{\mathbb{S}}_R \times \mathbb{S}$ and $k\Delta \geq T$. Substituting (6.100) and (6.101) into (6.102) yields

$$|\mathbb{E}F(X_{k+u}^{(x,i),\Delta}, r_{k+u}^{i,\Delta}) - \mathbb{E}F(X_k^{(x,i),\Delta}, r_k^{i,\Delta})| < \varepsilon \quad \text{for any } u \geq 1$$

whenever $k\Delta \geq T$. Since F is arbitrary, we have

$$d_{\mathbb{L}}(P_k^\Delta((x,i), \cdot \times \cdot), P_{k+u}^\Delta((x,i), \cdot \times \cdot)) < \varepsilon \quad \text{for any } u \geq 1$$

whenever $k\Delta \geq T$. Letting $u \to \infty$, we obtain (6.99). □

6.5.3 *Convergence of Stationary Distributions*

Corollary 6.21 shows that under Assumption 6.20 and condition (6.89), for each sufficiently small step size Δ, $\{Y_k^{(x,i),\Delta}\}_{k \geq 0}$ has a unique stationary distribution $\pi^\Delta(\cdot \times \cdot)$. The question is: Will the stationary distribution $\pi^\Delta(\cdot \times \cdot)$ converge weakly to a probability measure in $\mathcal{P}(\mathbb{R}^n \times \mathbb{S})$ as $\Delta \to 0$? If yes, one may further ask: What is the limit probability measure? In this section we will answer the questions positively and reveal that the limit probability measure is in fact the stationary distribution of the exact solution.

Let us recall some results from Section 5.6. Let $x(t)$ be the (exact) solution of equation (6.1) and let $Y(t) = (x(t), r(t))$. Then $Y(t)$ is a time homogeneous Markov process. If the process starts from $(z, i) \in \mathbb{R}^n \times \mathbb{S}$, we denote the process by $Y^{z,i}(t) = (y^{z,i}(t), r^i(t))$. Let $P_t((z,i), \cdot \times \cdot)$ be the transition probability function of the Markov process $Y(t)$. By the general results established in Section 5.6, we have

Theorem 6.22 *Under Assumption 6.20 and condition (6.89), the Markov process $Y(t)$ has a unique stationary distribution $\pi(\cdot \times \cdot) \in \mathcal{P}(\mathbb{R}^n \times \mathbb{S})$.*

To reveal the important relationship between π^Δ and π, let us establish another lemma.

Lemma 6.16 *Let Assumption 6.20 hold and fix any $(x,i) \in \mathbb{R}^n \times \mathbb{S}$. Then for any given $T > 0$ and $\varepsilon > 0$, there is a $\Delta^* > 0$, which is sufficiently*

small, such that

$$d_{\mathbb{L}}(P_k^\Delta((z,i),\cdot \times \cdot), P_{k\Delta}((z,i),\cdot \times \cdot)) < \varepsilon. \qquad (6.103)$$

provided $\Delta < \Delta^*$ and $k\Delta \leq T$.

Proof. Let $X^{(z,i),\Delta}(t)$ be the continuous approximate solution (recall (6.82)). Under Assumption 6.20, Theorem 4.1 shows that

$$\lim_{\Delta \to 0} \mathbb{E}\left[\sup_{0 \leq t \leq T} |X^{(z,i),\Delta}(t) - x^{(z,i)}(t)|^2\right] = 0. \qquad (6.104)$$

Hence there is a $\Delta^* > 0$, sufficiently small such that

$$\mathbb{E}(|X_k^{(z,i),\Delta} - x^{(z,i)}(k\Delta)|) < \varepsilon \qquad (6.105)$$

provided $\Delta < \Delta^*$ and $k\Delta \leq T$. Therefore, for any $F \in \mathbb{L}$,

$$|\mathbb{E}F(X_k^{(z,i),\Delta}, r_k^{i,\Delta}) - \mathbb{E}F(x^{(z,i)}(k\Delta), r^i(k\Delta))|$$
$$\leq \mathbb{E}(|X_k^{(z,i),\Delta} - y^{(z,i)}(k\Delta)|) < \varepsilon. \qquad (6.106)$$

The required assertion follows then. $\square$

We can now show that the numerical stationary distribution will weakly converge to the stationary distribution of the exact solution.

Theorem 6.23 *Under Assumption 6.20 and condition (6.89),*

$$\lim_{\Delta \to 0} d_{\mathbb{L}}(\pi^\Delta(\cdot \times \cdot), \pi(\cdot \times \cdot)) = 0. \qquad (6.107)$$

Proof. Fix any $(z,i) \in \mathbb{R}^n \times \mathbb{S}$ and let $\varepsilon > 0$ be arbitrary. By Theorem 6.22, there exists a $T_1 > 0$ such that

$$d_{\mathbb{L}}(P_{T_1}((z,i),\cdot \times \cdot), \pi(\cdot \times \cdot)) \leq \frac{\varepsilon}{3}. \qquad (6.108)$$

By Corollary 6.21, there is a pair of $\Delta_0 > 0$ and $T_2 > 0$ such that if $\Delta < \Delta_0$ and $k\Delta \geq T_2$, then

$$d_{\mathbb{L}}(P_k^\Delta((z,i),\cdot \times \cdot), \pi^\Delta(\cdot \times \cdot)) \leq \frac{\varepsilon}{3}. \qquad (6.109)$$

Set $T = T_1 \vee T_2$. By Lemma 6.16, there is a $\Delta^* > 0$ such that

$$d_{\mathbb{L}}(P_k^\Delta((z,i),\cdot \times \cdot), P_{k\Delta}((z,i),\cdot \times \cdot)) < \frac{\varepsilon}{3} \qquad (6.110)$$

provided $\Delta < \Delta^*$ and $k\Delta \leq T+1$. Now, for any $\Delta < \Delta_0 \wedge \Delta^*$, letting $k = [T/\Delta] + 1$ and using (6.108), (6.109) and (6.110) we derive that

$$d_{\mathbb{L}}(\pi^\Delta(\cdot \times \cdot), \pi(\cdot \times \cdot))$$
$$\leq d_{\mathbb{L}}(P_k^\Delta((z,i), \cdot \times \cdot), \pi^\Delta(\cdot \times \cdot)) + d_{\mathbb{L}}(P_{k\Delta}((z,i), \cdot \times \cdot), \pi(\cdot \times \cdot))$$
$$+ d_{\mathbb{L}}(P_k^\Delta((z,i), \cdot \times \cdot), P_{k\Delta}((z,i), \cdot \times \cdot))$$
$$< \frac{\varepsilon}{3} + \frac{\varepsilon}{3} + \frac{\varepsilon}{3} = \varepsilon,$$

as required. □

Let us make a remark to close our section. We observe that Theorem 6.22 is an existence theorem but does not provide a method to obtain the stationary distribution. In general, it is difficult to obtain the explicit stationary distribution $\pi(\cdot \times \cdot)$. However, Theorem 6.23 reveals that the numerical stationary distribution π^Δ will weakly converge to the stationary distribution π of the exact solution. In other words, this theorem provides us with a numerical method to obtain the approximate distribution for the stationary distribution π if the step size Δ is sufficiently small.

6.6 Asymptotic Stability in Distribution of the EM Method: Variable Step Sizes

Let $\Delta = \{\Delta_k\}_{k\geq 1}$ be a sequence of positive numbers such that

$$\lim_{k\to\infty} \Delta_k = 0 \quad \text{and} \quad \sum_{k=1}^{\infty} \Delta_k = \infty. \tag{6.111}$$

In general $\Delta_k \neq \Delta_l, k \neq l$, we shall call $\Delta = \{\Delta_k\}_{k\geq 1}$ the *variable step sizes*. Set $t_0 = 0$ and define $t_k = \sum_{l=1}^{k} \Delta_l$ for $k \geq 1$, $\{t_k\}_{k\geq 0}$ is a nonrandom partition of $[0, \infty)$. Define $r_k = r(t_k)$ for $k \geq 0$. Given initial data $(x, i) \in \mathbb{R}^n \times \mathbb{S}$, the EM approximate solution to the SDE (6.58) based on the variable step sizes is to compute the discrete approximations $X_k \approx x(t_k)$ by setting $X_0 = x, r_0 = i$ and forming

$$X_{k+1} = X_k + f(X_k, r_k)\Delta_{k+1} + g(X_k, r_k)\Delta B_{k+1}, \tag{6.112}$$

where $\Delta B_{k+1} = B(t_{k+1}) - B(t_k)$. Set $Y_k = (X_k, r_k)$. Then $\{Y_k\}_{k\geq 0}$ is a Markov process. To highlight the initial data we will use the notation $Y_k^{x,i} = (X_k^{x,i}, r_k^i)$. Let $P_k^\Delta((x,i), \cdot \times \cdot)$ be the probability measure induced by $Y_k^{x,i}$, namely

$$P_k^\Delta((x,i), A \times D) = \mathbb{P}\{Y_k^{x,i} \in A \times D\}, \quad \text{for } A \in \mathcal{B}^n, \ D \subset \mathbb{S}.$$

We can now state our main result of this section.

Theorem 6.24 *Let Assumption 6.20 and condition (6.89) hold. Then for any given variable step sizes $\{\Delta_k\}_{k \geq 1}$ satisfying (6.111), the probability measure $P_k^\Delta((x,i), \cdot \times \cdot)$ induced by $Y_k^{x,i}$ will converge weakly to the stationary distribution $\pi(\cdot \times \cdot)$ of the solution process $Y(t) = (x(t), r(t))$ as $k \to \infty$ for any initial data $(x,i) \in \mathbb{R}^n \times \mathbb{S}$.*

We will leave the proof to the reader as an exercise. Instead, we shall give a sufficient condition for condition (6.89) in terms of M-matrix.

Assumption 6.25 Assume that for each $j \in \mathbb{S}$, there is a pair of constants β_j and δ_j such that

$$(u-v)^T(f(u,j) - f(v,j)) \leq \beta_j |u-v|^2, \tag{6.113}$$

$$|g(u,j) - g(v,j)|^2 \leq \delta_j |u-v|^2 \tag{6.114}$$

for all $u, v \in \mathbb{R}^n$. Moreover,

$$\mathcal{A} := -\text{diag}(2\beta_1 + \delta_1, \cdots, 2\beta_N + \delta_N) - \Gamma \tag{6.115}$$

is a nonsingular M-matrix.

Lemma 6.17 *Assumption 6.25 implies condition (6.89).*

Proof. By Theorem 2.10 there is a vector $(q_1, \cdots, q_N)^T \gg 0$ such that

$$(\lambda_1, \cdots, \lambda_N)^T := \mathcal{A}(q_1, \cdots, q_N)^T \gg 0.$$

Set $\mu = \min_{1 \leq i \leq N} \lambda_i > 0$ and define the symmetric matrices $Q_j = q_j I$ for $j \in \mathbb{S}$, where I is the $n \times n$ identity matrix. By (6.113) and (6.114) we compute that

$$2(u-v)^T Q_j [f(u,j) - f(v,j)] + \text{trace}\Big[(g(u,j) - g(v,j))^T$$

$$Q_j(g(u,j) - g(v,j))\Big] + \sum_{l=1}^N \gamma_{jl}(u-v)^T Q_l(u-v)$$

$$= 2q_j(u-v)^T(f(u,j) - f(v,j) + q_j|g(u,j) - g(v,j)|^2$$
$$+ \sum_{l=1}^{N} \gamma_{jl} q_l |u-v|^2$$
$$\leq \left(2\beta_j q_j + \delta_j q_j + \sum_{l=1}^{N} \gamma_{jl} q_l\right)|u-v|^2$$
$$= -\lambda_j |u-v|^2 \leq -\mu|u-v|^2.$$

We hence conclude that Assumption 6.25 implies condition (6.89). □

Let us now discuss an example to illustrate this new technique of M-matrices.

Example 6.26 Let $B(t)$ be a scalar Brownian motion. Let $r(t)$ be a continuous-time Markov chain taking values in $S = \{1, 2, 3\}$ with the generator

$$\Gamma = \begin{bmatrix} -2 & 1 & 1 \\ 3 & -4 & 1 \\ 1 & 1 & -2 \end{bmatrix}.$$

Assume that $B(t)$ and $r(t)$ are independent. Consider a three-dimensional SDE

$$dx(t) = A(r(t))x(t)dt + g(x(t), r(t))dB(t) \tag{6.116}$$

on $t \geq 0$. Here

$$A(1) = A_1 = \begin{bmatrix} -2 & -1 & -2 \\ 2 & -2 & 1 \\ 1 & -2 & -3 \end{bmatrix}, \quad A(2) = A_2 = \begin{bmatrix} 0.5 & 1.0 & 0.5 \\ -0.8 & 0.5 & 1.0 \\ -0.7 & -0.9 & 0.2 \end{bmatrix},$$

$$A(3) = A_3 = \begin{bmatrix} -0.5 & -0.9 & -1.0 \\ 1.0 & -0.6 & -0.7 \\ 0.8 & 1.0 & -1.0 \end{bmatrix}.$$

Moreover, $g : \mathbb{R}^3 \times S \to \mathbb{R}^3$ satisfying

$$|g(u,j) - g(v,j)|^2 \leq \delta_j |u-v|^2, \quad (u,v,j) \in \mathbb{R}^3 \times \mathbb{R}^3 \times \mathbb{S},$$

where

$$\delta_1 = 0.5, \quad \delta_2 = 0.1, \quad \delta_3 = 0.3.$$

Defining $f(u,j) = A_j u$ for $(u,j) \in \mathbb{R}^3 \times \mathbb{S}$ we have

$$(u-v)^T[f(u,j) - f(v,j)] = \frac{1}{2}(u-v)^T(A_j + A_j^T)(u-v) \le \beta_j |u-v|^2,$$

where $\beta_j = \frac{1}{2}\lambda_{\max}(A_j + A_j^T)$. It is easy to compute

$$\beta_1 = -1.21925, \quad \beta_2 = 0.60359, \quad \beta_3 = -0.47534.$$

In other words, (6.113) and (6.114) hold with β_j and δ_j as above. So the matrix $\mathcal{A}$ defined by (6.115) becomes

$$\mathcal{A} = \mathrm{diag}(1.939, -1.30718, 0.65068) - \Gamma = \begin{bmatrix} 3.939 & -1 & -1 \\ -3 & 2.69372 & -1 \\ -1 & -1 & 2.65068 \end{bmatrix}.$$

It is easy to verify that all the leading principal minors of $\mathcal{A}$ are positive and hence $\mathcal{A}$ is a nonsingular M-matrix. We have therefore verified Assumption 6.25, which in turn, implies condition (6.89). Moreover, it is plain that Assumption 6.20 holds. By Theorem 6.24 and Lemma 6.17 we can conclude that equation (6.116) has a unique stationary distribution which can be obtained by the EM approximate scheme with variable step sizes.

6.7 Exercises

6.1 Prove (6.12) in detail.

6.2 Show that under Assumption 6.3, the solution (6.1) with initial data x_0 and r_0 satisfies, for any $T > 0$,

$$\sup_{0 \le t \le T} \mathbb{E}|x(t)|^2 \le 3\mathbb{E}|x_0|^2 e^{3(T+1)KT}.$$

6.3 Prove (6.14) in the same way as Theorem 4.1 was proved by using (6.9).

6.4 Prove Lemma 6.5.

6.5 Show

$$|(1-\theta)x + \theta y|^2 \le |x|^2 + |y|^2, \quad \forall x, y \in \mathbb{R}^n, \ \theta \in [0,1].$$

6.6 Prove (6.53).

6.7 Prove Lemma 6.14.

6.8 Prove Theorem 6.24 (you may refer to [Mao et al. (2005a)]).

Chapter 7

Stochastic Differential Delay Equations with Markovian Switching

7.1 Introduction

In many applications, one assumes that the system under consideration is governed by a principle of causality; that is, the future state of the system is independent of the past states and is determined solely by the present. However, under closer scrutiny, it becomes apparent that the principle of causality is often only a first approximation to the true situation and that a more realistic model would include some of the past states of the system. Stochastic functional differential equations give a mathematical formulation for such systems.

The simplest type of past dependence in a differential equation is that in which the past dependence is through the state variable but not the derivative of the state variable. The Lotka–Volterra model for the population growth of a single species is described by

$$\dot{x}(t) = x(t)[-\alpha + \beta x(t-\tau)]. \tag{7.1}$$

Under suitable assumptions, the equation

$$\dot{x}(t) = \sum_{i=1}^{k} A_i x(t-\tau_i) \tag{7.2}$$

is a suitable model for describing the mixing of a dye from a central tank as dyed water circulates through a number of pipes. Taking into account the transmission time in the triode oscillator, Rubanik in 1969 studied the van der Pol equation

$$\ddot{x}(t) + \alpha \dot{x}(t) - f(x(t-\tau))\dot{x}(t-\tau) + x(t) = 0 \tag{7.3}$$

with the delayed argument τ. The Hopfield model for a network with delays can be described by a differential delay equation

$$C_i \dot{u}_i(t) = -\frac{1}{R_i} u_i(t) + \sum_{j=1}^{n} T_{ij} g_j(u_j(t - \tau_j)), \quad 1 \le i \le n. \quad (7.4)$$

If we take into account the estimation error for system parameters as well as the environmental noise we are led to stochastic differential delay equations (SDDEs). For example, in equation (7.1) it is better to estimate parameter β as the point estimator $\bar{\beta}$ plus an error. But by the central limit theorem, the error may be described by a normally distributed random variable. That is

$$\beta = \bar{\beta} + error = \bar{\beta} + \sigma \dot{B}(t).$$

Substituting this into (7.1) gives

$$\dot{x}(t) = x(t)[-\alpha + (\bar{\beta} + \sigma \dot{B}(t))x(t - \tau)].$$

That is, in the differential form,

$$dx(t) = x(t)[(-\alpha + \bar{\beta} x(t - \tau))dt + \sigma x(t - \tau)dB(t)], \quad (7.5)$$

which is an SDDE.

Let us now take a further step to consider another type of random fluctuation. The population system may switch between (finite) regimes of environment, which differ by factors such as nutrition or as rain falls (see e.g. [Du et al. (2004); Slatkin (1978)]). The switching is without memory and the waiting time for the next switch has an exponential distribution. We can hence model the regime switching by a finite-state Markov chain $r(t)$. Under a different regime, the system parameters α, $\bar{\beta}$ and σ are different. As a result, system (7.5) becomes a more general equation

$$dx(t) = x(t)[(-\alpha(r(t)) + \bar{\beta}(r(t))x(t - \tau))dt + \sigma(r(t))x(t - \tau)dB(t)], \quad (7.6)$$

which is an SDDE with Markovian switching.

In general, an SDDE with Markovian switching has the form

$$dx(t) = f(x(t), x(t - \delta), r(t), t)dt + g(x(t), x(t - \delta), r(t), t)dB(t), \quad (7.7)$$

while the time delay δ may be a variable of time t. When we try to carry over the theory of stochastic differential equations (SDEs) to SDDEs, the following questions naturally arise:

- What is the initial-value problem for equation (7.7)?
- What are the conditions to guarantee the existence and uniqueness of the solution?
- What properties does the solution have?
- Is there any explicit solution or otherwise how can one obtain the approximate solution?

In this chapter we shall answer these questions one by one.

7.2 Stochastic Differential Delay Equations

To make the theory of SDDEs with Markovian switching more understandable, let us begin with SDDEs without Markovian switching.

Let the probability space $(\Omega, \mathcal{F}, \{\mathcal{F}_t\}_{t \geq 0}, \mathbb{P})$ and the m-dimensional Brownian motion $B(t) = (B_1(t), \cdots, B_m(t))^T$ be the same as before. Let $p > 0$ and $\tau > 0$. Denote by $C([-\tau, 0]; \mathbb{R}^n)$ the space of continuous functions $\varphi : [-\tau, 0] \to \mathbb{R}^n$ with norm $\|\varphi\| = \sup_{-\tau \leq \theta \leq 0} |\varphi(\theta)|$. Denote by $L^p_{\mathcal{F}_t}([-\tau, 0]; \mathbb{R}^n)$ the family of $\mathcal{F}_t$-measurable $C([-\tau, 0]; \mathbb{R}^n)$-valued random variables $\xi = \{\xi(\theta) : -\tau \leq \theta \leq 0\}$ such that $\mathbb{E}\|\xi\|^p < \infty$. If $x(t)$ is an $\mathbb{R}^n$-valued stochastic process on $t \in [-\tau, \infty)$, we let $x_t = \{x(t+\theta) : -\tau \leq \theta \leq 0\}$ for $t \geq 0$ so x_t is a $C([-\tau, 0]; \mathbb{R}^n)$-valued stochastic process.

Consider a nonlinear SDDE

$$dx(t) = f(x(t), x(t - \delta(t)), t)dt + g(x(t), x(t - \delta(t)), t)dB(t) \qquad (7.8)$$

on $t \in [0, T]$. Here $\delta : [0, T] \to [0, \tau]$ is a Borel measurable function which stands for the time lag, while

$$f : \mathbb{R}^n \times \mathbb{R}^n \times [0, T] \to \mathbb{R}^n \quad \text{and} \quad g : \mathbb{R}^n \times \mathbb{R}^n \times [0, T] \to \mathbb{R}^{n \times m}.$$

Moreover, to make notations simple, we let the initial time be 0 rather than t_0. In order to solve the equation we need to know the initial data and we assume that they are given by

$$\{x(\theta) : -\tau \leq \theta \leq 0\} = \xi \in L^2_{\mathcal{F}_0}([-\tau, 0]; \mathbb{R}^n). \qquad (7.9)$$

Moreover, we also require the coefficients f and g to be sufficiently smooth. The well-known conditions imposed for the existence and uniqueness of the solution are the Local Lipschitz condition and the linear growth condition which are stated as follows.

Assumption 7.1 (**The local Lipschitz condition**) For each integer $k \geq 1$ there is a positive constant H_k such that

$$|f(x,y,t) - f(\bar{x},\bar{y},t)|^2 \vee |g(x,y,t) - g(\bar{x},\bar{y},t)|^2 \leq H_k(|x-\bar{x}|^2 + |y-\bar{y}|^2)$$

for those $x, y, \bar{x}, \bar{y} \in \mathbb{R}^n$ with $|x| \vee |y| \vee |\bar{x}| \vee |\bar{y}| \leq k$ and any $t \in [0,T]$.

Assumption 7.2 (**The linear growth condition**) There is a constant $K > 0$ such that

$$|f(x,y,t)|^2 \vee |g(x,y,t)|^2 \leq K(1 + |x|^2 + |y|^2)$$

for all $(x,y,t) \in \mathbb{R}^n \times \mathbb{R}^n \times [0,T]$.

With these assumptions we can state the existence-and-uniqueness theorem.

Theorem 7.3 *Under Assumptions 7.1 and 7.2, equation (7.8) with the given initial data (7.9) has a unique continuous solution $x(t)$ on $t \in [-\tau, T]$. Moreover, the solution has the property that*

$$\mathbb{E}\left(\sup_{-\tau \leq t \leq T} |x(t)|^2\right) < \infty. \tag{7.10}$$

This theorem can be proved by the standard technique of Picard iterations but the details are left to the reader as an exercise.

However, when the time lag $\delta(t)$ is a constant, say $\delta(t) \equiv \tau$, the proof becomes much easier. In this case, the SDDE (7.8) become an SDDE with constant delay

$$dx(t) = f(x(t), x(t-\tau), t)dt + g(x(t), x(t-\tau), t)dB(t). \tag{7.11}$$

Note that on $[0, \tau]$, this SDDE become

$$dx(t) = f(x(t), \xi(t-\tau), t)dt + g(x(t), \xi(t-\tau), t)dB(t)$$

with the initial value $x(0) = \xi(0)$. But this is a stochastic differential equation (SDE) and it will have a unique solution if the linear growth condition holds while $f(x,y,t)$ and $g(x,y,t)$ are locally Lipschitz continuous in x only (not necessary in y). Moreover, the linear growth condition also guarantees the solution has the property that

$$\mathbb{E}\left(\sup_{-\tau \leq t \leq \tau} |x(t)|^2\right) < \infty.$$

Once the solution $x(t)$ on $[0,\tau]$ is known, we can proceed the arguments above on $[\tau, 2\tau]$, $[2\tau, 3\tau]$ etc, and hence obtain the solution on the entire

$[-\tau, T]$. This simple technique leads us to impose a weakly local Lipschitz condition:

Assumption 7.4 (**The weakly local Lipschitz condition**) For each integer $k \geq 1$ there is a positive constant $\bar{H}_k$ such that

$$|f(x,y,t) - f(\bar{x},y,t)|^2 \vee |g(x,y,t) - g(\bar{x},y,t)|^2 \leq \bar{H}_k(|x-\bar{x}|^2)$$

for those $x, \bar{x} \in \mathbb{R}^n$ with $|x| \vee |\bar{x}| \leq k$ and any $(y,t) \in \mathbb{R}^n \times [0,T]$.

The simple arguments above give the following result.

Theorem 7.5 *Under Assumptions 7.4 and 7.2, the SDDE (7.11) with the given initial data (7.9) has a unique continuous solution $x(t)$ on $t \in [-\tau, T]$. Moreover, the solution obeys property (7.10).*

This theorem shows that it is unnecessary to require the coefficients $f(x,y,t)$ and $g(x,y,t)$ to be locally Lipschitz continuous in y if the time delay is a positive constant. Let us discuss a number of examples to illustrate this simple but useful technique.

Example 7.6 Consider an SDDE

$$dx(t) = F(x(t-\tau),t)dt + G(x(t-\tau),t)dB(t) \quad \text{on } t \geq 0 \qquad (7.12)$$

with initial data $\{x(\theta) : -\tau \leq \theta \leq 0\} = \xi \in L^2_{\mathcal{F}_0}([-\tau,0];\mathbb{R}^n)$. where

$$F : \mathbb{R}^n \times \mathbb{R}_+ \to \mathbb{R}^n \quad \text{and} \quad G : \mathbb{R}^n \times \mathbb{R}_+ \to \mathbb{R}^{n \times m}$$

which satisfy the linear growth condition. For $t \in [0,\tau]$, we have the explicit solution

$$x(t) = \xi(0) + \int_0^t F(\xi(s-\tau),s)ds + \int_0^t G(\xi(s-\tau),s)dB(s).$$

It is also easy to show by the linear growth condition that

$$\{x(t) : 0 \leq t \leq \tau\} \in L^2_{\mathcal{F}_\tau}([-\tau,0];\mathbb{R}^n).$$

Then, for $t \in [\tau, 2\tau]$, we have

$$x(t) = x(\tau) + \int_\tau^t F(x(s-\tau),s)ds + \int_\tau^t G(x(s-\tau),s)dB(s).$$

Repeating this procedure over the intervals $[2\tau, 3\tau]$ *etc*, we obtain the explicit solution for the SDDE (7.12). Clearly, all we require here is the condition which guarantees that the integrals are well-defined, and the linear growth condition is sufficient.

Example 7.7 Consider a one-dimensional linear SDDE

$$dx(t) = [ax(t) + \bar{a}x(t-\tau)]dt$$
$$+ \sum_{k=1}^{m}[b_k x(t) + \bar{b}_k x(t-\tau)]dB_k(t) \quad \text{on } t \geq 0 \quad (7.13)$$

with initial data $\{x(\theta) : -\tau \leq \theta \leq 0\} = \xi \in L^2_{\mathcal{F}_0}([-\tau, 0]; \mathbb{R})$. On $t \in [0, \tau]$, the linear SDDE becomes a linear SDE

$$dx(t) = [ax(t) + \alpha_1(t)]dt + \sum_{k=1}^{m}[b_k x(t) + \beta_{k1}(t)]dB_k(t)$$

with initial value $x(0) = \xi(0)$, where

$$\alpha_1(t) = \bar{a}\xi(t-\tau), \quad \beta_{k1}(t) = \bar{b}_k \xi(t-\tau).$$

This linear SDE has the explicit solution

$$x(t) = \Psi_1(t)\bigg[\xi(0) + \int_0^t \Psi_1^{-1}(s)\bigg(\bar{a}\xi(s-\tau) - \sum_{k=1}^{m} b_k \bar{b}_k \xi(s-\tau)\bigg)ds$$
$$+ \sum_{k=1}^{m} \int_0^t \Psi_1^{-1}(s)\bar{b}_k \xi(s-\tau)dB_k(s)\bigg],$$

where

$$\Psi_1(t) = \exp\bigg[\bigg(a - \frac{1}{2}\sum_{k=1}^{m} b_k^2\bigg)t + \sum_{k=1}^{m} b_k B_k(t)\bigg].$$

Next, on $t \in [\tau, 2\tau]$, the linear SDDE becomes a linear SDE

$$dx(t) = [ax(t) + \alpha_2(t)]dt + \sum_{k=1}^{m}[b_k x(t) + \beta_{k2}(t)]dB_k(t)$$

with the initial value $x(\tau)$ at $t = \tau$ obtained above, where

$$\alpha_2(t) = \bar{a}x(t-\tau), \quad \beta_{k2}(t) = \bar{b}_k x(t-\tau).$$

This linear SDE has the explicit solution

$$x(t) = \Psi_2(t)\bigg[x(\tau) + \int_\tau^t \Psi_2^{-1}(s)\bigg(\bar{a}x(s-\tau) - \sum_{k=1}^{m} b_k \bar{b}_k x(s-\tau)\bigg)ds$$
$$+ \sum_{k=1}^{m} \int_\tau^t \Psi_2^{-1}(s)\bar{b}_k x(s-\tau)dB_k(s)\bigg],$$

where

$$\Psi_2(t) = \exp\left[\left(a - \frac{1}{2}\sum_{k=1}^{m} b_k^2\right)(t-\tau) + \sum_{k=1}^{m} b_k(B_k(t) - B_k(\tau))\right].$$

Repeating this procedure over the intervals $[2\tau, 3\tau]$ *etc*, we obtain the explicit solution for the SDDE (7.13).

7.3 SDDEs with Markovian Switching

Consider an n-dimensional SDDE with Markovian switching

$$dx(t) = f(x(t), x(t-\delta(t)), t, r(t))dt + g(x(t), x(t-\delta(t)), t, r(t))dB(t) \quad (7.14)$$

on $t \in [0, T]$ with initial data (7.9). Here $r(\cdot)$ is the same Markov chain as before with initial data $r(0) = r_0$ which is an $\mathcal{F}_0$-measurable $\mathbb{S}$-valued random variable, while

$$f : \mathbb{R}^n \times \mathbb{R}^n \times \mathbb{R}_+ \times \mathbb{S} \to \mathbb{R}^n \quad \text{and} \quad g : \mathbb{R}^n \times \mathbb{R}^n \times \mathbb{R}_+ \times \mathbb{S} \to \mathbb{R}^{n \times m}.$$

Naturally, we impose the following local Lipschitz and linear growth condition.

Assumption 7.8 (**The local Lipschitz condition**) For each integer $k \geq 1$ there is a positive constant H_k such that

$$|f(x,y,t,i) - f(\bar{x},\bar{y},t,i)|^2 \vee |g(x,y,t,i) - g(\bar{x},\bar{y},t,i)|^2 \leq H_k(|x-\bar{x}|^2 + |y-\bar{y}|^2)$$

for those $x, y, \bar{x}, \bar{y} \in \mathbb{R}^n$ with $|x| \vee |y| \vee |\bar{x}| \vee |\bar{y}| \leq k$ and any $(t,i) \in [0,T] \times \mathbb{S}$.

Assumption 7.9 (**The linear growth condition**) There is a constant $K > 0$ such that

$$|f(x,y,t,i)|^2 \vee |g(x,y,t,i)|^2 \leq K(1 + |x|^2 + |y|^2)$$

for all $(x, y, t, i) \in \mathbb{R}^n \times \mathbb{R}^n \times [0,T] \times \mathbb{S}$.

It is not surprising to have the following existence-and-uniqueness theorem.

Theorem 7.10 *Under Assumptions 7.8 and 7.9, equation (7.14) has a unique continuous solution $x(t)$ on $t \in [-\tau, T]$. Moreover, the solution obeys property (7.10).*

This theorem can be proved in the same way as Theorem 3.13 was proved by making use of the fact that almost every sample path of $r(\cdot)$ is a right-continuous step function with a finite number of jumps on $[0, T]$. The details are left to the reader as an exercise.

To establish more general existence-and-uniqueness theorems, let us introduce the concept of local solution.

Definition 7.11 Let σ_∞ be a stopping time such that $0 \leq \sigma_\infty \leq T$ a.s. An $\mathbb{R}^n$-valued $\mathcal{F}_t$-adapted continuous stochastic process $\{x(t) : -\tau \leq t \leq \sigma_\infty\}$ is called a *local solution* of the SDDE (7.14) if $x(t) = \xi(t)$ on $t \in [-\tau, 0]$ and, moreover, there is a nondecreasing sequence $\{\sigma_k\}_{k \geq 1}$ of stopping times such that $0 \leq \sigma_k \uparrow \sigma_\infty$ a.s. and

$$x(t) = x(0) + \int_0^{t \wedge \sigma_k} f(x(s), x(s - \delta(s)), s, r(s)) ds$$
$$+ \int_0^{t \wedge \sigma_k} g(x(s), x(s - \delta(s)), s, r(s)) dB(s)$$

holds for any $t \in [0, T]$ and $k \geq 1$ with probability 1. If, furthermore,

$$\limsup_{t \to \sigma_\infty} |x(t)| = \infty \text{ whenever } \sigma_\infty < T,$$

then it is called a *maximal local solution* and σ_∞ is called the *explosion time*. A maximal local solution $\{x(t) : -\tau \leq t \leq \sigma_\infty\}$ is said to be *unique* if any other maximal solution $\{\bar{x}(t) : -\tau \leq t \leq \bar{\sigma}_\infty\}$ is indistinguishable from it, that is, with probability 1, $\sigma_\infty = \bar{\sigma}_\infty$ and $x(t) = \bar{x}(t)$ for $-\tau \leq t \leq \sigma_\infty$.

Similar to SDEs, we have the following result on the maximal local solutions under the local Lipschitz condition without the linear growth condition.

Theorem 7.12 *Under Assumptions 7.8, there exists a unique maximal local solution to the SDDE (7.14).*

Proof. For each $k \geq 1$, define, for $x \in \mathbb{R}^n$,

$$x^{[k]} = \begin{cases} x & \text{if } |x| \leq k, \\ \frac{kx}{|x|} & \text{if } |x| > k. \end{cases}$$

Define then the truncation functions

$$f_k(x, y, t, i) = f_k(x^{[k]}, y^{[k]}, t, i), \quad g_k(x, y, t, i) = g_k(x^{[k]}, y^{[k]}, t, i)$$

Then f_k and g_k obey Assumptions 7.8 and 7.9. By Theorem 7.10, there is a unique solution $x_k(t)$ to the equation

$$dx_k(t) = f_k(x_k(t), x_k(t - \delta(t)), t, r(t))dt$$
$$+ g_k(x_k(t), x + k(t - \delta(t)), t, r(t))dB(t) \quad (7.15)$$

on $t \in [0, T]$ with initial data $x_k(t) = \xi(t)$ on $t \in [-\tau, 0]$. Define the stopping time

$$\sigma_k = T \wedge \inf\{t \in [t_0, T] : \|x_{k,t}\| \geq k\},$$

where $x_{k,t} = \{x_k(t + \theta) : -\tau \leq \theta \leq 0\}$. It is not difficult to show that

$$x_k(t) = x_{k+1}(t) \quad \text{if } -\tau \leq t \leq \sigma_k. \quad (7.16)$$

This implies that σ_k is increasing so has its limit $\sigma_\infty = \lim_{k \to \infty} \sigma_k$. Define $\{x(t) : -\tau \leq t \leq \sigma_\infty\}$ by $x(t) = \xi(t)$ on $t \in [-\tau, 0]$ and

$$x(t) = x_k(t), \quad t \in]]\sigma_{k-1}, \sigma_k]], \ k \geq 1,$$

where $\sigma_0 = 0$ and we set $x(\sigma_\infty) = \infty$ if $\sigma_\infty < T$. By (7.16), $x(t \wedge \sigma_k) = x_k(t \wedge \sigma_k)$. It therefore follows from (7.15) that

$$x(t \wedge \tau_k) = x(0) + \int_{t_0}^{t \wedge \sigma_k} f(x(s), x(s - \delta(s)), s, r(s))ds$$
$$+ \int_0^{t \wedge \sigma_k} g(x(s), x(s - \delta(s)), s, r(s))dB(s)$$

for any $t \in [0, T]$ and $k \geq 1$. It is also easy to see that if $\sigma_\infty < T$, then

$$\limsup_{t \to \sigma_\infty} |x(t)| = \limsup_{k \to \infty} |x(\sigma_k)| = \limsup_{k \to \infty} |x_k(\sigma_k)| = \infty.$$

Hence $\{x(t) : t_0 \leq t \leq \sigma_\infty\}$ is a maximal local solution. It is rather standard to prove the uniqueness so we leave the details as an exercise. □

To establish a more general result we need a new notation. If $V \in C^{2,1}(\mathbb{R}^n \times [0, T] \times \mathbb{S}; \mathbb{R}_+)$, define an operator LV from $\mathbb{R}^n \times \mathbb{R}^n \times [0, T] \times \mathbb{S}$ to $\mathbb{R}$ by

$$LV(x, y, t, i) = V_t(x, t, i) + V_x(x, t, i)f(x, y, t, i)$$
$$+ \frac{1}{2}\text{trace}[g^T(x, y, t, i)V_{xx}(x, t, i)g(x, y, t, i)] + \sum_{j=1}^N \gamma_{ij} V(x, t, j).$$

Let us emphasise that the operator LV is defined on $\mathbb{R}^n \times \mathbb{R}^n \times [0,T] \times \mathbb{S}$ although the function V is only defined on $\mathbb{R}^n \times [0,T] \times \mathbb{S}$.

Theorem 7.13 *Let Assumption 7.8 hold. Assume that there is a pair of functions $V \in C^{2,1}(\mathbb{R}^n \times [0,T] \times \mathbb{S}; \mathbb{R}_+)$ and $U \in C(\mathbb{R}^n; \mathbb{R}_+)$ as well as a constant $\alpha > 0$ such that*

$$\lim_{|x| \to \infty} U(x) = \infty, \tag{7.17}$$

$$U(x) \leq V(x,t,i), \tag{7.18}$$

$$LV(x,y,t,i) \leq \alpha(1 + U(x) + U(y)) \tag{7.19}$$

for all $x,y \in \mathbb{R}^n$, $t \in [0,T]$ and $i \in \mathbb{S}$. Assume furthermore that the initial data ξ obeys that $\sup_{-\tau \leq \theta \leq 0} \mathbb{E}U(\xi(\theta)) < \infty$ and $\mathbb{E}V(\xi(0),0,r_0) < \infty$. Then there exists a unique solution $x(t)$ to equation (7.14) on $[-\tau, T]$ which obeys

$$\sup_{-\tau \leq t \leq T} \mathbb{E}U(x(t)) < \infty. \tag{7.20}$$

Proof. By Theorem 7.12, Assumption 7.8 guarantees the existence of the unique maximal solution $x(t)$ on $[-\tau, \sigma_\infty]$, where σ_∞ is the explosion time. We need to show $\sigma_\infty = T$ a.s. If this is not true, then we can find a $\varepsilon > 0$ such that

$$\mathbb{P}\{\sigma_\infty < T\} > 2\varepsilon.$$

For each integer $k \geq 1$, define the stopping time

$$\sigma_k = T \wedge \inf\{t \in [0,T] : |x(t)| \geq k\}.$$

Since $\sigma_k \to \sigma_\infty$ almost surely, we can find a sufficiently large integer k_0 for

$$\mathbb{P}\{\sigma_k < T\} > \varepsilon, \quad \forall k \geq k_0. \tag{7.21}$$

Fix any $k \geq k_0$. For any $0 \leq t \leq T$, by the generalised Itô formula and the conditions, we compute

$$\mathbb{E}U(x(t \wedge \sigma_k)) \leq \mathbb{E}V(x(t \wedge \sigma_k), t \wedge \sigma_k, r(t \wedge \sigma_k))$$
$$= \mathbb{E}V(x(0), 0, r_0) + \mathbb{E}\int_0^{t \wedge \sigma_k} LV(x(s), x(s - \delta(s)), s, r(s))ds$$
$$\leq \mathbb{E}V(\xi(0), 0, r_0) + \alpha T + \alpha \mathbb{E}\int_0^{t \wedge \sigma_k} [U(x(s)) + U(x(s - \delta(s)))]ds$$

$$\leq \beta + \alpha \int_0^t [\mathbb{E}U(x(s \wedge \sigma_k)) + \mathbb{E}U(x((s-\delta(s)) \wedge \sigma_k))]ds$$

$$\leq \beta + 2\alpha \int_0^t \left(\sup_{-\tau \leq u \leq s} \mathbb{E}U(x(u \wedge \sigma_k)) \right) ds,$$

where $\beta = \mathbb{E}V(\xi(0), 0, r_0) + \alpha T$. Since the right-hand-side term is increasing in t, we must have

$$\sup_{0 \leq u \leq t} \mathbb{E}U(x(u \wedge \sigma_k)) \leq \beta + 2\alpha \int_0^t \left(\sup_{-\tau \leq u \leq s} \mathbb{E}U(x(u \wedge \sigma_k)) \right) ds.$$

So

$$\sup_{-\tau \leq u \leq t} \mathbb{E}U(x(u \wedge \sigma_k))$$

$$\leq \sup_{-\tau \leq u \leq 0} \mathbb{E}U(\xi(u)) + \sup_{0 \leq u \leq t} \mathbb{E}U(x(u \wedge \sigma_k))$$

$$\leq \beta_1 + 2\alpha \int_0^t \left(\sup_{-\tau \leq u \leq s} \mathbb{E}U(x(u \wedge \sigma_k)) \right) ds,$$

where $\beta_1 = \sup_{-\tau \leq u \leq 0} \mathbb{E}U(\xi(u)) + \beta$. The Gronwall inequality implies

$$\sup_{-\tau \leq u \leq T} \mathbb{E}U(x(u \wedge \sigma_k)) \leq \beta_1 e^{2\alpha T}. \tag{7.22}$$

In particular,

$$\mathbb{E}U(x(T \wedge \sigma_k)) \leq \beta_1 e^{2\alpha T}.$$

This implies

$$\mathbb{E}\Big(I_{\{\sigma_k < T\}} U(x(\sigma_k))\Big) \leq \beta_1 e^{2\alpha T}. \tag{7.23}$$

On the other hand, if we define

$$h_k = \inf\{U(x) : |x| \geq k\},$$

then $h_k \to \infty$ as $k \to \infty$ by (7.17). It now follows from (7.23) and (7.21) that

$$\beta_1 e^{2\alpha T} \geq h_k \mathbb{P}\{\sigma_k \leq T\} \geq \varepsilon h_k.$$

Letting $k \to \infty$ yields a contradiction so we must have $\sigma_\infty = T$ a.s. Finally, let $k \to \infty$ in (7.22) we obtain

$$\sup_{-\tau \leq u \leq T} \mathbb{E}U(x(u)) \leq \beta_1 e^{2\alpha T} < \infty.$$

which is the assertion (7.20). □

This theorem covers many important nonlinear SDDEs including the case specified in Exercise 7.5 and the case discussed in the following section.

7.4 Moment Properties

Let us proceed to discuss the moment properties of the solutions to the SDDE (7.14).

Theorem 7.14 *Let Assumption 7.8 hold. Assume that there is a positive constant K such that*

$$[x^T f(x,y,t,i)] \vee |g(x,y,t,i)|^2 \leq K(1+|x|^2+|y|^2) \qquad (7.24)$$

for all $(x,y,t,i) \in \mathbb{R}^n \times \mathbb{R}^n \times [0,T] \times \mathbb{S}$. Let $p \geq 2$ and assume the initial data $\xi \in L^p_{\mathcal{F}_0}([-\tau,0];\mathbb{R}^n)$. Then the solution $x(t)$ of equation (7.14) obeys

$$\mathbb{E}\left(\sup_{-\tau \leq t \leq T} |x(t)|^p\right) \leq 2^{\frac{1}{2}(p+2)}(1+\mathbb{E}\|\xi\|^p)e^{2Kp(10p+1)T}. \qquad (7.25)$$

Proof. Let $V(x,t,i) = U(x) = (1+|x|^2)^{\frac{1}{2}p}$. It is not difficult to show that

$$LV(x,y,t,i) \leq Kp(p+1)(U(x)+U(y)). \qquad (7.26)$$

By Theorem 7.13, equation (7.14) has a unique solution $x(t)$ on $[-\tau,T]$ which obeys that $\sup_{-\tau \leq t \leq T} \mathbb{E}|x(t)|^p < \infty$. Now, by the Itô formula, we can show that

$$\mathbb{E}\left(\sup_{0 \leq s \leq t} U(x(s))\right) \leq \mathbb{E}U(x(0)) + J(t)$$
$$+ Kp(p+1)\mathbb{E}\int_0^t [U(x(s))+U(x(s-\delta(s)))]ds, \qquad (7.27)$$

where

$$J(t) = p\mathbb{E}\left(\sup_{0 \leq s \leq t} \int_0^s (1+|x(s)|^2)^{\frac{1}{2}(p-2)}\right.$$
$$\left.\times x^T(u)g(x(u),x(u-\delta(u)),u,r(u))dB(u)\right).$$

But, by the Burkholder–Davis–Gundy inequality (see Exercise 2.5), we compute

$$J(t) \leq 3p\mathbb{E}\bigg(\int_0^t (1+|x(u)|^2)^{(p-1)}|g(x(u),x(u-\delta(u)),u,r(u))|^2 du\bigg)^{\frac{1}{2}}$$

$$\leq 3p\mathbb{E}\bigg(\sup_{0\leq u\leq t} U(x(u)) \int_0^t 2K[U(x(u))+U(x(u-\delta(u)))]du\bigg)^{\frac{1}{2}}$$

$$\leq \tfrac{1}{2}\mathbb{E}\bigg(\sup_{0\leq u\leq t} U(x(u))\bigg) + 9Kp^2\mathbb{E}\int_0^t [U(x(u))+U(x(u-\delta(u)))]du.$$

Substituting this into (7.27) gives

$$\mathbb{E}\bigg(\sup_{0\leq s\leq t} U(x(s))\bigg)$$
$$\leq 2\mathbb{E}U(x(0)) + 2Kp(10p+1)\int_0^t \mathbb{E}\bigg(\sup_{-\tau\leq u\leq s} U(x(u))\bigg)ds \quad (7.28)$$

Noting that $U(x) \leq 2^{\frac{1}{2}(p-2)}(1+|x|^p)$, we hence have

$$\mathbb{E}\bigg(\sup_{-\tau\leq s\leq t} U(x(s))\bigg)$$
$$\leq \mathbb{E}\bigg(\sup_{-\tau\leq s\leq 0} U(x(s))\bigg) + \mathbb{E}\bigg(\sup_{0\leq s\leq t} U(x(s))\bigg)$$
$$\leq 2^{\frac{1}{2}(p+2)}(1+\mathbb{E}\|\xi\|^p) + 2Kp(10p+1)\int_0^t \mathbb{E}\bigg(\sup_{-\tau\leq u\leq s} U(x(s))\bigg)ds.$$

Finally, the assertion (7.25) follows from the Gronwall inequality. □

If the linear growth condition, i.e. Assumption 7.9 holds, we compute

$$x^T f(x,y,t,i) \leq |x||f(x,y,t,i)| \leq \frac{\sqrt{K}}{2}|x|^2 + \frac{1}{2\sqrt{K}}|f(x,y,t,i)|^2$$
$$\leq \frac{\sqrt{K}}{2}|x|^2 + \frac{\sqrt{K}}{2}(1+|x|^2+|y|^2) \leq \sqrt{K}(1+|x|^2+|y|^2).$$

Without loss of generality, we may let $K \geq 1$ and observe that Assumption 7.9 implies (7.24). We hence have the following useful corollary.

Corollary 7.15 *Let Assumptions 7.8 and 7.9 hold with $K \geq 1$. Let $p \geq 2$ and assume the initial data $\xi \in L^p_{\mathcal{F}_0}([-\tau,0];\mathbb{R}^n)$. Then the solution of the SDDE (7.14) obeys (7.25).*

Let us now prove a theorem which reveals the important moment continuity property of the solution.

Theorem 7.16 *Let Assumptions 7.8 and 7.9 hold with $K \geq 1$. Let $p \geq 2$ and assume the initial data $\xi \in L^p_{\mathcal{F}_0}([-\tau,0];\mathbb{R}^n)$. Then the solution of the SDDE (7.14) obeys (7.25).*

$$\mathbb{E}|x(t) - x(s)|^p \leq C(t-s)^{p/2} \quad \forall 0 \leq s < t < T,$$

where C is a positive constant.

Proof. It is easy to show

$$\mathbb{E}|x(t) - x(s)|^p \leq 2^{p-1}\mathbb{E}\left|\int_s^t f(x(u), x(u-\delta(u)), u, r(u))du\right|^p$$
$$+ 2^{p-1}\mathbb{E}\left|\int_s^t g(x(u), x(u-\delta(u)), u, r(u))dB(r)\right|^p.$$

But, by the Hölder inequality and Corollary 7.15, we compute

$$\mathbb{E}\left|\int_s^t f(x(u), x(u-\delta(u)), u, r(u))du\right|^p$$
$$\leq (t-s)^{p-1}\mathbb{E}\int_s^t |f(x(u), x(u-\delta(u)), u, r(u))|^p dr$$
$$\leq (t-s)^{p-1}\mathbb{E}\int_s^t K^{p/2}[1 + |x(u)|^2 + |x(u-\delta(u))|^2]^{p/2}dr$$
$$\leq 3^{(p-2)/2}K^{p/2}(t-s)^p\left[1 + 2\mathbb{E}\left(\sup_{-\tau \leq u \leq T}|x(u)|^p\right)\right]$$
$$\leq C_1(t-s)^p.$$

Moreover, by Theorem 2.11, we can similarly compute

$$\mathbb{E}\left|\int_s^t g(x(u), x(u-\delta(u)), u, r(u))dB(r)\right|^p$$
$$\leq [p(p-1)/2]^{p/2}(t-s)^{(p-2)/2}\mathbb{E}\int_s^t |g(x(u), x(u-\delta(u)), u, r(u))|^p du$$
$$\leq C_2(t-s)^{p/2}.$$

Combining the three inequalities above yields the required assertion. □

7.5 Asymptotic Boundedness

If we let $T = \infty$, the SDDE (7.14) becomes an SDDE on $t \in \mathbb{R}_+$, namely

$$dx(t) = f(x(t), x(t-\delta(t)), t, r(t))dt + g(x(t), x(t-\delta(t)), t, r(t))dB(t). \quad (7.29)$$

We are interested in the properties of the solutions as $t \to \infty$.

Of course, δ is now defined on the entire $\mathbb{R}_+$, namely $\delta : \mathbb{R}_+ \to [0, \tau]$. Moreover, we assume that δ is differentiable and its derivative is bounded by a constant $\bar\delta \in [0, 1)$, that is

$$\dot\delta(t) \leq \bar\delta, \quad \forall t \geq 0. \quad (7.30)$$

In the case when $\delta(t) \equiv \tau$, we have $\bar\delta = 0$. Let $p > 0$ and let initial data $\{x(\theta) : -\tau \leq \theta \leq 0\} = \xi \in L^p_{\mathcal{F}_0}([-\tau, 0]; \mathbb{R}^n)$. We will fix initial state r_0 arbitrarily for the Markov chain. Assume that the SDDE (7.29) has a unique solution on $t \geq -\tau$ which is denoted by $x(t; \xi)$ to emphasise ξ but not r_0.

Definition 7.17 The SDDE (7.29) is said to be asymptotically bounded in pth moment if there is a positive constant C such that

$$\limsup_{t \to \infty} \mathbb{E}|x(t; \xi)|^p \leq C, \quad \forall \xi \in L^p_{\mathcal{F}_0}([-\tau, 0]; \mathbb{R}^n).$$

When $p = 2$ we say that the SDDE is asymptotically bounded in mean square.

Theorem 7.18 *Let (7.30) hold. Assume that there is a function $V \in C^{2,1}(\mathbb{R}^n \times \mathbb{R}_+ \times \mathbb{S}; \mathbb{R}_+)$ and positive constants $p, \alpha, c_1, c_2, \lambda_1, \lambda_2$ such that $\lambda_1 > \lambda_2/(1 - \bar\delta)$,*

$$c_1|x|^p \leq V(x, t, i) \leq c_2|x|^p \quad (7.31)$$

and

$$LV(x, y, t, i) \leq -\lambda_1|x|^p + \lambda_2|y|^2 + \alpha \quad (7.32)$$

for all $x, y \in \mathbb{R}^n$, $t \geq 0$ and $i \in \mathbb{S}$. Then the SDDE (7.29) is asymptotically bounded in pth moment. More precisely,

$$\limsup_{t \to \infty} \mathbb{E}|x(t; \xi)|^p \leq \frac{\alpha}{c_1 \lambda}, \quad \forall \xi \in L^p_{\mathcal{F}_0}([-\tau, 0]; \mathbb{R}^n), \quad (7.33)$$

where $\lambda \in (0, \lambda_1 - \lambda_2/(1-\bar{\delta}))$ is the unique root to the equation

$$\frac{\lambda_2 e^{\lambda \tau}}{1-\bar{\delta}} = \lambda_1 - \lambda c_2. \tag{7.34}$$

Proof. Fix any ξ and write $x(t;\xi) = x(t)$. Applying the generalised Itô formula to $e^{\lambda t} V(x(t), t, r(t))$ and then using conditions (7.31)–(7.32) we can show that

$$c_1 e^{\lambda t} \mathbb{E}|x(t)|^p \leq c_2 \mathbb{E}|\xi(0)|^p + \frac{\alpha}{\lambda} e^{\lambda t}$$
$$+ \mathbb{E} \int_0^t e^{\lambda s} \Big[-(\lambda_1 - \lambda c_2)|x(s)|^p + \lambda_2 |x(s-\delta(s))|^p \Big] ds. \tag{7.35}$$

But, by (7.30), we compute,

$$\int_0^t e^{\lambda s} |x(s-\delta(s))|^p ds \leq e^{\lambda \tau} \int_0^t e^{\lambda (s-\delta(s))} |x(s-\delta(s))|^p ds$$
$$\leq \frac{e^{\lambda \tau}}{1-\bar{\delta}} \int_{-\tau}^t e^{\lambda u} |x(u)|^p du.$$

Substituting this into (7.35) and then making use of (7.34) we obtain that

$$c_1 e^{\lambda t} \mathbb{E}|x(t)|^p \leq c_2 \mathbb{E}|\xi(0)|^p + \frac{\alpha}{\lambda} e^{\lambda t} + \frac{e^{\lambda \tau}}{1-\bar{\delta}} \mathbb{E} \int_{-\tau}^0 e^{\lambda u} \lambda_2 |x(u)|^p du$$
$$\leq c_2 \mathbb{E}|\xi(0)|^p + \frac{\alpha}{\lambda} e^{\lambda t} + \frac{\tau \lambda_2 e^{\lambda \tau}}{1-\bar{\delta}} \mathbb{E}\|\xi\|^p. \tag{7.36}$$

Dividing both sides by $c_1 e^{\lambda t}$ and then letting $t \to \infty$ we obtain the required assertion (7.33). □

The following result can be used much more easily.

Theorem 7.19 *Let (7.30) hold and $p \geq 2$, $\alpha > 0$. Assume that for each $i \in \mathbb{S}$, there is a pair of real numbers β_i and σ_i such that*

$$x^T f(x,y,t,i) + \frac{p-1}{2}|g(x,y,t,i)|^2 \leq \alpha + \beta_i |x|^2 + \sigma_i |y|^2 \tag{7.37}$$

for all $(x,y,t) \in \mathbb{R}^n \times \mathbb{R}^n \times \mathbb{R}_+$. Assume that

$$\mathcal{A} := -\mathrm{diag}(p\beta_1, \cdots, p\beta_N) - \Gamma$$

is a nonsingular M-matrix so (by Theorem 2.10)

$$(q_1, \cdots, q_N)^T := \mathcal{A}^{-1} \vec{1} \gg 0, \tag{7.38}$$

where $\vec{1} = (1, \cdots, 1)^T$. If
$$pq_i\sigma_i < 1 - \bar{\delta}, \quad \forall i \in \mathbb{S}, \tag{7.39}$$
then the SDDE (7.29) is asymptotically bounded in pth moment.

Proof. Define the function $V : \mathbb{R}^n \times \mathbb{R}_+ \times \mathbb{S} \to \mathbb{R}_+$ by
$$V(x, t, i) = q_i|x|^p.$$
Clearly V obeys (7.31) with $c_1 = \min_{i \in \mathbb{S}} q_i$ and $c_2 = \max_{1 \in \mathbb{S}} q_i$. To verify (7.32), we compute the operator LV from $\mathbb{R}^n \times \mathbb{R}^n \times \mathbb{R}_+ \times \mathbb{S}$ to $\mathbb{R}$ by condition (7.37) and (7.38) as follows:

$$\begin{aligned}
LV(x, y, t, i) &= pq_i|x|^{p-2}x^T f(x, y, t, i) + \tfrac{1}{2}pq_i|x|^{p-2}|g(x, y, t, i)|^2 \\
&\quad + \tfrac{1}{2}p(p-2)q_i|x|^{p-4}|x^T g(x, y, t, i)|^2 + \sum_{j=1}^N \gamma_{ij}q_j|x|^p \\
&\leq pq_i|x|^{p-2}\left[x^T f(x, y, t, i) + \frac{p-1}{2}|g(x, y, t, i)|^2\right] + \sum_{j=1}^N \gamma_{ij}q_j|x|^p \\
&\leq pq_i|x|^{p-2}(\alpha + \beta_i|x|^2 + \sigma_i|y|^2) + \sum_{j=1}^N \gamma_{ij}q_j|x|^p \\
&\leq \left(p\beta_i q_i + \sum_{i=1}^N \gamma_{ij}q_j\right)|x|^p + \alpha pq_i|x|^{p-2} + pq_i\sigma_i|x|^{p-2}|y| \\
&= -|x|^p + c_3|x|^{p-2} + \rho|x|^{p-2}|y|^2. \tag{7.40}
\end{aligned}$$

where $c_3 = \alpha p c_2$ and $\rho = 0 \vee \max_{i \in \mathbb{S}} pq_i\sigma_i$ and, by (7.39), $\rho < 1 - \bar{\delta}$. By the elementary inequality
$$a^c b^{1-c} \leq ac + b(1-c) \quad \forall a, b \geq 0, \ c \in [0, 1]$$
we have
$$|x|^{p-2}|y|^2 \leq \frac{p-2}{p}|x|^p + \frac{2}{p}|y|^p,$$
while, for any $\varepsilon > 0$
$$c_3|x|^{p-2} \leq (\varepsilon^{-(p-2)/2}c_3^{p/2})^{2/p}(\varepsilon|x|^p)^{(p-2)/p} \leq c_4 + \varepsilon|x|^p,$$
where $c_4 = \varepsilon^{-(p-2)/2}c_3^{p/2}$. Substituting this into (7.40) gives
$$LV(x, y, t, i) \leq -\lambda_1|x|^p + \lambda_2|y|^p + c_4.$$

where

$$\lambda_1 = 1 - \varepsilon - \frac{(p-2)\rho}{p}, \quad \lambda_2 = \frac{2\rho}{p}$$

Noting that

$$\frac{(p-2)\rho}{p} + \frac{\lambda_2}{1-\bar{\delta}} \le \frac{(p-2)\rho}{p(1-\bar{d}e)} + \frac{2\rho}{p(1-\bar{\delta})} = \frac{1}{p(1-\bar{\delta})} < 1,$$

we can choose ε sufficiently small for $\lambda_1 > \lambda_2/(1-\bar{\delta})$. Hence all the conditions of Theorem 7.18 have been verified and the conclusion follows of course. □

Let us now discuss an example to illustrate the theory.

Example 7.20 Let $B(t)$ be a scalar Brownian motion. Let $r(t)$ be a right-continuous Markov chain taking values in $\mathbb{S} = \{1, 2\}$ with generator

$$\Gamma = \begin{pmatrix} -\gamma_{12} & \gamma_{12} \\ \gamma_{21} & -\gamma_{21} \end{pmatrix},$$

where $\gamma_{12} > 0$ and $\gamma_{21} > 0$. Consider a scalar SDDE

$$dx(t) = f(x(t), r(t))dt + g(x(t-\tau), r(t))dB(t), \tag{7.41}$$

where, for $x, y \in \mathbb{R}$,

$$f(x, i) = \begin{cases} -x^3 & \text{if } i = 1, \\ x & \text{if } i = 2, \end{cases} \quad g(y, i) = \begin{cases} 1 + \theta_1^2 y & \text{if } i = 1, \\ \theta_2 y & \text{if } i = 2, \end{cases}$$

and both θ_1, θ_2 are real numbers. We assume that

$$\gamma_{21} > p \ge 2. \tag{7.42}$$

Compute, for $x, y \in \mathbb{R}$,

$$xf(x,1) + \frac{p-1}{2}|g(y,1)|^2 = -x^4 + \frac{p-1}{2}|1+\theta_1 y|^2 \le -x^4 + (p-1)(\theta_1^2 y^2 + 1).$$

But $-x^4 \le -2\rho x^2 + \rho^2$ for any $\rho > 0$. So

$$xf(x,1) + \frac{p-1}{2}|g(y,1)|^2 \le -2\rho x^4 + (p-1)\theta_1^2 y^2 + \rho^2 + p - 1.$$

Moreover,

$$xf(x,2) + \frac{p-1}{2}|g(y,2)|^2 = x^2 + \frac{p-1}{2}\theta_2^2 y^2.$$

The matrix $\mathcal{A}$ defined in Theorem 7.19 becomes

$$\mathcal{A} = \begin{pmatrix} 2p\rho + \gamma_{12} & -\gamma_{12} \\ -\gamma_{21} & -p + \gamma_{21} \end{pmatrix}.$$

Now, choose ρ sufficiently large for $\mathcal{A}$ to be a nonsingular M-matrix. Then

$$\mathcal{A}^{-1} = \frac{1}{H_\rho} \begin{pmatrix} -p + \gamma_{21} & \gamma_{12} \\ \gamma_{21} & 2p\rho + \gamma_{12} \end{pmatrix},$$

where

$$H_\rho = \det.(\mathcal{A}) = (\gamma_{12} + 2p\rho)(\gamma_{21} - p) - \gamma_{12}\gamma_{21} \to \infty \quad \text{as } \rho \to \infty.$$

Thus

$$\mathcal{A}^{-1}\vec{1} = \frac{1}{H_\rho} \begin{pmatrix} -p + \gamma_{21} + \gamma_{12} \\ \gamma_{21} + 2p\rho + \gamma_{12} \end{pmatrix}.$$

Noting $\bar{\delta} = 0$, condition (7.39) becomes

$$\frac{p(p-1)\theta_1^2}{H_\rho}(\gamma_{21} + \gamma_{12} - p) < 1, \quad \frac{p(p-1)\theta_2^2}{2H_\rho}(\gamma_{21} + \gamma_{12} + 2p\rho) < 1. \quad (7.43)$$

For any given θ_1, we can always make the first inequality above hold by choosing ρ sufficiently large. On the other hand, noting

$$\lim_{\rho \to \infty} \frac{p(p-1)\theta_2^2}{2H_\rho}(\gamma_{21} + \gamma_{12} + 2p\rho) = \frac{p(p-1)\theta_2^2}{\gamma_{21} - p},$$

we can make the second inequality in (7.43) hold if

$$\theta_2^2 < \frac{\gamma_{21} - p}{p(p-1)}. \quad (7.44)$$

By Theorem 7.19 we can therefore conclude that the SDDE (7.41) is asymptotically bounded in pth moment if conditions (7.42) and (7.44) are satisfied while there is no restriction on either θ_1 or γ_{12}.

7.6 Exponential Stability

Let us proceed to discuss the exponential stability of the SDDE (7.29). For the stability purpose, we impose the standing hypothesis in this section

$$f(0,0,t,i) \equiv 0, \quad g(0,0,t,i) \equiv 0 \quad \forall (t,i) \in \mathbb{R}_+ \times \mathbb{S},$$

whence the SDDE (7.29) admits a trivial solution $x(t) \equiv 0$.

Definition 7.21 The trivial solution of the SDDE (7.29), or simply, the SDDE (7.29) is said to be *exponentially stable in pth moment* if

$$\limsup_{t\to\infty} \frac{1}{t} \log(\mathbb{E}|x(t;\xi)|^p) < 0$$

for any $\xi \in L^p_{\mathcal{F}_0}([-\tau,0]; \mathbb{R}^n)$. When $p = 2$, it is said to be *exponentially stable in mean square*. It is said to be *almost surely exponentially stable* if

$$\limsup_{t\to\infty} \frac{1}{t} \log(|x(t;\xi)|) < 0 \quad a.s.$$

for any $\xi \in C([-\tau,0]; \mathbb{R}^n)$.

The following is a key criterion on the pth moment exponential stability.

Theorem 7.22 *Let (7.30) hold. Assume that there is a function $V \in C^{2,1}(\mathbb{R}^n \times \mathbb{R}_+ \times \mathbb{S}; \mathbb{R}_+)$ and positive constants $p, c_1, c_2, \lambda_1, \lambda_2$ such that $\lambda_1 > \lambda_2/(1-\bar{\delta})$,*

$$c_1|x|^p \leq V(x,t,i) \leq c_2|x|^p \tag{7.45}$$

and

$$LV(x,y,t,i) \leq -\lambda_1|x|^p + \lambda_2|y|^2 \tag{7.46}$$

for all $x, y \in \mathbb{R}^n$, $t \geq 0$ and $i \in \mathbb{S}$. Then the SDDE (7.29) is asymptotically stable in pth moment. More precisely,

$$\limsup_{t\to\infty} \frac{1}{t} \log(\mathbb{E}|x(t;\xi)|^p) \leq -\lambda, \quad \forall \xi \in L^p_{\mathcal{F}_0}([-\tau,0]; \mathbb{R}^n), \tag{7.47}$$

where $\lambda \in (0, \lambda_1 - \lambda_2/(1-\bar{\delta}))$ is the unique root to equation (7.34).

Proof. The proof of Theorem 7.19 still works except α there is now zero so the assertion (7.47) follows from (7.36).   □

Similarly, by setting $\alpha = 0$ in Theorem 7.19 and its proof, we obtain a useful criterion.

Theorem 7.23 *Let (7.30) hold and $p \geq 2$. Assume that for each $i \in \mathbb{S}$, there is a pair of real numbers β_i and σ_i such that*

$$x^T f(x,y,t,i) + \frac{p-1}{2}|g(x,y,t,i)|^2 \leq \beta_i|x|^2 + \sigma_i|y|^2 \tag{7.48}$$

for all $(x,y,t) \in \mathbb{R}^n \times \mathbb{R}^n \times \mathbb{R}_+$. Assume that

$$\mathcal{A} := -\text{diag}(p\beta_1, \cdots, p\beta_N) - \Gamma$$

is a nonsingular M-matrix so $(q_1, \cdots, q_N)^T := \mathcal{A}^{-1}\vec{1} \gg 0$. If

$$pq_i\sigma_i < 1 - \bar{\delta}, \quad \forall i \in \mathbb{S}, \tag{7.49}$$

then the SDDE (7.29) is asymptotically stable in pth moment.

The following theorem shows that the pth moment exponential stability implies the almost sure exponential stability.

Theorem 7.24 *Assume that there is a $K > 0$ such that*

$$|f(x,y,t,i)| \vee |g(x,y,t,i)| \leq K(|x| + |y|) \tag{7.50}$$

for all $(x,y,t,i) \in \mathbb{R}^n \times \mathbb{R}^n \times \mathbb{R}_+ \times \mathbb{S}$. Let $p \geq 1$, $\lambda > 0$ and the initial data $\xi \in C([-\tau,0];\mathbb{R}^n)$. If the solution of the SDDE (7.29) obeys

$$\limsup_{t \to \infty} \frac{1}{t} \log(\mathbb{E}|x(t;\xi)|^p) \leq -\lambda, \tag{7.51}$$

then

$$\limsup_{t \to \infty} \frac{1}{t} \log(|x(t;\xi)|) \leq -\frac{\lambda}{p} \quad a.s. \tag{7.52}$$

Proof. Write $x(t;\xi) = x(t)$. Let $\varepsilon \in (0, \lambda/2)$ be arbitrary. By (7.51) and Theorem 7.14, there is a positive constant κ such that

$$\mathbb{E}|x(t)|^p \leq \kappa e^{-(\lambda-\varepsilon)t} \quad \text{on } t \geq -\tau. \tag{7.53}$$

For each integer $k \geq 1$, compute

$$\mathbb{E}\left[\sup_{(k-1)\tau \leq t \leq k\tau} |x(t)|^p\right] \leq 3^{p-1}\mathbb{E}|x((k-1)\tau)|^p$$
$$+ 3^{p-1}\mathbb{E}\left(\int_{(k-1)\tau}^{k\tau} |f(x(s), x(s-\delta(s)), s, r(s))|ds\right)^p$$
$$+ 3^{p-1}\mathbb{E}\left[\sup_{(k-1)\tau \leq t \leq k\tau} \left|\int_{(k-1)\tau}^{t} g(x(s), x(s-\delta(s)), s, r(s))dB(s)\right|^p\right]. \tag{7.54}$$

By (7.53),

$$\mathbb{E}|x((k-1)\tau)|^p \leq \kappa e^{-(\lambda-\varepsilon)(k-1)\tau}. \tag{7.55}$$

By condition (7.50) and Hölder's inequality, compute that

$$\mathbb{E}\left(\int_{(k-1)\tau}^{k\tau} |f(x(s), x(s-\delta(s)), s, r(s))| ds\right)^p$$
$$\leq \tau^{p-1} \mathbb{E}\int_{(k-1)\tau}^{k\tau} |f(x(s), x(s-\delta(s)), s, r(s))|^p ds$$
$$\leq 2^{p-1}\tau^{p-1} K^p \int_{(k-1)\tau}^{k\tau} (\mathbb{E}|x(s)|^p + \mathbb{E}|x(s-\delta(s))|^p) ds$$
$$\leq (2K\tau)^p \kappa e^{-(\lambda-\varepsilon)(k-1)\tau}. \qquad (7.56)$$

Moreover, by the Burkholder–Davis–Gundy inequality, we compute

$$\mathbb{E}\left[\sup_{(k-1)\tau \leq t \leq k\tau} \left|\int_{(k-1)\tau}^{t} g(x(s), x(s-\delta(s)), s, r(s)) dB(s)\right|^p\right]$$
$$\leq C\mathbb{E}\left(\int_{(k-1)\tau}^{k\tau} [|x(s)|^2 + |x(s-\delta(s))|^2] ds\right)^{\frac{p}{2}}$$
$$\leq C\mathbb{E}\left(\left[\sup_{(k-2)\tau \leq t \leq k\tau} |x(t)|\right]\int_{(k-1)\tau}^{k\tau} [|x(s)| + |x(s-\delta(s))|] ds\right)^{\frac{p}{2}}$$
$$\leq \theta\mathbb{E}\left[\sup_{(k-2)\tau \leq t \leq k\tau} |x(t)|^p\right] + C\mathbb{E}\left(\int_{(k-1)\tau}^{k\tau} [|x(s)| + |x(s-\delta(s))|] ds\right)^p$$
$$\leq \theta\mathbb{E}\left[\sup_{(k-2)\tau \leq t \leq (k-1)\tau} |x(t)|^p\right]$$
$$+ \theta\mathbb{E}\left[\sup_{(k-1)\tau \leq t \leq k\tau} |x(t)|^p\right] + Ce^{-(\lambda-\varepsilon)(k-1)\tau}, \qquad (7.57)$$

where $C = C(p, \tau, K, \theta) > 0$ is a constant independent of k which may change line by line and $\theta \in (0, 3^{-(p-1)})$ is so small that

$$\frac{3^{p-1}\varepsilon}{1 - 3^{p-1}\varepsilon} \leq e^{-(\lambda-\varepsilon)\tau}.$$

Substituting (7.55)–(7.57) into (7.54) yields

$$\mathbb{E}\left[\sup_{(k-1)\tau \leq t \leq k\tau} |x(t)|^p\right]$$
$$\leq Ce^{-(\lambda-\varepsilon)(k-1)\tau} + e^{-(\lambda-\varepsilon)\tau}\mathbb{E}\left[\sup_{(k-2)\tau \leq t \leq (k-1)\tau} |x(t)|^p\right]. \qquad (7.58)$$

By induction, we have

$$\mathbb{E}\left[\sup_{(k-1)\tau \le t \le k\tau} |x(t)|^p\right] \le kCe^{-(\lambda-\varepsilon)(k-1)\tau} + \|\xi\|^p e^{-(\lambda-\varepsilon)k\tau}$$
$$\le C(k+1)e^{-(\lambda-\varepsilon)k\tau}. \tag{7.59}$$

Hence

$$\mathbb{P}\left\{\omega: \sup_{(k-1)\tau \le t \le k\tau} |x(t)| > e^{-(\lambda-2\varepsilon)k\tau/p}\right\} \le C(k+1)e^{-\varepsilon k\tau}.$$

In view of the well-known Borel–Cantelli lemma (see Lemma 1.2), we see that for almost all $\omega \in \Omega$,

$$\sup_{(k-1)\tau \le t \le k\tau} |x(t)| \le e^{-(\lambda-2\varepsilon)k\tau/p} \tag{7.60}$$

holds for all but finitely many k. Hence there exists a $k_0(\omega)$, for all $\omega \in \Omega$ excluding a $\mathbb{P}$-null set, for which (7.60) holds whenever $k \ge k_0$. Consequently, for almost all $\omega \in \Omega$,

$$\frac{1}{t}\log(|x(t)|) \le -\frac{(\lambda-2\varepsilon)k\tau}{pt} \le -\frac{\lambda-2\varepsilon}{p}$$

if $(k-1)\tau \le t \le k\tau$. Therefore

$$\limsup_{t\to\infty} \frac{1}{t}\log(|x(t)|) \le -\frac{\lambda-2\varepsilon}{p} \quad a.s.$$

and the required assertion (7.52) follows by letting $\varepsilon \to 0$. $\square$

Combining Theorems 7.22 and 7.24 together we obtain the following criterion on the almost sure exponential stability.

Theorem 7.25 *Under the conditions of Theorem 7.22, if, moreover, $p \ge 1$ and (7.50) holds, then the SDDE (7.29) is almost surely exponentially stable.*

Example 7.26 As an example, let us consider an n-dimensional linear SDDE

$$dx(t) = A(r(t))x(t)dt + D(r(t))x(t-\delta(t))dB(t) \quad \text{on } t \ge 0, \tag{7.61}$$

where B is a scalar Brownian motion and $A(i)$, $D(i)$, $i \in \mathbb{S}$, are $n \times n$ constant matrices. We will write $A(i) = A_i$ and $D(i) = D_i$. Accordingly, condition (7.48) becomes

$$x^T A_i x + \frac{p-1}{2}|D_i y|^2 \leq \tfrac{1}{2}\lambda_{\max}(A_i + A_i^T)|x|^2 + \tfrac{1}{2}(p-1)\|D_i\|^2|y|^2.$$

Assume that

$$\mathcal{A} := -\mathrm{diag}(\tfrac{1}{2}p\lambda_{\max}(A_1 + A_1^T), \cdots, \tfrac{1}{2}p\lambda_{\max}(A_N + A_N^T)) - \Gamma$$

is a nonsingular M-matrix and set

$$(q_1, \cdots, q_N)^T := \mathcal{A}^{-1}\vec{1} \gg 0.$$

If, moreover,

$$\tfrac{1}{2}p(p-1)q_i\|D_i\|^2 < 1 - \bar{\delta}, \quad \forall i \in \mathbb{S},$$

by Theorems 7.22 and 7.24, we can then conclude that the linear SDDE (7.61) is asymptotically stable in pth moment as well as almost surely.

7.7 Approximate Solutions

Most of SDDEs with Markovian switching do not have explicit solutions. Several numerical methods e.g. the Euler–Maruyama and the backward Euler have been developed to obtain approximate solutions to the SDDEs. Due to the page limit we will only discuss the most useful Euler–Maruyama method in this section. Moreover, to make our theory more understandable, we will only discuss the case of constant time lag. That is, we consider the SDDE with Markovian switching of the form

$$dx(t) = f(x(t), x(t-\tau), r(t))dt + g(x(t), x(t-\tau), r(t))dB(t) \qquad (7.62)$$

on $t \in [0, T]$ with initial data $\{x(\theta) : -\tau \leq \theta \leq 0\} = \xi \in L^2_{\mathcal{F}_0}([-\tau, 0]; \mathbb{R}^n)$.

Let us now define the Euler–Maruyama (EM) approximate solution to equation (7.62). To cope with the time lag, we choose a step size $\Delta > 0$ to be a fraction of τ, whence $\bar{k} := T/\Delta$ is a positive integer. Let $\{r_k^\Delta\}_{k \geq 0}$ be the same discrete Markov chain as defined in section 4.2. Let $t_k = k\Delta$ for $k \geq -\bar{k}$. Set $X_k = \xi(t_k)$ for $k = -\bar{k}, -\bar{k}+1, \cdots, 0$ and then compute the discrete approximations $X_k \approx x(t_k)$ for $k \geq 1$ by

$$X_{k+1} = X_k + f(X_k, X_{k-\bar{k}}, r_k^\Delta)\Delta + g(X_k, X_{k-\bar{k}}, r_k^\Delta)\Delta B_k, \qquad (7.63)$$

where $\Delta B_k = B(t_{k+1}) - B(t_k)$. Let

$$\bar{r}(t) = r_k^\Delta \quad \text{for } t \in [t_k, t_{k+1}), \ k \geq 0$$

and
$$\bar{X}(t) = X_k \quad \text{for } t \in [t_k, t_{k+1}), \ k \geq -\bar{k}.$$

Define the continuous EM approximate solution by setting $X(t) = \xi(t)$ for $t \in [-\tau, 0]$ and forming

$$X(t) = X_0 + \int_0^t f(\bar{X}(s), \bar{X}(s-\tau), \bar{r}(s)) ds$$
$$+ \int_0^t g(\bar{X}(s), \bar{X}(s-\tau), \bar{r}(s)) dB(s) \quad (7.64)$$

for $t \in [0, T]$. Note that $X(t_k) = \bar{X}(t_k) = X_k$ for $k \geq -\bar{k}$, that is $X(t)$ and $\bar{X}(t)$ coincide with the discrete solution at the gridpoints.

To show the convergence of the EM approximate solution to the true solution we impose the following global Lipschitz condition.

Assumption 7.27 There is a positive constant $\bar{K}$ such that

$$|f(x,y,i) - f(\bar{x},\bar{y},i)|^2 \vee |g(x,y,i) - g(\bar{x},\bar{y},i)|^2 \leq \bar{K}(|x-\bar{x}|^2 + |y-\bar{y}|^2)$$

for all $x, y, \bar{x}, \bar{y} \in \mathbb{R}^n$ and $i \in \mathbb{S}$.

Moreover, we will need another condition on the continuity of the initial data.

Assumption 7.28 There is a positive constant $\hat{K}$ such that the initial data x obeys

$$\mathbb{E}|x(u) - x(v)|^2 \leq \hat{K}(v-u), \quad -\tau \leq u < v \leq 0.$$

Let us first present a useful lemma.

Lemma 7.1 *Under Assumption 7.27, there is a constant H, which is dependent on $T, \bar{K}, \xi$ but independent of Δ, such that the exact solution and the EM approximate solution to equation (7.62) have the property that*

$$\mathbb{E}\left[\sup_{0 \leq t \leq T} |x(t)|^2\right] \vee \mathbb{E}\left[\sup_{0 \leq t \leq T} |X(t)|^2\right] \leq H. \quad (7.65)$$

Proof. We first observe that Assumption 7.27 implies

$$|f(x,y,i)|^2 \vee |g(x,y,i)|^2 \leq K(1 + |x|^2 + |y|^2) \quad (7.66)$$

for all $(x, y, i) \in \mathbb{R}^n \times \mathbb{R}^n \times \mathbb{S}$, where

$$K = 2\Big(\bar{K} \vee \max_{i \in \mathbb{S}}[|f(0,0,i)|^2 \vee |g(0,0,i)|^2]\Big).$$

Hence, it follows from Corollary 7.15 that the true solution obeys property (7.65). We therefore need to prove the assertion for the EM solution. It follows from (7.64) that, for $t \in [0, T]$,

$$|X(t)|^2 \leq 3\Big[|X_0|^p + T\int_0^t |f(\bar{X}(s), \bar{X}(s-\tau), \bar{r}(s))|^2 ds$$

$$+ |\int_0^t g(\bar{X}(s), \bar{X}(s-\tau), \bar{r}(s))dB(s)|^2\Big].$$

By the Doob martingale inequality and (7.66) we can then show that for any $0 \leq t_1 \leq T$,

$$\mathbb{E}\Big[\sup_{0 \leq t \leq t_1} |X(t)|^p\Big]$$

$$\leq 3\Big[\mathbb{E}|X_0|^2 + K(T+4)\int_0^{t_1}(1 + \mathbb{E}|\bar{X}(s)|^2 + \mathbb{E}|\bar{X}(s-\tau)|^2)ds\Big]$$

$$\leq 3\mathbb{E}\|\xi\|^2 + 3KT(T+4)(1 + \mathbb{E}\|\xi\|^2)$$

$$+ 6K(T+4)\int_0^{t_1} \mathbb{E}\Big[\sup_{0 \leq t \leq s} |X(r)|^2\Big] ds.$$

The Gronwall inequality then yields

$$\mathbb{E}\Big[\sup_{0 \leq t \leq T} |X(t)|^p\Big] \leq [3\mathbb{E}\|\xi\|^2 + 3KT(T+4)(1 + \mathbb{E}\|\xi\|^2)]e^{6KT(T+4)} := H$$

as required. □

The following theorem shows the strong convergence of the EM approximate solution to the the true solution.

Theorem 7.29 *Under Assumptions 7.27 and 7.28,*

$$\mathbb{E}\Big[\sup_{0 \leq t \leq T} |X(t) - x(t)|^2\Big] \leq C\Delta + o(\Delta), \qquad (7.67)$$

where C is a positive constant independent of Δ.

Proof. By the Hölder inequality and the Doob martingale inequality, it is not difficult to show that for $0 \leq t \leq T$,

$$\mathbb{E}\Big(\sup_{0 \leq s \leq t} |X(s) - x(s)|^2\Big)$$

$$\leq 2TE\int_0^t |f(\bar{X}(s), \bar{X}(s-\tau), \bar{r}(s)) - f(x(s), x(s-\tau), r(s))|^2 ds$$

$$+8\mathbb{E}\int_0^t |g(\bar{X}(s),\bar{X}(s-\tau),\bar{r}(s)) - g(x(s),x(s-\tau),r(s))|^2 ds. \quad (7.68)$$

By Assumption 7.27, compute

$$\mathbb{E}\int_0^t |f(\bar{X}(s),\bar{X}(s-\tau),\bar{r}(s)) - f(x(s),x(s-\tau),r(s))|^2 ds$$

$$\leq 2\mathbb{E}\int_0^t |f(\bar{X}(s),\bar{X}(s-\tau),r(s)) - f(x(s),x(s-\tau),r(s))|^2 ds$$

$$+ 2\mathbb{E}\int_0^t |f(\bar{X}(s),\bar{X}(s-\tau),\bar{r}(s)) - f(\bar{X}(s),\bar{X}(s-\tau),r(s))|^2 ds$$

$$\leq 2\bar{K}\int_0^t \left[\mathbb{E}|\bar{X}(s) - x(s)|^2 + \mathbb{E}|\bar{X}(s-\tau) - x(s-\tau)|^2\right] ds$$

$$+ 2\mathbb{E}\int_0^T |f(\bar{X}(s),\bar{X}(s-\tau),\bar{r}(s)) - f(\bar{X}(s),\bar{X}(s-\tau),r(s))|^2 ds. \quad (7.69)$$

Let j be the integer part of T/Δ. Then

$$\mathbb{E}\int_0^T |f(\bar{X}(s),\bar{X}(s-\tau),\bar{r}(s)) - f(\bar{X}(s),\bar{X}(s-\tau),r(s))|^2 ds$$

$$= \sum_{k=0}^j \mathbb{E}\int_{t_k}^{t_{k+1}} |f(X_k, X_{k-\bar{k}}, r(t_k)) - f(X_k, X_{k-\bar{k}}, r(s))|^2 ds \quad (7.70)$$

with t_{j+1} being now set to be T. In the remainder of the proof, C will be a positive constant independent of Δ which may change line by line. By (7.66) we compute

$$\mathbb{E}\int_{t_k}^{t_{k+1}} |f(X_k, X_{k-\bar{k}}, r(t_k)) - f(X_k, X_{k-\bar{k}}, r(s))|^2 ds$$

$$\leq 2\mathbb{E}\int_{t_k}^{t_{k+1}} \left[|f(X_k, X_{k-\bar{k}}, r(t_k))|^2 + |f(X_k, X_{k-\bar{k}}, r(s))|^2\right] I_{\{r(s)\neq r(t_k)\}} ds$$

$$\leq C\mathbb{E}\int_{t_k}^{t_{k+1}} \left[1 + |X_k|^2 + |X_{k-\bar{k}}|^2\right] I_{\{r(s)\neq r(t_k)\}} ds$$

$$\leq C\int_{t_k}^{t_{k+1}} \mathbb{E}\Big[\mathbb{E}[(1+|X_k|^2+|X_{k-\bar{k}}|^2)I_{\{r(s)\neq r(t_k)\}}|r(t_k)]\Big] ds$$

$$= C\int_{t_k}^{t_{k+1}} \mathbb{E}\Big[\mathbb{E}[(1+|X_k|^2+|X_{k-\bar{k}}|^2)|r(t_k)]\mathbb{E}[I_{\{r(s)\neq r(t_k)\}}|r(t_k)]\Big] ds,$$

where in the last step we use the fact that X_k and $X_{k-\bar{k}}$ are conditionally independent of $I_{\{r(s)\neq r(t_k)\}}$ given the σ-algebra generated by $r(t_k)$. But, by the Markov property,

$$\begin{aligned}
\mathbb{E}\big[I_{\{r(s)\neq r(t_k)\}}|r(t_k)\big] &= \sum_{i\in\mathbb{S}} I_{\{r(t_k)=i\}}\mathbb{P}(r(s)\neq i|r(t_k)=i) \\
&= \sum_{i\in\mathbb{S}} I_{\{r(t_k)=i\}} \sum_{j\neq i}(\gamma_{ij}(s-t_k)+o(s-t_k)) \\
&\leq \Big(\max_{1\leq i\leq N}(-\gamma_{ii})\Delta + o(\Delta)\Big)\sum_{i\in\mathbb{S}} I_{\{r(t_k)=i\}} \\
&\leq C\Delta + o(\Delta). \quad (7.71)
\end{aligned}$$

So, by Lemma 7.1,

$$\begin{aligned}
\mathbb{E}\int_{t_k}^{t_{k+1}} &|f(X_k, X_{k-\bar{k}}, r(t_k)) - f(X_k, X_{k-\bar{k}}, r(s))|^2 ds \\
&\leq (C\Delta + o(\Delta))\int_{t_k}^{t_{k+1}}[1+\mathbb{E}|X_k|^2 + \mathbb{E}|X_{k-\bar{k}}|^2]ds \\
&\leq \Delta(C\Delta + o(\Delta)).
\end{aligned}$$

Substituting this into (7.70) gives

$$\begin{aligned}
\mathbb{E}\int_0^T &|f(\bar{X}(s),\bar{X}(s-\tau),\bar{r}(s)) - f(\bar{X}(s),\bar{X}(s-\tau),r(s))|^2 ds \\
&\leq C\Delta + o(\Delta). \quad (7.72)
\end{aligned}$$

Moreover, by Assumption 7.28, we compute

$$\begin{aligned}
\int_0^t \mathbb{E}|\bar{X}(s-\tau) - x(s-\tau)|^2 ds &= \int_{-\tau}^{t-\tau} \mathbb{E}|\bar{X}(s) - x(s)|^2 ds \\
&\leq \int_{-\tau}^0 \mathbb{E}|\bar{X}(s) - \xi(s)|^2 ds + \int_0^t \mathbb{E}|\bar{X}(s) - x(s)|^2 ds \\
&= \sum_{k=-\bar{k}}^{-1}\int_{t_k}^{t_{k+1}} \mathbb{E}|\xi(t_k) - \xi(s)|^2 ds + \int_0^t \mathbb{E}|\bar{X}(s) - x(s)|^2 ds \\
&\leq \tau \hat{K}\Delta + \int_0^t \mathbb{E}|\bar{X}(s) - x(s)|^2 ds. \quad (7.73)
\end{aligned}$$

Combining (7.69), (7.72) and (7.73) together we obtain that

$$\mathbb{E}\int_0^t |f(\bar{X}(s), \bar{X}(s-\tau), \bar{r}(s)) - f(x(s), x(s-\tau), r(s))|^2 ds$$
$$\leq 4\bar{K}\int_0^t \mathbb{E}|\bar{X}(s) - x(s)|^2 ds + C\Delta + o(\Delta). \tag{7.74}$$

Similarly, we can show that

$$\mathbb{E}\int_0^t |g(\bar{X}(s), \bar{X}(s-\tau), \bar{r}(s)) - g(x(s), x(s-\tau), r(s))|^2 ds$$
$$\leq 4\bar{K}\int_0^t \mathbb{E}|\bar{X}(s) - x(s)|^2 ds + C\Delta + o(\Delta). \tag{7.75}$$

Substituting (7.74) and (7.75) into (7.68) yields that

$$\mathbb{E}\left(\sup_{0\leq s\leq t} |X(s) - x(s)|^2\right) \leq C\int_0^t \mathbb{E}|\bar{X}(s) - x(s)|^2 ds + C\Delta + o(\Delta). \tag{7.76}$$

Note

$$\mathbb{E}|\bar{X}(s) - x(s)|^2 \leq 2\mathbb{E}|X(s) - x(s)|^2 + 2\mathbb{E}|X(s) - \bar{X}(s)|^2. \tag{7.77}$$

But, for $s \in [0,t]$, let k_s be the integer part of s/Δ. It then follows from (7.64) and (7.66) as well as Lemma 7.1 that

$$\mathbb{E}|\bar{X}(s) - X(s)|^2 \leq C\mathbb{E}\left[(1 + |X_{k_s}|^2 + |X_{k_s - \bar{k}}|^2)(\Delta^2 + |B(s) - B(t_{k_s})|^2)\right]$$
$$\leq C\Delta. \tag{7.78}$$

Putting (7.77) and (7.78) into (7.76) we see that

$$\mathbb{E}\left(\sup_{0\leq s\leq t} |X(s) - x(s)|^2\right) \leq C\int_0^t \mathbb{E}|X(s) - x(s)|^2 ds + C(\Delta + o(\Delta))$$
$$\leq C\int_0^t \mathbb{E}\left(\sup_{0\leq r\leq s} |X(r) - x(r)|^2\right) ds + C\Delta + o(\Delta)$$

and the required result (7.67) follows from the Gronwall inequality. □

The techniques developed in Section 4.2 for SDEs can be used to deal with SDDEs under the local Lipschitz condition. The other numerical methods introduced in Chapter 4 for SDEs e.g. the backward Euler method can also be used to deal with SDDEs under more general conditions. But we have to leave the details to the reader due to the page limit.

7.8 Exercises

7.1 By referring e.g. [Mao (1997)], prove Theorem 7.3.

7.2 Prove Theorem 7.10.

7.3 Show (7.16).

7.4 Prove the uniqueness part of Theorem 7.12.

7.5 Consider the SDDE (7.14) on $t \geq 0$ with initial data (7.9). Let Assumption 7.8 hold. If, for some $K > 0$,

$$2x^T f(x, y, t, i) + |g(x, y, t, i)|^2 \leq K(1 + |x|^2 + |y|^2)$$

for all $(x, y, t, i) \in \mathbb{R}^n \times \mathbb{R}^n \times \mathbb{R}_+ \times \mathbb{S}$, show that equation (7.14) has a unique global solution on $[-\tau, \infty)$ and the solution has the property

$$\mathbb{E}|x(t)|^2 < \infty \quad \forall t \geq 0.$$

7.6 Show (7.26).

7.7 Show (7.35) in detail.

7.8 Let $B(t)$ be a scalar Brownian motion. Let $r(t)$ be a right-continuous Markov chain taking values in $\mathbb{S} = \{1, 2\}$ with generator

$$\Gamma = \begin{pmatrix} -\gamma_{12} & \gamma_{12} \\ 1 & -1 \end{pmatrix},$$

where $\gamma > 0$. Consider a linear SDDE

$$dx(t) = f(x(t), r(t))dt + g(x(t - \tau), r(t))dB(t),$$

where, for $x, y \in \mathbb{R}$,

$$f(x, i) = \begin{cases} 1 + x & \text{if } i = 1, \\ 1 - 4x & \text{if } i = 2, \end{cases} \quad g(y, i) = \begin{cases} 1 - y & \text{if } i = 1, \\ 1 + y & \text{if } i = 2. \end{cases}$$

Show that if $\gamma_{12} \geq 4$, the SDDE is asymptotically bounded in mean square.

7.9 Show that Theorem 7.24 still holds in the case when $p > 0$ and $\delta(t) \equiv \tau$. (You may refer to [Mao et al. (2000)].)

Chapter 8

Stochastic Functional Differential Equations with Markovian Switching

8.1 Introduction

In the previous chapter we discussed the situation where the underlying equation depends on its current state as well as a past one. In this chapter we will take a further step to consider a more general situation where the underlying equation depends on a finite-time segment of past states. That is, we consider a stochastic functional differential equation (SFDE) with Markovian switching of the form

$$dx(t) = f(x_t, t, r(t))dt + g(x_t, t, r(t))dB(t), \quad t \geq 0. \tag{8.1}$$

Here $r(t)$ and $B(t)$ are the same as before, $x_t = \{x(t+\theta) : -\tau \leq \theta \leq 0\}$ is the past segment while

$$f : C([-\tau, 0]; \mathbb{R}^n) \times \mathbb{R}_+ \times \mathbb{S} \to \mathbb{R}^n, \quad g : C([-\tau, 0]; \mathbb{R}^n) \times \mathbb{R}_+ \times \mathbb{S} \to \mathbb{R}^{n \times m}.$$

In this chapter we will discuss:

- the initial-value problem for equation (8.1);
- the existence and uniqueness of the solution;
- the properties of the solution.

In particular, we will develop the powerful Razumikhin technique to study the exponential stability.

8.2 Stochastic Functional Differential Equations

To make our theory more understandable, let us recall some fundamental concepts on SFDEs without Markovian switching. For a more detailed account the reader is referred to [Mao (1997)].

Consider the n-dimensional SFDE

$$dx(t) = f(x_t, t)dt + g(x_t, t)dB(t) \quad \text{on } t \geq 0, \tag{8.2}$$

where

$$f : C([-\tau, 0]; \mathbb{R}^n) \times \mathbb{R}_+ \to \mathbb{R}^n \quad \text{and} \quad g : C([-\tau, 0]; \mathbb{R}^n) \times \mathbb{R}_+ \to \mathbb{R}^{n \times m}.$$

The first question is the following: What is the initial-value problem for this equation? More specifically, what is the minimum amount of initial data that must be specified in order for equation (8.2) to define a stochastic process $x(t)$ on $t \geq 0$? A moment of reflection indicates that a stochastic process must be specified on the entire interval $[-\tau, 0]$. We therefore impose the initial data:

$$\{x(\theta) : -\tau \leq \theta \leq 0\} = \xi = \{\xi(\theta) : -\tau \leq \theta \leq 0\} \in L^2_{\mathcal{F}_0}([-\tau, 0]; \mathbb{R}^n). \tag{8.3}$$

The initial-value problem for equation (8.2) is now to find the solution satisfying the initial data (8.3). But, what is the solution?

Definition 8.1 An $\mathbb{R}^n$-valued stochastic process $x(t)$ on $t \geq -\tau$ is called a solution to equation (8.2) with initial data (8.3) if it has the following properties:

- it is continuous and $\{x_t\}_{t \geq 0}$ is $\mathcal{F}_t$-adapted $C([-\tau, 0]; \mathbb{R}^n)$-valued process;
- $\{f(x_t, t)\}_{t \geq 0} \in \mathcal{L}^1(\mathbb{R}_+; \mathbb{R}^n)$ and $\{g(x_t, t)\}_{t \geq 0} \in \mathcal{L}^2(\mathbb{R}_+; \mathbb{R}^{n \times m})$;
- (8.3) is satisfied and, for every $t \in \mathbb{R}_+$,

$$x(t) = \xi(0) + \int_0^t f(x_s, s)ds + \int_0^t g(x_s, s)dB(s) \quad a.s.$$

A solution $x(t)$ is said to be unique if any other solution $\bar{x}(t)$ is indistinguishable from it, that is

$$\mathbb{P}\{x(t) = \bar{x}(t) \text{ for all } -\tau \leq t < \infty\} = 1.$$

The following theorem is classical which shows that the local Lipschitz condition and the linear growth condition guarantee the existence and uniqueness of the solution to the SFDE.

Theorem 8.2 *Assume that for every integer $k \geq 1$, there exists a positive constant H_k such that, for all $t \geq 0$ and those $\varphi, \phi \in C([-\tau, 0]; \mathbb{R}^n)$ with $\|\varphi\| \vee \|\phi\| \leq k$,*

$$|f(\varphi, t) - f(\phi, t)|^2 \vee |g(\varphi, t) - g(\phi, t)|^2 \leq H_k \|\varphi - \phi\|^2.$$

Assume moreover that there is another positive constant H such that for all $(\varphi, t) \in C([-\tau, 0]; \mathbb{R}^n) \times \mathbb{R}_+$

$$|f(\varphi, t)|^2 \vee |g(\varphi, t)|^2 \leq H(1 + \|\varphi\|^2).$$

Then there exists a unique solution $x(t)$ to equation (8.2) with initial data (8.3). Moreover, the solution obeys

$$\mathbb{E}\left(\sup_{-\tau \leq t \leq T} |x(t)|^2\right) < \infty, \quad \forall T < 0.$$

8.3 SFDEs with Markovian Switching

Let us now return to the n-dimensional SFDE with Markovian switching of the form (8.1). We still impose the initial data (8.3).

Theorem 8.3 *Assume that for each $k = 1, 2, \cdots$, there is an $h_k > 0$ such that*

$$|f(\varphi_1, t, i) - f(\varphi_2, t, i)|^2 \vee |g(\varphi_1, t, i) - g(\varphi_2, t, i)|^2 \leq h_k \|\varphi_1 - \varphi_2\|^2$$

for all $t \geq 0$, $i \in \mathbb{S}$ and those $\varphi_1, \varphi_2 \in C([-\tau, 0]; \mathbb{R}^n)$ with $\|\varphi_1\| \vee \|\varphi_2\| \leq k$. Assume moreover that there is an $h > 0$ such that

$$|f(\varphi, t, i)|^2 \vee |g(\varphi, t, i)|^2 \leq h(1 + \|\varphi\|^2)$$

for all $\varphi \in C([-\tau, 0]; \mathbb{R}^n)$, $t \geq 0$ and $i \in \mathbb{S}$. Then there exists a unique solution $x(t)$ to equation (8.1) with initial data (8.3). Moreover, the solution obeys

$$\mathbb{E}\left[\sup_{-\tau \leq t \leq T} |x(t)|^2\right] < \infty, \quad \forall T > 0. \tag{8.4}$$

Proof. It is known that there is a sequence $\{\tau_k\}_{k \geq 0}$ of stopping times such that $0 = \tau_0 < \tau_1 < \cdots < \tau_k \to \infty$ and $r(t)$ is constant on every interval $[\tau_k, \tau_{k+1})$, i.e. for every $k \geq 0$

$$r(t) = r(\tau_k) \quad \text{on } \tau_k \leq t < \tau_{k+1}.$$

Let $T > 0$ arbitrary. We first consider equation (8.1) on $t \in [0, T \wedge \tau_1]$ which becomes

$$dx(t) = f(x_t, t, r(0))dt + g(x_t, t, r(0))dB(t) \tag{8.5}$$

with initial data $x_0 = \xi$. This is an SFDE without Markovian switching. By Theorem 8.2, it has a unique continuous solution on $[-\tau, T \wedge \tau_1]$ which obeys

$$\mathbb{E}\left[\sup_{-\tau \leq t \leq T \wedge \tau_1} |x(t)|^2\right] < \infty.$$

We next consider equation (8.1) on $t \in [T \wedge \tau_1, T \wedge \tau_2]$ which becomes

$$dx(t) = f(x_t, t, r(\tau_1))dt + g(x_t, t, r(\tau_1))dB(t) \tag{8.6}$$

with initial data $x_{T \wedge \tau_1}$ given by the solution of equation (8.5). By Theorem 8.2 again, we know equation (8.6) has a unique continuous solution on $[T \wedge \tau_1 - \tau, T \wedge \tau_2]$. Repeating this procedure we see that equation (8.1) has a unique solution $x(t)$ on $t \in [-\tau, T]$. As $T > 0$ is arbitrary, it must have a unique solution $x(t)$ on $t \geq -\tau$.

To show (8.4), it is not difficult to compute that, for $0 \leq t \leq T$,

$$\mathbb{E}\left[\sup_{0 \leq s \leq t} |x(s)|^2\right] \leq 3\mathbb{E}|\xi(0)|^2 + 3h(T+4)\mathbb{E}\int_0^t (1 + \|x_s\|^2)ds. \tag{8.7}$$

Consequently

$$\mathbb{E}\left[\sup_{-\tau \leq s \leq t} |x(s)|^2\right] \leq \mathbb{E}\|\xi\|^2 + \mathbb{E}\left[\sup_{0 \leq s \leq t} |x(s)|^2\right]$$

$$\leq 4\mathbb{E}\|\xi\|^2 + 3hT(T+4) + 3h(T+4)\int_0^t \mathbb{E}\left[\sup_{-\tau \leq s \leq u} |x(s)|^2\right]du.$$

An application of the Gronwall inequality implies

$$\mathbb{E}\left[\sup_{-\tau \leq s \leq T} |x(s)|^2\right] \leq e^{3hT(T+4)}\left[4\mathbb{E}\|\xi\|^2 + +3hT(T+4)\right]$$

and the required assertion (8.4) follows. □

If we strengthen the initial data by assuming $\xi \in L^p_{\mathcal{F}_0}([-\tau, 0]; \mathbb{R}^n)$ for some $p > 2$, we can show that

$$\mathbb{E}\left[\sup_{-\tau \leq t \leq T} |x(t)|^p\right] < \infty, \quad \forall T > 0. \tag{8.8}$$

In particular, if the initial data $\xi \in C^b_{\mathcal{F}_0}([-\tau, 0]; \mathbb{R}^n)$ (i.e. ξ is a bounded $\mathcal{F}_0$-measurable $C([-\tau, 0]; \mathbb{R}^n)$-valued random variable), then (8.8) holds for any $p > 0$. For convenience we will restrict initial data $\xi \in C^b_{\mathcal{F}_0}([-\tau, 0]; \mathbb{R}^n)$

in the remainder of this chapter so that the solution will have finite moment of any order.

8.4 Boundedness

To establish more properties, let us introduce an important operator associated with the SFDE (8.1). For a given $V \in C^{2,1}(\mathbb{R}^n \times [-\tau, \infty) \times \mathbb{S}; \mathbb{R}_+)$, define an operator $\mathcal{L}V$ from $C([-\tau, 0]; \mathbb{R}^n) \times \mathbb{R}_+ \times \mathbb{S}$ to $\mathbb{R}$ by

$$\mathcal{L}V(\varphi, t, i) = V_t(\varphi(0), t, i) + V_x(\varphi(0), t, i) f(\varphi, t, i)$$
$$+ \frac{1}{2}\text{trace}\left[g^T(\varphi, t, i) V_{xx}(\varphi(0), t, i) g(\varphi, t, i)\right]$$
$$+ \sum_{j=1}^{N} \gamma_{ij} V(\varphi(0), t, j). \qquad (8.9)$$

It should be emphasised that the operator $\mathcal{L}V$ (thought as a single notation rather than $\mathcal{L}$ acting on V) is defined on $C([-\tau, 0]; \mathbb{R}^n) \times \mathbb{R}_+ \times \mathbb{S}$, although V is defined on $\mathbb{R}^n \times [-\tau, \infty) \times \mathbb{S}$. Let us also introduce a new notation $\mathcal{W}([-\tau, 0]; \mathbb{R}_+)$ which denotes the family of Borel-measurable functions $w : [-\tau, 0] \to \mathbb{R}_+$ such that $\int_{-\tau}^{0} w(u) du = 1$. Such a function is sometimes called a *weighting function*.

Theorem 8.4 *Assume that there are functions* $V \in C^{2,1}(\mathbb{R}^n \times [-\tau, \infty) \times \mathbb{S}; \mathbb{R}_+)$ *and* $w \in \mathcal{W}([-\tau, 0]; \mathbb{R}_+)$ *and positive constants* $p, \alpha, c_1, c_2, \lambda_1, \lambda_2$ *such that* $\lambda_1 > \lambda_2$,

$$c_1 |x|^p \leq V(x, t, i) \leq c_2 |x|^p \qquad (8.10)$$

for $(x, t, i) \in \mathbb{R}^n \times [-\tau, \infty) \times \mathbb{S}$ *and*

$$\mathcal{L}V(\varphi, t, i) \leq \alpha - \lambda_1 |\varphi(0)|^p + \lambda_2 \int_{-\tau}^{0} w(u) |\varphi(u)|^p du \qquad (8.11)$$

for $(\varphi, t, i) \in C([-\tau, 0]; \mathbb{R}^n) \times \mathbb{R}_+ \times \mathbb{S}$. *Then for any initial data* $\xi \in C^b_{\mathcal{F}_0}([-\tau, 0]; \mathbb{R}^n)$, *the solution of the SFDE (8.1) obeys*

$$\limsup_{t \to \infty} \mathbb{E}|x(t)|^p \leq \frac{\alpha}{c_1 \lambda}, \qquad (8.12)$$

where $\lambda \in (0, \lambda_1 - \lambda_2)$ *is the unique root to the equation*

$$\lambda_2 e^{\lambda \tau} = \lambda_1 - \lambda c_2. \qquad (8.13)$$

In this case, the solution of the SFDE (8.1) is said to be asymptotically bounded in pth moment.

Proof. Applying the generalised Itô formula to $e^{\lambda t}V(x(t), t, r(t))$ and then using conditions (8.10) and (8.11) we can show that, for $t \geq 0$,

$$c_1 e^{\lambda t}\mathbb{E}|x(t)|^p \leq c_2\mathbb{E}|\xi(0)|^p + \frac{\alpha}{\lambda}e^{\lambda t} - (\lambda_1 - \lambda c_2)\mathbb{E}\int_0^t e^{\lambda s}|x(s)|^p ds$$
$$+ \lambda_2 \mathbb{E}\int_0^t e^{\lambda s}\left(\int_{-\tau}^0 w(u)|x(s+u)|^p du\right) ds. \tag{8.14}$$

Compute

$$\int_0^t e^{\lambda s}\left(\int_{-\tau}^0 w(u)|x(s+u)|^p du\right) ds = \int_{-\tau}^0 w(u)\left(\int_0^t e^{\lambda s}|x(s+u)|^p ds\right) du$$
$$= \int_{-\tau}^0 w(u)e^{-\lambda u}\left(\int_0^t e^{\lambda(s+u)}|x(s+u)|^p ds\right) du$$
$$\leq e^{\lambda \tau}\int_{-\tau}^0 w(u)\left(\int_{-\tau}^t e^{\lambda s}|x(s)|^p ds\right) du$$
$$\leq \tau e^{\lambda \tau}\|\xi\|^p + e^{\lambda \tau}\int_0^t e^{\lambda s}|x(s)|^p ds.$$

Substituting this into (8.14) and then making use of (8.13) we obtain that

$$c_1 e^{\lambda t}\mathbb{E}|x(t)|^p \leq (c_2 + \lambda_2 \tau e^{\lambda \tau})\mathbb{E}\|\xi\|^p + \frac{\alpha}{\lambda}e^{\lambda t}. \tag{8.15}$$

Dividing both sides by $c_1 e^{\lambda t}$ and then letting $t \to \infty$ we obtain the required assertion (8.12). □

The following result is in terms of M-matrices which can be used much more easily.

Theorem 8.5 *Let $p \geq 2$, $\alpha > 0$ and $w \in \mathcal{W}([-\tau, 0]; \mathbb{R}_+)$. Assume that for each $i \in \mathbb{S}$, there is a pair of real numbers β_i and σ_i such that*

$$\varphi^T(0)f(\varphi, t, i) + \frac{p-1}{2}|g(\varphi, t, i)|^2 \leq \alpha + \beta_i|\varphi(0)|^2 + \sigma_i \int_{-\tau}^0 w(u)|\varphi(u)|^2 du \tag{8.16}$$

for all $(\varphi, t) \in C([-\tau, 0]; \mathbb{R}^n) \times \mathbb{R}_+$. Assume that

$$\mathcal{A} := -\text{diag}(p\beta_1, \cdots, p\beta_N) - \Gamma$$

is a nonsingular M-matrix so (by Theorem 2.10)

$$(q_1, \cdots, q_N)^T := \mathcal{A}^{-1}\vec{1} \gg 0, \tag{8.17}$$

where $\vec{1} = (1, \cdots, 1)^T$. If

$$pq_i\sigma_i < 1, \quad \forall i \in \mathbb{S}, \tag{8.18}$$

then the solution of the SFDE (8.1) is asymptotically bounded in pth moment.

Proof. Define a $C^{2,1}$-function $V : \mathbb{R}^n \times [-\tau, \infty) \times \mathbb{S} \to \mathbb{R}_+$ by

$$V(x, t, i) = q_i|x|^p.$$

Clearly V obeys (8.10) with $c_1 = \min_{i \in \mathbb{S}} q_i$ and $c_2 = \max_{i \in \mathbb{S}} q_i$. To verify (8.11), we compute the operator $\mathcal{L}V$ from $C([-\tau, 0]; \mathbb{R}^n) \times \mathbb{R}_+ \times \mathbb{S}$ to $\mathbb{R}$ by condition (8.16) and (8.18) as follows:

$$\mathcal{L}V(\varphi, t, i) = pq_i|\varphi(0)|^{p-2}\varphi^T(0)f(\varphi, t, i) + \tfrac{1}{2}pq_i|\varphi(0)|^{p-2}|g(\varphi, t, i)|^2$$

$$+ \tfrac{1}{2}p(p-2)q_i|\varphi(0)|^{p-4}|\varphi^T(0)g(\varphi, t, i)|^2 + \sum_{j=1}^{N}\gamma_{ij}q_j|\varphi(0)|^p$$

$$\leq pq_i|x|^{p-2}\left[\varphi^T(0)f(\varphi, t, i) + \frac{p-1}{2}|g(\varphi, t, i)|^2\right] + \sum_{j=1}^{N}\gamma_{ij}q_j|\varphi(0)|^p$$

$$\leq \left(p\beta_i q_i + \sum_{i=1}^{N}\gamma_{ij}q_j\right)|\varphi(0)|^p + \alpha pq_i|\varphi(0)|^{p-2}$$

$$+ pq_i\sigma_i|\varphi(0)|^{p-2}\int_{-\tau}^{0} w(u)|\varphi(u)|^2 du$$

$$\leq -|\varphi(0)|^p + c_3|\varphi(0)|^{p-2} + \rho\int_{-\tau}^{0} w(u)|\varphi(0)|^{p-2}|\varphi(u)|^2 du, \tag{8.19}$$

where $c_3 = \alpha p c_2$ and $\rho = 0 \vee \max_{i \in \mathbb{S}} pq_i\sigma_i$ and, by (8.18), $\rho < 1$. Note that

$$|\varphi(0)|^{p-2}|\varphi(u)|^2 \leq \frac{p-2}{p}|\varphi(0)|^p + \frac{2}{p}|\varphi(u)|^p,$$

while, for any $\varepsilon > 0$

$$c_3|\varphi(0)|^{p-2} \leq (\varepsilon^{-(p-2)/2}c_3^{p/2})^{2/p}(\varepsilon|\varphi(0)|^p)^{(p-2)/p} \leq c_4 + \varepsilon|\varphi(0)|^p,$$

where $c_4 = \varepsilon^{-(p-2)/2} c_3^{p/2}$. Substituting this into (8.19) gives

$$\mathcal{L}V(\varphi,t,i) \le c_4 - \lambda_1|\varphi(0)|^p + \lambda_2 \int_{-\tau}^0 w(u)|\varphi(u)|^p du,$$

where

$$\lambda_1 = 1 - \varepsilon - \frac{(p-2)\rho}{p}, \quad \lambda_2 = \frac{2\rho}{p}$$

Choosing $\rho = \frac{1}{2}(1-\rho)$ gives $\lambda_1 - \lambda_2 = \frac{1}{2}(1-\rho) > 0$. Hence all the conditions of Theorem 8.4 have been verified and the conclusion of this theorem follows of course. $\square$

8.5 Asymptotic Stability

Let us now proceed to discuss asymptotic stability of the SFDE (8.1). We shall always fix the Markov chain $r(t)$ but let the initial data ξ vary in $C_{\mathcal{F}_0}^b([-\tau,0]; \mathbb{R}^n)$. The solution of equation (8.1) is denoted by $x(t;\xi)$. For the purpose of stability we assume in this section as well as the next one that $f(0,t,i) \equiv 0$ and $g(0,t,i) \equiv 0$. So equation (8.1) admits a trivial solution $x(t;0) \equiv 0$.

Definition 8.6 The trivial solution of the SFDE (8.1), or simply, the SFDE (8.1) is said to be *exponentially stable in pth moment* if

$$\limsup_{t\to\infty} \frac{1}{t} \log(\mathbb{E}|x(t;\xi)|^p) < 0$$

for any $\xi \in C_{\mathcal{F}_0}^b([-\tau,0]; \mathbb{R}^n)$. When $p=2$, it is said to be *exponentially stable in mean square*. It is said to be *almost surely exponentially stable* if

$$\limsup_{t\to\infty} \frac{1}{t} \log(|x(t;\xi)|) < 0 \quad a.s.$$

It is easy to observe that Theorems 8.4 and 8.5 will become the criteria on the pth moment exponential stability if α there becomes zero. For example, we have:

Theorem 8.7 *Assume that there are functions $V \in C^{2,1}(\mathbb{R}^n \times [-\tau, \infty) \times \mathbb{S}; \mathbb{R}_+)$ and $w \in \mathcal{W}([-\tau,0]; \mathbb{R}_+)$ as well as positive constants p, c_1, c_2 and*

$\lambda_1 > \lambda_2$ such that (8.10) holds and

$$\mathcal{L}V(\varphi,t,i) \leq -\lambda_1|\varphi(0)|^p + \lambda_2 \int_{-\tau}^{0} w(u)|\varphi(u)|^p du$$

for $(\varphi,t,i) \in C([-\tau,0];\mathbb{R}^n) \times \mathbb{R}_+ \times \mathbb{S}$. Then for any $\xi \in C_{\mathcal{F}_0}^b([-\tau,0];\mathbb{R}^n)$, the solution of the SFDE (8.1) obeys

$$\mathbb{E}|x(t)|^p \leq c\mathbb{E}\|\xi\|^p e^{-\lambda t} \qquad (8.20)$$

where $c = (c_2 + \lambda_2\tau e^{\lambda\tau})/c_1$ and $\lambda \in (0,\lambda_1 - \lambda_2)$ is the unique root to equation (8.13).

The following theorem gives a condition under which the pth moment exponential stability implies the almost sure exponential stability.

Theorem 8.8 Let $p \geq 1$ and assume that there is a constant $K > 0$ such that for all $\phi \in L_{\mathcal{F}_t}^p([-\tau,0];\mathbb{R}^n)$, $t \geq 0$ and $i \in \mathbb{S}$,

$$\mathbb{E}\big[|f(\phi,t,i)|^p + |g(\phi,t,i)|^p\big] \leq K \sup_{-\tau \leq \theta \leq 0} \mathbb{E}|\phi(\theta)|^p. \qquad (8.21)$$

Then (8.20) implies that

$$\limsup_{t\to\infty} \frac{1}{t} \log|x(t;\xi)| \leq -\frac{\lambda}{p} \quad a.s. \qquad (8.22)$$

for all $\xi \in C_{\mathcal{F}_0}^b([-\tau,0];\mathbb{R}^n)$.

Proof. Fix any $\xi \in C_{\mathcal{F}_0}^b([-\tau,0];\mathbb{R}^n)$ and write $x(t;\xi) = x(t)$. For each integer $k \geq 2$, compute

$$\mathbb{E}\|x_{k\tau}\|^p = \mathbb{E}\Big(\sup_{0\leq h\leq \tau} |x((k-1)\tau + h)|^p\Big)$$
$$\leq 3^{p-1}\bigg(\mathbb{E}|x((k-1)\tau)|^p + \mathbb{E}\bigg[\int_{(k-1)\tau}^{k\tau} |f(x_s,s,r(s))|ds\bigg]^p$$
$$+ \mathbb{E}\bigg[\sup_{0\leq h\leq\tau}\bigg|\int_{(k-1)\tau}^{(k-1)\tau+h} g(x_s,s,r(s))dB(s)\bigg|^p\bigg]\bigg). \qquad (8.23)$$

By Hölder's inequality,

$$\mathbb{E}\bigg[\int_{(k-1)\tau}^{k\tau} |f(x_s,s,r(s))|ds\bigg]^p \leq \tau^{p-1}\int_t^{t+\tau} \mathbb{E}|f(x_s,s,r(s))|^p ds. \qquad (8.24)$$

But, by (8.21) and (8.20), we can derive that

$$\mathbb{E}|f(x_s, s, r(s))|^p \leq \sum_{1 \leq i \leq N} \mathbb{E}|f(x_s, s, i)|^p$$

$$\leq NK \sup_{-\tau \leq \theta \leq 0} \mathbb{E}|x(s+\theta)|^p \leq cNK\mathbb{E}\|\xi\|^p e^{-\lambda(s-\tau)}.$$

Substituting this into (8.24) yields

$$\mathbb{E}\left[\int_{(k-1)\tau}^{k\tau} |f(x_s, s, r(s))|ds\right]^p \leq cNK\tau^p \mathbb{E}\|\xi\|^p e^{-(k-2)\tau\lambda}. \tag{8.25}$$

On the other hand, by the Burkholder–Davis–Gundy inequality, we have that

$$J := \mathbb{E}\left[\sup_{0 \leq h \leq \tau} \left|\int_{(k-1)\tau}^{(k-1)\tau+h} g(x_s, s, r(s))dB(s)\right|^p\right]$$

$$\leq C_p E\left(\int_{(k-1)\tau}^{k\tau} |g(x_s, s, r(s))|^2 ds\right)^{p/2}$$

$$\leq C_p \mathbb{E}\left[\left(\sup_{(k-1)\tau \leq s \leq k\tau} |g(x_s, s, r(s))|\right) \int_{(k-1)\tau}^{k\tau} |g(x_s, s, r(s))|ds\right]^{p/2}, \tag{8.26}$$

where C_p is a positive constant dependent of p only. Let $\varepsilon \in (0, 1/3^{p-1}KN)$ be sufficiently small for

$$\frac{3^{p-1}KN\varepsilon}{1 - 3^{p-1}KN\varepsilon} < e^{-\lambda\tau}. \tag{8.27}$$

Using the elementary inequality $|ab| \leq \varepsilon a^2 + b^2/4\varepsilon$ we derive from (8.26) that

$$J \leq \varepsilon \mathbb{E}\left(\sup_{(k-1)\tau \leq s \leq k\tau} |g(x_s, s, r(s))|^p\right) \tag{8.28}$$

$$+ \frac{C_p^2}{4\varepsilon} \mathbb{E}\left[\int_{(k-1)\tau}^{k\tau} |g(x_s, s, r(s))|ds\right]^p.$$

In the same way as (8.25) was proved we can show that

$$\mathbb{E}\left[\int_{(k-1)\tau}^{k\tau} |g(x_s, s, r(s))|ds\right]^p \leq cNK\tau^p \mathbb{E}\|\xi\|^p e^{-(k-2)\tau\lambda}. \tag{8.29}$$

Note also from condition (8.21) that

$$|g(x_s, s, r(s))|^p \leq \sum_{i=1}^{N} |g(x_s, s, i)|^p \leq KN\|x_s\|^p. \tag{8.30}$$

Consequently

$$\mathbb{E}\left(\sup_{(k-1)\tau \leq s \leq k\tau} |g(x_s, s, r(s))|^p\right) \leq KN\left[\mathbb{E}\|x_{(k-1)\tau}\|^p + \mathbb{E}\|x_{k\tau}\|^p\right]. \tag{8.31}$$

Substituting (8.29) and (8.31) into (8.28) yields

$$J \leq \varepsilon KN\left[\mathbb{E}\|x_{(k-1)\tau}\|^p + \mathbb{E}\|x_{k\tau}\|^p\right]$$
$$+ \frac{cNKC_p^2\tau^p}{4\varepsilon}\mathbb{E}\|\xi\|^p e^{-(k-2)\tau\lambda}. \tag{8.32}$$

Making use of (8.27), we can now substitute (8.20), (8.25) and (8.32) into (8.23) to obtain that

$$\mathbb{E}\|x_{k\tau}\|^p \leq e^{-\lambda\tau}\mathbb{E}\|x_{(k-1)\tau}\|^p + Ce^{-(k-2)\tau\lambda}, \tag{8.33}$$

where C is a constant independent of k. By induction we can easily show from (8.33) that

$$\mathbb{E}\|x_{k\tau}\|^p \leq \left(kCe^{2\lambda\tau} + \mathbb{E}\|\xi\|^p\right)e^{-k\tau\lambda}. \tag{8.34}$$

Finally we can show the required assertion (8.22) from (8.34) in the same way as in the proof of Theorem 7.24 (see Exercise 8.4).

8.6 Razumikhin-Type Theorems on Stability

In this section, we will develop a new method, known as the Razumikhin technique, to study the pth moment exponential stability for SFDEs.

Theorem 8.9 *Let λ, p, c_1, c_2 be all positive numbers and $q > 1$. Assume that there exists a function $V \in C^{2,1}(\mathbb{R}^n \times [-\tau, \infty) \times \mathbb{S}; \mathbb{R}_+)$ such that*

$$c_1|x|^p \leq V(x, t, i) \leq c_2|x|^p \quad \forall (x, t, i) \in \mathbb{R}^n \times [-\tau, \infty) \times \mathbb{S}, \tag{8.35}$$

and also for all $t \geq 0$

$$\mathbb{E}\left[\max_{1 \leq i \leq N} \mathcal{L}V(\phi, t, i)\right] \leq -\lambda \mathbb{E}\left[\max_{1 \leq i \leq N} V(\phi(0), t, i)\right] \tag{8.36}$$

provided $\phi = \{\phi(\theta) : -\tau \leq \theta \leq 0\} \in L^p_{\mathcal{F}_t}([-\tau, 0]; \mathbb{R}^n)$ satisfying

$$\mathbb{E}\left[\min_{1 \leq i \leq N} V(\phi(\theta), t+\theta, i)\right] < q\mathbb{E}\left[\max_{1 \leq i \leq N} V(\phi(0), t, i)\right], \quad \forall \theta \in [-\tau, 0]. \tag{8.37}$$

Then, for any $\xi \in C^b_{\mathcal{F}_0}([-\tau, 0]; \mathbb{R}^n)$, the solution of equation (8.1) obeys

$$\mathbb{E}|x(t;\xi)|^p \leq \frac{c_2}{c_1}\mathbb{E}\|\xi\|^p e^{-\gamma t} \quad \text{on } t \geq 0, \tag{8.38}$$

where $\gamma = \min\{\lambda, \log(q)/\tau\}$.

Proof. Fix any $\xi \in C^b_{\mathcal{F}_0}([-\tau, 0]; \mathbb{R}^n)$ and write $x(t;\xi) = x(t)$. Extend $r(t)$ to $[-\tau, 0)$ by setting $r(t) = r(0)$. Recalling the facts that $x(t)$ is continuous, $\mathbb{E}(\sup_{-\tau \leq s \leq t} |x(s)|^p) < \infty$ for all $t \geq 0$ and $r(t)$ is right continuous, we see easily that $\mathbb{E}V(x(t), t, r(t))$ is right continuous on $t \geq -\tau$. Let $\varepsilon \in (0, \gamma)$ be arbitrary and set $\bar{\gamma} = \gamma - \varepsilon$. Define

$$U(t) = \sup_{-\tau \leq \theta \leq 0}\left[e^{\bar{\gamma}(t+\theta)}\mathbb{E}V(x(t+\theta), t+\theta, r(t+\theta))\right] \quad \text{for } t \geq 0.$$

We claim that

$$D_+U(t) := \limsup_{h \to 0+} \frac{U(t+h) - U(t)}{h} \leq 0 \quad \text{for all } t \geq 0. \tag{8.39}$$

It is easy to observe that for each $t \geq 0$ (which is fixed for the moment), either $U(t) > e^{\bar{\gamma}t}\mathbb{E}V(x(t), t, r(t))$ or $U(t) = e^{\bar{\gamma}t}\mathbb{E}V(x(t), t, r(t))$. In the case of the former, if follows from the right continuity of $\mathbb{E}V(x(\cdot), \cdot, r(\cdot))$ that for all $h > 0$ sufficiently small

$$U(t) > e^{\bar{\gamma}(t+h)}\mathbb{E}V(x(t+h), t+h, r(t+h)),$$

hence

$$U(t+h) \leq U(t) \quad \text{and} \quad D_+U(t) \leq 0.$$

In the other case, i.e. $U(t) = e^{\bar{\gamma}t}\mathbb{E}V(x(t), t, r(t))$, we have

$$e^{\bar{\gamma}(t+\theta)}\mathbb{E}V(x(t+\theta), t+\theta, r(t+\theta)) \leq e^{\bar{\gamma}t}\mathbb{E}V(x(t)), t, r(t))$$

for all $-\tau \leq \theta \leq 0$. So

$$\mathbb{E}V(x(t+\theta), t+\theta, r(t+\theta)) \leq e^{-\bar{\gamma}\theta}\mathbb{E}V(x(t), t, r(t)) \tag{8.40}$$

$$\leq e^{\bar{\gamma}\tau}\mathbb{E}V(x(t), t, r(t))$$

for all $-\tau \leq \theta \leq 0$. Note that either $\mathbb{E}V(x(t),t,r(t)) = 0$ or $\mathbb{E}V(x(t),t,r(t)) > 0$. In the former case, (8.40) and (8.35) yield that $x(t+\theta) = 0$ a.s. for all $-\tau \leq \theta \leq 0$. Recalling the fact that $f(0,t,i) \equiv 0$ and $g(0,t,i) \equiv 0$, one sees that $x(t+h) = 0$ a.s. for all $h > 0$ hence $U(t+h) = 0$ and $D_+U(t) = 0$. On the other hand, in the case of $\mathbb{E}V(x(t),t,r(t)) > 0$, (8.40) implies

$$\mathbb{E}V(x(t+\theta),t+\theta,r(t+\theta)) < q\mathbb{E}V(x(t),t,r(t))$$

for all $-\tau \leq \theta \leq 0$, since $e^{\bar{\gamma}\tau} < q$. Consequently

$$\mathbb{E}\left[\min_{1 \leq i \leq N} V(x(t+\theta),t+\theta,i)\right] < q\mathbb{E}\left[\max_{1 \leq i \leq N} V(x(t),t,i)\right]$$

for all $-\tau \leq \theta \leq 0$. In other words, $x_t \in L^p_{\mathcal{F}_t}([-\tau,0];\mathbb{R}^n)$ satisfying (8.37). Thus, by condition (8.36), we have

$$\mathbb{E}\left[\max_{1 \leq i \leq N} \mathcal{L}V(x_t,t,i)\right] \leq -\lambda\mathbb{E}\left[\max_{1 \leq i \leq N} V(x(t),t,i)\right],$$

which implies

$$\mathbb{E}\mathcal{L}V(x_t,t,r(t)) \leq -\lambda\mathbb{E}V(x(t),t,r(t)). \tag{8.41}$$

This gives

$$\bar{\gamma}\mathbb{E}V(x(t),t,r(t)) + \mathbb{E}\mathcal{L}V(x_t,t,r(t)) \leq -(\lambda - \bar{\gamma})\mathbb{E}V(x(t),t,r(t)) < 0.$$

By the right continuity of the processes involved we hence see that for all $h > 0$ sufficiently small

$$\bar{\gamma}\mathbb{E}V(x(s),s,r(s)) + \mathbb{E}\mathcal{L}V(x_s,s,r(s)) \leq 0 \quad \text{if } t \leq s \leq t+h.$$

By Itô's formula, we can then derive that

$$e^{\bar{\gamma}(t+h)}\mathbb{E}V(x(t+h),t+h,r(t+h)) - e^{\bar{\gamma}t}\mathbb{E}V(x(t),t,r(t))$$

$$= \int_t^{t+h} e^{\bar{\gamma}s}\Big[\bar{\gamma}\mathbb{E}V(x(s),s,r(s)) + \mathbb{E}\mathcal{L}V(x_s,s,r(s))\Big]ds \leq 0.$$

That is

$$e^{\bar{\gamma}(t+h)}\mathbb{E}V(x(t+h),t+h,r(t+h)) \leq e^{\bar{\gamma}t}\mathbb{E}V(x(t),t,r(t)).$$

So it must hold that $U(t+h) = U(t)$ for all $h > 0$ sufficiently small, and hence $D_+U(t) = 0$. Inequality (8.39) has therefore been proved. It now follows from (8.39) immediately that

$$U(t) \leq U(0) \quad \text{for all } t \geq 0.$$

By the definition of $U(t)$ and condition (8.35) one sees

$$\mathbb{E}|x(t)|^p \leq \frac{c_2}{c_1}\mathbb{E}\|\xi\|^p e^{-\bar{\gamma}t} = \frac{c_2}{c_1}\mathbb{E}\|\xi\|^p e^{-(\gamma-\varepsilon)t}.$$

Since ε is arbitrary, the required inequality (8.38) must hold. □

Theorem 8.9 will play a key role in the next chapter but meanwhile let us turn to discuss a number of examples to illustrate the theory established above.

8.7 Examples

In the following examples we shall omit mentioning the initial data.

Example 8.10 Consider a linear SFDE with Markovian switching of the form

$$dx(t) = [a(r(t)) + A(r(t))x(t)]dt + g(x_t, r(t))dB(t) \qquad (8.42)$$

on $t \geq 0$. Here $B(t)$ is a scalar Brownian motion, $a : \mathbb{S} \to \mathbb{R}^n$, $A : \mathbb{S} \to \mathbb{R}^{n\times n}$ and we shall write $a(i) = a_i$ and $A(i) = A_i$, while $g : C([-\tau, 0]; \mathbb{R}^n) \times \mathbb{S} \to \mathbb{R}^n$ is defined by

$$g(\varphi, i) = \int_{-\tau}^{0} D_i\varphi(u)du$$

with D_i's being all $n \times n$ matrices. Assume that there are symmetric positive-definite matrices Q_i, $1 \leq i \leq N$, such that

$$-\lambda_1 := \max_{1\leq i\leq N}\lambda_{\max}\left(Q_iA_i + A_i^TQ_i + \sum_{j=1}^{N}\gamma_{ij}Q_j\right) < 0. \qquad (8.43)$$

In order to apply Theorem 8.4 to establish a condition on D_i for the SFDE (8.42) to be asymptotically bounded in mean square, we define a $C^{2,1}$-function $V : \mathbb{R}^n \times [-\tau, \infty) \times \mathbb{S} \to \mathbb{R}$ by $V(x,t,i) = x^TQ_ix$. Then, the operator $\mathcal{L}V : C([-\tau, \mathbb{R}^n) \times \mathbb{R}_+ \times \mathbb{S} \to \mathbb{R}$, associated with equation (8.42), has the form

$$\begin{aligned}\mathcal{L}V(\varphi, t, i) &= 2a_i^T \varphi(0) + \varphi^T(0)(Q_i A_i + A_i^T Q_i)\varphi(0) \\ &\quad + g^T(\varphi, i)Q_i g(\varphi, i) + \sum_{j=1}^N \gamma_{ij} \varphi^T(0) Q_j \varphi(0) \\ &= 2a_i^T \varphi(0) + \varphi^T(0)\Big(Q_i A_i + A_i^T Q_i + \sum_{j=1}^N \gamma_{ij} Q_j\Big)\varphi(0) \\ &\quad + \Big(\int_{-\tau}^0 \varphi(u) du\Big)^T D_i^T Q_i D_i \Big(\int_{-\tau}^0 \varphi(u) du\Big) \\ &\leq 2a|\varphi(0)|^2 - \lambda_1 |\varphi(0)|^2 + \tau \lambda_2 \int_{-\tau}^0 |\varphi(u)|^2 du,\end{aligned}$$

where $a = \max_{1 \leq i \leq N} |a_i|$ and $\lambda_2 = \max_{1 \leq i \leq N} \|D_i^T Q_i D_i\|$. Let us now assume that

$$\tau^2 \lambda_2 < \lambda_1. \tag{8.44}$$

Choose an $\varepsilon \in (0, \lambda_1 - \tau^2 \lambda_2)$. Then

$$\mathcal{L}V(\varphi, t, i) \leq \frac{a^2}{\varepsilon} - (\lambda_1 - \varepsilon)|\varphi(0)|^2 + \tau^2 \lambda_2 \int_{-\tau}^0 w(u) |\varphi(u)|^2 du,$$

where $w(u) = 1/\tau$ for $u \in [-\tau, 0]$ which is in $\mathcal{W}([-\tau, 0]; \mathbb{R}_+)$. By Theorem 8.4 we can then conclude that under conditions (8.43) and (8.44) the SFDE (8.42) is asymptotically bounded in mean square.

Example 8.11 If $a_i = 0$ for all $i \in \mathbb{S}$, the SFDE (8.42) becomes

$$dx(t) = A(r(t))x(t)dt + g(x_t, r(t))dB(t). \tag{8.45}$$

Under conditions (8.43) and (8.44), we have

$$\mathcal{L}V(\varphi, t, i) \leq -\lambda_1 |\varphi(0)|^2 + \tau^2 \lambda_2 \int_{-\tau}^0 w(u)|\varphi(u)|^2 du,$$

whence by Theorem 8.7, equation (8.45) is exponentially stable in mean square. To show the almost sure exponential stability, we verify condition

(8.21) with $p = 2$: For $\phi \in L^2_{\mathcal{F}_t}([-\tau, 0]; \mathbb{R}^n)$ and $i \in \mathbb{S}$

$$\mathbb{E}\big[|A_i\phi(0)|^2 + |g(\phi, i)|^2\big]$$
$$\leq \|A_i\|^2 \mathbb{E}|\phi(0)|^2 + \tau\|B_i\|^2 \int_{-\tau}^0 \mathbb{E}|\phi(u)|^2 du$$
$$\leq \left[\max_{1 \leq i \leq N}(\|A_i\|^2 + \tau^2\|B_i\|^2)\right] \sup_{-\tau \leq u \leq 0} \mathbb{E}\|\phi(u)\|^2.$$

Hence, by Theorem 8.8, equation (8.45) is also almost surely exponentially stable.

Example 8.12 Let $\delta : \mathbb{R}_+ \to [0, \tau]$ be a Borel-measurable function. Consider a linear SDDE with Markovian switching of the form

$$dx(t) = A(r(t))x(t)dt + D(r(t))x(t - \delta(t))dB(t), \qquad (8.46)$$

where B, r, A, D are the same as in Example 8.10. The stability theory developed in Chapter 7 requires that the time lag function $\delta(\cdot)$ be differentiable and its derivative be bounded above by a positive constant less than 1 (see condition (7.30)) and hence it cannot be applied to the SDDE (8.46). However, we shall now demonstrate that the Razumikhin-type theorem developed in this chapter can be applied to it. First of all, if we define $f, g : C([-\tau, 0]; \mathbb{R}^n) \times \mathbb{R}_+ \times \mathbb{S} \to \mathbb{R}^n$ by

$$f(\varphi, t, i) = A_i\varphi(0) \quad \text{and} \quad g(\varphi, t, i) = D_i\varphi(-\delta(t)),$$

then equation (8.46) can be written in the form of (8.1). We still assume that condition (8.43) holds and let $V(x, t, i) = x^T Q_i x$ as before. Then, the operator $\mathcal{L}V : C([-\tau, \mathbb{R}^n) \times \mathbb{R}_+ \times \mathbb{S} \to \mathbb{R}$, associated with equation (8.46), has the form

$$\mathcal{L}V(\varphi, t, i) = \varphi^T(0)(Q_i A_i + A_i^T Q_i)\varphi(0)$$
$$+ \varphi^T(-\delta(t))D_i^T Q_i D_i \varphi(-\delta(t)) + \sum_{j=1}^N \gamma_{ij}\varphi^T(0)Q_j\varphi(0)$$
$$\leq -\lambda_1|\varphi(0)|^2 + \lambda_2|\varphi(-\delta(t))|^2, \qquad (8.47)$$

where $\lambda_2 = \max_{1 \leq i \leq N} \|D_i^T Q_i D_i\|$. Set $\lambda_3 = \min_{1 \leq i \leq N} \lambda_{\min}(Q_i)$ and $\lambda_4 = \max_{1 \leq i \leq N} \lambda_{\max}(Q_i)$ and assume

$$\lambda_1 > \frac{\lambda_2 \lambda_4}{\lambda_3}. \qquad (8.48)$$

Choose $q > 1$ for $\lambda_1 > q\lambda_2\lambda_4/\lambda_3$. Then, for any $t \geq 0$ and $\phi \in L^2_{\mathcal{F}_t}([-\tau,0];\mathbb{R}^n)$ satisfying

$$\mathbb{E}\left[\min_{1\leq i \leq N} \phi^T(\theta)Q_i\phi(\theta)\right] < q\mathbb{E}\left[\max_{1\leq i \leq N} \phi^T(0)Q_i\phi(0)\right]$$

on $-\tau \leq \theta \leq 0$, we have

$$\lambda_3\mathbb{E}|\phi(\theta)|^2 < q\lambda_4\mathbb{E}|\phi(0)|^2 \quad \text{on} \ -\tau \leq \theta \leq 0$$

and hence, by (8.47),

$$\mathbb{E}\left[\max_{1\leq i \leq N} \mathcal{L}V(\phi,t,i)\right] \leq -\left(\lambda_1 - \frac{q\lambda_2\lambda_4}{\lambda_3}\right)\mathbb{E}|\phi(0)|^2.$$

By Theorem 8.9 we can conclude that under conditions (8.43) and (8.48), equation (8.46) is mean square exponentially stable. Moreover, by Theorem 8.8, it is also almost surely exponentially stable.

8.8 Exercises

8.1 Show (8.7).
8.2 Under the conditions of Theorem 8.3 show (8.8) if the initial data $\xi \in L^p_{\mathcal{F}_0}([-\tau,0];\mathbb{R}^n)$ for some $p > 2$.
8.3 By condition (8.21) show (8.30).
8.4 Show (8.22) from (8.34) by the Borel–Cantelli argument.
8.5 Let $B(t)$ be a scalar Brownian motion. Let $r(t)$ be a right-continuous Markov chain taking values in $\mathbb{S} = \{1,2\}$ with generator

$$\Gamma = (\gamma_{ij})_{2\times 2} = \begin{pmatrix} -1 & 1 \\ 1 & -1 \end{pmatrix}.$$

Assume that $B(t)$ and $r(t)$ are independent. Let $\delta : R_+ \to [0,\tau]$ be Borel measurable. Consider a one-dimensional stochastic differential delay equation with Markovian switching of the form

$$dx(t) = \alpha(r(t))x(t)dt + \sigma(x(t-\delta(t)),t,r(t))dB(t) \quad (8.49)$$

on $t \geq 0$, where

$$\alpha(1) = \frac{1}{4}, \quad \alpha(2) = -3,$$

while $\sigma : \mathbb{R} \times \mathbb{R}_+ \times \mathbb{S} \to \mathbb{R}$ satisfying
$$|\sigma(y,t,1)| \leq \frac{|y|}{8}, \quad |\sigma(y,t,2)| \leq \frac{|y|}{4}.$$
Let $p \in [2, 2.5]$. Show that equation (8.49) is exponentially stable in pth moment.

Chapter 9

Stochastic Interval Systems with Markovian Switching

9.1 Introduction

As mentioned before, the hybrid systems driven by continuous-time Markov chains have been used to model many practical systems where they may experience abrupt changes in their structure and parameters. One of the important issues in the study of hybrid systems is the automatic control, with consequent emphasis being placed on the analysis of stability.

An important class of hybrid systems is the jump linear systems,

$$\dot{x}(t) = A(r(t))x(t), \qquad (9.1)$$

where a part of the state $x(t)$ takes values in $\mathbb{R}^n$ while another part of the state $r(t)$ is a Markov chain taking values in $\mathbb{S}$. However, the abrupt changes of structure and parameters in the hybrid systems are usually caused by phenomena such as component failures or repairs, changing subsystem interconnections, and abrupt environmental disturbances. When we model such systems it is therefore necessary to take parameter uncertainty and environmental noise as well as time delay into account. In system (9.1), if we consider the effect of time delay, the system should be described by

$$\dot{x}(t) = A(r(t))x(t) + G(r(t))x(t-\delta). \qquad (9.2)$$

If we also take the environmental noise into account, the system might become a linear SDDE with Markovian switching

$$dx(t) = [A(r(t))x(t) + G(r(t))x(t-\delta)]dt$$
$$+ [C(r(t))x(t) + D(r(t))x(t-\delta)]dB(t), \qquad (9.3)$$

where $B(t)$ is a scalar Brownian motion. This is a special case of the SDDEs discussed in Chapter 7.

However, when we estimate systems parameter matrices $A(\cdot)$, $G(\cdot)$ etc, there is always some uncertainty and error. If we take this uncertainty and error into account further we arrive at a stochastic delay interval system (SDIS) with Markovian switching

$$\begin{aligned}dx(t) = &\Big[(A(r(t)) + \Delta A(r(t)))x(t) \\ &+ (G(r(t)) + \Delta G(r(t)))x(t-\delta)\Big]dt \\ &+ \Big[(C(r(t)) + \Delta C(r(t)))x(t) \\ &+ (D(r(t)) + \Delta D(r(t)))x(t-\delta)\Big]dB(t).\end{aligned} \qquad (9.4)$$

Such systems take all the features of interval systems, Itô equations, hybrid systems as well as time-lag into account so they are very advanced. The theory developed in this chapter is applicable in many different and complicated situations and hence the importance of this chapter is clear.

9.2 Interval Matrices

For $A = (a_{ij})_{n \times m} \in \mathbb{R}^{n \times m}$ and $C = (c_{ij})_{n \times m} \in \mathbb{R}^{n \times m}$ satisfying

$$a_{ij} \le c_{ij} \quad \forall 1 \le i \le n, \ 1 \le j \le m,$$

the $n \times m$ interval matrix $[A, C]$ is defined by

$$[A, C] = \{(d_{ij})_{n \times m} \in \mathbb{R}^{n \times m} : a_{ij} \le d_{ij} \le c_{ij} \ \forall 1 \le i \le n, \ 1 \le j \le m\}.$$

For $A, \bar{A} \in \mathbb{R}^{n \times m}$, where $\bar{A} = (\bar{a}_{ij})_{n \times m}$ is a non-negative matrix (element-wise, namely $\bar{a}_{ij} \ge 0$ for $\forall i, j$), we use the notation $[A \pm \bar{A}]$ to denote the interval matrix $[A - \bar{A}, A + \bar{A}]$. In fact, any interval matrix $[C_1, C_2]$ has a unique representation of the form $[A \pm \bar{A}]$, where $A = (1/2)(C_1 + C_2)$ and $\bar{A} = (1/2)(C_2 - C_1)$.

Let us establish a couple of inequalities on interval matrices.

Lemma 9.1 *If* $\Delta A \in [-\bar{A}, \bar{A}]$, *where* $\bar{A} \in \mathbb{R}^{n \times m}$ *is a non-negative matrix. Then* $\|\Delta A\| \le \|\bar{A}\|$.

Proof. Let $\Delta A = (\Delta a_{ij})_{n \times m}$ and $\bar{A} = (\bar{a}_{ij})_{n \times m}$. Then

$$|\Delta a_{ij}| \le \bar{a}_{ij}, \quad \forall i, j.$$

Now, for any $x = (x_1, \cdots, x_m)^T \in \mathbb{R}^m$ with $|x| = 1$, we compute

$$|\Delta Ax|^2 = \sum_{i=1}^{n}\Big(\sum_{j=1}^{m}\Delta a_{ij}x_j\Big)^2 \leq \sum_{i=1}^{n}\Big(\sum_{j=1}^{m}|\Delta a_{ij}||x_j|\Big)^2$$

$$\leq \sum_{i=1}^{n}\Big(\sum_{j=1}^{m}|\bar{a}_{ij}||x_j|\Big)^2 = |\bar{A}(|x_1|,\cdots,|x_m|)^T|^2$$

$$\leq \Big(\|\bar{A}\|\sqrt{\sum_{j=1}^{m}|x_j|^2}\Big)^2 = \|\bar{A}\|^2.$$

Hence

$$\|\Delta A\| = \sup_{x \in \mathbb{R}^m, |x|=1} |\Delta Ax|^2 \leq \|\bar{A}\|^2$$

as required. $\square$

Let us recall that if A, G are both symmetric matrices with the same dimensions, by $G \leq A$ or $A \geq G$ we mean that $A - G$ is non-negative definite. In particular, for any $m \times m$ non-negative matrix A, we always have that $A \leq \|A\|I_m$, where I_m is the $m \times m$ identity matrix.

Lemma 9.2 *Let $[A \pm \bar{A}]$ be an $n \times m$ interval matrix and $\Delta A \in [-\bar{A}, \bar{A}]$. Let $Q \in \mathbb{R}^{n \times n}$ be a non-negative symmetric matrix. Then*

$$(A + \Delta A)^T Q (A + \Delta A)^T$$
$$\leq \frac{\|A\| + \|\bar{A}\|}{\|A\|} A^T Q A + \|Q\|\|\bar{A}\|(\|A\| + \|\bar{A}\|)I_m, \quad (9.5)$$

where we set $\|A\|^{-1} = \infty$ when $\|A\| = 0$.

Proof. Clearly, (9.5) holds if either $\|A\| = 0$ or $\|\bar{A}\| = 0$. Otherwise, by Lemma 9.1,

$$(A + \Delta A)^T Q (A + \Delta A)^T$$
$$\leq A^T Q A + \Delta A^T Q A + A^T Q \Delta A + \Delta A^T Q \Delta A$$
$$\leq A^T Q A + \Delta A^T Q A + A^T Q \Delta A + \|Q\|\|\Delta A\|^2 I_m. \quad (9.6)$$

But for any $\varepsilon > 0$,

$$0 \leq (\varepsilon A - \varepsilon^{-1}\Delta A)^T Q (\varepsilon A - \varepsilon^{-1}\Delta A),$$

namely

$$0 \leq \varepsilon^2 A^T Q A - \Delta A^T Q A - A^T Q \Delta A + \varepsilon^{-2} \Delta A^T Q \Delta A.$$

This implies

$$\Delta A^T Q A + A^T Q \Delta A \leq \varepsilon^2 A^T Q A + \varepsilon^{-2} \|Q\| \, \|\bar{A}\|^2 I_m.$$

Choosing $\varepsilon = \sqrt{\|\bar{A}\|/\|A\|}$ gives

$$\Delta A^T Q A + A^T Q \Delta A \leq \frac{\|\bar{A}\|}{\|A\|} A^T Q A + \|Q\| \, \|A\| \, \|\bar{A}\| I_m.$$

Substituting this into (9.6) yields the assertion (9.5). □

9.3 SDISs with Markovian Switching

As in Chapter 7, we let $\delta : \mathbb{R}_+ \to [0, \tau]$ stand for the time lag. Given, for each $i \in \mathbb{S}$, the $n \times n$ interval matrices

$$[A_i \pm \bar{A}_i], \quad [G_i \pm \bar{G}_i], \quad [C_i \pm \bar{C}_i], \quad [D_i \pm \bar{D}_i],$$

let us consider an n-dimensional SDIS with Markovian switching of the form

$$dx(t) = \Big[(A_r + \Delta A_r)x(t) + (G_r + \Delta G_r)x(t-\delta)\Big] dt$$
$$+ \Big[(C_r + \Delta C_r)x(t) + (D_r + \Delta D_r)x(t-\delta)\Big] dB(t) \quad (9.7)$$

on $t \geq 0$ with initial data $\{x(\theta) : -\tau \leq \theta \leq 0\} = \xi \in L^2_{\mathcal{F}_0}([-\tau, 0]; \mathbb{R}^n)$. Here $r = r(t)$ is the same Markov chain as before but $B(t)$ is a scalar Brownian motion, while

$$\Delta A_i \in [-\bar{A}_i, \bar{A}_i], \quad \Delta G_i \in [-\bar{G}_i, \ \bar{G}_i], \quad \Delta C_i \in [-\bar{C}_i, \bar{C}_i], \quad \Delta D_i \in [-\bar{D}_i, \bar{D}_i]$$

and throughout this chapter, unless otherwise necessary, we drop t from $r(t)$ and $\delta(t)$ for the sake of simplicity.

It should be pointed out that the theory developed in this chapter can be generalised to cope with the more general SDIS with Markovian switching driven by multi-dimensional Brownian motion

$$dx(t) = \Big[(A_r + \Delta A_r)x(t) + (G_r + \Delta G_r)x(t-\delta)\Big]dt$$
$$+ \sum_{k=1}^{m} \Big[(C_{r,k} + \Delta C_{r,k})x(t)$$
$$+ (D_{r,k} + \Delta D_{r,k})x(t-\delta)\Big]dB_k(t). \qquad (9.8)$$

The reason we concentrate on equation (9.7) rather than (9.8) is to avoid the notations becoming too complicated. Once the theory developed in this chapter is understood, the reader should be able to cope with equation (9.8) without any difficulty.

If we define $f, g : \mathbb{R}^n \times \mathbb{R}^n \times \mathbb{S} \to \mathbb{R}^n$ by

$$f(x, y, i) = (A_i + \Delta A_i)x + (G_i + \Delta G_i)y$$

and

$$g(x, y, i) = (C_i + \Delta C_i)x + (D_i + \Delta D_i)y,$$

the SDIS (9.7) can be written as

$$dx(t) = f(x(t), x(t-\delta), r)dt + g(x(t), x(t-\delta), r)dB(t)$$

which is in the form of the SDDE discussed in Chapter 7 so the theory established there can be applied to equation (9.7). For example, we observe that equation (9.7) has a unique solution, denoted by $x(t;\xi)$, on $t \in [-\tau, \infty)$ which obeys

$$\mathbb{E}\Big(\sup_{-\tau \leq t \leq T} |x(t;\xi)|^2\Big) < \infty \quad \forall T > 0.$$

To apply Theorems 7.23 and 7.24, we compute by Lemma 9.1 that for $(x, y, i) \in \mathbb{R}^n \times \mathbb{R}^n \times \mathbb{S}$,

$$x^T f(x, y, i) + \tfrac{1}{2}|g(x, y, i)|^2$$
$$= x^T(A_i + \Delta A_i)x + x^T(G_i + \Delta G_i)y$$
$$+ \tfrac{1}{2}|(C_i + \Delta C_i)x + (D_i + \Delta D_i)y|^2$$
$$\leq \tfrac{1}{2}x^T(A_i + A_i^T)x + \|\Delta A_i\| |x|^2 + (\|G_i\| + \|\Delta G_i\|)|x| |y|$$
$$+ (\|C_i\| + \|\Delta C_i\|)^2 |x|^2 + (\|D_i\| + \|\Delta D_i\|)^2 |y|^2$$
$$\leq \beta_i |x|^2 + \sigma_i^2 |y|^2,$$

where

$$\beta_i = \tfrac{1}{2}\lambda_{\max}(A_i + A_i^T) + \|\bar{A}_i\| + \tfrac{1}{2}(\|G_i\| + \|\bar{G}_i\|) + (\|C_i\| + \|\bar{C}_i\|)^2$$

and

$$\sigma_i = \tfrac{1}{2}(\|G_i\| + \|\bar{G}_i\|) + (\|D_i\| + \|\bar{D}_i\|)^2.$$

By Theorems 7.23 and 7.24 we then obtain the following useful stability criterion.

Theorem 9.1 *Assume that δ is differentiable and its derivative is bounded by a constant less than one, namely*

$$\frac{d\delta(t)}{dt} \leq \bar{\delta} < 1 \quad \forall t \geq 0. \tag{9.9}$$

Assume that $\mathcal{A} := -\mathrm{diag}(2\beta_1, \cdots, 2\beta_N) - \Gamma$ is a nonsingular M-matrix so (by Theorem 2.10) $(q_1, \cdots, q_N)^T := \mathcal{A}^{-1}\vec{1} \gg 0$. If

$$2q_i\sigma_i < 1 - \bar{\delta}, \quad \forall i \in \mathbb{S},$$

then the SDIS (9.7) is both mean square and almost surely exponentially stable.

Let us discuss two examples to illustrate this theorem. The examples demonstrate that Theorem 9.1 can be used in two ways: (i) when the system parameter matrices and the bounds for the uncertainties are all known it can be used to verify whether the underlying system is stable; (ii) when some bounds for the uncertainties are unknown it can be used to estimate these bounds so that the underlying system will remain stable if the uncertainties can be controlled within the bounds.

Example 9.2 Consider a linear jump system which is described by the linear differential equation with Markovian switching

$$\dot{x}(t) = A_r x(t), \quad t \geq 0, \tag{9.10}$$

where a part of the state $x(t)$ take values in $\mathbb{R}^3$ while the other part of the state $r(t)$ is a right-continuous Markov chain taking values in $\mathbb{S} = \{1, 2, 3\}$ with generator

$$\Gamma = \begin{pmatrix} -2 & 1 & 1 \\ 3 & -4 & 1 \\ 1 & 1 & -2 \end{pmatrix}.$$

Moreover, we specify the system matrices as follows

$$A_1 = \begin{pmatrix} -2 & 1 & -2 \\ 2 & -2 & 1 \\ 1 & -2 & -3 \end{pmatrix}, \quad A_2 = \begin{pmatrix} 0.5 & 1 & 0.5 \\ -0.8 & 0.5 & 1 \\ -0.7 & -0.9 & 0.2 \end{pmatrix},$$

$$A_3 = \begin{pmatrix} -0.5 & -0.9 & -1 \\ 1 & -0.6 & -0.7 \\ 0.8 & 1 & -1 \end{pmatrix}.$$

Given these parameters, by Corollary 5.17, we can show (see Exercise 9.1) that equation (9.10) is exponentially stable in mean square (and hence almost surely as well) if

$$\tilde{\mathcal{A}} := \text{diag}(-\lambda_1, -\lambda_2, -\lambda_3) - \Gamma$$

is a nonsingular M-matrix, where

$$\lambda_i = \lambda_{\max}(A_i + A_i^T), \quad i = 1, 2, 3.$$

To verify this is the case, we compute

$$\lambda_1 = -2.4385, \quad \lambda_2 = 1.20718, \quad \lambda_3 = -0.95067.$$

So the matrix $\tilde{\mathcal{A}}$ becomes

$$\tilde{\mathcal{A}} = \begin{pmatrix} 4.4385 & -1 & -1 \\ -3 & 2.79282 & -1 \\ -1 & -1 & 2.95067 \end{pmatrix}.$$

Using Maple we can easily compute the inverse matrix of $\tilde{\mathcal{A}}$:

$$\tilde{\mathcal{A}}^{-1} = \begin{pmatrix} 0.43902 & 0.23954 & 0.22997 \\ 0.59735 & 0.73344 & 0.45101 \\ 0.35123 & 0.32975 & 0.56969 \end{pmatrix}.$$

This implies, by Theorem 2.10, that $\tilde{\mathcal{A}}$ is a nonsingular M-matrix. We can therefore conclude that equation (9.10) with the parameters specified above is indeed exponentially stable in mean square.

However, as explained before, the abrupt changes of structure and parameters in the hybrid systems are usually caused by phenomena such as component failures or repairs, changing subsystem interconnections, abrupt environmental disturbances and the effects of time delay. The question is: *If we take parameter uncertainty and environmental noise as well as time* delay into account, what extent of the uncertainty and environmental noise can the system tolerate so that it remains stable?

To explain how we can apply Theorem 9.1 to answer the question, assume the system under the environmental noise *etc*, is described by a three-dimensional SDIS with Markovian switching of the form

$$dx(t) = [A_r x(t) + \Delta G_r x(t-\delta)]dt$$
$$+ \left[\Delta C_r x(t) + \Delta D_r x(t-\delta)\right]dB(t). \quad (9.11)$$

on $t \geq 0$, where A_r's are the same as above and

$$\Delta G_i \in [-\bar{G}_i, \bar{G}_i], \quad \Delta C_i \in [-\bar{C}_i, \bar{C}_i], \quad \Delta D_i \in [-\bar{D}_i, \bar{D}_i].$$

For illustration, we let $\delta = \delta(t) = 0.1 \sin^2 t$ and assume that the bounds for the uncertainties are known as follows

$$\|\bar{G}_1\| = \|\bar{G}_2\| = \|\bar{G}_3\| = 0.1,$$
$$\|\bar{C}_1\| = \|\bar{C}_2\| = \|\bar{C}_3\| = 0.2, \quad (9.12)$$
$$\|\bar{D}_1\| = \|\bar{D}_2\| = \|\bar{D}_3\| = 0.3.$$

So the matrix $\mathcal{A}$ defined in Theorem 9.1 becomes

$$\mathcal{A} = \begin{pmatrix} 4.4585 & -1 & -1 \\ -3 & 2.61282 & -1 \\ -1 & -1 & 2.77067 \end{pmatrix},$$

while the parameters $\sigma_1 = \sigma_2 = \sigma_3 = 0.14$. Compute

$$\mathcal{A}^{-1} = \begin{pmatrix} 0.48393 & 0.29246 & 0.28022 \\ 0.72226 & 0.88056 & 0.57850 \\ 0.43534 & 0.42337 & 0.67085 \end{pmatrix},$$

which implies, by Theorem 2.10, that $\mathcal{A}$ is a nonsingular M-matrix. Also compute

$$(q_1, q_2, q_3)^T = \mathcal{A}^{-1}\vec{1} = (1.05661, 2.18132, 1.52957)^T.$$

On the other hand, noting $d\delta(t)/dt = 0.2 \sin t \cos t \leq 0.1 = \bar{\delta}$, we observe that

$$2q_1\sigma_1 = 0.296, \quad 2q_2\sigma_2 = 0.611, \quad 2q_3\sigma_3 = 0.428$$

which are all less than $1 - \bar{\delta} = 0.9$. Therefore, by Theorem 9.1 we can conclude that the SDIS (9.11) is mean square as well as almost surely exponentially stable under the bounds for uncertainties specified by (9.12).

Example 9.3 However, we may not know all the bounds for the uncertainties. For example, we may only know the following bounds

$$\|\bar{G}_1\| = \|\bar{G}_2\| = \|\bar{G}_3\| = 0.05, \quad \|\bar{C}_1\| = \|\bar{C}_2\| = \|\bar{C}_3\| = 0.1 \quad (9.13)$$

but the bounds for ΔD_i's are unknown. In this case, if we can estimate the bounds for ΔD_i's so that the SDIS (9.11) will still be exponentially stable in mean square, we then know that the system will be stable as long as we can control the uncertainties ΔD_i's within the bounds. To explain how our theory can be applied to solve this problem, we note that the matrix $\mathcal{A}$ becomes

$$\mathcal{A} = \begin{pmatrix} 4.3685 & -1 & -1 \\ -3 & 2.72282 & -1 \\ -1 & -1 & 2.88067 \end{pmatrix},$$

while

$$\sigma_1 = 0.025 + \|\bar{D}_1\|^2, \ \sigma_2 = 0.025 + \|\bar{D}_2\|^2, \ \sigma_3 = 0.025 + \|\bar{D}_3\|^2.$$

Compute

$$\mathcal{A}^{-1} = \begin{pmatrix} 0.47095 & 0.26706 & 0.25619 \\ 0.66354 & 0.79720 & 0.50708 \\ 0.39383 & 0.36945 & 0.61211 \end{pmatrix}$$

whence $\mathcal{A}$ is a nonsingular M-matrix. Compute then

$$\mathcal{A}^{-1}\vec{1} = (0.9942, 1.9678, 1.3754)^T.$$

For equation (9.11) to be exponentially stable in mean sqaure, it is sufficient to require

$$0.9942(0.05 + 2\|\bar{D}_1\|^2) < 0.9,$$
$$1.9678(0.05 + 2\|\bar{D}_2\|^2) < 0.9,$$
$$1.3754(0.05 + 2\|\bar{D}_3\|^2) < 0.9.$$

Solving these inequalities gives

$$\|\bar{D}_1\| < 0.6539, \quad \|\bar{D}_2\| < 0.4513, \quad \|\bar{D}_3\| < 0.5497. \quad (9.14)$$

By Theorem 9.1, we can therefore conclude that the SDIS (9.11) is exponentially stable in mean square as long as (9.13) and (9.14) are satisfied.

9.4 Razumikhin Technology on SDISs

However, condition (9.9) is sometimes restrictive. In this section we will remove this condition. In other words, *we will no longer require $\delta(t)$ be differentiable in this section.*

The method used is the Razumikhin technology developed in the previous chapter. To see this more clearly, let us define f, $g : C([-\tau, 0]; \mathbb{R}^n) \times \mathbb{R}_+ \times \mathbb{S} \to \mathbb{R}^n$ by

$$f(\varphi, t, i) = (A_i + \Delta A_i)\varphi(0) + (G_i + \Delta G_i)\varphi(-\delta(t))$$

and

$$g(\varphi, t, i) = (C_i + \Delta C_i)\varphi(0) + (D_i + \Delta D_i)\varphi(-\delta(t)).$$

Then SDIS (9.7) can be written as

$$dx(t) = f(x_t, t, r(t))dt + g(x_t, t, r(t))dB(t)$$

which is in the form of the SFDE discussed in Chapter 8 so the theory established there, in particular, the Razumikhin technology can be applied to equation (9.7).

9.4.1 Delay Independent Criteria

The following stability criterion is delay independent as the only condition (9.15) imposed is independent of the time lag.

Theorem 9.4 *Assume that there exist symmetric positive-definite $n \times n$-matrices Q_i, $1 \le i \le N$, such that*

$$\max_{1 \le i \le N} \left[\lambda_{\max}\left(Q_i A_i + A_i^T Q_i + \sum_{j=1}^{N} \gamma_{ij} Q_j \right) \right.$$

$$+ \|Q_i\| \left(2a_i + c_i^2 + (g_i + c_i d_i)\sqrt{\frac{\alpha_2}{\alpha_1}} \right) \right]$$

$$+ \max_{1 \le i \le N} \left[\|Q_i\| \left(\frac{d_i^2 \alpha_2}{\alpha_1} + (g_i + c_i d_i)\sqrt{\frac{\alpha_2}{\alpha_1}} \right) \right] < 0, \quad (9.15)$$

where

$$\alpha_1 = \min_{1 \le i \le N} \lambda_{\min}(Q_i), \quad \alpha_2 = \max_{1 \le i \le N} \lambda_{\max}(Q_i),$$

$$a_i = \|\bar{A}_i\|, \quad g_i = \|G_i\| + \|\bar{G}_i\|, \quad c_i = \|C_i\| + \|\bar{C}_i\|, \quad d_i = \|D_i\| + \|\bar{D}_i\|.$$

Then the trivial solution of equation (9.7) is exponentially stable in mean square. More precisely, for all $\xi \in L^2_{\mathcal{F}_0}([-\tau,0];\mathbb{R}^n)$, the solution of equation (9.7) has the property that

$$\limsup_{t \to \infty} \frac{1}{t} \log(\mathbb{E}|x(t;\xi)|^2) \le -\frac{\log(q)}{\tau}, \qquad (9.16)$$

where $q > 1$ is the unique root to the following equation

$$\max_{1 \le i \le N} \left[\lambda_{\max}\left(Q_i A_i + A_i^T Q_i + \sum_{j=1}^N \gamma_{ij} Q_j\right) \right.$$
$$+ \|Q_i\|\left(2a_i + c_i^2 + (g_i + c_i d_i)\sqrt{\frac{q\alpha_2}{\alpha_1}}\right)\right]$$
$$+ \max_{1 \le i \le N}\left[\|Q_i\|\left(\frac{d_i^2 q\alpha_2}{\alpha_1} + (g_i + c_i d_i)\sqrt{\frac{q\alpha_2}{\alpha_1}}\right)\right] = -\alpha_2 \frac{\log(q)}{\tau}. \qquad (9.17)$$

Proof. The proof is an application of Theorem 8.9 with $p = 2$. For this purpose, we define $V \in C^{2,1}(\mathbb{R}^n \times [-\tau,\infty) \times \mathbb{S}; \mathbb{R}_+)$ by

$$V(x,t,i) = x^T Q_i x.$$

Since V defined is independent of t, we shall write $V(x,t,i) = V(x,i)$. Accordingly, the operator $\mathcal{L}V : C([-\tau,0];\mathbb{R}^n) \times \mathbb{R}_+ \times \mathbb{S} \to \mathbb{R}$ associated with equation (9.7) becomes

$$\mathcal{L}V(\varphi,t,i) = 2\varphi^T(0)Q_i\Big[(A_i + \Delta A_i)\varphi(0) + (G_i + \Delta G_i)\varphi(-\delta(t))\Big]$$
$$+ \Big[(C_i + \Delta C_i)\varphi(0) + (D_i + \Delta D_i)\varphi(-\delta(t))\Big]^T$$
$$\times Q_i\Big[(C_i + \Delta C_i)\varphi(0) + (D_i + \Delta D_i)\varphi(-\delta(t))\Big]$$
$$+ \sum_{j=1}^N \gamma_{ij}\varphi^T(0)Q_j\varphi(0). \qquad (9.18)$$

Clearly
$$\alpha_1 |x|^2 \le V(x,i) \le \alpha_2 |x|^2.$$

If we can show that
$$\mathbb{E}\left[\max_{1 \le i \le N} \mathcal{L}V(\phi,t,i)\right] \le -\frac{\log(q)}{\tau}\mathbb{E}\left[\max_{1 \le i \le N} V(\phi(0),i)\right] \qquad (9.19)$$

for all $t \geq 0$ and those $\phi = \{\phi(\theta) : -\tau \leq \theta \leq 0\} \in L^2_{\mathcal{F}_t}([-\tau,0];\mathbb{R}^n)$ satisfying

$$\mathbb{E}\left[\min_{1\leq i\leq N} V(\phi(\theta),i)\right] < q\mathbb{E}\left[\max_{1\leq i\leq N} V(\phi(0),i)\right], \quad \forall \theta \in [-\tau,0], \qquad (9.20)$$

then the required assertion (9.16) follows from Theorem 8.9. To show (9.19) under (9.20), we compute by Lemma 9.1 that for $(\varphi,t,i) \in C([-\tau,0];\mathbb{R}^n) \times \mathbb{R}_+ \times \mathbb{S}$,

$$2\varphi^T(0)Q_i\Big[(A_i + \Delta A_i)\varphi(0) + (G_i + \Delta G_i)\varphi(-\delta(t))\Big]$$
$$\leq \varphi^T(0)(Q_i A_i + A_i^T Q_i)\varphi(0) + 2\|Q_i\|\|\Delta A_i\||\varphi(0)|^2$$
$$+ 2|\varphi^T(0)|\|Q_i\|(\|G_i\| + \|\Delta G_i\|)|\varphi(-\delta(t))|$$
$$\leq \varphi^T(0)(Q_i A_i + A_i^T Q_i)\varphi(0) + \|Q_i\|\left(2a_i + g_i\sqrt{\frac{q\alpha_2}{\alpha_1}}\right)|\varphi(0)|^2$$
$$+ \|Q_i\|g_i\sqrt{\frac{\alpha_1}{q\alpha_2}}\,|\varphi(-\delta(t))|^2, \qquad (9.21)$$

and

$$\Big[(C_i + \Delta C_i)\varphi(0) + (D_i + \Delta D_i)\varphi(-\delta(t))\Big]^T$$
$$\times Q_i\Big[(C_i + \Delta C_i)\varphi(0) + (D_i + \Delta D_i)\varphi(-\delta(t))\Big]$$
$$\leq \|Q_i\|(c_i|\varphi(0)| + d_i|\varphi(-\delta(t))|)^2$$
$$\leq \|Q_i\|\left[\left(c_i^2 + c_i d_i\sqrt{\frac{q\alpha_2}{\alpha_1}}\right)|\varphi(0)|^2\right.$$
$$\left.+ \left(d_i^2 + c_i d_i\sqrt{\frac{\alpha_1}{q\alpha_2}}\right)|\varphi(-\delta(t))|^2\right]. \qquad (9.22)$$

Substituting (9.21) and (9.22) into (9.18) yields

$$\mathcal{L}V(\varphi,t,i) \leq \varphi^T(0)\Big(Q_i A_i + A_i^T Q_i + \sum_{j=1}^N \gamma_{ij} Q_j\Big)\varphi(0)$$
$$+ \|Q_i\|\left(2a_i + c_i^2 + (g_i + c_i d_i)\sqrt{\frac{q\alpha_2}{\alpha_1}}\right)|\varphi(0)|^2$$
$$+ \|Q_i\|\left(d_i^2 + (g_i + c_i d_i)\sqrt{\frac{\alpha_1}{q\alpha_2}}\right)|\varphi(-\delta(t))|^2$$
$$\leq \lambda_1|\varphi(0)|^2 + \lambda_2|\varphi(-\delta(t))|^2, \qquad (9.23)$$

where

$$\lambda_1 = \max_{1 \leq i \leq N} \left[\lambda_{\max}\left(Q_i A_i + A_i^T Q_i + \sum_{j=1}^{N} \gamma_{ij} Q_j\right) \right.$$
$$\left. + \|Q_i\|\left(2a_i + c_i^2 + (g_i + c_i d_i)\sqrt{\frac{q\alpha_2}{\alpha_1}}\right) \right]$$

and

$$\lambda_2 = \max_{1 \leq i \leq N} \left[\|Q_i\|\left(d_i^2 + (g_i + c_i d_i)\sqrt{\frac{\alpha_1}{q\alpha_2}}\right) \right].$$

Now, for any $t \geq 0$ and any $\phi = \{\phi(\theta) : -\tau \leq \theta \leq 0\} \in L^2_{\mathcal{F}_t}([-\tau, 0]; \mathbb{R}^n)$ satisfying (9.20), we have

$$\mathbb{E}|\phi(\theta)|^2 < \frac{q\alpha_2}{\alpha_1} \mathbb{E}|\phi(0)|^2, \quad \forall \theta \in [-\tau, 0].$$

Making use of this inequality and (9.23), we have

$$\mathbb{E}\left[\max_{1 \leq i \leq N} \mathcal{L}V(\phi, t, i)\right] \leq \lambda_1 \mathbb{E}|\phi(0)|^2 + \lambda_2 \mathbb{E}|\phi(-\delta(t))|^2$$
$$\leq \left(\lambda_1 + \frac{\lambda_2 q \alpha_2}{\alpha_1}\right) \mathbb{E}|\phi(0)|^2. \quad (9.24)$$

By (9.17) we note that $\lambda_1 + \frac{\lambda_2 q \alpha_2}{\alpha_1} < 0$ so

$$\mathbb{E}\left[\max_{1 \leq i \leq N} \mathcal{L}V(\phi, t, i)\right] \leq \frac{1}{\alpha_2}\left(\lambda_1 + \frac{\lambda_2 q \alpha_2}{\alpha_1}\right) \mathbb{E}\left[\max_{1 \leq i \leq N} V(\phi(0), i)\right]. \quad (9.25)$$

But by (9.17),

$$\frac{1}{\alpha_2}\left(\lambda_1 + \frac{\lambda_2 q \alpha_2}{\alpha_1}\right) = -\frac{\log(q)}{\tau}.$$

Hence

$$\mathbb{E}\left[\max_{1 \leq i \leq N} \mathcal{L}V(\phi, t, i)\right] \leq -\frac{\log(q)}{\tau} \mathbb{E}\left[\max_{1 \leq i \leq N} V(\phi(0), i)\right],$$

which is the desired inequality (9.19). The proof is therefore complete. □

Let us now apply Theorem 9.4 to establish a very useful criterion in terms of an M-matrix which can be verified much more easily.

Theorem 9.5 *Let a_i, g_i, c_i and d_i be the same as defined in Theorem 9.4. Define the diagonal matrix*

$$K = \mathrm{diag}\Big(-\lambda_{\max}(A_1 + A_1^T) - 2a_1 - c_1^2,$$
$$\cdots, -\lambda_{\max}(A_N + A_N^T) - 2a_N - c_N^2\Big).$$

If $K - \Gamma$ is a nonsingular matrix, then

$$\vec{q} = (q_1, \cdots, q_N)^T := (K - \Gamma)^{-1}\vec{1} \gg 0, \tag{9.26}$$

where $\vec{1} = (1, \cdots, 1)^T$. If, moreover, g_i and d_i are sufficiently small such that

$$\max_{1 \le i \le N}\Big[q_i(g_i + c_i d_i)\sqrt{\frac{\alpha_2}{\alpha_1}}\Big] + \max_{1 \le i \le N}\Big[q_i\Big(\frac{d_i^2 \alpha_2}{\alpha_1} + (g_i + c_i d_i)\sqrt{\frac{\alpha_2}{\alpha_1}}\Big)\Big] < 1, \tag{9.27}$$

where $\alpha_1 = \min_{1 \le i \le N} q_i$ and $\alpha_2 = \max_{1 \le i \le N} q_i$, then equation (9.7) is exponentially stable in mean square. More precisely, for every $\xi \in L^2_{\mathcal{F}_0}([-\tau, 0]; \mathbb{R}^n)$, the solution of equation (9.7) has the property that

$$\limsup_{t \to \infty} \frac{1}{t} \log(\mathbb{E}|x(t;\xi)|^2) \le -\frac{\log(q)}{\tau}, \tag{9.28}$$

where $q > 1$ is the unique root to the following equation

$$\max_{1 \le i \le N}\Big[q_i(g_i + c_i d_i)\sqrt{\frac{q\alpha_2}{\alpha_1}}\Big]$$
$$+ \max_{1 \le i \le N}\Big[q_i\Big(\frac{d_i^2 q\alpha_2}{\alpha_1} + (g_i + c_i d_i)\sqrt{\frac{q\alpha_2}{\alpha_1}}\Big)\Big] = 1 - \alpha_2 \frac{\log(q)}{\tau}. \tag{9.29}$$

Proof. By Theorem 2.10, we observe that all the elements of $(K - \Gamma)^{-1}$ are non-negative. Since $(K - \Gamma)^{-1}$ is invertible, each of its rows must have at least one non-zero, and hence positive element. We must therefore have (9.26), namely all q_i's are positive. Note

$$(-K + \Gamma)\vec{q} = -\vec{1},$$

that is,

$$\lambda_{\max}(A_i + A_i^T)q_i + \sum_{j=1}^{N} \gamma_{ij} q_j + (2a_i + c_i^2)q_i = -1, \quad 1 \le i \le N. \tag{9.30}$$

To apply Theorem 9.4, we let $Q_i = q_i I$ for $i \in S$, where I is the $n \times n$-identity matrix. Hence, α_1 and α_2 defined in this theorem coincide with those defined in Theorem 9.4. Then, by (9.30) and (9.27), we compute

$$\max_{1 \le i \le N} \left[\lambda_{\max}\left(Q_i A_i + A_i^T Q_i + \sum_{j=1}^N \gamma_{ij} Q_j\right) \right.$$
$$+ \left. \|Q_i\|\left(2a_i + c_i^2 + (g_i + c_i d_i)\sqrt{\frac{\alpha_2}{\alpha_1}}\right) \right]$$
$$+ \max_{1 \le i \le N}\left[\|Q_i\|\left(\frac{d_i^2 \alpha_2}{\alpha_1} + (g_i + c_i d_i)\sqrt{\frac{\alpha_2}{\alpha_1}}\right)\right]$$

$$= \max_{1 \le i \le N} \left[\lambda_{\max}(A_i + A_i^T)q_i + \sum_{j=1}^N \gamma_{ij} q_j \right.$$
$$+ q_i\left(2a_i + c_i^2 + (g_i + c_i d_i)\sqrt{\frac{\alpha_2}{\alpha_1}}\right)\Big]$$
$$+ \max_{1 \le i \le N}\left[q_i\left(\frac{d_i^2 \alpha_2}{\alpha_1} + (g_i + c_i d_i)\sqrt{\frac{\alpha_2}{\alpha_1}}\right)\right]$$

$$= \max_{1 \le i \le N} \left[\lambda_{\max}(A_i + A_i^T)q_i + \sum_{j=1}^N \gamma_{ij} q_j \right.$$
$$+ (2a_i + c_i^2)q_i + q_i(g_i + c_i d_i)\sqrt{\frac{\alpha_2}{\alpha_1}}\Big]$$
$$+ \max_{1 \le i \le N}\left[q_i\left(\frac{d_i^2 \alpha_2}{\alpha_1} + (g_i + c_i d_i)\sqrt{\frac{\alpha_2}{\alpha_1}}\right)\right]$$

$$= -1 + \max_{1 \le i \le N}\left[q_i(g_i + c_i d_i)\sqrt{\frac{\alpha_2}{\alpha_1}}\right]$$
$$+ \max_{1 \le i \le N}\left[q_i\left(\frac{d_i^2 \alpha_2}{\alpha_1} + (g_i + c_i d_i)\sqrt{\frac{\alpha_2}{\alpha_1}}\right)\right]$$
$$< 0.$$

That is, condition (9.15) of Theorem 9.4 is fulfilled. The required assertion (9.28) follows therefore from Theorem 9.4. □

It is interesting to compare Theorem 9.4 with Theorem 9.5. Theorem 9.4 depends on the choices of matrices Q_i while Theorem 9.5 only requires to verify the specified matrix $K - \Gamma$ to be an M-matrix. Theorem 9.4 is more general while Theorem 9.5 is much simpler. The advantage of Theorem 9.4 is that it gives more flexibility in applications but the drawback is that one needs to construct a number of matrices required, while Theorem 9.5 is user friendly but it covers less situation as Theorem 9.4 does.

Two stability criteria established above are independent of the time delay $\delta(t)$ as long as it is a bounded function of t. It does not matter if the bound τ of $\delta(t)$ is large or small. Such criteria require in general that the corresponding non-delay stochastic interval system (SIS)

$$dx(t) = (A_r + \Delta A_r)x(t)dt + (C_r + \Delta C_r)x(t)dB(t)$$

be stable (in fact, we will see in the next section that under the conditions of either Theorem 9.4 or 9.5, this SIS is exponentially stable in mean square), while the delay part is treated as a stochastic perturbation to the stable system. In other words, the delay part is regarded as a destabilising effect.

9.4.2 Delay Dependent Criteria

However, it is known that the delay part could have a stabilising effect as well. In what follows we will make use of the stablising effect of the delay part to establish delay dependent criteria.

Write equation (9.7) as

$$\begin{aligned}dx(t) = &\Big[(A_r + G_r + \Delta A_r)x(t) \\ &- G_r[x(t) - x(t-\delta)] + \Delta G_r x(t-\delta))\Big]dt \\ &+ \Big[(C_r + \Delta C_r)x(t) + (D_r + \Delta D_r)x(t-\delta)\Big]dB(t) \end{aligned} \quad (9.31)$$

on $t \geq \tau$ with

$$\begin{aligned} &x(t) - x(t-\delta(t)) \\ &= \int_{t-\delta(t)}^{t} \Big[(A_r + \Delta A_r)x(s) + (G_r + \Delta G_r)x(s-\delta(s))\Big]ds \\ &+ \int_{t-\delta(t)}^{t} \Big[(C_r + \Delta C_r)x(s) + (D_r + \Delta D_r)x(s-\delta(s))\Big]dB(s). \end{aligned} \quad (9.32)$$

Define $C([-2\tau, 0]; \mathbb{R}^n)$ and $L^2_{\mathcal{F}_t}([-2\tau, 0]; \mathbb{R}^n)$ similarly as before. Denote by $L^2(\Omega; \mathbb{R}^n)$ the family of all $\mathbb{R}^n$-valued random variables X such that $\mathbb{E}|X|^2 < \infty$. For each $t \geq \tau$, define an operator $F(t, \cdot) : L^2_{\mathcal{F}_t}([-2\tau, 0]; \mathbb{R}^n) \to$

$L^2(\Omega; \mathbb{R}^n)$ by

$$F(t, \phi)$$
$$= \int_{t-\delta(t)}^{t} \Big[(A_r + \Delta A_r)\phi(s-t) + (G_r + \Delta G_r)\phi(s - \delta(s) - t)\Big] ds$$
$$+ \int_{t-\delta(t)}^{t} \Big[(C_r + \Delta C_r)\phi(s-t)$$
$$+ (D_r + \Delta D_r)\phi(s - \delta(s) - t)\Big] dB(s). \qquad (9.33)$$

Moreover, let $\hat{x}(t) = \{x(t + \theta) : -2\tau \leq \theta \leq 0\}$ on $t \geq \tau$ which is regarded as a $C([-2\tau, 0]; \mathbb{R}^n)$-valued stochastic process. With these new notations, we note that

$$x(t) - x(t - \delta(t)) = F(t, \hat{x}(t))$$

whence, via (9.31), equation (9.7) can be regarded as the equation

$$dx(t) = \Big[(A_r + G_r + \Delta A_r)x(t) - G_r F(t, \hat{x}(t)) + \Delta G_r x(t - \delta)\Big] dt$$
$$+ \Big[(C_r + \Delta C_r)x(t) + (D_r + \Delta D_r)x(t - \delta)\Big] dB(t) \qquad (9.34)$$

on $t \geq \tau$ with initial data $x(t) = \xi(t)$ for $t \in [-\tau, 0]$ and $x(t) = x(t; \xi)$ for $t \in [0, \tau]$. It should be pointed out that equation (9.34) has time lag 2τ instead of τ. Let us prepare another lemma.

Lemma 9.3 *The operator F defined by (9.33) has the property that*

$$\mathbb{E}|F(t, \phi)|^2 \leq K_\tau \sup_{-2\tau \leq \theta \leq 0} \mathbb{E}|\phi(\theta)|^2$$

for all $t \geq \tau$ and $\phi \in L^2_{\mathcal{F}_t}([-2\tau, 0]; \mathbb{R}^n)$, where $K_\tau = 4\tau^2(\bar{a} + \bar{b}) + 4\tau(\bar{c} + \bar{d})$ in which

$$\bar{a} = \max_{1 \leq i \leq N}(\|A_i\| + \|\bar{A}_i\|)^2, \quad \bar{b} = \max_{1 \leq i \leq N}(\|G_i\| + \|\bar{G}_i\|)^2,$$

$$\bar{c} = \max_{1 \leq i \leq N}(\|C_i\| + \|\bar{C}_i\|)^2, \quad \bar{d} = \max_{1 \leq i \leq N}(\|D_i\| + \|\bar{D}_i\|)^2.$$

Proof. Compute from (9.33) that

$$\mathbb{E}|F(t,\phi)|^2 \leq 2\mathbb{E}\left|\int_{t-\delta(t)}^{t}\left[(A_{r(s)}+\Delta A_{r(s)})\phi(s-t)+(G_{r(s)}\right.\right.$$
$$\left.\left.+\Delta G_{r(s)})\phi(s-\delta(s)-t)\right]ds\right|^2$$
$$+2\mathbb{E}\left|\int_{t-\delta(t)}^{t}\left[(C_{r(s)}+\Delta C_{r(s)})\phi(s-t)\right.\right.$$
$$\left.\left.+(D_{r(s)}+\Delta D_{r(s)})\phi(s-\delta(s)-t)\right]dB(s)\right|^2$$
$$\leq 2\tau\mathbb{E}\int_{t-\delta(t)}^{t}\left|(A_{r(s)}+\Delta A_{r(s)})\phi(s-t)\right.$$
$$\left.+(G_{r(s)}+\Delta G_{r(s)})\phi(s-\delta(s)-t)\right|^2 ds$$
$$+2\int_{t-\delta(t)}^{t}\left|(C_{r(s)}+\Delta C_{r(s)})\phi(s-t)\right.$$
$$\left.+(D_{r(s)}+\Delta D_{r(s)})\phi(s-\delta(s)-t)\right|^2 ds$$
$$\leq 4\tau\mathbb{E}\int_{t-\tau}^{t}\left[\bar{a}|\phi(s-t)|^2+\bar{b}|\phi(s-\delta(s)-t)|^2\right]ds$$
$$+4\int_{t-\tau}^{t}\left[\bar{c}|\phi(s-t)|^2+\bar{d}|\phi(s-\delta(s)-t)|^2\right]ds$$
$$\leq K_{\tau}\sup_{-2\tau\leq\theta\leq 0}\mathbb{E}|\phi(\theta)|^2$$

as required. □

Theorem 9.6 *Assume that there exist symmetric positive-definite $n\times n$-matrices Q_i, $1\leq i\leq N$, such that*

$$\max_{1\leq i\leq N}\left[\lambda_{\max}\left(Q_i(A_i+G_i)+(A_i+G_i)^T Q_i+\sum_{j=1}^{N}\gamma_{ij}Q_j\right)\right.$$
$$\left.+\|Q_i\|\left(2a_i+c_i^2+[c_i d_i+\|\bar{G}_i\|+\|G_i\|\sqrt{K_\tau}\,]\sqrt{\frac{\alpha_2}{\alpha_1}}\,\right)\right]$$
$$+\max_{1\leq i\leq N}\left[\|Q_i\|\left(\frac{d_i^2\alpha_2}{\alpha_1}+(\|\bar{G}_i\|+c_i d_i)\sqrt{\frac{\alpha_2}{\alpha_1}}\,\right)\right]$$
$$+\max_{1\leq i\leq N}\left[\|Q_i\|\|G_i\|\sqrt{\frac{\alpha_2 K_\tau}{\alpha_1}}\,\right]<0, \tag{9.35}$$

where $\alpha_1, \alpha_2, a_i, c_i, d_i$ and K_τ are the same as defined in Theorem 9.4 and Lemma 9.3, respectively. Then the trivial solution of equation (9.7) is exponentially stable in mean square. More precisely, for all $\xi \in L^2_{\mathcal{F}_0}([-\tau, 0]; \mathbb{R}^n)$, the solution of equation (9.7) has the property that

$$\limsup_{t \to \infty} \frac{1}{t} \log(\mathbb{E}|x(t;\xi)|^2) \leq -\frac{\log(q)}{\tau}, \qquad (9.36)$$

where $q > 1$ is the unique root to the following equation

$$\max_{1 \leq i \leq N} \left[\lambda_{\max}\left(Q_i(A_i + G_i) + (A_i + G_i)^T Q_i + \sum_{j=1}^{N} \gamma_{ij} Q_j \right) \right.$$
$$\left. + \|Q_i\| \left(2a_i + c_i^2 + [c_i d_i + \|\bar{G}_i\| + \|G_i\|\sqrt{K_\tau}] \sqrt{\frac{q\alpha_2}{\alpha_1}} \right) \right]$$
$$+ \max_{1 \leq i \leq N} \left[\|Q_i\| \left(\frac{d_i^2 q \alpha_2}{\alpha_1} + (\|\bar{G}_i\| + c_i d_i) \sqrt{\frac{q\alpha_2}{\alpha_1}} \right) \right]$$
$$+ \max_{1 \leq i \leq N} \left[\|Q_i\| \|G_i\| \sqrt{\frac{q\alpha_2 K_\tau}{\alpha_1}} \right] = -\alpha_2 \frac{\log(q)}{\tau}. \qquad (9.37)$$

Proof. The proof is an application of Theorem 8.9 to equation (9.34), which is an alternative form of equation (9.7). For this purpose, we still define $V \in C^{2,1}(\mathbb{R}^n \times [-\tau, \infty) \times \mathbb{S}; \mathbb{R}_+)$ by $V(x, t, i) = x^T Q_i x$, and again write $V(x, t, i) = V(x, i)$. Accordingly, the operator $\mathcal{L}V : C([-2\tau, 0]; \mathbb{R}^n) \times \mathbb{R}_+ \times \mathbb{S} \to R$ associated with equation (9.34) becomes

$$\mathcal{L}V(\varphi, t, i) = 2\varphi^T(0) Q_i \Big[(A_i + G_i + \Delta A_i)\varphi(0)$$
$$- G_i F(t, \varphi) + \Delta G_i \varphi(-\delta(t)) \Big]$$
$$+ \Big[(C_i + \Delta C_i)\varphi(0) + (D_i + \Delta D_i)\varphi(-\delta(t)) \Big]^T$$
$$\times Q_i \Big[(C_i + \Delta C_i)\varphi(0) + (D_i + \Delta D_i)\varphi(-\delta(t)) \Big]$$
$$+ \sum_{j=1}^{N} \gamma_{ij} \varphi^T(0) Q_j \varphi(0). \qquad (9.38)$$

If we can show that

$$\mathbb{E}\left[\max_{1 \leq i \leq N} \mathcal{L}V(\phi, t, i) \right] \leq -\frac{\log(q)}{\tau} \mathbb{E}\left[\max_{1 \leq i \leq N} V(\phi(0), i) \right] \qquad (9.39)$$

for all $t \geq \tau$ and those $\phi = \{\phi(\theta) : -2\tau \leq \theta \leq 0\} \in L^2_{\mathcal{F}_t}([-2\tau, 0]; \mathbb{R}^n)$ satisfying

$$\mathbb{E}\left[\min_{1\leq i\leq N} V(\phi(\theta), i)\right] < q\mathbb{E}\left[\max_{1\leq i\leq N} V(\phi(0), i)\right], \quad \forall \theta \in [-2\tau, 0], \quad (9.40)$$

then the required assertion (9.36) follows from Theorem 8.9. To show (9.39) under (9.40), we note that for $(\varphi, t, i) \in C([-2\tau, 0]; \mathbb{R}^n) \times \mathbb{R}_+ \times \mathbb{S}$, (9.22) still holds and we compute that

$$2\varphi^T(0)Q_i\Big[(A_i + G_i + \Delta A_i)\varphi(0) - G_i F(t, \varphi) + \Delta G_i \varphi(-\delta(t))\Big]$$
$$\leq \varphi^T(0)[Q_i(A_i + G_i) + (A_i + G_i)^T Q_i]\varphi(0) + 2\|Q_i\|\|\Delta A_i\|\|\varphi(0)\|^2$$
$$+ 2|\varphi^T(0)|\|Q_i\|(\|G_i\||F(t,\varphi)| + \|\Delta G_i\||\varphi(-\delta(t))|)$$
$$\leq \varphi^T(0)[Q_i(A_i + G_i) + (A_i + G_i)^T Q_i]\varphi(0)$$
$$+ \|Q_i\|\Big(2a_i + [\|\bar{G}_i\| + \|G_i\|\sqrt{K_\tau}\,]\sqrt{\frac{q\alpha_2}{\alpha_1}}\,\Big)|\varphi(0)|^2$$
$$+ \|Q_i\|\|\bar{G}_i\|\sqrt{\frac{\alpha_1}{q\alpha_2}}\,|\varphi(-\delta(t))|^2 + \|Q_i\|\|G_i\|\sqrt{\frac{\alpha_1}{K_\tau q\alpha_2}}\,|F(t,\varphi)|^2. \quad (9.41)$$

Substituting (9.41) and (9.22) into (9.38) yields

$$\mathcal{L}V(\varphi, t, i) \leq \varphi^T(0)\Big(Q_i(A_i + G_i) + (A_i + G_i)^T Q_i + \sum_{j=1}^N \gamma_{ij} Q_j\Big)\varphi(0)$$
$$+ \|Q_i\|\Big(2a_i + c_i^2 + [c_i d_i + \|\bar{G}_i\| + \|G_i\|\sqrt{K_\tau}\,]\sqrt{\frac{q\alpha_2}{\alpha_1}}\,\Big)|\varphi(0)|^2$$
$$+ \|Q_i\|\Big(d_i^2 + (\|\bar{G}_i\| + c_i d_i)\sqrt{\frac{\alpha_1}{q\alpha_2}}\,\Big)|\varphi(-\delta(t))|^2$$
$$+ \|Q_i\|\|G_i\|\sqrt{\frac{\alpha_1}{K_\tau q\alpha_2}}\,|F(t,\varphi)|^2$$
$$\leq \beta_1 |\varphi(0)|^2 + \beta_2 |\varphi(-\delta(t))|^2 + \beta_3 |F(t,\varphi)|^2, \quad (9.42)$$

where

$$\beta_1 = \max_{1\leq i\leq N}\Big[\lambda_{\max}\Big(Q_i(A_i + G_i) + (A_i + G_i)^T Q_i + \sum_{j=1}^N \gamma_{ij} Q_j\Big)$$
$$+ \|Q_i\|\Big(2a_i + c_i^2 + [c_i d_i + \|\bar{G}_i\| + \|G_i\|\sqrt{K_\tau}\,]\sqrt{\frac{q\alpha_2}{\alpha_1}}\,\Big)\Big],$$

$$\beta_2 = \max_{1\leq i\leq N}\left[\|Q_i\|\left(d_i^2 + (\|\bar{G}_i\| + c_i d_i)\sqrt{\frac{\alpha_1}{q\alpha_2}}\right)\right],$$

$$\beta_3 = \max_{1\leq i\leq N}\left[\|Q_i\|\|G_i\|\sqrt{\frac{\alpha_1}{K_\tau q\alpha_2}}\right].$$

Now, for any $t \geq \tau$ and any $\phi = \{\phi(\theta) : -2\tau \leq \theta \leq 0\} \in L^2_{\mathcal{F}_t}([-2\tau, 0]; \mathbb{R}^n)$ satisfying (9.40), we have

$$\mathbb{E}|\phi(\theta)|^2 < \frac{q\alpha_2}{\alpha_1}\mathbb{E}|\phi(0)|^2, \quad \forall \theta \in [-2\tau, 0].$$

Using this inequality and (9.42) and applying Lemma 9.3, we have

$$\mathbb{E}\left[\max_{1\leq i\leq N}\mathcal{L}V(\phi,t,i)\right] \leq \beta_1 \mathbb{E}|\phi(0)|^2 + \beta_2 \mathbb{E}|\phi(-\delta(t))|^2 + \beta_3 \mathbb{E}|F(t,\phi)|^2$$

$$\leq \left(\beta_1 + (\beta_2 + K_\tau \beta_2)\frac{q\alpha_2}{\alpha_1}\right)\mathbb{E}|\phi(0)|^2. \quad (9.43)$$

By (9.37) we note that $\beta_1 + (\beta_2 + K_\tau \beta_2)\frac{q\alpha_2}{\alpha_1} < 0$ so

$$\mathbb{E}\left[\max_{1\leq i\leq N}\mathcal{L}V(\phi,t,i)\right] \leq \frac{1}{\alpha_2}\left(\beta_1 + (\beta_2 + K_\tau \beta_2)\frac{q\alpha_2}{\alpha_1}\right)\mathbb{E}\left[\max_{1\leq i\leq N} V(\phi(0),i)\right].$$

But by (9.37),

$$\frac{1}{\alpha_2}\left(\beta_1 + (\beta_2 + K_\tau \beta_2)\frac{q\alpha_2}{\alpha_1}\right) = -\frac{\log(q)}{\tau}.$$

Hence

$$\mathbb{E}\left[\max_{1\leq i\leq N}\mathcal{L}V(\phi,t,i)\right] \leq -\frac{\log(q)}{\tau}\mathbb{E}\left[\max_{1\leq i\leq N} V(\phi(0),i)\right],$$

which is the desired inequality (9.39). □

The following is a new delay dependent criterion in terms of an M-matrix.

Theorem 9.7 *Let a_i, c_i and d_i be the same as defined in Theorem 9.4. Define the matrix*

$$H = \mathrm{diag}\Big(-\lambda_{\max}(A_1 + G_1 + A_1^T + G_1^T) - 2a_1 - c_1^2,$$
$$\cdots, -\lambda_{\max}(A_N + G_N + A_N^T + G_N^T) - 2a_N - c_N^2\Big).$$

If $H - \Gamma$ is a nonsingular matrix, then

$$\vec{q} = (q_1, \cdots, q_N)^T := (H - \Gamma)^{-1}\vec{1} \gg 0, \quad (9.44)$$

where $\vec{1} = (1, \cdots, 1)^T$. If, moreover, $\bar{G}_i, d_i$ and τ are sufficiently small such that

$$\max_{1 \leq i \leq N} \left[q_i[c_i d_i + \|\bar{G}_i\| + \|G_i\|\sqrt{K_\tau}\,] \sqrt{\frac{\alpha_2}{\alpha_1}}\,\right]$$
$$+ \max_{1 \leq i \leq N} \left[q_i \Big(\frac{d_i^2 \alpha_2}{\alpha_1} + (\|\bar{G}_i\| + c_i d_i)\sqrt{\frac{\alpha_2}{\alpha_1}}\,\Big)\right]$$
$$+ \max_{1 \leq i \leq N} \left[q_i \|G_i\| \sqrt{\frac{\alpha_2 K_\tau}{\alpha_1}}\,\right] < 1, \qquad (9.45)$$

where K_τ is the same as defined in Lemma 9.3, $\alpha_1 = \min_{1 \leq i \leq N} q_i$ and $\alpha_2 = \max_{1 \leq i \leq N} q_i$, then, for every $\xi \in L^2_{\mathcal{F}_0}([-\tau, 0]; \mathbb{R}^n)$, the solution of equation (9.7) has the property that

$$\limsup_{t \to \infty} \frac{1}{t} \log(\mathbb{E}|x(t;\xi)|^2) \leq -\frac{\log(q)}{\tau},$$

where $q > 1$ is the unique root to the following equation

$$\max_{1 \leq i \leq N} \left[q_i[c_i d_i + \|\bar{G}_i\| + \|G_i\|\sqrt{K_\tau}\,] \sqrt{\frac{q\alpha_2}{\alpha_1}}\,\right]$$
$$+ \max_{1 \leq i \leq N} \left[q_i \Big(\frac{d_i^2 q\alpha_2}{\alpha_1} + (\|\bar{G}_i\| + c_i d_i)\sqrt{\frac{q\alpha_2}{\alpha_1}}\,\Big)\right]$$
$$+ \max_{1 \leq i \leq N} \left[q_i \|G_i\| \sqrt{\frac{q\alpha_2 K_\tau}{\alpha_1}}\,\right] = 1 - \alpha_2 \frac{\log(q)}{\tau}. \qquad (9.46)$$

Proof. We observe that (9.44) follows from Theorem 2.10 and (9.44) gives that

$$\lambda_{\max}(A_i + G_i + A_i^T + G_i^T)q_i + \sum_{j=1}^N \gamma_{ij} q_j + (2a_i + c_i^2)q_i = -1 \qquad (9.47)$$

for all $i \in \mathbb{S}$. To apply Theorem 9.6, we let $Q_i = q_i I$ for $i \in \mathbb{S}$, where I is the $n \times n$-identity matrix. Hence, α_1 and α_2 defined in this theorem coincide with those defined in Theorem 9.4. Then, by (9.47) and (9.44), we

compute

$$\max_{1\leq i\leq N}\left[\lambda_{\max}\Big(Q_i(A_i+G_i)+(A_i+G_i)^T Q_i+\sum_{j=1}^{N}\gamma_{ij}Q_j\Big)\right.$$

$$\left.+\|Q_i\|\Big(2a_i+c_i^2+[c_i d_i+\|\bar{G}_i\|+\|G_i\|\sqrt{K_\tau}\,]\sqrt{\frac{\alpha_2}{\alpha_1}}\,\Big)\right]$$

$$+\max_{1\leq i\leq N}\left[\|Q_i\|\Big(\frac{d_i^2\alpha_1}{\alpha_2}+(\|\bar{G}_i\|+c_i d_i)\sqrt{\frac{\alpha_2}{\alpha_1}}\,\Big)\right]$$

$$+\max_{1\leq i\leq N}\left[\|Q_i\|\|G_i\|\sqrt{\frac{\alpha_2 K_\tau}{\alpha_1}}\,\right]$$

$$\leq \max_{1\leq i\leq N}\left[\lambda_{\max}(A_i+G_i+A_i^T+G_i^T)q_i+\sum_{j=1}^{N}\gamma_{ij}q_j+(2a_i+c_i^2)q_i\right]$$

$$+\max_{1\leq i\leq N}\left[q_i[c_i d_i+\|\bar{G}_i\|+\|G_i\|\sqrt{K_\tau}\,]\sqrt{\frac{\alpha_2}{\alpha_1}}\,\right]$$

$$+\max_{1\leq i\leq N}\left[q_i\Big(\frac{d_i^2\alpha_2}{\alpha_1}+(\|\bar{G}_i\|+c_i d_i)\sqrt{\frac{\alpha_2}{\alpha_1}}\,\Big)\right]+\max_{1\leq i\leq N}\left[q_i\|G_i\|\sqrt{\frac{\alpha_2 K_\tau}{\alpha_1}}\,\right]$$

$$< 0.$$

That is, condition (9.35) of Theorem 9.6 is fulfilled. Therefore, the assertion follows from Theorem 9.6. □

We will see in the next section that the condition that $H-\Gamma$ is a nonsingular M-matrix guarantees the mean-square exponential stability of the hybrid stochastic interval system (without delay)

$$dx(t)=(A_r+G_r+\Delta A_r)x(t)dt+(C_r+\Delta C_r)x(t)dB(t). \qquad (9.48)$$

Theorem 9.7 can therefore be interpreted as that if the non-delay interval system (9.48) is exponentially stable in mean square (in the sense that $H-\Gamma$ is a nonsingular M-matrix) and the matrices $\bar{G}_i, D_i, \bar{D}_i$ as well as the delay $\delta(t)$ are sufficiently small (in the sense that (9.45) holds), then the delay system (9.7) will remain to be exponentially stable in mean square.

9.4.3 *Examples*

Let us now discuss some examples to illustrate our theory. The examples demonstrate that our theory can be used in two ways: (i) when the system parameter matrices and the bounds for the uncertainties are all known our theory can be used to verify whether the underlying system is stable; (ii) when some bounds for the uncertainties are unknown our theory can be used to estimate these bounds so that the underlying system will remain stable should we be able to control the uncertainties within the bounds.

Example 9.8 Let $r(t)$, $t \geq 0$, be a right-continuous Markov chain taking values in $\mathbb{S} = \{1, 2\}$ with the generator

$$\Gamma = \begin{pmatrix} -6 & 6 \\ 1 & -1 \end{pmatrix}.$$

Consider a scalar SDIS with Markovian switching

$$dx(t) = (G_r + \Delta G_r)x(t - \delta(t))dt + \Delta D_r x(t - \delta(t))dB(t), \quad (9.49)$$

where $\delta : \mathbb{R}_+ \to [0, \tau]$ and

$$G_1 = 1, \quad G_2 = -4, \quad \Delta G_i, \Delta D_i \in [-\varepsilon, \varepsilon] \text{ for } i = 1, 2$$

with $\varepsilon > 0$. We want to obtain the bounds for τ and ε so that equation (9.49) will be exponentially stable in mean square. It is observed that the theorems established in the previous section can not be applied to this example, even if the time delay $\delta(t)$ is of constant or differentiable. The new contributions of this section are therefore more clear. To apply Theorem 9.7, we note that the matrix H defined there becomes

$$H = \begin{pmatrix} -2 & 0 \\ 0 & 8 \end{pmatrix},$$

whence

$$H - \Gamma = \begin{pmatrix} 4 & -6 \\ -1 & 9 \end{pmatrix}.$$

Compute

$$(H - \Gamma)^{-1} = \begin{pmatrix} 3/10 & 1/5 \\ 1/30 & 2/15 \end{pmatrix}.$$

By Theorem 2.10, $H - \Gamma$ is a nonsingular M-matrix. Compute

$$\vec{q} = (q_1, q_2)^T = (H - \Gamma)^{-1}\vec{1} = (1/2, 1/6)^T,$$

namely $q_1 = 1/2$ and $q_2 = 1/6$. Consequently, $\alpha_1 = 1/6$, $\alpha_2 = 1/2$ and $\alpha_2/\alpha_1 = 3$. Recalling the definition of K_τ in Lemma 9.3, we compute

$$K_\tau = 4\tau^2(4 + \varepsilon)^2 + 4\tau\varepsilon^2.$$

It is therefore easy to verify that condition (9.45) becomes

$$\max\{1.5(\varepsilon+\sqrt{K_\tau}), 0.5\varepsilon+2\sqrt{K_\tau}\}+(1.5+0.5\sqrt{3})\varepsilon+\frac{2}{\sqrt{3}}\sqrt{K_\tau}<1. \quad (9.50)$$

A sufficient condition for this is if

$$(3+0.5\sqrt{3})\varepsilon+\left(4+\frac{4}{\sqrt{3}}\right)\sqrt{\tau^2(4+\varepsilon)^2+\tau\varepsilon^2}<1,$$

which gives the bounds for ε and τ simultaneously, namely

$$\varepsilon<\frac{1}{3+0.5\sqrt{3}} \quad \text{and} \quad \tau<\frac{-\varepsilon^2+\sqrt{\varepsilon^4+4(4+\varepsilon)^2\bar{\varepsilon}^2}}{2(4+\varepsilon)^2}, \quad (9.51)$$

where

$$\bar{\varepsilon}=\frac{\sqrt{3}[1-(3+0.5\sqrt{3})\varepsilon]}{4(1+\sqrt{3})}.$$

We therefore conclude by Theorem 9.7 that if ε and τ are sufficiently small for (9.51) to hold, then equation (9.49) is exponentially stable in mean square. For instance, if $\varepsilon<0.1$, then (9.51) gives $\tau<0.1358$.

Example 9.9 Let $r(t)$ be a right-continuous Markov chain taking values in $\mathbb{S}=\{1,2,3\}$ with the generator

$$\Gamma=\begin{pmatrix}-2 & 1 & 1 \\ 3 & -4 & 1 \\ 1 & 1 & -2\end{pmatrix}.$$

Consider the three-dimensional SDIS with Markovian switching

$$dx(t)=[A_r x(t)+\Delta G_r x(t-\delta)]dt$$
$$+[\Delta C_r x(t)+\Delta D_r x(t-\delta)]dB(t), \quad (9.52)$$

where

$$A_1=\begin{pmatrix}-2 & 1 & -2 \\ 2 & -2 & 1 \\ 1 & -2 & -3\end{pmatrix}, \quad A_2=\begin{pmatrix}0.5 & 1 & 0.5 \\ -0.8 & 0.5 & 1 \\ -0.7 & -0.9 & 0.2\end{pmatrix},$$

$$A_3=\begin{pmatrix}-0.5 & -0.9 & -1 \\ 1 & -0.6 & -0.7 \\ 0.8 & 1 & -1\end{pmatrix},$$

and
$$\Delta G_i \in [-\bar{G}_i, \bar{G}_i], \quad \Delta C_i \in [-\bar{C}_i, \bar{C}_i], \quad \Delta D_i \in [-\bar{D}_i, \bar{D}_i]$$
for $i = 1, 2, 3$ with
$$\begin{aligned} \|\bar{G}_1\| &= 0.15, \quad \|\bar{G}_2\| = 0.05, \quad \|\bar{G}_3\| = 0.1, \\ \|\bar{C}_1\| &= \|\bar{C}_2\| = \|\bar{C}_3\| = 0.2, \\ \|\bar{D}_1\| &= \|\bar{D}_2\| = \|\bar{D}_3\| = 0.3. \end{aligned} \quad (9.53)$$

It is easy to compute
$$\lambda_{\max}(A_i + A_i^T) = \begin{cases} -2.43850 & \text{if } i = 1, \\ 1.20718 & \text{if } i = 2, \\ -0.95067 & \text{if } i = 3. \end{cases}$$

So the matrix K defined in Theorem 9.5 becomes
$$K = \text{diag}(2.39850, -1.24718, 0.91077).$$

Hence
$$K - \Gamma = \begin{pmatrix} 9.A.39850 & -1 & -1 \\ -3 & 2.75282 & -1 \\ -1 & -1 & 2.91077 \end{pmatrix}.$$

Compute
$$(K - \Gamma)^{-1} = \begin{pmatrix} 0.36625 & 0.20424 & 0.19599 \\ 0.50826 & 0.69851 & 0.41459 \\ 0.30044 & 0.31014 & 0.55332 \end{pmatrix}.$$

By Theorem 2.10, $K - \Gamma$ is a nonsingular M-matrix. Then
$$\vec{q} = (q_1, q_2, q_3)^T = (K - \Gamma)^{-1}\vec{1} = (0.76648, 1.62138, 1.16391)^T,$$
giving $\alpha_1 = 0.76648$, $\alpha_2 = 1.62138$ and $\alpha_2/\alpha_1 = 2.11536$. Compute the right hand side of (9.27)
$$= \max\{0.23410, 0.25969, 0.27085\} + \max\{0.38003, 0.57871, 0.49244\}$$
$$= 0.83956 < 1,$$

namely (9.27) is fulfilled. By Theorem 9.5 we can therefore conclude that equation (9.52) is exponentially stable in mean square, and this is independent of τ. However, to obtain the upper bound for the second Lyapunov exponent, we need to specify τ. For example, we let $\tau = 0.1$. Then, it is easy to show that equation (9.29) becomes

$$0.53054\sqrt{q} + 0.30902q = 1 - 16.2138\log(q),$$

which has a unique root $q = 1.0096$ on $(1, \infty)$. Hence the solution of equation (9.52) has the property

$$\limsup_{t\to\infty} \frac{1}{t}\log(\mathbb{E}|x(t;\xi)|^2) \leq -0.09555.$$

Example 9.10 Let us continue to discuss Example 9.9 to illustrate another way in which our theory can be used. In practice, we may not know all the bounds for the uncertainties. For example, we may only know the following bounds

$$\|\bar{C}_1\| = \|\bar{C}_2\| = \|\bar{C}_3\| = 0.2, \quad \|\bar{D}_1\| = \|\bar{D}_2\| = \|\bar{D}_3\| = 0.3 \quad (9.54)$$

but the bounds for ΔG_i's are unknown. In this case, if we can estimate the bounds for ΔG_i's so that equation (9.52) will still be exponentially stable in mean square, we then know that the system will be stable as long as we can control the uncertainties ΔG_i's within the bounds. To explain how Theorem 9.5 can be applied to solve this problem, we note that condition (9.27) becomes

$$\max\{1.115\|G_1\| + 0.067,\ 2.359\|G_2\| + 0.142,\ 1.693\|G_3\| + 0.102\}$$
$$+ \max\{0.146 + 1.115\|G_1\| + 0.067,\ 0.309 + 2.359\|G_2\| + 0.142,$$
$$0.222 + 1.693\|G_3\| + 0.102\} < 1.$$

This holds if

$$2\max\{1.115\|G_1\| + 0.067,\ 2.359\|G_2\| + 0.142,\ 1.693\|G_3\| + 0.102\}$$
$$< 1 - 0.309 = 0.691.$$

This yields

$$1.115\|G_1\| + 0.067 < 0.345,$$
$$2.359\|G_2\| + 0.142 < 0.345,$$
$$1.693\|G_3\| + 0.102 < 0.345,$$

namely

$$\|G_1\| < 0.250, \quad \|G_2\| < 0.086, \quad \|G_3\| < 0.144. \quad (9.55)$$

By Theorem 9.5 we can therefore conclude that equation (9.52) is exponentially stable in mean square as long as (9.54) and (9.55) are satisfied.

9.5 SISs with Markovian Switching

Let us now proceed to discuss a couple of special cases of the SDIS (9.7). First of all, if $G_i = \bar{G}_i = D_i = \bar{D}_i = 0$ for all $i \in \mathbb{S}$ (whence $\Delta G_i = \Delta D_i = 0$), then (9.7) becomes

$$dx(t) = (A_{r(t)} + \Delta A_{r(t)})x(t)dt + (C_{r(t)} + \Delta C_{r(t)})x(t)dB(t). \quad (9.56)$$

This is a non-delay stochastic interval system (SIS) with Markovian switching. It is sufficient to consider initial value $x(0) = x_0 \in \mathbb{R}^n$ rather than in $L^2_{\mathcal{F}_0}(\Omega; \mathbb{R}^n)$ due to the Markov property. Accordingly, we observe that condition (9.27) holds automatically as the parameters $g_i = d_i = 0$. The following useful corollary hence follows from Theorem 9.5.

Corollary 9.11 *Let* $\lambda_i = \lambda_{\max}(A_i + A_i^T) + 2\|\bar{A}_i\| + (\|C_i\| + \|\bar{C}_i\|)^2$ *for* $i \in \mathbb{S}$. *If*

$$-\mathrm{diag}(\lambda_1, \cdots, \lambda_N) - \Gamma$$

is a nonsingular M-matrix, then the SIS (9.56) is exponentially stable in mean square (and hence almost surely as well).

Of course, this corollary can be proved directly by using Theorems 7.23 and 7.24 and Lemma 9.1, but we leave it as an exercise.

Theorem 9.5 can therefore be interpreted as that if the non-delay SIS (9.56) is exponentially stable in mean square (in the sense that $K - \Gamma$ is a nonsingular M-matrix) and the matrices $G_i, \bar{G}_i$ etc, of the delay terms are sufficiently small (in the sense that (9.27) holds), then the SDIS (9.7) will remain to be exponentially stable in mean square.

So far in this chapter we have discussed mainly the exponential stability in mean square, from which follows the almost sure exponential stability. Let us now discuss the almost sure stability directly. For this purpose, we assume that that the Markov chain is *irreducible*. In this case, the Markov chain has a unique stationary (probability) distribution $\pi = (\pi_1, \pi_2, \cdots, \pi_N) \in \mathbb{R}^{1 \times N}$ which can be determined by solving the following linear equation

$$\pi \Gamma = 0 \quad (9.57)$$

subject to

$$\sum_{j=1}^{N} \pi_j = 1 \quad \text{and} \quad \pi_j > 0 \quad \forall j \in \mathbb{S}.$$

Theorem 9.12 *Assume that the Markov chain is irreducible and its stationary distribution $\pi = (\pi_1, \pi_2, \cdots, \pi_N)$ is given by (9.57). Assume that for every $i \in \mathbb{S}$, $C_i + C_i^T$ is either non-negative or non-positive definite. Set*

$$\begin{aligned}\gamma_i &= \tfrac{1}{2}\lambda_{\max}(A_i + A_i^T) + \|\bar{A}\| + \tfrac{1}{2}(\|C_i\| + \|\bar{C}_i\|)^2 \\ &\quad + \|\bar{C}_i\|[|\lambda_{\max}(C_i + C_i^T)| \vee |\lambda_{\min}(C_i + C_i^T)|] \\ &\quad - \tfrac{1}{4}\big[|\lambda_{\max}(C_i + C_i^T)| \wedge |\lambda_{\min}(C_i + C_i^T)|\big]^2.\end{aligned} \quad (9.58)$$

Then the solution of the SIS (9.56) with initial value $x(0) = x_0 \neq 0$ obeys

$$\limsup_{t \to \infty} \frac{1}{t} \log(|x(t)|) \le \sum_{i=1}^{N} \pi_i \gamma_i \quad a.s. \quad (9.59)$$

In particular, if $\sum_{i=1}^{N} \pi_i \gamma_i < 0$, then the SIS (9.56) is almost surely exponentially stable.

Proof. By Lemma 5.1, $x(t) \neq 0$ for all $t \geq 0$ with probability 1. We can therefore apply the Itô formula to $\log(|x(t)|^2)$ to obtain that

$$\begin{aligned}\log(|x(t)|^2) &= \log(|x_0|^2) + M(t) \\ &\quad + \int_0^t 2\Big(\frac{x^T(s)(A_{r(s)} + \Delta A_{r(s)})x(s) + \tfrac{1}{2}|(C_{r(s)} + \Delta C_{r(s)})x(s)|^2}{|x(s)|^2} \\ &\qquad - \frac{|x^T(s)(C_{r(s)} + \Delta C_{r(s)})x(s)|^2}{|x(s)|^4} \Big) ds,\end{aligned}$$

where

$$M(t) = \int_0^t \frac{2x^T(s)(C_{r(s)} + \Delta C_{r(s)})x(s)}{|x(s)|^2} \, dB(s).$$

For $x \neq 0$ and $i \in \mathbb{S}$, it is easy to see that

$$\begin{aligned}&\frac{x^T(A_i + \Delta A_i)x + \tfrac{1}{2}|(C_i + \Delta C_i)x|^2}{|x|^2} \\ &\quad \le \tfrac{1}{2}\lambda_{\max}(A_i + A_i^T) + \|\bar{A}\| + \tfrac{1}{2}(\|C_i\| + \|\bar{C}_i\|)^2\end{aligned}$$

while it can be shown (as an exercise)

$$-\frac{|x^T(C_i + \Delta C_i)x|^2}{|x|^4} \le \|\bar{C}_i\|\left[|\lambda_{\max}(C_i + C_i^T)| \vee |\lambda_{\min}(C_i + C_i^T)|\right]$$
$$-\tfrac{1}{4}\left[|\lambda_{\max}(C_i + C_i^T)| \wedge |\lambda_{\min}(C_i + C_i^T)|\right]^2. \quad (9.60)$$

Hence

$$\log(|x(t)|^2) \le \log(|x_0|^2) + M(t) + \int_0^t 2\gamma_{r(s)} ds.$$

Dividing both sides by $2t$ and noting that $M(t)/t \to 0$ almost surely as $t \to \infty$ (by Theorem 1.6) we obtain

$$\limsup_{t\to\infty} \frac{1}{t}\log(|x(t)|) \le \limsup_{t\to\infty} \frac{1}{t}\int_0^t \gamma_{r(s)} ds \quad a.s.$$

But, by the ergodic property of the Markov chain, we have

$$\limsup_{t\to\infty} \frac{1}{t}\int_0^t \gamma_{r(s)} ds = \lim_{t\to\infty} \frac{1}{t}\int_0^t \gamma_{r(s)} ds = \sum_{i=1}^N \pi_i \gamma_i \quad a.s.$$

whence the required assertion (9.59) follows. □

The technique used in the proof above can also be applied to prove the following result on instability, though the details are left as an exercise.

Theorem 9.13 *Assume that the Markov chain is irreducible and its stationary distribution* $\pi = (\pi_1, \pi_2, \cdots, \pi_N)$ *is given by (9.57). For every* $i \in \mathbb{S}$, *set*

$$\rho_i = \tfrac{1}{2}\lambda_{\min}(A_i + A_i^T) - \|\bar{A}\| + \tfrac{1}{2}\Big(0 \vee \left[(\lambda_{\min}(C_i^T C_i) - 2\|C_i\|\,\|\bar{C}_i\|\right]\Big)$$
$$+ \Big(\|\bar{C}_i\| + \tfrac{1}{2}\left[|\lambda_{\max}(C_i + C_i^T)| \vee |\lambda_{\min}(C_i + C_i^T)|\right]\Big)^2. \quad (9.61)$$

Then the solution of the SIS (9.56) with initial value $x(0) = x_0 \ne 0$ *obeys*

$$\limsup_{t\to\infty} \frac{1}{t}\log(|x(t)|) \ge \sum_{i=1}^N \pi_i \rho_i \quad a.s. \quad (9.62)$$

In particular, if $\sum_{i=1}^N \pi_i \rho_i > 0$, *then the SIS (9.56) is almost surely exponentially unstable.*

Let us now consider a special but important case. When $C_i = \Delta C_i = 0$ for every $i \in \mathbb{S}$, equation (9.56) becomes

$$\frac{dx(t)}{dt} = (A_{r(t)} + \Delta A_{r(t)})x(t). \tag{9.63}$$

This is a jump interval system (JIS). The following corollaries follow from the corresponding results above.

Corollary 9.14 *Let $\lambda_i = \lambda_{\max}(A_i + A_i^T) + 2\|\bar{A}_i\|$ for $i \in \mathbb{S}$. If*

$$-\mathrm{diag}(\lambda_1, \cdots, \lambda_N) - \Gamma$$

is a nonsingular M-matrix, then the JIS (9.63) is exponentially stable in mean square (and hence almost surely as well).

Corollary 9.15 *Assume that the Markov chain is irreducible and its stationary distribution $\pi = (\pi_1, \pi_2, \cdots, \pi_N)$ is given by (9.57). For every $i \in \mathbb{S}$, set $\gamma_i = \frac{1}{2}\lambda_{\max}(A_i + A_i^T) + \|\bar{A}\|$. Then the solution of the JIS (9.63) with initial value $x(0) = x_0 \neq 0$ obeys*

$$\limsup_{t \to \infty} \frac{1}{t}\log(|x(t)|) \leq \sum_{i=1}^{N} \pi_i \gamma_i \quad a.s.$$

In particular, if $\sum_{i=1}^{N} \pi_i \gamma_i < 0$, then the JIS (9.63) is almost surely exponentially stable.

Corollary 9.16 *Assume that the Markov chain is irreducible and its stationary distribution $\pi = (\pi_1, \pi_2, \cdots, \pi_N)$ is given by (9.57). For every $i \in \mathbb{S}$, set $\rho_i = \frac{1}{2}\lambda_{\min}(A_i + A_i^T) - \|\bar{A}\|$. Then the solution of the JIS (9.63) with initial value $x(0) = x_0 \neq 0$ obeys*

$$\limsup_{t \to \infty} \frac{1}{t}\log(|x(t)|) \geq \sum_{i=1}^{N} \pi_i \rho_i \quad a.s.$$

In particular, if $\sum_{i=1}^{N} \pi_i \rho_i > 0$, then the JIS (9.63) is almost surely exponentially unstable.

9.6 Exercises

9.1 In Example 9.2, show by Corollary 5.17 that equation (9.10) is exponentially stable in mean square if $\tilde{\mathcal{A}}$ defined there is a nonsingular M-matrix.

9.2 Let $r(t)$, $t \geq 0$, be a right-continuous Markov chain taking values in $\mathbb{S} = \{1, 2\}$ with the generator
$$\Gamma = \begin{pmatrix} -8 & 8 \\ 1 & -1 \end{pmatrix}.$$
Consider a scalar SDIS with Markovian switching
$$dx(t) = (G_r + \Delta G_r)x(t - \delta(t))dt + \Delta D_r x(t - \delta(t))dB(t),$$
where $\delta : \mathbb{R}_+ \to [0, \tau]$ and
$$G_1 = 2, \quad G_2 = -3, \quad \Delta G_i, \Delta D_i \in [-\varepsilon, \varepsilon] \text{ for } i = 1, 2$$
with $\varepsilon > 0$. Give bounds for τ and ε so that the equation will be exponentially stable in mean square.

9.3 Prove Corollary 9.11 by using Theorems 7.23 and 7.24 and Lemma 9.1.

9.4 Show (9.60).

9.5 Prove Theorem 9.13.

9.6 Let $r(t)$, $t \geq 0$, be a right-continuous Markov chain taking values in $\mathbb{S} = \{1, 2\}$ with the generator
$$\Gamma = \begin{pmatrix} -1 & 1 \\ \lambda & -\lambda \end{pmatrix},$$
where $\lambda > 0$. Consider a scalar SIS with Markovian switching of the form
$$dx(t) = (A_{r(t)} + \Delta A_{r(t)})x(t)dt + (C_{r(t)} + \Delta C_{r(t)})dB(t),$$
where $A_1 = 1$, $A_2 = 2$, $C_1 = 3$, $C_2 = 1$ and $\Delta A_i, \Delta C_i \in [-0.1, 0.1]$ for $i = 1, 2$. Give a value λ_1 such that when $\lambda > \lambda_1$ then the SIS is almost surely exponentially stable. Give an alternative value λ_2 such that when $\lambda < \lambda_2$ then the SIS is almost surely exponentially unstable.

Chapter 10

Applications

10.1 Introduction

SDEs with Markovian switching discussed in the text take all the features of Itô equations, Markovian switching, interval systems as well as time-lag into account. To demonstrate that the theory developed in the previous chapters will be applicable in different and complicated situations in many branches of science and industry, we will discuss a number of important applications including population dynamics, financial modelling, stochastic stabilisation and stochastic neural networks in this chapter. All of them are currently hot topics in research.

10.2 Stochastic Population Dynamics

Single-species deterministic population dynamics can often be described by the ordinary differential equation $\dot{x} = f(x)$, and to avoid an explosion (i.e. infinite population size at a finite time) $f(x)$ has to satisfy certain conditions. Consider, for example, the one-dimensional logistic (i.e. quadratic) equation

$$\dot{x}(t) = x(t)[b + ax(t)] \qquad (10.1)$$

on $t \geq 0$ with initial value $x(0) = x_0 > 0$. Since here the variable $x(t)$ denotes population size, only positive solutions are of interest. For parameters $a < 0$ and $b > 0$, equation (10.1) has the global solution

$$x(t) = \frac{b}{-a + e^{-bt}(b + ax_0)/x_0} \qquad (t \geq 0),$$

which is not only positive and bounded but also has the asymptotic property that $\lim_{t\to\infty} x(t) = b/|a|$. In contrast, if we now let $a > 0$, whilst retaining $b > 0$, then equation (10.1) has only the local solution

$$x(t) = \frac{b}{-a + e^{-bt}(b + ax_0)/x_0} \quad (0 \le t < T),$$

which explodes to infinity at the finite time

$$T = -\frac{1}{b} \log\left(\frac{ax_0}{b + ax_0}\right).$$

However, given that population systems are often subject to environmental noise, it is important to discover whether the presence of such noise affects this result. Suppose that the parameter a is stochastically perturbed, with

$$a \to a + \sigma \dot{B}(t)$$

where $\dot{B}(t)$ is white noise (i.e. $B(t)$ is a scalar Brownian motion) and $\sigma > 0$ represents the intensity of the noise. Then this environmentally perturbed system may be described by the Itô equation

$$dx(t) = x(t)[(b + ax(t))dt + \sigma x(t)dB(t)]. \tag{10.2}$$

It is already known (see [Mao et al. (2002a)]) that with probability one the solution of equation (10.2) can no longer explode in a finite time if $a > 0$. In summary, when $a > 0$ and $\sigma = 0$ the solution explodes at the finite time $t = T$; whilst conversely, no matter how small $\sigma > 0$, the solution will not explode in a finite time. In other words, stochastic environmental noise suppresses deterministic explosion.

Let us now take a further step to consider another type of random fluctuation. The population system may switch between (finite) regimes of environment, which differ by factors such as nutrition or as rain falls (see e.g. [Du et al. (2004); Slatkin (1978)]). The switching is without memory and the waiting time for the next switch has an exponential distribution. We can hence model the regime switching by a finite-state Markov chain $r(t)$. Under different regime, the system parameters b, a and σ are different. As a result, system (10.2) becomes a more general equation

$$dx(t) = x(t)[(b(r(t)) + a(r(t))x(t))dt + \sigma(r(t))x(t)dB(t)], \tag{10.3}$$

which is an SDE with Markovian switching.

More generally, consider the Lotka–Volterra model with Markovian switching for a system with n interacting components, namely

$$dx(t) = \text{diag}(x_1(t), \cdots, x_n(t))$$
$$\times [(b(r(t)) + A(r(t))x(t))dt + \sigma(r(t))x(t)dB(t)] \quad (10.4)$$

on $t \geq 0$ with initial value $x(0) = x_0 \in \mathbb{R}_+^n = \{x \in \mathbb{R}^n : x_i > 0 \text{ for all } 1 \leq i \leq n\}$. Here $r(t)$ is the same Markov chain as before, $B(t)$ is a scalar Brownian motion, $b : \mathbb{S} \to \mathbb{R}^n$ and $A, \sigma : \mathbb{S} \to \mathbb{R}^{n \times n}$. More precisely, for each $u \in \mathbb{S}$,

$$b(u) = (b_1(u), \cdots, b_n(u))^T, \quad A(u) = (a_{ij}(u))_{n \times n}, \quad \sigma = (\sigma_{ij}(u))_{n \times n}.$$

Since our purpose is to discover the effect of environmental noise, we naturally impose the following simple hypothesis on the noise intensities.

Assumption 10.1 For each $u \in \mathbb{S}$, $\sigma_{ii}(u) > 0$ if $1 \leq i \leq n$ whilst $\sigma_{ij}(u) \geq 0$ if $i \neq j$.

10.2.1 Global Positive Solutions

As the ith state $x_i(t)$ of equation (10.4) is the size of the ith component in the system, it should be non-negative. Moreover, the coefficients of equation (10.4) do not satisfy the linear growth condition, though they are locally Lipschitz continuous, so the solution of equation (10.4) may explode at a finite time. However we shall show that under the simple Assumption 10.1 the solution of equation (10.4) is positive and global. This result reveals the important property that the environmental noise suppresses the explosion.

Theorem 10.2 *Under Assumption 10.1, for any system parameters $b(\cdot)$, $A(\cdot)$, and any given initial value $x_0 \in \mathbb{R}_+^n$, there is a unique solution $x(t)$ to equation (10.4) on $t \geq 0$ and the solution will remain in $\mathbb{R}_+^n$ with probability 1, namely $x(t) \in \mathbb{R}_+^n$ for all $t \geq 0$ almost surely.*

Proof. Since the coefficients of the equation are locally Lipschitz continuous, for any given initial value $x_0 \in \mathbb{R}_+^n$ there is a unique maximal local solution $x(t)$ on $t \in [0, \tau_e)$, where τ_e is the explosion time (see Chapter 3). To show this solution is global, we need to show that $\tau_e = \infty$ a.s. Let $k_0 > 0$ be sufficiently large for every component of x_0 lying within the interval $[1/k_0, k_0]$. For each integer $k \geq k_0$, define the stopping time

$$\tau_k = \inf\{t \in [0, \tau_e) : x_i(t) \notin (1/k, k) \text{ for some } i = 1, \cdots, n\},$$

Clearly, τ_k is increasing as $k \to \infty$. Set $\tau_\infty = \lim_{k\to\infty} \tau_k$, whence $\tau_\infty \leq \tau_e$ a.s. If we can show that $\tau_\infty = \infty$ a.s., then $\tau_e = \infty$ a.s. and $x(t) \in \mathbb{R}_+^n$ a.s. for all $t \geq 0$. In other words, to complete the proof all we need to show is that $\tau_\infty = \infty$ a.s. For if this statement is false, then there is a pair of constants $T > 0$ and $\varepsilon \in (0,1)$ such that

$$\mathbb{P}\{\tau_\infty \leq T\} > \varepsilon.$$

Hence there is an integer $k_1 \geq k_0$ such that

$$\mathbb{P}\{\tau_k \leq T\} \geq \varepsilon \quad \text{for all } k \geq k_1. \tag{10.5}$$

Define a C^2-function $V : \mathbb{R}_+^n \to \mathbb{R}_+$ by

$$V(x) = \sum_{i=1}^n \left[\sqrt{x_i} - 1 - 0.5 \log(x_i)\right].$$

The non-negativity of this function can be seen from

$$\sqrt{y} - 1 - 0.5 \log(y) \geq 0 \quad \text{on } y > 0.$$

If $x(t) \in \mathbb{R}_+^n$, the Itô formula shows that

$$\begin{aligned}
&dV(x(t)) \\
&= \sum_{i=1}^n \left\{ 0.5(x_i^{-0.5} - x_i^{-1}) x_i \left[\left(b_i + \sum_{j=1}^n a_{ij} x_j\right) dt + \sum_{j=1}^n \sigma_{ij} x_j dB(t)\right] \right. \\
&\quad + 0.5(-0.25 x_i^{-1.5} + 0.5 x_i^{-2}) x_i^2 \left[\sum_{j=1}^n \sigma_{ij} x_j\right]^2 dt \bigg\} \\
&= \sum_{i=1}^n \left\{ 0.5(x_i^{0.5} - 1)\left(b_i + \sum_{j=1}^n a_{ij} x_j\right) + (0.25 - 0.125 x_i^{0.5})\left[\sum_{j=1}^n \sigma_{ij} x_j\right]^2 \right\} dt \\
&\quad + \sum_{i=1}^n \sum_{j=1}^n 0.5(x_i^{0.5} - 1)\sigma_{ij} x_j dB(t),
\end{aligned}$$

where we drop t from $x(t)$ and $r(t)$ from $b_i(r(t))$ etc. Compute

$$\sum_{i=1}^n (x_i^{0.5} - 1)\left(b_i + \sum_{j=1}^n a_{ij}x_j\right)$$

$$\leq \sum_{i=1}^n |b_i|(x_i^{0.5} + 1) + \sum_{i=1}^n \sum_{j=1}^n |a_{ij}|x_j + \sum_{i=1}^n \sum_{j=1}^n |a_{ij}|x_i^{0.5}x_j$$

$$\leq \sum_{i=1}^n |b_i|(x_i^{0.5} + 1) + \sum_{j=1}^n \sum_{i=1}^n |a_{ij}|x_j + \sum_{i=1}^n \sum_{j=1}^n 0.5|a_{ij}|(x_i + x_j^2)$$

$$= \sum_{i=1}^n \left(|b_i|(1 + x_i^{0.5}) + \sum_{j=1}^n (|a_{ji}| + 0.5|a_{ij}|)x_i + 0.5\sum_{j=1}^n |a_{ji}|x_i^2\right)$$

and

$$\sum_{i=1}^n \left[\sum_{j=1}^n \sigma_{ij}x_j\right]^2 \leq \sum_{i=1}^n \left[\sum_{j=1}^n \sigma_{ij}^2 \sum_{j=1}^n x_j^2\right] = |\sigma|^2 \sum_{i=1}^n x_i^2.$$

Moreover, by Assumption 10.1,

$$\sum_{i=1}^n x_i^{0.5}\left[\sum_{j=1}^n \sigma_{ij}x_j\right]^2 \geq \sum_{i=1}^n \sigma_{ii}^2 x_i^{2.5}.$$

So

$$\sum_{i=1}^n \left\{0.5(x_i^{0.5} - 1)\left(b_i + \sum_{j=1}^n a_{ij}x_j\right) + (0.25 - 0.125x_i^{0.5})\left[\sum_{j=1}^n \sigma_{ij}x_j\right]^2\right\}$$

$$\leq \sum_{i=1}^n \left\{0.5|b_i|(1 + x_i^{0.5}) + \sum_{j=1}^n (0.5|a_{ji}| + 0.25|a_{ij}|)x_i\right.$$

$$\left. + 0.25\left(\sum_{j=1}^n |a_{ji}| + |\sigma|^2\right)x_i^2 - 0.125\sigma_{ii}^2 x_i^{2.5}\right\},$$

which is bounded, say by K, in $\mathbb{R}_+^n \times \mathbb{S}$. We therefore obtain

$$dV(x(t)) \leq K dt + \sum_{i=1}^n \sum_{j=1}^n 0.5(x_i^{0.5} - 1)\sigma_{ij}x_j dB(t)$$

as long as $x(t) \in \mathbb{R}_+^n$. Whence integrating both sides from 0 to $\tau_k \wedge T$, and then taking expectations, yields

$$\mathbb{E}V(x(\tau_k \wedge T)) \leq V(x_0) + K\mathbb{E}(\tau_k \wedge T) \leq V(x_0) + KT. \qquad (10.6)$$

Set $\Omega_k = \{\tau_k \leq T\}$ for $k \geq k_1$ and, by (10.5), $\mathbb{P}(\Omega_k) \geq \varepsilon$. Note that for every $\omega \in \Omega_k$, there is some i such that $x_i(\tau_k, \omega)$ equals either k or $1/k$, and hence $V(x(\tau_k, \omega))$ is no less than either

$$\sqrt{k} - 1 - 0.5 \log(k)$$

or

$$\sqrt{1/k} - 1 - 0.5 \log(1/k) = \sqrt{1/k} - 1 + 0.5 \log(k).$$

Consequently,

$$V(x(\tau_k, \omega)) \geq \left[\sqrt{k} - 1 - 0.5 \log(k)\right] \wedge \left[0.5 \log(k) - 1\right].$$

It then follows from (10.6) that

$$V(x_0) + KT \geq \mathbb{E}\left[I_{\Omega_k}(\omega) V(x(\tau_k, \omega))\right]$$

$$\geq \varepsilon \left(\left[\sqrt{k} - 1 - 0.5 \log(k)\right] \wedge \left[0.5 \log(k) - 1\right]\right).$$

Letting $k \to \infty$ leads to the contradiction

$$\infty > V(x_0) + KT = \infty,$$

so we must therefore have $\tau_\infty = \infty$ a.s. □

Theorem 10.2 shows that under the simple Assumption 10.1 the solutions of equation (10.4) will remain in the positive cone $\mathbb{R}^n_+$. This nice positive property provides us with a great opportunity to construct different types of Lyapunov functions to discuss how the solutions vary in $\mathbb{R}^n_+$ in more detail.

10.2.2 Ultimate Boundedness

The non-explosion property in a population dynamical system is often not good enough but the property of ultimate boundedness is more desired. Let us now give the definition of stochastically ultimate boundedness.

Definition 10.3 Equation (10.4) is said to be *stochastically ultimately bounded* if for any $\varepsilon \in (0, 1)$, there is a positive constant $H = H(\varepsilon)$ such that for any initial value $x_0 \in \mathbb{R}^n_+$, the solution $x(t)$ of equation (10.4) has the property that

$$\limsup_{t \to \infty} \mathbb{P}\{|x(t)| \leq H\} \geq 1 - \varepsilon. \tag{10.7}$$

Let us present a useful lemma from which the stochastically ultimate boundedness will follow directly.

Lemma 10.1 *Let Assumption 10.1 hold and $\theta \in (0,1)$. Then there is a positive constant $K = K(\theta)$, which is independent of the initial data $x_0 \in \mathbb{R}_+^n$, such that the solution $x(t)$ of equation (10.4) has the property that*

$$\limsup_{t \to \infty} \mathbb{E}|x(t)|^\theta \leq K. \tag{10.8}$$

Proof. Define

$$V(x) = \sum_{i=1}^n x_i^\theta \quad \text{for } x \in \mathbb{R}_+^n.$$

By the Itô formula, we have

$$dV(x(t)) = LV(x(t), r(t))dt + \Big(\sum_{i=1}^n \theta x_i^\theta(t) \sum_{j=1}^n \sigma_{ij}(r(t)) x_j(t)\Big) dB(t), \tag{10.9}$$

where $LV : \mathbb{R}_+^n \times \mathbb{S} \to \mathbb{R}$ is defined by

$$LV(x, u) = \sum_{i=1}^n \theta x_i^\theta \Big[b_i(u) + \sum_{j=1}^n a_{ij}(u) x_j\Big] - \frac{\theta(1-\theta)}{2} \sum_{i=1}^n x_i^\theta \Big[\sum_{j=1}^n \sigma_{ij}(u) x_j\Big]^2.$$

By Assumption 10.1, we have

$$LV(x, u) \leq \sum_{i=1}^n \theta x_i^\theta \Big[b_i(u) + \sum_{j=1}^n a_{ij} x_j\Big] - \frac{\theta(1-\theta)}{2} \sum_{i=1}^n \sigma_{ii}^2(u) x_i^{2+\theta}$$
$$= F(x, u) - V(x).$$

where

$$F(x, u) = V(x) + \sum_{i=1}^n \theta x_i^\theta \Big[b_i(u) + \sum_{j=1}^n a_{ij} x_j\Big] - \frac{\theta(1-\theta)}{2} \sum_{i=1}^n \sigma_{ii}^2(u) x_i^{2+\theta}.$$

Note that $F(x, u)$ is bounded in $\mathbb{R}_+^n \times \mathbb{S}$, namely

$$K_1 := \sup_{(x,u) \in \mathbb{R}_+^n \times \mathbb{S}} F(x, u) < \infty.$$

We therefore have

$$LV(x, u) \leq K_1 - V(x).$$

On the other hand, we have
$$|x|^2 \le n \max_{1\le i\le n} x_i^2$$
so
$$|x|^\theta \le n^{\theta/2} \max_{1\le i\le n} x_i^\theta \le n^{\theta/2} V(x).$$
Applying Theorem 5.2, we obtain
$$\limsup_{t\to\infty} \mathbb{E}|x(t)|^\theta \le n^{\theta/2} K_1$$
and the assertion (10.8) follows by setting $K = n^{\theta/2} K_1$. □

Theorem 10.4 *Under Assumption 10.1, equation (10.4) is stochastically ultimately bounded.*

Proof. By Lemma 10.1, there is a $K > 0$ such that
$$\limsup_{t\to\infty} \mathbb{E}(\sqrt{|x(t)|}) \le K.$$
Now, for any $\varepsilon > 0$, let $H = K^2/\varepsilon^2$. Then by Chebyshev's inequality,
$$\mathbb{P}\{|x(t)| > H\} \le \frac{\mathbb{E}(\sqrt{|x(t)|})}{\sqrt{H}}.$$
Hence
$$\limsup_{t\to\infty} \mathbb{P}\{|x(t)| > H\} \le \frac{K}{\sqrt{H}} = \varepsilon.$$
This implies
$$\limsup_{t\to\infty} \mathbb{P}\{|x(t)| \le H\} \ge 1 - \varepsilon$$
as required. □

10.2.3 Moment Average in Time

The result in the previous sub-section shows that the solutions of equation (10.4) will be stochastically ultimately bounded. That is, the solutions will be ultimately bounded with large probability. The following result shows that the average in time of the second moment of the solutions will be bounded.

Theorem 10.5 *Under Assumption 10.1, there is a positive constant K, which is independent of the initial value $x_0 \in \mathbb{R}_+^n$, such that the solution $x(t)$ of equation (10.4) has the property that*

$$\limsup_{T \to \infty} \frac{1}{T} \int_0^T \mathbb{E}|x(t)|^2 dt \leq K. \tag{10.10}$$

Proof. Define

$$V(x) = \sum_{i=1}^n \sqrt{x_i} \quad \text{for } x \in \mathbb{R}_+^n.$$

By the Itô formula, we have

$$dV(x(t)) = LV(x(t), r(t))dt + \Big(\sum_{i=1}^n \tfrac{1}{2}\sqrt{x_i(t)} \sum_{j=1}^n \sigma_{ij}(r(t))x_j(t) \Big) dB(t), \tag{10.11}$$

where $LV : \mathbb{R}_+^n \times \mathbb{S} \to \mathbb{R}$ is defined by

$$LV(x, u) = \sum_{i=1}^n \tfrac{1}{2}\sqrt{x_i}\Big[b_i(u) + \sum_{j=1}^n a_{ij}(u)x_j\Big] - \frac{1}{8}\sum_{i=1}^n \sqrt{x_i}\Big[\sum_{j=1}^n \sigma_{ij}(u)x_j\Big]^2.$$

By Assumption 10.1,

$$LV(x, u) \leq \sum_{i=1}^n \tfrac{1}{2}\sqrt{x_i}\Big[b_i(u) + \sum_{j=1}^n a_{ij}(u)x_j\Big] - \frac{1}{8}\sum_{i=1}^n \sigma_{ii}^2(u)x_i^{2.5}$$

$$= F_1(x, u) - |x|^2,$$

where

$$F_1(x, u) = |x|^2 + \sum_{i=1}^n \tfrac{1}{2}\sqrt{x_i}\Big[b_i(u) + \sum_{j=1}^n a_{ij}(u)x_j\Big] - \frac{1}{8}\sum_{i=1}^n \sigma_{ii}^2(u)x_i^{2.5}.$$

Note that $F_1(x, u)$ is bounded in $\mathbb{R}_+^n \times \mathbb{S}$, namely

$$K := \sup_{(x,u) \in \mathbb{R}_+^n \times \mathbb{S}} F_1(x, u) < \infty.$$

We therefore have

$$LV(x, u) \leq K - |x|^2. \tag{10.12}$$

Now, for each sufficiently large integer k, let τ_k be the same stopping time as defined in the proof of Theorem 10.2. Moreover, let T be any positive number. Integrating both sides of (10.11) from 0 to $\tau_k \wedge T$, using (10.12) and then taking expectations, we obtain that

$$0 \le V(x(0)) + K\mathbb{E}(\tau_k \wedge T) - \mathbb{E}\int_0^{\tau_k \wedge T} |x(t)|^2 dt.$$

Letting $k \to \infty$ yields

$$\mathbb{E}\int_0^T |x(t)|^2 dt \le V(x(0)) + KT.$$

Dividing both sides by T and then letting $T \to \infty$ we get

$$\limsup_{T \to \infty} \frac{1}{T} \int_0^T \mathbb{E}|x(t)|^2 dt \le K$$

as required. $\square$

10.3 Stochastic Financial Modelling

In the well-known Black–Scholes model, the asset price is described by a geometric Brownian motion

$$dX(t) = \mu X(t)dt + \nu X(t)dB_1(t), \qquad (10.13)$$

where $B_1(t)$ is a scalar Brownian motion, μ is the rate of return of the underlying asset and ν is the volatility. In this classical model, [Black and Scholes (1973)] assumed that the rate of return and the volatility are constants. However, it has been proved by many authors that the volatility is itself an Itô process in many situations. For instance, [Hull and White (1987)] assume that the instantaneous variance, $V = \nu^2$, obeys another geometric Brownian motion

$$dV(t) = \alpha V(t)dt + \beta V(t)dB_2(t), \qquad (10.14)$$

where α, β are constants while $B_2(t)$ is another Brownian motion and $B_1(t)$ and $B_2(t)$ have correlation ρ. [Heston (1993)] assumes that the variance V obeys the mean-reverting square root process

$$dV(t) = \alpha(\lambda - V(t))dt + \beta\sqrt{V(t)}dB_2(t) \qquad (10.15)$$

while the mean-reverting process

$$dV(t) = \alpha(\lambda - V(t))dt + \beta V(t)dB_2(t) \qquad (10.16)$$

is also proposed as the volatility process by others. In particular, [Lewis (2000)] proposes the mean-reverting θ-process

$$dV(t) = \alpha(\lambda - V(t))dt + \beta V^\theta(t)dB_2(t), \tag{10.17}$$

where $\theta \geq \frac{1}{2}$. This process unifies processes (10.15) and (10.16).

On the other hand, the rate of return, μ is not a constant either and there is strong evidence to indicate that it is a Markov jump process (see e.g. [Buffington and Elliott (2002); Yin and Zhou (2004)]). Of course, when the rate jumps, the volatility will jump accordingly. For example, the hybrid geometric Brownian motion

$$dX(t) = \mu(r(t))X(t)dt + \nu(r(t))X(t)dB_1(t) \tag{10.18}$$

has been proposed by several authors. Here $r(t)$ is a Markov chain with a finite state space $\mathbb{S}$ and μ, ν are mappings from $\mathbb{S}$ to $\mathbb{R}_+$. Equation (10.18) is also known as the geometric Brownian motion under regime-switching. We observe that in this model, the volatility is also assumed to obey a Markov jump process. Recalling the stochastic volatility models mentioned above, we may more reasonably assume that the volatility process obeys an SDE under regime-switching, for example, the hybrid mean-reverting θ-process

$$dV(t) = \alpha(r(t))(\lambda(r(t)) - V(t))dt + \beta(r(t))V^\theta(t)dB_2(t). \tag{10.19}$$

Such stochastic models under regime-switching have recently been developed to model various financial quantities, e.g. option pricing, stock returns, portfolio optimisation. In particular, the mean-reverting square root process under regime-switching or, more generally, equation (10.19) has found its considerable use as a model for volatility and interest rate. In general, SDEs under regime-switching have no explicit solutions so numerical methods for approximations have become one of the powerful techniques in valuation of financial quantities e.g. option price. In this section we will concentrate on the Euler–Maruyama (EM) scheme for the typical hybrid mean-reverting θ-process (10.19) but the theory established here can certainly be developed to cope with other SDEs under regime-switching in finance.

10.3.1 Non-Negative Solutions

Let $B(t)$ be a scalar Brownian motion and $r(t)$ be the same Markov chain as before. Of course, they are independent. Let $\theta \in [\frac{1}{2}, 1]$. Consider the mean-reverting θ-process under regime-switching of the form

$$dS(t) = \lambda(r(t))[\mu(r(t)) - S(t)]dt + \sigma(r(t))S^\theta(t)dB(t), \quad t \geq 0, \quad (10.20)$$

with initial data $S(0) = S_0 > 0$ and $r(0) = i_0 \in \mathbb{S}$. Here $\lambda(i), \mu(i)$, $\sigma(i), i \in \mathbb{S}$ are positive constants. Since equation (10.20) is mainly used to model stochastic volatility or interest rate or an asset price, it is critical the solution $S(t)$ will never become negative. The following lemma reveals this non-negative property.

Lemma 10.2 *For given any initial data $S(0) = S_0 > 0$ and $r(0) = i_0 \in \mathbb{S}$, the solution $S(t)$ of equation (10.20) will never become negative with probability 1.*

Proof. Clearly, the statement of the lemma is equivalent to that the solution of equation

$$dS(t) = \lambda(r(t))[\mu(r(t)) - S(t)]dt + \sigma(r(t))|S(t)|^\theta dB(t), \quad t \geq 0, \quad (10.21)$$

will never become negative with probability 1 for any initial data $S(0) = S_0 > 0$ and $r(0) = i_0 \in \mathbb{S}$. To show this, let $a_0 = 1$ and, for each integer $k = 1, 2, \cdots$,

$$a_k = \begin{cases} e^{-k(k+1)}, & \text{if } \theta = \frac{1}{2}, \\ \left[\frac{(2\theta-1)k(k+1)}{2}\right]^{\frac{1}{1-2\theta}}, & \text{if } \frac{1}{2} < \theta \leq 1, \end{cases}$$

so that

$$\int_{a_k}^{a_{k-1}} \frac{du}{u^{2\theta}} = k.$$

For each $k = 1, 2, \cdots$, there clearly exists a continuous function $\psi_k(u)$ with support in (a_k, a_{k-1}) such that

$$0 \leq \psi_k(u) \leq \frac{2}{ku^{2\theta}} \quad \text{for } a_k < u < a_{k-1}$$

and $\int_{a_k}^{a_{k-1}} \psi_k(u)du = 1$. Define $\varphi_k(x) = 0$ for $x \geq 0$ and

$$\varphi_k(x) = \int_0^{-x} dy \int_0^y \psi_k(u)du \quad \text{for } x < 0.$$

Then $\varphi_k \in C^2(\mathbb{R}, \mathbb{R})$ and has the following properties:

(i) $-1 \leq \varphi_k'(x) \leq 0$, for $a_k < x < a_{k-1}$ or otherwise $\varphi_k'(x) = 0$;
(ii) $|\varphi_k''(x)| \leq \frac{2}{k|x|^{2\theta}}$ for $a_k < x < a_{k-1}$ or otherwise $\varphi_k'(x) = 0$;

(iii) $|x| - a_{k-1} \leq \varphi_k(x) \leq |x|$ for all $x \in \mathbb{R}$.

Let $\bar\lambda = \max_{i\in\mathbb{S}} \lambda(i), \bar\mu = \max_{i\in\mathbb{S}} \mu(i)$ and $\bar\sigma = \max_{i\in\mathbb{S}} \sigma(i)$. Now for any $t \geq 0$, by the Itô formula, we can derive that

$$\mathbb{E}\varphi_k(S(t)) = \varphi_k(S_0) + \mathbb{E}\int_0^t \Big[\lambda(r(u))(\mu(r(u)) - S(u))\varphi'_k(S(u))$$
$$+ \frac{\sigma^2(r(u))}{2}|S(u)|^{2\theta}\varphi''_k(S(r(u)))\Big]du$$
$$\leq \frac{\bar\sigma^2 t}{k}. \qquad (10.22)$$

Hence

$$-a_{k-1} \leq \mathbb{E}S^-(t) - a_{k-1} \leq \frac{\bar\sigma^2 t}{k},$$

where $S^-(t) = -S(t)$ if $S(t) < 0$ or otherwise $S^-(t) = 0$. Letting $k \to \infty$ we get that $\mathbb{E}S^-(t) = 0$ for all $t \geq 0$. This implies that $S(t) \geq 0$ for all $t \geq 0$ with probability 1 as required. □

10.3.2 The EM Approximations

As equation (10.20) does not have an explicit solution, we now use the EM method discussed in Chapter 4 to obtain the approximate solution.

Given a stepsize $\Delta > 0$, let $t_k = k\Delta$ for $k = 0, 1, 2, \cdots$. Let $r_k^\Delta = r(t_k)$. As shown in Chapter 4, $\{r_k^\Delta\}_{k\geq 0}$ is a discrete-time Markov chain. The EM method is to compute the discrete approximations $s_k \approx S(t_k)$ by setting $s_0 = S_0$, $r_0^\Delta = i_0$ and forming

$$s_{k+1} = s_k + \lambda(r_k^\Delta)(\mu(r_k^\Delta) - s_k)\Delta + \sigma(r_k^\Delta)|s_k|^\theta \Delta B_k, \quad k = 0, 1, 2, \cdots, \qquad (10.23)$$

where $\Delta B_k = B(t_{k+1}) - B(t_k)$. Let

$$\bar s(t) = s_k, \quad \bar r(t) = r_k^\Delta \quad \text{for } t \in [t_k, t_{k+1}), \ k = 0, 1, 2, \cdots \qquad (10.24)$$

and define the continuous EM approximate solution by

$$s(t) = s_0 + \int_0^t \lambda(\bar r(u))[\mu(\bar r(u)) - \bar s(u)]du + \int_0^t \sigma(\bar r(u))|\bar s(u)|^\theta dB(u). \qquad (10.25)$$

Note that $s(t_k) = \bar s(t_k) = s_k$, that is $s(t)$ and $\bar s(t)$ coincide with the discrete approximate solution at the gridpoints.

The theorems established in Chapter 4 cannot be applied to equation (10.20) to conclude the convergence of the EM approximate solution to the true solution, because the diffusion coefficient $\sigma(i)s^\theta$ does not obey the conditions imposed there (its derivative tends to infinity as $s \downarrow 0$ if $\frac{1}{2} \leq \theta < 1$). We therefore have to develop new techniques to show the convergence.

First of all, we observe that the coefficients of equation (10.20) satisfy the linear growth condition. By Lemma 4.1, we have the following moment bound.

Lemma 10.3 *Let $S(t)$ be the solution of equation (10.20). Then for any $p \geq 1$ there is a constant K, which is dependent on only p, T, S_0 but independent of Δ, such that the exact solution and the EM approximate solution to equation (10.20) have the property that*

$$\mathbb{E}\left[\sup_{0 \leq t \leq T} |S(t)|^p\right] \vee \mathbb{E}\left[\sup_{0 \leq t \leq T} |s(t)|^p\right] \leq K. \tag{10.26}$$

From this the following useful result follows easily.

Lemma 10.4 *There is a constant C, which is independent of Δ, such that*

$$\mathbb{E}|s(t) - \bar{s}(t)|^{2\theta} \leq C\Delta^\theta, \quad \forall t \in [0, T]. \tag{10.27}$$

Proof. From now on, C used in the proofs below will be a generic positive number independent of Δ but may have a different value where it appears.

For any $t \in [0, T]$, let k_t be the integer part of t/Δ. By Lemma 10.3 we then derive that

$$\mathbb{E}|\bar{s}(t) - s(t)|^2 \leq 4(\bar{\lambda} \vee \bar{\lambda}\bar{\mu} \vee \bar{\sigma})\mathbb{E}\left[(1 + |s_{k_t}|^2)(\Delta^2 + |B(t) - B(k_t\Delta)|^2)\right]$$
$$\leq C\Delta.$$

So, since $\frac{1}{2} \leq \theta \leq 1$, by the Hölder inequality we get

$$\mathbb{E}|\bar{s}(t) - s(t)|^{2\theta} \leq [\mathbb{E}|\bar{s}(t) - s(t)|^2]^\theta \leq C\Delta^\theta$$

as required. $\square$

The following theorem reveals the convergence in L^1.

Theorem 10.6 *For each integer $k = 1, 2, \cdots$,*

$$\sup_{0 \le t \le T} \mathbb{E}|S(t) - s(t)|$$
$$\le e^{\bar{\lambda} T} \left[e^{-k(k-1)/2} + \frac{4\bar{\sigma}^2 T}{k} + \left(\frac{1}{k a_k^{2\theta}} + 1 \right) (C\Delta^\theta + o(\Delta)) \right],$$

where C is a constant which is independent of the step-size Δ and $\bar{\lambda}$, $\bar{\sigma}$ have been defined in the proof of Lemma 10.2.

Proof. Note that

$$S(t) - s(t) = \int_0^t \Big[\lambda(r(u))\mu(r(u)) - \lambda(\bar{r}(u))\mu(\bar{r}(u)) $$
$$- \lambda(r(u))S(u) + \lambda(\bar{r}(u))\bar{s}(u) \Big] du$$
$$+ \int_0^t \Big[\sigma(r(u))|S(u)|^\theta - \sigma(\bar{r}(u))|\bar{s}(u)|^\theta \Big] dB(u).$$

Let φ_k be the same as defined in the proof of Lemma 10.2. Applying the Itô formula gives

$$\mathbb{E}\varphi_k(S(t) - s(t))$$
$$= \mathbb{E} \int_0^t \varphi_k'(S(u) - s(u)) \Big[\lambda(r(u))\mu(r(u)) - \lambda(\bar{r}(u))\mu(\bar{r}(u))$$
$$- \lambda(r(u))S(u) + \lambda(\bar{r}(u))\bar{s}(u) \Big] du$$
$$+ \frac{1}{2}\mathbb{E} \int_0^t \varphi_k''(S(u) - s(u)) \Big[\sigma(r(u))|S(u)|^\theta - \sigma(\bar{r}(u))|\bar{s}(u)|^\theta \Big]^2 du$$
$$=: I(t) + \frac{1}{2} J(t)$$

By property (i) of φ_k

$$|I(t)| \le \mathbb{E} \int_0^t \Big| \varphi_k'(S(u) - s(u)) \Big[\lambda(r(u))\mu(r(u)) - \lambda(\bar{r}(u))\mu(\bar{r}(u))$$
$$- \lambda(r(u))S(u) + \lambda(\bar{r}(u))\bar{s}(u) \Big] \Big| du$$
$$\le \mathbb{E} \int_0^t |\lambda(r(u))\mu(r(u)) - \lambda(\bar{r}(u))\mu(\bar{r}(u))| du$$
$$+ \mathbb{E} \int_0^t |\lambda(r(u))S(u) - \lambda(\bar{r}(u))\bar{s}(u)| du. \qquad (10.28)$$

Let j be the integer part of T/Δ. Then

$$\mathbb{E}\int_0^T |\lambda(r(u))\mu(r(u)) - \lambda(\bar{r}(u))\mu(\bar{r}(u))|du$$
$$= \sum_{k=0}^n \mathbb{E}\int_{t_k}^{t_{k+1}} |\lambda(r(u))\mu(r(u)) - \lambda(\bar{r}(u))\mu(\bar{r}(u))|du \qquad (10.29)$$

with t_{j+1} being now set to be T. Compute

$$\mathbb{E}\int_{t_k}^{t_{k+1}} |\lambda(r(u))\mu(r(u)) - \lambda(\bar{r}(u))\mu(\bar{r}(u))|du$$
$$\leq 2\bar{\lambda}\bar{\mu}\int_{t_k}^{t_{k+1}} \mathbb{P}(r(u) \neq r(t_k))du$$
$$= 2\bar{\lambda}\bar{\mu}\int_{t_k}^{t_{k+1}} \sum_{i\in\mathbb{S}} \mathbb{P}(r(t_k) = i)\mathbb{P}(r(u) \neq i | r(t_k) = i)du$$
$$= 2\bar{\lambda}\bar{\mu}\int_{t_k}^{t_{k+1}} \sum_{i\in\mathbb{S}} \mathbb{P}(r(t_k) = i)\sum_{j\neq i}(\gamma_{ij}(u - t_k) + o(u - T_k))$$
$$\leq 2\bar{\lambda}\bar{\mu}[\max_{1\leq i\leq N}(-\gamma_{ii})\Delta + o(\Delta)]\Delta. \qquad (10.30)$$

Therefore

$$\mathbb{E}\int_0^T |\lambda(r(u))\mu(r(u)) - \lambda(\bar{r}(u))\mu(\bar{r}(u))|du$$
$$\leq 2\bar{\lambda}\bar{\mu}[\max_{1\leq i\leq N}(-\gamma_{ii})\Delta + o(\Delta)]. \qquad (10.31)$$

On the other hand

$$\mathbb{E}\int_0^t |\lambda(r(u))S(u) - \lambda(\bar{r}(u))\bar{s}(u)|du$$
$$\leq \mathbb{E}\int_0^t |\lambda(r(u)) - \lambda(\bar{r}(u))||\bar{s}(u)|du$$
$$+ \mathbb{E}\int_0^t \lambda(r(u))|S(u) - \bar{s}(u)|du. \qquad (10.32)$$

But

$$\mathbb{E}\int_0^t |\lambda(r(u)) - \lambda(\bar{r}(u))||\bar{s}(u)|du$$

$$= \sum_{k=0}^{j} \int_{t_k}^{t_{k+1}} \mathbb{E}\left[\mathbb{E}[|\lambda(r(u)) - \lambda(r(t_k))||s_k||I_{\{r(u)\neq r(t_k)\}}]\right] du$$

$$= \sum_{k=0}^{j} \int_{t_k}^{t_{k+1}} \mathbb{E}\left[\mathbb{E}[|\lambda(r(u)) - \lambda(r(t_k))||I_{\{r(u)\neq r(t_k)\}}]\mathbb{E}[|s_k||I_{\{r(u)\neq r(t_k)\}}]\right],$$

where in the last step we use the fact that s_k and $I_{\{r(u)\neq r(t_k)\}}$ are conditionally independent with respect to the σ-algebra generated by $r(t_k)$. In the same way as in (10.30), we have

$$\mathbb{E}\int_0^t |\lambda(r(u)) - \lambda(\bar{r}(u))||\bar{s}(u)|du$$
$$\leq 2\bar{\lambda}\bar{\mu}[\max_{1\leq i\leq N}(-\gamma_{ii})\Delta + o(\Delta)]\int_0^T \mathbb{E}|\bar{s}(u)|du.$$

So, by Lemma 10.3,

$$\mathbb{E}\int_0^t |\lambda(r(u)) - \lambda(\bar{r}(u))\bar{s}(u)|du$$
$$\leq 2(1+K)\bar{\lambda}\bar{\mu}[\max_{1\leq i\leq N}(-\gamma_{ii})\Delta + o(\Delta)]. \tag{10.33}$$

Substitute (10.33) to (10.32) and use Lemma 10.4, we obtain

$$\mathbb{E}\int_0^t |\lambda(r(u))S(u) - \lambda(\bar{r}(u))\bar{s}(u)|du$$
$$\leq C\Delta + o(\Delta) + \bar{\lambda}\mathbb{E}\int_0^t |S(u) - \bar{s}(u)|du$$
$$\leq C\Delta + o(\Delta) + \bar{\lambda}\mathbb{E}\int_0^t |S(u) - s(u)|du, \tag{10.34}$$

where C is a positive constant independent of Δ and it may change line by line. This, together with (10.31), yields

$$|I(t)| \leq C\Delta + o(\Delta) + \bar{\lambda}\mathbb{E}\int_0^t |S(u) - s(u)|du. \tag{10.35}$$

In the following, we will estimate $J(t)$. Note

$$|J(t)| \leq 2\bar{\sigma}^2 \mathbb{E} \int_0^t |\varphi_k''(S(u) - s(u))| \left(|S(u)|^\theta - |\bar{s}(u)|^\theta\right)^2 du$$
$$+ 2\mathbb{E} \int_0^t |\varphi_k''(S(u) - s(u))| [\sigma(r(u)) - \sigma(\bar{r}(u))]^2 |S(u)|^{2\theta} du. \quad (10.36)$$

Using property (ii) of φ_k and Lemma 10.4, we have

$$\mathbb{E} \int_0^t |\varphi_k''(S(u) - s(u))| \left(|S(u)|^\theta - |\bar{s}(u)|^\theta\right)^2 du$$
$$\leq \mathbb{E} \int_0^t |\varphi_k''(S(u) - s(u))| |S(u) - \bar{s}(u)|^{2\theta} du$$
$$\leq 2^{2\theta-1} \mathbb{E} \int_0^t |\varphi_k''(S(u) - s(u))| |S(u) - s(u)|^{2\theta} du$$
$$+ 2^{2\theta-1} \mathbb{E} \int_0^t |\varphi_k''(S(u) - s(u))| |s(u) - \bar{s}(u)|^{2\theta} du$$
$$\leq 2\mathbb{E} \int_0^t \frac{2}{k} I_{\{a_k < |S(u) - s(u)| < a_{k-1}\}} du + 2 \int_0^t \frac{2}{ka_k^{2\theta}} \mathbb{E} |s(u) - \bar{s}(u)|^{2\theta} du$$
$$\leq \frac{4T}{k} + \frac{C\Delta^\theta}{ka_k^{2\theta}}. \quad (10.37)$$

In the same way as (10.33) was proved, we can show that

$$\mathbb{E} \int_0^t |\varphi_k''(S(u) - s(u))| [\sigma(r(u)) - \sigma(\bar{r}(u))]^2 |S(u)|^{2\theta} du$$
$$\leq \mathbb{E} \int_0^t \frac{2}{ka_k^{2\theta}} [\sigma(r(u)) - \sigma(\bar{r}(u))]^2 |S(u)|^{2\theta} du \leq \frac{C\Delta + o(\Delta)}{ka_k^{2\theta}}. \quad (10.38)$$

Substituting (10.38) and (10.37) to (10.36), we have

$$|J(t)| \leq \frac{8\bar{\sigma}^2 T}{k} + \frac{C\Delta^\theta + o(\Delta)}{ka_k^{2\theta}}. \quad (10.39)$$

Therefore

$$\mathbb{E}\varphi_k(S(t) - s(t)) \leq \frac{4\bar{\sigma}^2 T}{k} + \frac{C\Delta^\theta + o(\Delta)}{ka_k^{2\theta}} + C\Delta$$
$$+ o(\Delta) + 2\bar{\lambda}\mathbb{E} \int_0^t |S(u) - s(u)| du.$$

Noting
$$\mathbb{E}\varphi_k(S(t) - s(t)) \geq \mathbb{E}|S(t) - s(t)| - a_{k-1},$$
gives
$$\mathbb{E}|S(t) - s(t)| \leq a_{k-1} + \frac{4\bar{\sigma}^2 T}{k} + \left[\frac{1}{ka_k^{2\theta}} + 1\right](C\Delta^\theta + o(\Delta))$$
$$+ \bar{\lambda}\int_0^t \mathbb{E}|S(u) - s(u)|du.$$

Finally, the required assertion follows from the Gronwall inequality. □

Next, we derive a bound for a stronger form of the error. This version uses an L^2-distance and places the supremum over time inside the expectation operation. The result below involves the L^1 error which is explicitly bounded in Theorem 10.6 and hence is also computable.

Theorem 10.7
$$\mathbb{E}\left[\sup_{0\leq t\leq T}(S(t) - s(t))^2\right]$$
$$\leq e^{(8\bar{\sigma}^2 + 2\bar{\lambda}^2)T^2}\left(C\Delta + o(\Delta) + 8\bar{\sigma}^2 T \sup_{0\leq u\leq T}\mathbb{E}|S(u) - s(u)|\right). \quad (10.40)$$

Proof. For any $0 \leq t \leq T$, using the Hölder inequality we have
$$(S(t) - s(t))^2$$
$$\leq T\int_0^t [\lambda(r(u))\mu(r(u)) - \lambda(\bar{r}(u))\mu(\bar{r}(u)) - \lambda(r(u))S(u) + \lambda(\bar{r}(u))\bar{s}(u)]^2 du$$
$$+ \left(\int_0^t [\sigma(r(u))|S(u)|^\theta - \sigma(\bar{r}(u))|\bar{s}(u)|^\theta]\,dB(u)\right)^2. \quad (10.41)$$

In the same way as (10.31) and (10.34) were proved, we derive
$$\mathbb{E}\int_0^t [\lambda(r(u))\mu(r(u)) - \lambda(\bar{r}(u))\mu(\bar{r}(u)) - \lambda(r(u))S(u) + \lambda(\bar{r}(u))\bar{s}(u)]^2 du$$
$$\leq 2\mathbb{E}\int_0^t [\lambda(r(u))\mu(r(u)) - \lambda(\bar{r}(u))\mu(\bar{r}(u))]^2 du$$
$$+ 2\mathbb{E}\int_0^t [\lambda(r(u))S(u) - \lambda(\bar{r}(u))\bar{s}(u)]^2 du$$
$$\leq C\Delta + o(\Delta) + 2\bar{\lambda}^2 \mathbb{E}\int_0^t (S(u) - s(u))^2 du. \quad (10.42)$$

Using the Doob martingale inequality, we find that for any $t_1 \in [0, T]$

$$\mathbb{E}\left[\sup_{0\le t\le t_1}\left(\int_0^t [\sigma(r(u))|S(u)|^\theta - \sigma(\bar{r}(u))|\bar{s}(u)|^\theta]\,dB(u)\right)^2\right]$$

$$\le 4\mathbb{E}\int_0^{t_1}[\sigma(r(u))|S(u)|^\theta - \sigma(\bar{r}(u))|\bar{s}(u)|^\theta]^2\,du$$

$$\le C\Delta + o(\Delta) + 8\bar{\sigma}^2\mathbb{E}\int_0^{t_1}|S(u) - s(u)|^{2\theta}\,du$$

$$\le C\Delta + o(\Delta) + 8\bar{\sigma}^2\mathbb{E}\int_0^{t_1}|S(u) - s(u)|\,du$$

$$+ 8\bar{\sigma}^2\mathbb{E}\int_0^{t_1}|S(u) - s(u)|^2\,du. \tag{10.43}$$

Therefore

$$\mathbb{E}\left[\sup_{0\le t\le t_1}(S(t) - s(t))^2\right]$$

$$\le C\Delta + o(\Delta) + (8\bar{\sigma}^2 + 2\bar{\lambda}^2)\mathbb{E}\int_0^{t_1}(S(u) - s(u))^2\,du$$

$$+ 8\bar{\sigma}^2\mathbb{E}\int_0^{t_1}|S(u) - s(u)|\,du$$

$$\le C\Delta + o(\Delta) + (8\bar{\sigma}^2 + 2\bar{\lambda}^2)\int_0^{t_1}\mathbb{E}\left[\sup_{0\le u\le v}(S(u) - s(u))^2\right]dv$$

$$+ 8\bar{\sigma}^2 T\sup_{0\le u\le T}\mathbb{E}|S(u) - s(u)|. \tag{10.44}$$

An application of the Gronwall inequality completes the proof. $\square$

10.3.3 Stochastic Volatility Model

Let us now consider the Heston stochastic volatility model under regime-switching, namely

$$dX(t) = \lambda_1(r(t))[\mu_1(r(t)) - X(t)]dt + \sigma_1(r(t))X(t)\sqrt{V(t)}\,dB_1(t), \tag{10.45}$$
$$dV(t) = \lambda_2(r(t))[\mu_2(r(t)) - V(t)]dt + \sigma_2(r(t))V^\theta(t)\,dB_2(t) \tag{10.46}$$

on $0 \le t \le T$. Here $V(t)$ is the volatility that feeds into the asset price $X(t)$. The Brownian motions $B_1(t)$ and $B_2(t)$ may be correlated. Naturally, we assume the initial values $X(0)$ and $V(0)$ are both positive constants. Moreover, λ_1, σ_1 etc, are all mappings from $\mathbb{S}$ to $\mathbb{R}_+$.

We begin with a lemma showing that the positivity in the initial data leads to the positive solution $X(t)$.

Lemma 10.5 *If $V(t)$, $t \in [0,T]$ is given by (10.46), then*

$$\mathbb{P}(X(t) > 0 \text{ for all } 0 \leq t \leq T) = 1. \tag{10.47}$$

Proof. $X(t)$ can be expressed explicitly as (see Exercise 10.2)

$$X(t) = \Psi(t)\Big(X(0) + \int_0^t \frac{\lambda_1(r(s))\mu_1(r(s))}{\Psi(s)} ds\Big), \tag{10.48}$$

where

$$\Psi(t) = \exp\Big(\int_0^t \big[-\lambda_1(r(s)) - \tfrac{1}{2}\sigma_1^2(r(s))V(s)\big]ds + \int_0^t \sigma_1(r(s))\sqrt{V(s)}dB_1(s)\Big).$$

The assertion follows clearly. $\square$

Applying the EM method to equation (10.46) gives

$$v_{k+1} = v_k + \lambda_1(r_k^\Delta)(\mu_1(r_k^\Delta) - v_k)\Delta + \sigma_1(r_k^\Delta)|v_k|^\theta \Delta B_{2,k}, \tag{10.49}$$

where $\Delta B_{2,k} = B_2(t_{k+1}) - B_2(t_k)$, while applying the EM method to equation (10.45) gives

$$x_{k+1} = x_k + \lambda_2(r_k^\Delta)(\mu_2(r_k^\Delta) - x_k)\Delta + \sigma_2(r_k^\Delta)x_k\sqrt{|v_k|}\Delta B_{1,k}, \tag{10.50}$$

where $\Delta B_{1,k} = B_1(t_{k+1}) - B_1(t_k)$. Let

$$\bar{x}(t) = x_k, \quad \bar{v}(t) = v_k \quad \bar{r}(t) = r_k^\Delta \quad \text{for } t \in [t_k, t_{k+1}), \tag{10.51}$$

and define the continuous EM approximate solution by

$$x(t) = x_0 + \int_0^t \lambda_1(\bar{r}(u))[\mu_1(\bar{r}(u)) - \bar{x}(u)]du$$
$$+ \int_0^t \sigma_1(\bar{r}(u))\bar{x}(u)\sqrt{|\bar{v}(u)|}dB_1(u), \tag{10.52}$$

$$v(t) = v_0 + \int_0^t \lambda_2(\bar{r}(u))[\mu_2(\bar{r}(u)) - \bar{v}(u)]du$$
$$+ \int_0^t \sigma_2(\bar{r}(u))|\bar{v}(u)|^\theta dB_2(u). \tag{10.53}$$

Lemma 10.6 *For any given pair of positive numbers p and q, define the stopping time*

$$\tau_{pq} = \inf\{t \geq 0 : X(t) > p \text{ or } |v(t)| > q\}.$$

Then

$$\lim_{\Delta \to 0} \mathbb{E}\left(\sup_{0 \leq t \leq T} |X(t \wedge \tau_{pq}) - x(t \wedge \tau_{pq})|^2\right) = 0. \qquad (10.54)$$

Proof. Fix p and q arbitrarily and write $\tau_{pq} = \tau$. For any $0 \leq t_1 \leq T$,

$$X(t_1 \wedge \tau) - x(t_1 \wedge \tau)$$
$$= \int_0^{t_1 \wedge \tau} [\lambda_1(r(u))\mu_1(r(u)) - \lambda_1(\bar{r}(u))\mu_1(\bar{r}(u))]du$$
$$- \int_0^{t_1 \wedge \tau} [\lambda_1(r(u))X(u) - \lambda_1(\bar{r}(u))\bar{x}(u)]du$$
$$+ \int_0^{t_1 \wedge \tau} [\sigma_1(r(u))X(u)\sqrt{V(u)} - \sigma_1(\bar{r}(u))\bar{x}(u)\sqrt{|\bar{v}(u)|}]dB_1(u)$$
$$= \int_0^{t_1 \wedge \tau} [\lambda_1(r(u))\mu_1(r(u)) - \lambda_1(\bar{r}(u))\mu_1(\bar{r}(u))]du$$
$$- \int_0^{t_1 \wedge \tau} X(u)[\lambda_1(r(u)) - \lambda_1(\bar{r}(u))]du - \int_0^{t_1 \wedge T} \lambda_1(\bar{r}(u))[X(u) - \bar{x}(u)]du$$
$$+ \int_0^{t_1 \wedge \tau} X(u)\sqrt{V(u)}[\sigma_1(r(u)) - \sigma_1(\bar{r}(u))]dB_1(u)$$
$$+ \int_0^{t_1 \wedge \tau} \sigma_1(\bar{r}(u))\sqrt{\bar{v}(u)}[X(u) - \bar{x}(u)]dB_1(u)$$
$$+ \int_0^{t_1 \wedge \tau} \sigma_1(\bar{r}(u))X(u)[\sqrt{V(u)} - \sqrt{|\bar{v}(u)|}]dB_1(u).$$

By the Hölder inequality and the Doob martingale inequality, we have

$$\mathbb{E}\left(\sup_{0 \leq t_1 \leq t} |X(t \wedge \tau) - x(t \wedge \tau)|^2\right)$$
$$\leq 32t\mathbb{E}\int_0^{t \wedge \tau} [\lambda_1(r(u))\mu_1(r(u)) - \lambda_1(\bar{r}(u))\mu_1(\bar{r}(u))]^2 du$$
$$+ 32t\mathbb{E}\int_0^{t \wedge \tau} X^2(u)[\lambda_1(r(u)) - \lambda_1(\bar{r}(u))]^2 du$$
$$+ 32t\mathbb{E}\int_0^{t \wedge \tau} \lambda_1^2(\bar{r}(u))[X(u) - \bar{x}(u)]^2 du$$

$$+128\mathbb{E}\int_0^{t\wedge\tau}(X(u)\sqrt{V(u)})^2[\sigma_1(r(u))-\sigma_1(\bar{r}(u))]^2du$$

$$+128\mathbb{E}\int_0^{t\wedge\tau}(\sigma_1(\bar{r}(u))\sqrt{\bar{v}(u)})^2[X(u)-\bar{x}(u)]^2du$$

$$+128\mathbb{E}\int_0^{t\wedge\tau}(\sigma_1(\bar{r}(u))X(u))^2[\sqrt{V(u)}-\sqrt{|\bar{v}(u)|}]^2du. \quad (10.55)$$

Using Lemma 10.3 and Lemma 10.4, the definition of τ and the techniques of the proof of (10.33), we derive that

$$\mathbb{E}\left(\sup_{0\leq t_1\leq t}|X(t_1\wedge\tau)-x(t_1\wedge\tau)|^2\right)$$
$$\leq 32[t+p^2t+4p^2K^2+4\bar{\sigma}_1^2p^2](C\Delta+o(\Delta))$$
$$+32[t\bar{\lambda}_1^2+4\bar{s}_1q]\mathbb{E}\int_0^{t\wedge\tau}[X(u)-\bar{x}(u)]^2du$$
$$\leq C_1\Delta+o(\Delta)+C_2\mathbb{E}\int_0^{t\wedge\tau}[X(u)-\bar{x}(u)]^2du$$
$$\leq C_1\Delta+o(\Delta)+C_2\mathbb{E}\int_0^{t\wedge\tau}[X(u)-x(u)]^2du$$
$$+C_2\mathbb{E}\int_0^{t\wedge\tau}[x(u)-\bar{x}(u)]^2du, \quad (10.56)$$

where C_1, C_2 and the following $C_3, \cdots$ are positive constants which may change line by line. For $0\leq u\leq t\wedge\tau$, let $[u/\Delta]$ be the integer part of u/Δ. Then

$$x(u)-\bar{x}(u)=\int_{[u/\Delta]\Delta}^u \lambda_1(\bar{r}(u))[\mu_1(\bar{r}(u))-\bar{x}(u)]du$$
$$+\int_{[u/\Delta]\Delta}^u \sigma_1(\bar{r}(u))\bar{x}(u)\sqrt{|\bar{v}(u)|}dB_1(u),$$

which yields

$$|x(u)-\bar{x}(u)|^2\leq 4\bar{\lambda}_1^2(\bar{\mu}_1^2+p^2)\Delta^2+2\bar{\sigma}_1^2p^2q(B_1(u)-B_1([u/\Delta]\Delta))^2.$$

Therefore

$$\mathbb{E}\int_0^{t\wedge\tau}[x(u)-\bar{x}(u)]^2du\leq C_3\Delta. \quad (10.57)$$

By (10.56) and (10.57), we have

$$\mathbb{E}\left(\sup_{0\leq t_1\leq t}|X(t_1\wedge\tau)-x(t_1\wedge\tau)|^2\right)$$
$$\leq C_4\Delta+o(\Delta)+C_5\int_0^{t\wedge\tau}\mathbb{E}\left(\sup_{0\leq t_1\leq u}|X(t_1)-x(t_1)|^2\right)du.$$

By the well-known Gronwall inequality,

$$\mathbb{E}\left(\sup_{0\leq t_1\leq t}|X(t_1\wedge\tau)-x(t_1\wedge\tau)|^2\right)\leq [C_4\Delta+o(\Delta)]e^{C_5 T}.$$

The required assertion (10.54) follows by letting $\Delta\to 0$. □

Lemma 10.7 *The continuous EM approximate solution (10.53) obeys*

$$\mathbb{E}\left(\sup_{0\leq t\leq T}|v(t)|\right)\leq (1+2v_0+2\bar{\lambda}_2\bar{\mu}_2 T)e^{(2\bar{\lambda}_2+9\bar{\sigma}_2^2)T}\quad\forall T>0, \quad (10.58)$$

where $\bar{\lambda}_2=\max_{i\in\mathbb{S}}\lambda_2(i)$ and $\bar{\mu}_2$ and $\bar{\sigma}_2$ are defined similarly.

Proof. By the well-known Burkholder–Davis–Gunday inequality, we derive from (10.53) that for $0\leq t\leq T$,

$$\mathbb{E}\left(\sup_{0\leq t_1\leq t}|v(t_1)|\right)\leq v_0+\int_0^{t_1}|\lambda_2(\bar{r}(u))[\mu_2(\bar{r}(u))-\bar{v}(u)]|du$$
$$+3\mathbb{E}\left[\int_0^t\left(\sigma_2(\bar{r}(u))|\bar{v}(u)|^\theta\right)^2 du\right]^{\frac{1}{2}}$$
$$\leq v_0+\bar{\lambda}_2\bar{\mu}_2 T+\bar{\lambda}_2\int_0^t\mathbb{E}|\bar{v}(u)|du$$
$$+3\bar{\sigma}_2\mathbb{E}\left[\left(\sup_{0\leq t_1\leq t}|v(t_1)|^{2\theta-1}\right)\int_0^t|\bar{v}(u)|du\right]^{\frac{1}{2}}$$
$$\leq v_0+\bar{\lambda}_2\bar{\mu}_2 T+\bar{\lambda}_2\int_0^t\mathbb{E}|\bar{v}(u)|du$$
$$+\frac{1}{2}\mathbb{E}\left(\sup_{0\leq t_1\leq t}|v(t_1)|^{2\theta-1}\right)+\frac{9}{2}\bar{\sigma}_2^2\mathbb{E}\int_0^t|\bar{v}(u)|du$$
$$\leq v_0+\bar{\lambda}_2\bar{\mu}_2 T+\frac{1}{2}\left[1+\mathbb{E}\left(\sup_{0\leq t_1\leq t}|v(t_1)|\right)\right]$$
$$+\left(\bar{\lambda}_2+\frac{9}{2}\bar{\sigma}_2^2\right)\int_0^t\mathbb{E}\left(\sup_{0\leq t_1\leq u}|v(t_1)|\right)du$$

This yields

$$\mathbb{E}\Big(\sup_{0\le t_1\le t}|v(t_1)|\Big)$$
$$\le 1+2v_0+2\bar{\lambda}_2\bar{\mu}_2 T+(2\bar{\lambda}_2+9\bar{\sigma}_2^2)\int_0^t \mathbb{E}\Big(\sup_{0\le t_1\le u}|v(t_1)|\Big)du$$

An application of the Gronwall inequality implies assertion (10.58). □

10.3.4 Options Under Stochastic Volatility

The EM method established above provides us with a numerical scheme to carry out the Monte Carlo simulation for the option price if the underlying asset price follows the Heston model under regime-switching (10.45) and (10.46).

Let K be the exercise price. Define the payoff for the European put option

$$\mathcal{P}=\mathbb{E}[(K-X(T))^+].$$

Accordingly, the payoff based on the numerical method (10.51) is

$$\mathcal{P}_\Delta=\mathbb{E}[(K-\bar{x}(T))^+].$$

Theorem 10.8 *In the notation above,*

$$\lim_{\Delta\to 0}|\mathcal{P}_\Delta-\mathcal{P}|=0.$$

Proof. For $p,q>0$, set

$$A_q=\{X(t)\le q,\ 0\le t\le T\}\quad\text{and}\quad G_q=\{|v(t)|\le q,\ 0\le t\le T\}.$$

Given any $\varepsilon>0$, by Lemmas 10.5 and 10.7 we can find p,q sufficiently large for

$$\mathbb{P}(A_p^c\cup G_q^c)\le\frac{\varepsilon}{4K}.$$

Compute

$$|\mathcal{P} - \mathcal{P}_\Delta| \leq \mathbb{E}|(K - X(T))^+ - (K - \bar{x}(T))^+|$$
$$= \mathbb{E}\Big(|(K - X(T))^+ - (K - \bar{x}(T))^+|I_{A_p \cap G_q}\Big)$$
$$+ \mathbb{E}\Big(|(K - X(T))^+ - (K - \bar{x}(T))^+|I_{A_p^c \cup G_q^c}\Big)$$
$$\leq \mathbb{E}\Big(|X(T) - \bar{x}(T)|I_{A_p \cap G_q}\Big) + 2K\mathbb{P}(A_p^c \cup G_q^c)$$
$$\leq \mathbb{E}\Big(|X(T) - \bar{x}(T)|I_{(\tau_{pq} > T)}\Big) + \frac{\varepsilon}{2}$$
$$\leq \mathbb{E}|X(T \wedge \tau_{pq}) - x(T \wedge \tau_{pq})| + \mathbb{E}\Big(|x(T) - \bar{x}(T)|I_{(\tau_{pq} > T)}\Big)$$
$$+ \frac{\varepsilon}{2}, \tag{10.59}$$

where τ_{pq} is the stopping time defined in Lemma 10.6. But, let j be the integer part of T/Δ and define $\bar{\lambda}_1 = \max_{i \in S} \lambda_1(i)$ and $\bar{\mu}_1, \bar{\sigma}_1$ similarly. Then

$$\mathbb{E}\Big(|x(T) - \bar{x}(T)|I_{(\tau_{pq} > T)}\Big)$$
$$\leq \mathbb{E}\Big(|\bar{\lambda}_1(\bar{\mu}_1 + p)(T - j\Delta) + \bar{\sigma}_1\sqrt{\bar{q}}\, p(B_1(T) - B_1(j\Delta))|I_{(\tau_{pq} > T)}\Big)$$
$$\leq \bar{\lambda}_1(\bar{\mu}_1 + p)\Delta + \bar{\sigma}_1\sqrt{\bar{q}}\, p\sqrt{\Delta}.$$

This, together with Lemma 10.6, shows that there is a $\Delta^* > 0$ such that for all $\Delta < \Delta^*$,

$$\mathbb{E}|X(T \wedge \tau_{pq}) - x(T \wedge \tau_{pq})| + \mathbb{E}\Big(|x(T) - \bar{x}(T)|I_{(\tau_{pq} > T)}\Big) < \frac{\varepsilon}{2}.$$

In view of (10.59), this completes the proof. □

Further more, let us consider the more complicated barrier option which is a path-dependent option. Let K be the exercise price and $b(> K)$ be a barrier. For the Heston model under regime-switching (10.45) and (10.46), the payoff for the barrier European put option is given by

$$U = \mathbb{E}\left[(X(T) - K)^+ I_{\{0 \leq X(t) \leq b,\ 0 \leq t \leq T\}}\right],$$

while the payoff based on the numerical method (10.51) is

$$U_\Delta = \mathbb{E}\left[(\bar{x}(T) - K)^+ I_{\{0 \leq \bar{x}(t) \leq b,\ 0 \leq t \leq T\}}\right].$$

Theorem 10.9 *In the notation above,*

$$\lim_{\Delta \to 0} |U_\Delta - U| = 0.$$

Proof. Let $A := \{0 \le X(t) \le b,\ 0 \le t \le T\}$ and $A_\Delta := \{0 \le \bar{x}(t) \le b,\ 0 \le t \le T\}$. It can be shown (see Exercise 10.3) that

$$|U - U_\Delta| \le \mathbb{E}\left(|X(T) - \bar{x}(T)|I_{A \cap A_\Delta}\right) + (b - K)\mathbb{P}(A \cap A_\Delta^c)$$
$$+ (b - K)\mathbb{P}(A^c \cap A_\Delta). \qquad (10.60)$$

Now, for $p,\ q > b$, set

$$A_p := \{X(t) \le p,\ 0 \le t \le T\} \quad \text{and} \quad G_q := \{|v(t)| \le q,\ 0 \le t \le T\}.$$

Let τ_{pq} be the same stopping time as defined in Lemma 10.6. Let $\varepsilon > 0$ be arbitrary. By Lemmas 10.5 and 10.7, we can find a pair of p and q sufficiently large for

$$\mathbb{P}(A_p^c \cup G_q^c) \le \frac{\varepsilon}{2(1 \vee b)}. \qquad (10.61)$$

Compute

$$\mathbb{E}\left(|X(T) - \bar{x}(T)|I_{A \cap A_\Delta}\right)$$
$$= \mathbb{E}\left(|X(T) - \bar{x}(T)|I_{A \cap A_\Delta \cap A_p \cap G_q}\right) + \mathbb{E}\left(|X(T) - \bar{x}(T)|I_{A \cap A_\Delta \cap (A_p^c \cup G_q^c)}\right)$$
$$\le \mathbb{E}\left(|X(T) - \bar{x}(T)|I_{A_p \cap G_q}\right) + b\mathbb{P}(A_p^c \cup G_q^c)$$
$$\le \mathbb{E}\left(|X(T) - \bar{x}(T)|I_{\{\tau_{pq} > T\}}\right) + \frac{\varepsilon}{2}.$$

Recalling the proof of Theorem 10.8, we observe that there is a $\Delta^* = \Delta^*(\varepsilon) > 0$ such that for any $\Delta < \Delta^*$,

$$\mathbb{E}\left(|X(T) - \bar{x}(T)|I_{\{\tau_{pq} > T\}}\right) < \frac{\varepsilon}{2},$$

whence

$$\mathbb{E}\left(|X(T) - \bar{x}(T)|I_{A \cap A_\Delta}\right) < \varepsilon.$$

In other words, we have shown that

$$\lim_{\Delta \to 0} \mathbb{E}\left(|X(T) - \bar{x}(T)|I_{A \cap A_\Delta}\right) = 0. \qquad (10.62)$$

Let us now begin to show that $\mathbb{P}(A \cap A_\Delta^c) \to 0$ as $\Delta \to 0$. Using (10.61), we note that

$$\mathbb{P}(A \cap A_\Delta^c) \le \mathbb{P}(A \cap A_\Delta^c \cap A_p \cap G_q) + \frac{\varepsilon}{2}. \qquad (10.63)$$

On the other hand, for any sufficiently small δ, we have

$$\begin{aligned}
A &= \{\sup_{0\leq t\leq T} X(t) \leq b\} \\
&= \{\sup_{0\leq t\leq T} X(t) \leq b-\delta\} \cup \{b-\delta < \sup_{0\leq t\leq T} X(t) \leq b\} \\
&\subseteq \{\sup_{0\leq k\Delta\leq T} X(k\Delta) \leq b-\delta\} \cup \{b-\delta < \sup_{0\leq t\leq T} X(t) \leq b\} \\
&=: \bar{A}_1 \cup \bar{A}_2.
\end{aligned}$$

Hence,

$$\begin{aligned}
A \cap A_\Delta^c \cap A_p \cap G_q &\subseteq \left[\bar{A}_1 \cap A_\Delta^c \cap A_p \cap G_q\right] \cup \bar{A}_2 \\
&\subseteq \left[\{\sup_{0\leq k\Delta\leq T} |X(k\Delta) - \bar{x}(k\Delta)| \geq \delta\} \cap \{\tau_{pq} > T\}\right] \cup \bar{A}_2 \\
&\subseteq \{\sup_{0\leq t\leq T} |X(t\wedge \tau_{pq}) - \bar{x}(t\wedge \tau_{pq})| \geq \delta\} \cup \bar{A}_2
\end{aligned}$$

So,

$$\begin{aligned}
&\mathbb{P}(A \cap A_\Delta^c \cap A_p \cap G_q) \\
&\leq \mathbb{P}(\sup_{0\leq t\leq T} |X(t\wedge \tau_{pq}) - \bar{x}(t\wedge \tau_{pq})| \geq \delta) + \mathbb{P}(\bar{A}_2) \\
&\leq \frac{1}{\delta^2}\mathbb{E}\left(\sup_{0\leq t\leq T} |X(t\wedge \tau_{pq}) - \bar{x}(t\wedge \tau_{pq})|^2\right) + \mathbb{P}(\bar{A}_2).
\end{aligned}$$

Now, for any $\varepsilon > 0$, we may choose δ so small that

$$\mathbb{P}(\bar{A}_2) < \frac{\varepsilon}{4}$$

and then, by Lemma 10.6, choose Δ so small that

$$\frac{1}{\delta^2}\mathbb{E}\left(\sup_{0\leq t\leq T} |X(t\wedge \tau_{pq}) - \bar{x}(t\wedge \tau_{pq})|^2\right) < \frac{\varepsilon}{4}$$

whence

$$\mathbb{P}(A \cap A_\Delta^c \cap A_p \cap G_q) < \frac{\varepsilon}{2}.$$

Substituting this into (10.63) we obtain

$$\mathbb{P}(A \cap A_\Delta^c) < \varepsilon$$

as long as Δ is sufficiently small. This shows that

$$\lim_{\Delta \to 0} \mathbb{P}(A \cap A_\Delta^c) = 0. \qquad (10.64)$$

Similarly, we can show (the details are left as an exercise) that

$$\lim_{\Delta \to 0} \mathbb{P}(A^c \cap A_\Delta) = 0. \tag{10.65}$$

Substituting (10.62), (10.64) and (10.65) into (10.60), we finally obtain

$$\lim_{\Delta \to 0} |U - U_\Delta| = 0,$$

as required. □

10.4 Stochastic Stabilisation and Destabilisation

Let us now begin to discuss the stochastic stabilisation for the hybrid ordinary differential equation (ODE)

$$\dot{x}(t) = f(x(t), t, r(t)). \tag{10.66}$$

In its operation, the hybrid system will switch from one mode to another according to the law of Markov chain. It happens often that the system is observable only when it operates in some modes but not all. Accordingly, let us decompose $\mathbb{S}$ into two subsets $\mathbb{S}_1$ and $\mathbb{S}_2$, namely $\mathbb{S} = \mathbb{S}_1 \cup \mathbb{S}_2$, where for each mode $i \in \mathbb{S}_1$ the ODE is not observable and hence cannot be stabilised by feedback control, but it can be stabilised for each $i \in \mathbb{S}_2$. The question is: Can we stabilise the hybrid ODE (10.66) if we can only control the partial system?

More precisely, let us consider the controlled stochastic system

$$dx(t) = f(x(t), t, r(t))dt + u(t, r(t))dB(t), \tag{10.67}$$

where $u(t, i) \equiv 0$ for $i \in \mathbb{S}_1$ while $u(t, i) = u(x(t), i)$ is a feedback control for $i \in \mathbb{S}_2$. Our aim is to design the control $u(x(t), i)$ for $i \in \mathbb{S}_2$ only so that the controlled system (10.67) is stabilised. In this section we only consider the linear feedback control of the form

$$u(x, i) = (G_{1,i}x, G_{2,i}x, \cdots, G_{m,i}x). \tag{10.68}$$

Thus the controlled system (10.67) becomes

$$dx(t) = f(x(t), t, r(t))dt + \sum_{k=1}^{m} G_{k,r(t)}x(t)dB_k(t), \tag{10.69}$$

where $G_{k,i} = 0$ whenever $i \in \mathbb{S}_1$ while the other $G_{k,i}$'s are all $n \times n$ matrices to be designed in order to make the controlled system (10.69) become stable.

Clearly not any given hybrid ODE (10.66) can be stabilised by stochastic control and we need to impose some conditions on it. To proceed, we impose the following simple assumption.

Assumption 10.10 There is a positive constant K such that
$$|f(x,t,i)| \leq K|x| \quad \forall (x,t,i) \in \mathbb{R}^n \times \mathbb{R}_+ \times \mathbb{S}.$$

Moreover, the Markov chain $r(t)$ is irreducible.

In other words, we require the coefficient of the given ODE be bounded by a linear function and this is reasonable since we use the linear feedback control of form (10.68). The irreducibility is also essential because the ODE is not observable in any mode in $\mathbb{S}_1$ but the irreducibility guarantees it will switch to a mode in $\mathbb{S}_2$ where it becomes observable. It is known that the irreducibility implies that the Markov chain has a unique stationary (probability) distribution $\pi = (\pi_1, \pi_2, \cdots, \pi_N) \in \mathbb{R}^{1 \times N}$ which can be determined by solving the following linear equation

$$\pi \Gamma = 0 \tag{10.70}$$

subject to

$$\sum_{j=1}^{N} \pi_j = 1 \quad \text{and} \quad \pi_j > 0 \quad \forall j \in \mathbb{S}.$$

Theorem 10.11 *Let Assumption 10.10 hold. Assume that for each $i \in \mathbb{S}_2$, the matrices $G_{k,i}$ in the controller have the property that*

$$\sum_{k=1}^{m} |G_{k,i}x|^2 \leq a_i|x|^2 \quad \text{and} \quad \sum_{k=1}^{m} |x^T G_{k,i}x|^2 \geq b_i|x|^4, \quad \forall x \in \mathbb{R}^n \tag{10.71}$$

where a_i and b_i are some non-negative constants. Then for any initial value $x_0 \in \mathbb{R}^n$ the solution of the controlled system (10.69) has the property that

$$\limsup_{t \to \infty} \frac{1}{t} \log(|x(t)|) \leq K + \sum_{i \in \mathbb{S}_2} \pi_i(0.5a_i - b_i) \quad a.s. \tag{10.72}$$

In particular, if $K + \sum_{i \in \mathbb{S}_2} \pi_i(0.5a_i - b_i) < 0$ then the controlled system (10.69) is almost surely exponentially stable.

Proof. Clearly, (10.72) holds when $x_0 = 0$. Fix any $x_0 \neq 0$. By Lemma 5.1, the solution $x(t)$ will never reach zero with probability 1. We can then apply the Itô formula to obtain that

$$d[\log(|x(t)|^2)] = \frac{2x^T(t)}{|x(t)|^2}\left[f(x(t),t,r(t))dt + \sum_{k=1}^{m} G_{k,r(t)}x(t)dB_k(t)\right]$$
$$+ \left[\frac{\sum_{k=1}^{m}|G_{k,r(t)}x(t)|^2}{|x(t)|^2} - \frac{2\sum_{k=1}^{m}|x^T(t)G_{k,r(t)}x(t)|^2}{|x(t)|^4}\right]dt.$$

Setting $a_i = b_i = 0$ for $i \in \mathbb{S}_1$ and using (10.71) as well as Assumption 10.10, we obtain that

$$\log(|x(t)|^2) \le \log(|x_0|^2) + \int_0^t \left[2K + a_{r(s)} - 2b_{r(s)}\right]ds + M(t), \quad (10.73)$$

where

$$M(t) = \sum_{k=1}^{m}\int_0^t \frac{2x^T(s)G_{k,r(s)}x(s)}{|x(s)|^2}dB_k(s)$$

which is a continuous martingale vanishing at $t = 0$. The quadratic variation of the martingale is given by

$$\langle M(t), M(t)\rangle = \sum_{k=1}^{m}\int_0^t \frac{4|x^T(s)G_{k,r(s)}x(s)|^2}{|x(s)|^4}ds \le 4t \max_{i\in\mathbb{S}_2}\left(\sum_{k=1}^{m}\|G_{k,i}\|^2\right).$$

By Theorem 1.6,

$$\lim_{t\to\infty} \frac{M(t)}{t} = 0 \quad a.s.$$

Moreover, by the ergodic property of the Markov chain, we have

$$\lim_{t\to\infty}\frac{1}{t}\int_0^t \left[2K + a_{r(s)} - 2b_{r(s)}\right]ds = 2K + \sum_{i\in\mathbb{S}_2}\pi_i(a_i - 2b_i) \quad a.s.$$

It then follows from (10.73) that

$$\limsup_{t\to\infty} \frac{1}{t}\log(|x(t)|) \le K + \sum_{i\in\mathbb{S}_2}\pi_i(0.5a_i - b_i) \quad a.s.$$

as required. □

Theorem 10.11 ensures that there are many choices for the matrices $G_{k,i}$ in order to stabilise the given hybrid system (10.66). For example, let

$$G_{k,i} = \theta_{k,i}I, \quad 1 \le k \le m, \; i \in \mathbb{S}_2,$$

where I is the $n \times n$ identity matrix and $\theta_{k,i}$ are constants. Then the controlled system (10.69) becomes

$$dx(t) = f(x(t), t, r(t))dt + \sum_{k=1}^{m} \theta_{k,r(t)} x(t) dB_k(t), \qquad (10.74)$$

where we set $\theta_{k,i} = 0$ for $i \in \mathbb{S}_1$ and $1 \leq k \leq m$. Note in this case that for $i \in \mathbb{S}_2$ and $x \in \mathbb{R}^n$,

$$\sum_{k=1}^{m} |G_{k,i} x|^2 = \Big(\sum_{k=1}^{m} \theta_{k,i}^2\Big) |x|^2 \quad \text{and} \quad \sum_{k=1}^{m} |x^T G_{k,i} x|^2 = \Big(\sum_{k=1}^{m} \theta_{k,i}^2\Big) |x|^4.$$

By Theorem 10.11 we can conclude that the solution of the controlled system (10.74) satisfies

$$\limsup_{t \to \infty} \frac{1}{t} \log(|x(t)|) \leq K - 0.5 \sum_{i \in \mathbb{S}_2} \pi_i \Big(\sum_{k=1}^{m} \theta_{k,i}^2\Big) \quad a.s.$$

Recalling that the stationary probabilities $\pi_i > 0$ for all $i \in \mathbb{S}$, given any $K > 0$, one can always choose the constants $\theta_{k,i}$ ($i \in \mathbb{S}_2$) sufficiently large for

$$0.5 \sum_{i \in \mathbb{S}_2} \pi_i \Big(\sum_{k=1}^{m} \theta_{k,i}^2\Big) > K$$

in order to make the controlled system (10.74) become stable.

As one more example, for each pair of $i \in \mathbb{S}_2$ and $1 \leq k \leq m$, choose a symmetric positive definite matrix $D_{k,i}$ such that

$$x^T D_{k,i} x \geq \frac{3}{4} \|D_{k,i}\| |x|^2.$$

Obviously, there are many such matrices. Let θ be a constant and $G_{k,i} = \theta D_{k,i}$. Then the controlled system (10.69) becomes

$$dx(t) = f(x(t), t, r(t))dt + \sum_{k=1}^{m} \theta D_{k,r(t)} x(t) dB_k(t), \qquad (10.75)$$

where we set $D_{k,i} = 0$ for $i \in \mathbb{S}_1$ and $1 \leq k \leq m$. Note that for each $i \in \mathbb{S}_2$,

$$\sum_{k=1}^{m} |G_{k,i} x|^2 \leq \theta^2 \Big(\sum_{k=1}^{m} \|D_{k,i}\|^2\Big) |x|^2$$

and
$$\sum_{k=1}^{m}|x^T G_{k,i}x|^2 \geq \frac{9\theta^2}{16}\Big(\sum_{k=1}^{m}\|D_{k,i}\|^2\Big)|x|^4$$

for all $x \in \mathbb{R}^n$. By Theorem 10.11, the solution of the controlled system (10.75) has the property that

$$\limsup_{t\to\infty}\frac{1}{t}\log(|x(t)|) \leq K - \frac{\theta^2}{16}\sum_{i\in\mathbb{S}_2}\pi_i\Big(\sum_{k=1}^{m}\|D_{k,i}\|^2\Big) \quad a.s.$$

If we choose θ sufficiently large such that

$$\theta^2 > \frac{16K}{\sum_{i\in\mathbb{S}_2}\pi_i\Big(\sum_{k=1}^{m}\|D_{k,i}\|^2\Big)},$$

then the controlled system (10.75) is almost surely asymptotically stable. Summarising the discussions above we can state a general theorem.

Theorem 10.12 *Given any nonlinear hybrid system (10.66) satisfying Assumption 10.10, one can always design a linear controller $u(x,i)$ of the form (10.68) for the partial modes $i \in \mathbb{S}_2$ so that the controlled system (10.69) becomes stable.*

Let us now turn to consider the opposite problem—stochastic destabilisation. More precisely, given a nonlinear stable hybrid system (10.66), can we design a linear controller $u(x,i)$ of the form (10.68) for those modes $i \in \mathbb{S}_2$ only so that the controlled system (10.69) become unstable? To answer this question positively, let us establish a new result.

Theorem 10.13 *Let Assumption 10.10 hold. Assume that for each $i \in \mathbb{S}_2$, the matrices $G_{k,i}$ in the controller (10.68) satisfy*

$$\sum_{k=1}^{m}|G_{k,i}x|^2 \geq a_i|x|^2 \quad \text{and} \quad \sum_{k=1}^{m}|x^T G_{k,i}x|^2 \leq b_i|x|^4, \quad \forall x \in \mathbb{R}^n \quad (10.76)$$

where a_i and b_i are some non-negative constants. Then for any initial value $x_0 \neq 0$, the solution of the controlled system (10.69) satisfies

$$\limsup_{t\to\infty}\frac{1}{t}\log|x(t)| \geq -K + \sum_{i\in\mathbb{S}_2}\pi_i(0.5a_i - b_i) \quad a.s. \quad (10.77)$$

In particular, if $\sum_{i\in\mathbb{S}_2}\pi_i(0.5a_i-b_i) > K$ then the controlled system (10.69) is almost surely exponentially unstable.

This theorem can be proved in the same way as Theorem 10.11 was proved and the details are left as an exercise. The question now becomes: Can we find matrices $G_{k,i}$ so that $\sum_{i \in \mathbb{S}_2} \pi_i(0.5a_i - b_i) > K$? We shall show that this is possible if the dimension of the state space is greater than or equal to 2.

First, let the dimension of the state space n be an even number. For each $i \in \mathbb{S}_2$, let θ_i be a constant and define

$$G_{1,i} = \begin{bmatrix} 0 & \theta_i & & & \\ -\theta_i & 0 & & & \\ & & \ddots & & \\ & & & 0 & \theta_i \\ & & & -\theta_i & 0 \end{bmatrix},$$

but set $G_{k,i} = 0$ for $2 \le k \le m$. The controlled system (10.69) becomes

$$dx(t) = f(x(t), t, r(t))dt + \theta_{r(t)} \begin{bmatrix} x_2(t) \\ -x_1(t) \\ \vdots \\ x_n(t) \\ -x_{n-1}(t) \end{bmatrix} dB_1(t), \qquad (10.78)$$

where we set $\theta_i = 0$ for $i \in \mathbb{S}_1$. Note that for each $i \in \mathbb{S}_2$,

$$\sum_{k=1}^m |G_{k,i}x|^2 = |G_{1,i}x|^2 = \theta_i^2 |x|^2 \quad \text{and} \quad \sum_{k=1}^m |x^T G_{k,i}x|^2 = |x^T G_{1,i}x|^2 = 0.$$

Hence, by Theorem 10.13, the solution of the controlled system (10.78) has the property that

$$\limsup_{t \to \infty} \frac{1}{t} \log |x(t)| \ge -K + \sum_{i \in \mathbb{S}_2} 0.5\pi_i \theta_i^2 \quad a.s. \qquad (10.79)$$

if the initial value $x_0 \ne 0$. Clearly we can choose θ_i ($i \in \mathbb{S}_2$) sufficiently large for $\sum_{i \in \mathbb{S}_2} 0.5\pi_i \theta_i^2 > K$ so that the controlled system (10.78) becomes unstable.

We next consider the case when the dimension of the state space n is an odd number and $n \ge 3$. Let the dimension of the Brownian motion $m \ge 2$.

For each $i \in \mathbb{S}_2$, let θ_i be a constant. Define

$$G_{1,i} = \begin{bmatrix} 0 & \theta_i & & & & \\ -\theta_i & 0 & & & & \\ & & \ddots & & & \\ & & & 0 & \theta_i & \\ & & & -\theta_i & 0 & \\ & & & & & 0 \end{bmatrix},$$

$$G_{2,i} = \begin{bmatrix} 0 & & & & & \\ & 0 & \theta_i & & & \\ & -\theta_i & 0 & & & \\ & & & \ddots & & \\ & & & & 0 & \theta_i \\ & & & & -\theta_i & 0 \end{bmatrix}$$

but set $G_{k,i} = 0$ for $2 < k \leq m$. So the controlled system (10.69) becomes

$$dx(t) = f(x(t), t, r(t))dt + \theta_{r(t)} \begin{bmatrix} x_2(t) \\ -x_1(t) \\ \vdots \\ x_{n-1}(t) \\ -x_{n-2}(t) \\ 0 \end{bmatrix} dB_1(t)$$

$$+ \theta_{r(t)} \begin{bmatrix} 0 \\ x_3(t) \\ -x_2(t) \\ \vdots \\ x_n(t) \\ -x_{n-2}(t) \end{bmatrix} dB_2(t), \qquad (10.80)$$

where we set $\theta_i = 0$ for $i \in \mathbb{S}_1$. Note that for each $i \in \mathbb{S}_2$,

$$\sum_{k=1}^{m} |G_{k,i}x|^2 = |G_{1,i}x|^2 + |G_{2,i}x|^2$$
$$= \theta_i^2(x_1^2 + \cdots + x_{n-1}^2) + \theta_i^2(x_2^2 + \cdots + x_n^2) \geq \theta_i^2|x|^2$$

and

$$\sum_{k=1}^{m} |x^T G_{k,i} x|^2 = |x^T G_{1,i} x|^2 + |x^T G_{2,i} x|^2 = 0.$$

Hence, by Theorem 10.13, the solution of the controlled system (10.80) has the property that

$$\limsup_{t \to \infty} \frac{1}{t} \log |x(t)| \geq -K + \sum_{i \in \mathbb{S}_2} 0.5 \pi_i \theta_i^2 \quad a.s.$$

for any initial value $x_0 \neq 0$. Clearly we can choose θ_i ($i \in \mathbb{S}_2$) sufficiently large for $\sum_{i \in \mathbb{S}_2} 0.5 \pi_i \theta_i^2 > K$ so that the controlled system (10.80) becomes unstable. Summarising these results, we state a general theorem in what follows.

Theorem 10.14 *Given any n-dimensional nonlinear hybrid system (10.66), one can always design a linear controller $u(x, i)$ of the form (10.68) for the partial modes $i \in \mathbb{S}_2$ so that the controlled system (10.69) become unstable provided Assumption 10.10 is satisfied and the dimension $n \geq 2$.*

Naturally one may ask what happens if the dimension $n = 1$. To answer this let us consider the scalar jump linear system

$$\dot{x}(t) = a_{r(t)} x(t). \tag{10.81}$$

Assume that this system is almost surely asymptotically stable. Clearly, this is equivalent to the condition that

$$\sum_{i=1}^{N} \pi_i a_i < 0. \tag{10.82}$$

For each mode $i \in \mathbb{S}_2$ we use the feedback controller

$$u(x, i) = (b_{1,i} x, b_{2,i} x, \cdots, b_{k,i} x)$$

while we set $u(x, i) = 0$, namely $b_{k,i} = 0$, for $i \in \mathbb{S}_1$. Then the controlled stochastic system is

$$dx(t) = a_{r(t)} x(t) dt + \sum_{k=1}^{m} b_{k,r(t)} x(t) dB_k(t). \tag{10.83}$$

By Theorems 10.11 and 10.13, we observe that if the initial value $x_0 \neq 0$, the solution of this equation has the property that

$$\lim_{t \to \infty} \frac{1}{t} \log(|x(t)|) = \sum_{i=1}^{N} \pi_i a_i - \frac{1}{2} \sum_{i \in \mathbb{S}_2} \sum_{k=1}^{m} \pi_i b_{k,i}^2 \quad a.s. \quad (10.84)$$

which is always negative under condition (10.82). That is, the controlled system remains stable. In other words, the one-dimensional stable system (10.81) cannot be destabilised by Brownian motions if the controller is of the linear form.

10.5 Stochastic Neural Networks

Since [Hopfield (1982)] initiated the study of neural networks, theoretical understanding of neural network dynamics has advanced greatly and we here mention [Hopfield (1984); Hopfield and Tank (1986); Denker (1986)] among the others. Much of the current interest in artificial networks stems not only from their richness as a theoretical model of collective dynamics but also from the promise they have shown as a practical tool for performing parallel computation. In performing the computation, there are various stochastic perturbations to the networks and it is important to understand how these perturbations affect the networks. Especially, it is very critical to know whether the networks are stable or not under the perturbations. Although the stability of neural networks had been studied to a great deal, the stochastic effects to the stability problem were not investigated until 1996 by [Liao and Mao (1996a); Liao and Mao (1996b)] and the main aim of this section is to introduce the study in this new direction.

The neural network proposed by [Hopfield (1982)] can be described by an ordinary differential equation of the form

$$C_i \dot{u}_i(t) = -\frac{1}{R_i} u_i(t) + \sum_{j=1}^{n} T_{ij} g_j(u_j(t)), \quad 1 \leq i \leq n, \quad (10.85)$$

on $t \geq 0$. The variable $u_i(t)$ represents the voltage on the input of the ith neuron. Each neuron is characterized by an input capacitance C_i and a transfer function $g_i(u)$. The connection matrix element T_{ij} has a value either $+1/R_{ij}$ or $-1/R_{ij}$ depending on whether the noninverting or inverting output of the jth neuron is connected to the input of the ith neuron through a resistance R_{ij}. The parallel resistance at the input of the ith

neuron is
$$R_i = \frac{1}{\sum_{j=1}^{n} |T_{ij}|}.$$

The nonlinear transfer function $g_i(u)$ is sigmoidal, saturating at ± 1 with maximum slope at $u = 0$. In terms of mathematics, $g_i(u)$ is a nondecreasing Lipschitz continuous function with properties that

$$ug_i(u) \geq 0 \quad \text{and} \quad |g_i(u)| \leq 1 \wedge (\beta_i|u|) \quad \text{on} \; -\infty < u < \infty, \qquad (10.86)$$

where β_i is the slope of $g_i(u)$ at $u = 0$ and is supposed to be positive and finite. By defining

$$f_i = \frac{1}{C_i R_i} \quad \text{and} \quad a_{ij} = \frac{T_{ij}}{C_i}$$

equation (10.85) can be re-written as

$$\dot{u}(t) = -Fu(t) + Ag(u(t)), \qquad (10.87)$$

where

$$u(t) = (u_1(t), \cdots, u_n(t))^T, \quad F = \text{diag}(f_1, \cdots, f_n),$$
$$A = (a_{ij})_{n \times n}, \quad g(u) = (g_1(u_1), \cdots, g_n(u_n))^T.$$

It is useful to note that

$$f_i = \sum_{j=1}^{n} |a_{ij}|, \quad 1 \leq i \leq n. \qquad (10.88)$$

In practice neural networks are subject to various types of noise e.g. white and colour. Many authors e.g. [Liao and Mao (1996a)] have taken the white noise into account and described the stochastically perturbed network by an SDE

$$dx(t) = [-Fx(t) + Ag(x(t))]dt + \sigma(x(t))dB(t) \qquad (10.89)$$

on $t \geq 0$. Here $B(t)$ is an m-dimensional Brownian motion as before and and $\sigma : \mathbb{R}^n \to \mathbb{R}^{n \times m}$. However, many neural networks may experience abrupt changes in their structure and parameters caused by phenomena such as component failures or repairs, changing subsystem interconnections, and abrupt environmental disturbances. In this situation the stochastic neural network (10.89) becomes an SDE with Markovian switching

$$dx(t) = [-F(r(t))x(t) + A(r(t))g(x(t))]dt + \sigma(x(t), r(t))dB(t) \qquad (10.90)$$

on $t \geq 0$. Here $r(t)$ is the same Markov chain as before and for each $k \in \mathbb{S}$,

$$F(k) = F_k = \text{diag}(f_1^k, \cdots, f_n^k), \quad A(k) = A_k = (a_{ij}^k)_{n \times n}$$

while $\sigma : \mathbb{R}^n \times \mathbb{S} \to \mathbb{R}^{n \times m}$. Recalling (10.88) we observe that for each $k \in \mathbb{S}$,

$$f_i^k = \sum_{j=1}^n |a_{ij}^k|, \quad 1 \leq i \leq n. \tag{10.91}$$

We always assume that for each $k \in \mathbb{S}$, $\sigma(\cdot, k)$ is locally Lipschitz continuous and satisfies the linear growth condition as well. By the theory of Chapter 3, we know that given any initial data $x(0) = x_0 \in \mathbb{R}^n$ and $r(0) = r_0 \in \mathbb{S}$, equation (10.90) has a unique global solution on $t \geq 0$.

Due to the page limit we will only discuss the stability of the neural network. For this purpose, we assume that $\sigma(0, k) = 0$ for all $k \in \mathbb{S}$. So equation (10.90) admits a trivial solution $x(t) \equiv 0$. Moreover, by Lemma 5.1, we know that if the initial value $x_0 \neq 0$, the solution will never be zero with probability 1, that is $x(t) \neq 0$ for all $t \geq 0$ a.s. As another standing hypothesis, we assume that the Markov chain $r(t)$ is irreducible so it has a unique stationary (probability) distribution $\pi = (\pi_1, \pi_2, \cdots, \pi_N) \in \mathbb{R}^{1 \times N}$ which can be determined by solving equation (10.70).

Theorem 10.15 *Assume that there exists a symmetric positive definite matrix $Q = (q_{ij})_{n \times n}$, and a pair of constants $\mu_k \in \mathbb{R}$ and $\rho_k \geq 0$ for each $k \in \mathbb{S}$, such that*

$$2x^T Q[-F_k x + A_k g(x)] + \text{trace}[\sigma^T(x, k) Q \sigma(x, k)] \leq \mu_k x^T Q x \tag{10.92}$$

and

$$|x^T Q \sigma(x, k)|^2 \geq \rho_k (x^T Q x)^2 \tag{10.93}$$

for all $x \in \mathbb{R}^n$. Then the solution of equation (10.90) has the property that

$$\limsup_{t \to \infty} \frac{1}{t} \log(|x(t)|) \leq \sum_{k=1}^N \pi_k(\tfrac{1}{2}\mu_k - \rho_k) \quad a.s. \tag{10.94}$$

In particular, if $\sum_{k=1}^N \pi_k(\tfrac{1}{2}\mu_k - \rho_k) < 0$, then the stochastic neural network (10.90) is almost surely exponentially stable.

Proof. Clearly, (10.94) holds when $x_0 = 0$. Fix any $x_0 \neq 0$. The Itô formula shows

$$\log(x^T(t)Qx(t)) = \log(x_0^T Q x_0) + M(t)$$
$$+ \int_0^t \Big[\frac{1}{x^T(s)Qx(s)} \Big(2x^T(s)Q[-F(r(s))x(s) + A(r(s))g(x(s))]$$
$$+ \text{trace}[\sigma^T(x(s),r(s))Q\sigma(x(s),r(s))]\Big)$$
$$- \frac{4|x^T(s)Q\sigma(x(s),r(s))|^2}{(x^T(s)Qx(s))^2}\Big]ds,$$

where

$$M(t) = \int_0^t \frac{2x^T(s)Q\sigma(x(s),r(s))}{x^T(s)Qx(s)}dB(s)$$

which is a continuous martingale vanishing at $t = 0$. By conditions (10.92) and (10.93), we get that

$$\log(x^T(t)Qx(t)) \leq \log(x_0^T Q x_0) + M(t) + \int_0^t [\mu_{r(s)} - 2r_{r(s)}]ds. \quad (10.95)$$

By Theorem 1.6, it is easy to see that

$$\lim_{t \to \infty} \frac{M(t)}{t} = 0 \quad a.s.$$

Moreover, by the ergodic property of the Markov chain, we have

$$\lim_{t \to \infty} \frac{1}{t} \int_0^t [\mu_{r(s)} - 2\rho_{r(s)}]ds = \sum_{k=1}^N \pi_k(\mu_k - 2\rho_k) \quad a.s.$$

It then follows from (10.95) that

$$\limsup_{t \to \infty} \frac{1}{t} \log(|x(t)|) = \frac{1}{2} \limsup_{t \to \infty} \frac{1}{t} \log(x^T(t)Qx(t))$$
$$\leq \sum_{k=1}^N \pi_k(\tfrac{1}{2}\mu_k - \rho_k) \quad a.s.$$

as required. □

Making use of the special features of the neural network e.g. (10.86) and (10.91) we can establish a number of specified stability criteria.

Theorem 10.16 *Let (10.86) hold. Assume that there exists a positive definite diagonal matrix $Q = \mathrm{diag}(q_1, q_2, \cdots, q_n)$, and a pair of non-negative constants $\bar{\mu}_k$ and ρ_k for each $k \in \mathbb{S}$, such that*

$$\mathrm{trace}[\sigma^T(x,k)Q\sigma(x,k)] \leq \bar{\mu}_k x^T Q x$$

and

$$|x^T Q \sigma(x)|^2 \geq \rho_k (x^T Q x)^2$$

for all $x \in \mathbb{R}^n$. Let $H_k = (h_{ij}^k)_{n \times n}$ be the symmetric matrix defined by

$$h_{ij}^k = \begin{cases} 2q_i[-f_i^k + (0 \vee a_{ii}^k)\beta_i] & \text{for } i = j, \\ q_i |a_{ij}^k| \beta_j + q_j |a_{ji}^k| \beta_i & \text{for } i \neq j. \end{cases}$$

Set

$$\lambda_k = \begin{cases} \dfrac{\lambda_{\max}(H_k)}{\min_{1 \leq i \leq n} q_i} & \text{if } \lambda_{\max}(H_k) \geq 0, \\ \dfrac{\lambda_{\max}(H_k)}{\max_{1 \leq i \leq n} q_i} & \text{if } \lambda_{\max}(H_k) < 0. \end{cases}$$

Then the solution of equation (10.90) has the property that

$$\limsup_{t \to \infty} \frac{1}{t} \log(|x(t)|) \leq \sum_{k=1}^{N} \pi_k \left(\tfrac{1}{2}[\lambda_k + \bar{\mu}_k] - \rho_k \right) \quad a.s.$$

Proof. Compute, by (10.86),

$$2x^T Q A_k g(x) = 2 \sum_{i,j=1}^{n} x_i q_i a_{ij}^k g_j(x_j)$$

$$\leq 2 \sum_i q_i (0 \vee a_{ii}^k) x_i g_i(x_i) + 2 \sum_{i \neq j} |x_i| q_i |a_{ij}^k| \beta_j |x_j|$$

$$\leq 2 \sum_i q_i (0 \vee a_{ii}^k) \beta_i x_i^2 + \sum_{i \neq j} |x_i|(q_i |a_{ij}^k| \beta_j + q_j |a_{ji}^k| \beta_i)|x_j|.$$

Hence

$$2x^T Q[-F_k x + A_k g(x)] \leq (|x_1|, \cdots, |x_n|) H_k (|x_1|, \cdots, |x_n|)^T$$
$$\leq \lambda_{\max}(H_k) |x|^2 \leq \lambda_k x^T Q x.$$

Therefore the conclusion follows from Theorem 10.15 immediately. □

Theorem 10.17 *Let (10.86) and (10.91) hold. Assume that there exist positive numbers* $q_1, q_2, \cdots, q_n$ *such that*

$$\beta_j^2 \sum_{i=1}^n q_i \, [0 \vee \text{sign}(a_{ii}^k)]^{\delta_{ij}} \, |a_{ij}^k| \le q_j f_j^k, \quad 1 \le j \le n, \; k \in \mathbb{S},$$

where δ_{ij} *is the Dirac delta function, i.e.*

$$\delta_{ij} = \begin{cases} 1 & \text{for } i = j, \\ 0 & \text{for } i \ne j. \end{cases}$$

Assume moreover that

$$\text{trace}[\sigma^T(x,k) Q \sigma(x,k)] \le \bar{\mu}_k x^T Q x$$

and

$$|x^T Q \sigma(x,k)|^2 \ge \rho_k (x^T Q x)^2$$

for all $x \in \mathbb{R}^n$ *and* $k \in \mathbb{S}$, *where* $Q = \text{diag}(q_1, q_2, \cdots, q_n)$, $\bar{\mu}_k$ *and* ρ_k *are non-negative constants. Then the solution of equation (10.90) satisfies*

$$\limsup_{t \to \infty} \frac{1}{t} \log(|x(t)|) \le \sum_{k=1}^N \pi_k \bigl(\tfrac{1}{2}\bar{\mu}_k - \rho_k\bigr) \quad a.s.$$

Proof. Compute, by the conditions,

$$2x^T Q A_k g(x) = 2 \sum_{i,j=1}^n x_i \, q_i \, a_{ij}^k \, g_j(x_j)$$

$$\le 2 \sum_{i,j=1}^n |x_i| \, q_i \, [0 \vee \text{sign}(a_{ii}^k)]^{\delta_{ij}} \, |a_{ij}^k| \, \beta_j |x_j|$$

$$\le \sum_{i,j=1}^n q_i \, [0 \vee \text{sign}(a_{ii}^k)]^{\delta_{ij}} \, |a_{ij}^k| (x_i^2 + \beta_j^2 x_j^2)$$

$$\le \sum_{i=1}^n q_i \Bigl(\sum_{j=1}^n |a_{ij}^k|\Bigr) x_i^2 + \sum_{j=1}^n \Bigl(\beta_j^2 \sum_{i=1}^n q_i \, [0 \vee \text{sign}(a_{ii}^k)]^{\delta_{ij}} \, |a_{ij}|\Bigr) x_j^2$$

$$\le \sum_{i=1}^n q_i f_i^k x_i^2 + \sum_{j=1}^n q_j f_j^k x_j^2 = 2x^T Q F_k x.$$

Hence

$$2x^T Q[-F_k x + A_k g(x)] + \text{trace}[\sigma^T(x) Q \sigma(x)] \le \bar{\mu} x^T Q x$$

and the conclusion follows from Theorem 10.15. □

Theorem 10.18 *Let (10.86) and (10.91) hold. Assume that the network is symmetric in the sense that for each $k \in \mathbb{S}$,*

$$|a_{ij}^k| = |a_{ji}^k| \quad \text{for all } 1 \leq i, j \leq n.$$

Assume also that for each $k \in \mathbb{S}$, there is a pair of non-negative constants $\bar{\mu}_k$ and ρ_k such that

$$|\sigma(x,k)|^2 \leq \bar{\mu}_k |x|^2 \quad \text{and} \quad |x^T \sigma(x,k)|^2 \geq \rho_k |x|^4$$

for all $x \in \mathbb{R}^n$. Then the solution of equation (10.90) has the property that

$$\limsup_{t \to \infty} \frac{1}{t} \log(|x(t)|) \leq \sum_{k=1}^N \pi_k \left(\tfrac{1}{2}[\bar{\mu}_k + \lambda_k] - \rho_k \right) \quad a.s.$$

where

$$\lambda_k = \begin{cases} -2\hat{f}_k(1 - \check{\beta}) & \text{if } \check{\beta} \leq 1, \\ 2\check{f}_k(\check{\beta} - 1) & \text{if } \check{\beta} > 1, \end{cases}$$

in which

$$\check{\beta} = \max_{1 \leq i \leq n} \beta_i, \quad \check{f}_k = \max_{1 \leq i \leq n} f_i^k, \quad \hat{f}_k = \min_{1 \leq i \leq n} f_i^k.$$

Proof. Compute

$$2x^T A_k g(x) = 2 \sum_{i,j=1}^n x_i \, a_{ij}^k \, g_j(x_j)$$

$$\leq 2 \sum_{i,j=1}^n |x_i| \, |a_{ij}^k| \, \beta_j \, |x_j| \leq \check{\beta} \sum_{i,j=1}^n |a_{ij}^k|(x_i^2 + x_j^2)$$

$$= \check{\beta} \left[\sum_{i=1}^n \left(\sum_{j=1}^n |a_{ij}^k| \right) x_i^2 + \sum_{j=1}^n \left(\sum_{i=1}^n |a_{ji}^k| \right) x_j^2 \right]$$

$$= \check{\beta} \left[\sum_{i=1}^n f_i^k x_i^2 + \sum_{j=1}^n f_j^k x_j^2 \right] = 2\check{\beta} x^T F_k x.$$

Hence

$$2x^T[-F_k x + A_k g(x)] \leq -2(1 - \check{\beta})x^T F_k x \leq \lambda_k |x|^2.$$

So the conclusion follows from Theorem 10.15 with Q being the identity matrix. □

10.6 Exercises

10.1 Using the same notation as in the proof of Lemma 10.1, show that
$$\limsup_{t\to\infty} \mathbb{E}V(x(t)) \le K_1.$$

10.2 By the well-known variation-of-constants formula (see e.g. [Mao (1997), Theorem 3.1 on page 96]), show (10.48).

10.3 Show (10.60).

10.4 Show (10.65).

10.5 Prove Theorem 10.13.

10.6 Show (10.84) directly rather than by Theorems 10.11 and 10.13.

Bibliographical Notes

Chapter 1: The material in this chapter is classical and we refer to [Arnold (1974); Doob (1953); Friedman (1975); Gihman and Skorohod (1972); Liptser and Shiryayev (1986); Mao (1997)] etc.

Chapter 2: Most of the material in this chapter is classical and we refer to [Berman and Plemmons (1994); Boyd et al. (1994); Mao (1991); Mao (1997)] while some results in Section 2.4 are based on [Bahar and Mao (2004b)].

Chapter 3: Most of the material in this chapter is essentially based on [Mao (1997); Mao (1999a)] while the proofs on the Markov property of the solutions presented in Section 3.7 are new.

Chapter 4: The material in Section 4.3 is from [Yuan and Mao (2004b)]. The results in Sections 4.4–4.6 are new, although in the case of SDEs (without Markovian switching) these results can be found in [Higham et al. (2002); Higham et al. (2003); Mao (1997); Mao (2003)].

Chapter 5: The material in this chapter is essentially based on [Mao (1999a); Mao (2002b); Yuan and Mao (2004b); Yuan and Mao (2004c)].

Chapter 6: The results in Sections 6.2–6.4 are new but the corresponding results for SDEs without Markovian switching can be found in [Higham et al. (2003)]. The results in Sections 6.5 and 6.6 are from [Mao et al. (2005a); Yuan and Mao (2004c)].

Chapter 7: Most of the material in this chapter is from [Mao (2000b); Mao (2002c); Mao et al. (2000)] while Section 7.7 is based on [Mao and Sabanis (2003)].

Chapter 8: This chapter is essentially based on [Mao (1999b)].

Chapter 9: The results of this chapter are essentially due to [Mao (2002d); Mao *et al.* (2006)].

Chapter 10: The results of this chapter are essentially new although the corresponding results for SDEs without Markovian switching can be found in [Bahar and Mao (2004a); Bahar and Mao (2004b); Higham and Mao (2005); Liao and Mao (1996a); Mao (1997); Mao *et al.* (2002a)].

Bibliography

Anderson, W.J. (1991). Continuous-Time Markov Chains, Springer, New York.

Arnold, L. (1974). Stochastic Differential Equations: Theory and Applications, John Wiley and Sons.

Arnold, L. (1998). Random Dynamical System, Springer-Verlag.

Athans, M. (1987). Command and control (C2) theory: A challenge to control science, *IEEE Trans. Automat. Contr.*, **32**, pp. 286–293.

Bahar, A. and Mao, X. (2004a). Stochastic delay population dynamics, *Journal of International Applied Math.*, **11**, 4, pp. 377–400.

Bahar, A. and Mao, X. (2004b) Stochastic delay Lotka-Volterra model, *J. Math. Anal. Appl.*, **292**, pp. 364–380.

Baker, C.T.H.B. and Buckwar, E. (2001). Exponential stability in p-th mean of solutions, and of convergent Euler-type solutions, to stochastic delay differential equations, *Numerical Analysis Report*, **390**. Univ. Manchester.

Basak, G.K., Bisi, A. and Ghosh, M.K. (1996). Stability of a random diffusion with linear drift, *J. Math. Anal. Appl.* **202**, pp. 604–622.

Baxendale, P. and Henning, E.M. (1993). Stabilization of a linear system, *Random Comput. Dyn.*, **1**, 4, pp. 395–421.

Beckenbach, E.F. and Bellman, R. (1961). Inequalities, Springer-Verlag.

Bell, D.R. and Mohammed, S.E.A. (1989). On the solution of stochastic differential equations via small delays, *Stochastics and Stochastics Reports*, **29**, pp. 293–299.

Bellman, R. and Cooke, K.L. (1963). Differential–Difference Equations, Academic Press.

Berman, A. and Plemmons, R. (1994). Nonnegative Matrices in the Mathematical Sciences, SIAM, Philadelphia.

Bihari, I. (1957). A generalization of a lemma of Bellman and its alication to uniqueness problem of differential equations, *Acta Math. Acad. Sci. Hungar.*, **7**, pp. 71–94.

Black, F. and Scholes, M. (1973). The prices of options and corporate liabilities, *J. Political Economy*, **81**, pp. 637–654.

Blythe, S., Mao, X. and Liao, X.X. (2001). Stability of stochastic delay neural networks, *Journal of The Franklin Institute*, **338**, 4, pp. 481–495.

Blythe, S., Mao, X. and Shah, A. (2001). Razumikhin-type theorems on stability of stochastic neural networks with delays, *Sto. Anal. Appl.*, **19**, 1, pp. 85–101.

Bouks, E.K., Control of system with controlled jump Markov disturbances, *Control Theory and Advanced Technology*, **9**, pp. 577–597.

Boyd, S., El Ghaoui, L., Feron, E. and Balakrishnan, V. (1994). Linear Matrix Inequalities in System and Control Theory, SIAM, Philadelphia.

Bucy, H.J. (1965). Stability and positive supermartingales, *J. Differential Equation*, **1**, pp. 151–155.

Buffington, J. and Elliott, R.J. (2002). American options with regime switching, *International Journal of Theoretical and Applied Finance*, **5**, pp. 497–514.

Burrage, K. and Butcher, J.C. (1979). Stability criteria for implicit Runge–Kutta methods, *SIAM J. Numer. Anal.*, **16**, pp. 46–57.

Butcher, J.C. (1975). A stability property of implicit Runge–Kutta methods, *BIT*, **15**, pp. 358–361.

Caraballo, T. Liu, K. and Mao, X. (2001). On stabilization of partial differential equations by noise, *Nagoya Mathematical Journal*, **161**, pp. 155–170.

Carmona, R. and Nulart, D. (1990). Nonlinear Stochastic Integrators, Equations and Flows, Gordon and Breach.

Chow, P. (1982). Stability of nonlinear stochastic evolution equations, *J. Math. Anal. Appl.*, **89**, pp. 400–419.

Coddington, R.F. and Levinson, N. (1955). Theory of Ordinary Differential Equations, McGraw-Hill.

Costa, O. L. V. and Boukas, K. (1998). Necessary and sufficient condition for robust stability and stabilizability of continuous-time linear systems with Markovian jumps, *J. Optimization Theory Appl.*, **99**, pp. 1155–1167.

Cox, J.C., Ingersoll, J. and Ross, S.A. (1985). A theory of the term structure of interest rates, *Econometrica* **53**, pp. 385–407.

Curtain, R.F. and Pritchard, A.J. (1978). Infinite Dimensional Linear System Theory, *Lecture Notes in Control and Information Sciences*, **8**, Springer.

Da Prato, G. and Zabczyk, J. (1992). Stochastic Equations in Infinite Dimensions, Cambridge University Press.

Davis, M. (1994). Linear Stochastic Systems, Chapman and Hall.

Dekker, K. and Verwer, J.G. (1984). Stability of Runge–Kutta Methods for Stiff Nonlinear Equations, North Holland, Amsterdam.

Denker, J.S. (1986). Neural Networks for Computing, Proceedings of the Conference on Neural Networks for Computing (Snowbird, UT, 1986). AIP, New York.

Doob, J.L. (1953). Stochastic Processes, John Wiley.

Dragan, V. and Morozan, T. (2002). Stability and robust stabilization to linear stochastic systems described by differential equations with Markovian jumping and multiplicative white noise, *Stoch. Anal. Appl.*, **20**, pp. 33–92.

Driver, R.D. (1963). A functional differential system of neutral type arising in a two-body problem of classical electrodynamics, in *Nonlinear Differential Equations and Nonlinear Mechanics*, Academic Press, pp. 474–484.

Du, N.H., Kon, R., Sato, K. and Takeuchi, Y. (2004). Dynamical behaviour of

Lotka-Volterra competition systems: non autonomous bistable case and the effect of telegraph noise, *J. Comput. Appl. Math.*, **170**, pp. 399–422.
Dunkel, G. (1968). Single-species model for population growth depending on past history, *Lecture Notes in Math.*, **60**, pp. 92–99.
Dynkin, E.B. (1965). Markov Processes, Vol.1 and 2, Springer.
Elliott, R.J. (1982). Stochastic Calculus and Aplications, Springer.
Elworthy, K.D. (1982). Stochastic Differential Equations on Manifolds, *London Math. Society, Lect. Notes Series*, **70**, Cambridge University Press.
Ergen, W.K. (1954). Kinetics of the circulating fuel nuclear reaction, *J. Appl. Phys.*, **25**, pp. 702–711.
Feng, X., Loparo, K.A., Ji, Y. and Chizeck, H.J. (1992). Stochastic stability properties of jump linear systems, *IEEE Trans. Automat. Control*, **37**, pp. 38–53.
Freidlin, M.I. (1985). Functional Integration and Partial Differential Equations, Princeton University Press.
Freidlin, M.I. and Wentzell, A.D. (1984). Random Perturbations of Dynamical Systems, Springer.
Friedman, A. (1975). Stochastic Differential Equations and Applications, Vol.1 and 2, Academic Press.
Ghosh, M. K., Arapostahis, A. and Marcus, S. I. (1993). Optimal control of switching diffusions with applications to flexible manufacturing systems, *SIAM Journal on Control and Optimization*, **31**, pp. 1183-1204.
Ghosh, M. K., Arapostahis, A. and Marcus, S. I. (1997). Ergodic control of switching diffusions, *SIAM Journal on Control and Optimization*, **35**, pp. 1952-1988.
Gihman, I.I. and Skorohod, A.V. (1972). Stochastic Differential Equations, Springer.
Girsanov, I.V. (1962). An example of nonuniqueness of the solution of K. Itô's stochastic equations, *Teoriya veroyatnostey i yeye primeneniya*, **7**, pp. 336–342.
Glasserman, P. (2004). Monte Carlo Methods in Financial Engineering, Springer.
Gyöngy, I. (1998). A note on Euler's approximations, *Potential Analysis*, **8**, pp. 205–216.
Gyöngy, I. and Krylov, N. (1996). Existence of strong solutions for Itô's stochastic equations via approximations, *Probab. Theory Relat. Fields*, **105**, pp. 143–158.
Hahn, W. (1967). Stability of Motion, Springer.
Hairer, E. and Wanner, G. (1996). Solving Ordinary Differential Equations II , Stiff and Differential-Algebraic Problems, Springer Verlag, Berlin, 2nd ed.
Hairer, E., Norsett, S.P. and Wanner, G. (1993). Solving Ordinary Differential Equations I: Nonstiff Problems, 2nd Ed., Springer, Berlin.
Hale, J.K. and Lunel, S.M.V. (1993). Introduction to Functional Differential Equations, Springer.
Halmos, P.R. (1974). Measure Theory, Springer.
Has'minskii, R.Z. (1967). Necessary and sufficient conditions for the asymptotic stability of linear stochastic systems, *Theory Prob. Appl.*, **12**, pp. 144–147.
Has'minskii, R.Z. (1980). Stochastic Stability of Differential Equations, Alphen:

Sijtjoff and Noordhoff (translation of the Russian edition, Moscow, Nauka 1969).

Heston, S.I. (1993). A closed-form solution for options with stochastic volatility with applications to bond and currency options, *Review of Financial Studies*, **6**, pp. 327–434.

Higham, D.J. (2000). Mean-square and asymptotic stability of the stochastic theta method, *SIAM J. Numer. Anal.*, **38**, pp. 753–769.

Higham, D.J. and Mao, X. (2005). Convergence of Monte Carlo simulations involving the mean-reverting square root process, *Journal of Computational Finance*, **8**, 3, pp. 35–62.

Higham, D.J., Mao, X. and Stuart, A.M. (2002). Strong convergence of Euler-type methods for nonlinear stochastic differential equations, *SIAM Journal on Numerical Analysis*, **40**, 3, pp. 1041–1063.

Higham, D.J., Mao, X. and Stuart, A.M. (2003). Exponential mean-square stability of numerical solutions to stochastic differential equations, *LMS J. Comput. Math.*, **6**, pp. 297–313.

Hopfield, J.J. (1982). Neural networks and physical systems with emergent collect computational abilities, *Proc. Natl. Acad. Sci. USA*, **79**, pp. 2554–2558.

Hopfield, J.J. (1984). Neurons with graded response have collective computational properties like those of two-state neurons, *Proc. Natl. Acad. Sci. USA*, **81**, pp. 3088–3092.

Hopfield, J.J. and Tank, D.W. (1986). Computing with neural circuits, *Model Science*, **233**, pp. 3088–3092.

Hu, Y. (1996). Semi-implicit Euler-Maruyama scheme for stiff stochastic equations, in *Stochastic Analysis and Related Topics V: The Silvri Workshop*, H. Koerezlioglu, ed., vol. **38**, Boston MA, Birkhauser, Progr. Probab., pp. 183–302.

Hu, S., Liao, X.X. and Mao, X. (2003). Stochastic Hopfield neural networks, *Journal of Physics A: Mathematical and General*, **36**, pp. 2235–2249.

Hull, J. and White, A. (1987). The pricing of options on assets with stochastic volatilities, *Journal of Finance*, **42**, pp. 281–300.

Ikeda, N. and Watanabe, S. (1981). Stochastic Differential Equations and Diffusion Processes, North-Holland.

Itô, K. (1942). Differential equations determining Markov processes, *Zenkoku Shijo Suguku Danwakai*, **1077**, pp. 1352–1400.

Ji, Y. and Chizeck, H.J. (1990). Controllability, stabilizability and continuous-time Markovian jump linear quadratic control, *IEEE Trans. Automat. Control*, **35**, pp. 777–788.

Kaneko, T. and Nakao, S. (1988). A note on approximation for stochastic differential equations, *Séminaire de Probabilités*, **22**, Lect. Notes Math., **1321**, pp. 155–162.

Karatzas, I. and Shreve, S.E. (1988). Brownian Motion and Stochastic Calculus, Springer.

Kazangey, T. and Sworder, D.D. (1971). Effective federal policies for regulating residential housing, Proc. Summer Computer Simulation Conference, 1971, pp. 1120–1128.

Kloeden, P.E. and Platen, E. (1992). Numerical Solution of Stochastic Differential Equations, Springer-Verlag.

Kolmanovskii, V., Koroleva, N., Maizenberg, T., Mao, X. and Matasov, A. (2003). Neutral stochastic differential delay equations with Markovian switching, *Sto. Anal. Appl.*, **21**, 4, pp. 819–847.

Kolmanovskii, V.B. and Myshkis, A. (1992). Applied Theory of Functional Differential Equations, Kluwer Academic Publishers.

Kolmanovskii, V.B. and Nosov, V.R. (1986). Stability of Functional Differential Equations, Academic Press.

Kolmanovskii, V.B. and Nosov, V.R. (1981). Stability and and Periodic Modes of Control Systems with Aftereffect, Nauka, Moscow.

Krasovskii, N. (1963). Stability of Motion, Standford University Press.

Krylov, N.V. (1980). Controlled Diffusion Processes, Springer.

Kunita, H. (1990). Stochastic Flows and Stochastic Differential Equations, Cambridge University Press.

Küchler, U. and Platen, E. (2000). Strong discrete time approximation of stochastic differential equations with time delay, *Mathematics and Computers in Simulation*, **54**, pp. 189–205.

Kushner, H.J. (1967). Stochastic Stability and Control, Academic Press.

Ladde, G.S. and Lakshmikantham, V. (1980). Random Differential Inequalities, Academic Press.

Lakshmikantham, V., Leeda, S. and Martynyuk, A.A. (1989). Stability Analysis of Nonlinear Systems, Marcel Dekker.

Lewin, M. (1986). On the boundedness recurrence and stability of solutions of an Itô equation perturbed by a Markov chain, *Stoch. Anal. Appl.*, **4**, pp. 431–487.

Lewis, A.L. (2000). Option Valuation under Stochastic Volatility : with Mathematica Code, Finance Press.

Liao, X.X. and Mao, X. (1996a). Exponential stability and instability of stochastic neural networks, *Sto. Anal. Appl.*, **14**, 2, pp. 165–185.

Liao, X.X. and Mao, X. (1996b). Stability of stochastic neural networks, Neural, *Parallel and Scientific Computations*, **4**, 205–224.

Liao, X.X. and Mao, X. (1997). Stability of large-scale neutral-type stochastic differential equations, *Dynamics of Continuous, Discrete and Impulsive Systems*, **3**, 1, pp. 43–56.

Liao, X.X. and Mao,X. (2000). Exponential stability of stochastic delay interval systems, *Systems and Control Letters*, **40**, pp. 171–181.

Liao, X., Mao, X., Wang, J. and Zeng, Z. (2004). Algebraic conditions of stability for the Hopfield neural networks, *Science in China Ser. F Information Sciences*, **47**, pp. 113–125.

Liptser, R.Sh. and Shiryayev, A.N. (1986). Theorey of Martingales, Klumer Academic Publishers.

Lyapunov, A.M. (1892). Probleme general de la stabilite du mouvement, *Comm. Soc. Math. Kharkov*, **2**, pp. 265–272.

Mao, X. (1991). Stability of Stochastic Differential Equations with Respect to

Semimartingales, Pitman Research Notes in Mathematics Series 251, Longman Scientific and Technical.

Mao, X. (1994). Exponential Stability of Stochastic Differential Equations, Marcel Dekker.

Mao, X. (1996a). Razumikhin-type theorems on exponential stability of stochastic functional differential equations, *Sto. Proc. Their Appl.*, **65**, pp. 233–250.

Mao, X. (1996b). Stochastic self-stabilization, *Stochastics and Stochastics Reports*, **57**, pp. 57–70.

Mao, X. (1997). Stochastic Differential Equations and Applications, Horwood.

Mao, X., Stability of stochastic differential equations with Markovian switching, *Sto. Proce. Their Appl.*, **79**, pp. 45–67.

Mao, X. (1999b). Stochastic functional differential equations with Markovian switching, *Functional Differential Equations*, **6**, pp. 375–396.

Mao, X. (2000a). Asymptotic properties of neutral stochastic differential delay equations, *Stochastics and Stochastics Reports*, **68**, pp. 273–295.

Mao, X. (2000b). Robustness of stability of stochastic differential delay equations with Markovian switching, *Stability and Control: Theory and Applications*, **3**, 1, pp. 48–61.

Mao, X. (2000c). The LaSalle-type theorems for stochastic functional differential equations, *Nonlinear Studies*, **7**,2, pp. 307–328.

Mao, X. (2001a). Some contributions to stochastic asymptotic stability and boundedness via multiple Lyapunov functions, *J. Math. Anal. Appl.*, **260**, pp. 325–340.

Mao, X. (2001b). Attraction, stability and boundedness for stochastic differential delay equations, *Nonlinear Analysis: Theory, Methods & Applications*, **47**, pp. 4795–4806.

Mao, X. (2002a). A note on the LaSalle-type theorems for stochastic differential delay equations, *J. Math. Anal. Appl.*, **268**, pp. 125–142.

Mao, X. (2002b). Asymptotic stability for stochastic differential equations with Markov switching, *WSEAS Transactions on Circuits*, **1**, 1, pp. 68–73.

Mao, X. (2002c). Asymptotic stability for stochastic differential delay equations with Markovian switching, *Functional Differential Equations*, **9**, 1-2, pp. 201–220.

Mao, X. (2002d). Exponential stability of stochastic delay interval systems with Markovian switching, *IEEE Trans. Automat. Contr.*, **47**, 10, pp. 1604-1612.

Mao, X. (2003). Numerical solutions of stochastic functional differential equations, *LMS Journal of Computation and Mathematics*, **6**, pp. 141–161.

Mao, X., Lam, J., Xu, S. and Gao, H. (2006). Razumikhin method and exponential stability of hybrid stochastic delay interval systems, *Journal of Mathematical Analysis and Applications*, **314**, pp. 45–66.

Mao, X., Marion, G. and Renshaw, E. (2002a). Environmental noise supresses explosion in population dynamics, *Stochastic Processes and Their Applications*, **97**, pp. 95–110.

Mao, X., Marion, G. and Renshaw, E. (2002b), Convergence of the Euler shceme for a class of stochastic differential equations, *International Mathematical Journal*, **1**, 1, pp. 9–22.

Mao, X., Matasov, A. and Piunovskiy, A. (2000). Stochastic differential delay equations with Markovian switching, *Bernoulli*, **6**, 1, pp. 73–90.

Mao, X. and Sabanis, S. (2003). Numerical solutions of stochastic differential delay equations under local Lipschitz condition, *Journal of Computational and Applied Mathematics*, **151**, pp. 215–227.

Mao, X., Sabanis, S. and Renshaw, R. (2003). Asymptotic behaviour of the stochastic Lotka-Volterra model, *J. Math. Anal. Appl.*, **287**, 1, pp. 141–156.

Mao, X. and Shaikhet, L. (2000). Delay-dependent stability criteria for stochastic differential delay equations with Markovian switching, *Stability and Control: Theory and Applications*, **3**, 2, pp. 87–101.

Mao, X., Yuan, C. and Yin, G. (2005). Numerical method for stationary distribution of stochastic differential equations with Markovian switching, *Journal of Computational and Applied Mathematics*, **174**, pp. 1–27.

Mao, X., Yuan, C. and Zou, J. (2005). Stochastic differential delay equations of population dynamics, *Journal of Math. Anal. Appl.*, **304**, pp. 296–320.

Mariton, M. (1990). Jump Linear Systems in Automatic Control, Marcel Dekker.

Mattingly, J., Stuart, A.M. and Higham, D.J. (2002). Ergodicity for SDEs and approximations: locally Lipschitz vector fields and degenerate noise, *Stoch. Proc. and Appl.*, **101**, pp.185–232.

McKean, H.P. (1969). Stochastic Integrals, Academic Press.

Mcshane, E.J. (1974). Stochastic Calculus and Stochastic Models, Academic Press.

Merton, R.C. (1973). The theory of rational optional pricing, *Bell Journal of Economics and Management Science*, **4**, pp. 141–183.

Métivier, M. (1982). Semimartingales, Wslter de Gruyter.

Meyer, P.A. (1972). Martingales and Stochastic Integrals I, *Lect. Notes in Math.*, **284**, Springer.

Mizel, V.J. and Trutzer, V. (1984). Stochastic hereditary equations: existence and asymptotic stability, *J. Integral Equations*, **7**, pp. 1–72.

Mohammed, S–E. A. (1984). Stochastic Functional Differential Equations, Longman Scientific and Technical.

Mohammed, S–E. A., Scheutzow, M. and WeizsaÉcher, H. V. (1986). Hyperbolic state space decomposition for a linear stochastic delay equations, *SIAM J. on Control and Optimization*, **24**, 3, pp. 543–551.

Morozan, T. (1998). Parametrized Riccati equations for controlled linear differential systems with jump Markov perturbations, *Stoch. Anal. Appl.*, **16**, pp. 661–682.

Neftci, S.N. (1996). An Introduction to the Mathematics of Financial Derivatives, Academic Press.

Øksendal, B. (1995). Stochastic Differential Equations: An Introduction with Applications, 5th Ed., Springer.

Oseledec, V.I. (1968). A multiplicative ergodic theorem: Laypunov characteristic numbers for dynamical systems, *Trans. Moscow Math. Soc.*, **19**, pp. 197–231.

Pan, G. and Bar-Shalom, Y. (1996). Stabilization of jump linear Gaussian systems

without mode observations, *Int. J. Contr.*, **64**, pp. 631–661.
Razumikhin, B.S. (1956). On the stability of systems with a delay, *Prikl. Mat. Meh.*, **20**, pp. 500–512.
Razumikhin, B.S. (1960). Application of Lyapunov's method to problems in the stability of systems with a delay, *Automat. i Telemeh.*, **21** , pp. 740–749.
Rogers, L.C.G. and Williams, D. (1987). Diffusions, Markov Processes and Martingales, Vol.2, John Wiley and Sons.
Saito, Y. and Mitsui, T. (1996). Stability analysis of numerical schemes for stochastic differential equations, *SIAM J. Numer. Anal.*, **33**, pp. 2254–2267.
Schurz, H. (1997). Stability, Stationarity, and Boundedness of some Implicit Numerical Methods for Stochastic Differential Equations and Applications, Logos Verlag, Berlin.
Shaikhet, L. (1996). Stability of stochastic hereditary systems with Markov switching, *Theory of Stochastic Processes*, **2**, 18, pp. 180–184.
Skorohod, A.V. (1989). Asymptotic Methods in the Theory of Stochastic Differential Equations, American Mathematical Society.
Slatkin, M. (1978). The dynamics of a population in a Markovian environment, *Ecology*, **59**, pp. 249–256.
Souza de C. E. and Fragoso, M. D. (1993). H_∞ control for linear systems with Markovian jumping parameters, *Control Theory and Advanced Technology*, **9**, pp. 457–466.
Stuart, A.M. and Humphries, A.R. (1996). Dynamical Systems and Numerical Analysis, Cambridge University Press, Cambridge.
Sworder, D.D. and Robinson, V.G. (1973). Feedback regulators for jump parameter systems with state and control depend transition rates, *IEEE Trans. Automat. Control*, **18**, pp. 355–360.
Truman, A. (1986). An introduction to the stochastic mechanics of stationary states with alications, in *From Local Times to Global Geometry, Control and Physics* edited by K. D. Elworthy, *Pitman Research Notes in Math.*, **150**, pp. 329–344.
Tsokos, C.P. and Padgett, W.J. (1974). Random Integral Equations with Applications to Life Sciences and Engineering, Academic Press.
Uhlenbeck, G.E. and Ornstein, L.S. (1930). On the theory of Brownian motion, *Phys. Rev.*, **36**, pp. 362–271.
Volterra, V. (1928). Sur la théorie mathématique des phénomènes héréditaires, *J. Math. Pures Appl.*, **7**, pp. 249–298.
Watanabe, S. and Yamada, T. (1971). On the uniqueness of solutions of stochastic differential equations II, *J. Math. Kyoto University*, **11**, pp. 553–563.
Willsky, A.S. and Rogers, B.C. (1979). Stochastic stability research for complex power systems, *DOE Contract, LIDS, MIT, Rep.*, ET-76-C-01-2295.
Wonham, W.M., Random differential equations in control theory, *Probabilistic Methods in Applied Mathematics*, **2**, pp. 131–212.
Wright, E.M. (1961). A functional equation in the heuristic theory of primes, *Mathematical Gazette*, **45**, pp. 15–16.
Yamada, T. (1981). On the successive approximation of solutions of stochastic

differential equations, *J. Math. Kyoto University*, **21**, pp. 501–515.

Yamada, T. and Watanabe, S. (1971). On the uniqueness of solutions of stochastic differential equations, *J. Math. Kyoto University*, **11**, pp. 155–167.

Yan, J.A. (1981). Introduction to Martingales and Stochastic Integrals, Shanghai Science and Technology Press.

Yin, G.G. and Zhang, Q. (1998). Continuous-Time Markov Chains and Applications: A Singular Perturbations Approach, Springer-Verlag, New York.

Yin, G.G. and Zhang, Q. (2005). Discrete-Time Markov Chains, Springer-Verlag, New York.

Yin, G.G. and Zhou, X.Y. (2004). Markowitzs mean-variance portfolio selection with regime switching: from discrete-time models to their continuous-time limits, *IEEE Transactions on Automatic Control*, **49**, pp. 349–360.

Yuan, C. (2004). Numerical Solutions and Stability of Stochastic Differential Equations with Markovian Switching, PhD Thesis, University of Strathclyde.

Yuan, C. and Lygeros, J. (2004a). Asymptotic stability and boundedness of delay switching diffusions, J. In R. Alur, G. Pappas Eds., *Hybrid Systems: Computation and Control, 7th International Workshop, HSCC 2004, Philadelphia, USA*, **2993**, pp. 646-659, Springer LNCS.

Yuan, C. and Lygeros, J. (2004b). Stabilization of a class of stochastic differential equations with Markovian switching, *16th International Symposium on Mathematical Theory of Networks and Systems*, Belgium, MTNS2004.

Yuan, C. and Lygeros, J. (2004c). Invariant measure of stochastic hybrid processes with jumps, *43th Conference of Decision and Control*, Bahamas, CDC2004.

Yuan, C. and Mao, X. (2003a). Asymptotic stability in distribution of stochastic differential equations with Markovian switching, *Sto. Proce. Their Appl.*, **103**, pp. 277–291.

Yuan, C. and Mao, X. (2003b). Asymptotic stability and boundedness of stochastic differential equations with respect to semimartingales, *Sto. Anal. Appl.*, **21**, 3, pp. 737–751.

Yuan, C. and Mao, X. (2004a). Robust stability and controllability of stochastic differential delay equations with Markovian switching, *Automatica*, **40**, 3, pp. 343–354.

Yuan, C. and Mao, X. (2004b) Convergence of the Euler-Maruyama method for stochastic differential equations with Markovian switching, *Mathematics and Computers in Simulation*, **64**, pp. 223–235.

Yuan, C. and Mao, X. (2004c). Stability in distribution of numerical solutions for stochastic differential equations, *Sto. Anal. Appl.*, **22**, 5, pp. 1133–1150.

Yuan, C. and Mao, X. (2005). Stationary Distributions of Euler-Maruyama-Type Stochastic Difference Equations with Markovian Switching and Their Convergence, *Journal of Difference Equations and Applications*, **11**, pp. 29–48.

Yuan, C., Zou, J. and Mao, X. (2003). Stability in distribution of stochastic differential delay equations with Markovian switching, *Systems and Control Letters*, **50**, pp. 195–207.

Index

adapted process 13
asymptotic stability 193
augmentation 20

Bihari's inequality 55
Black–Scholes model 360
Borel
  measurable 5
  set 5
  σ-algebra 5
Borel–Cantelli's lemma 10
Brownian motion 19
Burkholder–Davis–Gundy's
  inequality 70

cadlag process 13
Caratheodory's approximation 126
Euler–Maruyama's
  approximation 111
Chebyshev's inequality 7
complement 4
completion 6
conditional expectation 11
conditional probability 11
continuous process 13
convergence
  almost sure $\sim$ 8
  with probability one 8
  stochastic $\sim$ 8
  in probability 8
  in pth moment 8
  in distribution 8

convergence theorem
  bounded $\sim$ 9
  dominated $\sim$ 9
  monotonic $\sim$ 8
covariance matrix 7

delay equation 271
destabilisation 383
differential equation 2
diffusion operator 41
Dirac delta function 23
distribution 7
Doob's martingale
  convergence theorem 17
  inequality 18
Doob's stopping theorem 15
Doob–Dynkin's lemma 6

equilibrium state 164
ergodic property 184
exit time 14
expectation 6
explosion time 91
exponential instability 383
exponential martingale formula 42
exponential martingale inequality 74
exponential stability
  almost sure $\sim$ 166
  moment $\sim$ 166

Feller property 106
finite variation process 13

filtration 12
   natural $\sim$ 19
   right continuous $\sim$ 12
flow property 78
functional differential equation 301
$\mathcal{F}$-measurable 5
   set 5

geometric Brownian motion 3
   with Markovian switching 4
global solution 94
Gronwall's inequality 54

Hölder's inequality 7

increasing process 13
indicator function 5
indefinite integral 30
independent increments 19
independent
   random variables 9
   sets 9
   σ-algebras 9
indistinguishable processes 14
inferior limit of sets 10
integrable process 13
integrable random variable 6
integration by parts formula 41
Itô's formula 39
   generalised $\sim$ 48
Itô's stochastic integral 25
Itô's process 39

Jensen's inequality 54
joint quadratic variation 15

Kolmogorov–Chapman's equation 44

Laplace operator 43
law of the iterated logarithm 22
left continuous process 13
Lévy's theorem 23
linear growth condition 81
linear matrix inequality 62
linear stochastic differential

equation 180
Lipschitz condition 81
local Lipschitz condition 91
local martingale 16
Lotka–Volterra model 353
Lyapunov exponent
   moment $\sim$ 101
   sample $\sim$ 102
Lyapunov inequality 62
Lyapunov method 155
Lyapunov-type function 122

Markov chain 45
   stable state of $\sim$ 45
   generator of $\sim$ 46
   finite $\sim$ 46
Markovian switching 3
Markov process 43
   homogeneous $\sim$ 44
   strong $\sim$ 45
Markov property 43
   strong $\sim$ 43
martingale 15
matrix
   induced operator norm of $\sim$ 59
   trace norm of $\sim$ 59
   eigenvalue of $\sim$ 59
   spectrum of $\sim$ 59
   symmetric $\sim$ 59
   positive-definite $\sim$ 59
   non-negative-definite $\sim$ 59
M-matrix 67
   nonsingular $\sim$ 67
mean-reverting
   process 360
   square root process 360
   θ-process 361
measurable process 13
measurable space 5
Minkovskii's inequality 7
modification 14
moment 6
monotone condition 93

neural network 387
   stochastic $\sim$ 388

option 375
    barrier $\sim$ 376
    European put $\sim$ 375
Ornstein–Uhlenbeck process 160

payoff 375
Picard's iterations 83
Poisson random measure 46
population dynamics 351
    stochastic $\sim$ 352
predictable process 13
probability measure 6
probability space 6
    complete $\sim$ 6
progressive process 13
progressively measurable 13

quadratic variation 15

Radon–Nikodym's theorem 11
random variable 5
Razumikhin theorem 311
Riccati equation 63
right continuous process 13

sample path 13
Schur's complements 64
semigroup property 78
simple process 25
solution 78
    global $\sim$ 94
    local $\sim$ 90
    maximal local $\sim$ 91
square integrable martingale 15
stability in probability 204
stabilization 379
state space 12
stationary distribution 183

stationary increments 19
stationary state 164
step function 11
step process 25
stochastic asymptotic stability 204
    in the large 204
stochastic differential 39
stochastic interval 14
stochastic process 12
stochastic stability 204
stochastic volatility model 370
stopped process 15
stopping time 14
strong law of large numbers 16
submartingale 16
    inequality 17
superior limit of sets 10
supermartingale 17
    inequality 17
    convergence theorem 17
σ-algebra 4
    generated 5

transformation formula 7
transition probability 44
    standard $\sim$ 45
trivial solution 164

uncorrelated 7
uniformly integrable 17
usual conditions 12

variance 6
version 14
volatility 360

weighting function 305